Nonlinear Dynamics and Chaos

Nonlinear Dynamics and Chaos

Second Edition

J. M. T. Thompson, FRS
University College London, UK

H. B. Stewart
The Ross Institute, New York, USA

JOHN WILEY & SONS, LTD

Baffins Lane, Chichester,
West Sussex, PO19 1UD, England

National 01243 779777
International (+44) 1243 779777

e-mail (for orders and customer service enquiries): cs-books@wiley.co.uk
Visit our Home Page on http://www.wiley.co.uk or http://www.wiley.com

Other Wiley Editorial Offices

John Wiley & Sons, Inc., 605 Third Avenue,
New York, NY 10158-0012, USA

WILEY-VCH Verlag GmbH
Pappelallee 3, D-69469 Weinheim, Germany

John Wiley & Sons Australia, Ltd
33 Park Road, Milton, Queensland 4064, Australia

John Wiley & Sons (Canada) Ltd, 22 Worcester Road
Rexdale, Ontario, M9W 1L1, Canada

John Wiley & Sons (Asia) Pte Ltd, 2 Clementi Loop #02-01,
Jin Xing Distripark, Singapore 129809

Library in Congress Cataloging-in-Publication Data

Thompson, J. M. T.
Nonlinear dynamics and chaos / J.M.T. Thompson, H.B. Stewart.—2nd ed.
p. cm.
Includes bibliographical references and index.
ISBN 0–471–87645–3 (cloth : alk. paper)—ISBN 0–471–87684–4 (pbk. : alk. paper)
1. Dynamics. 2. Nonlinear theories. 3. Chaotic behavior in systems. I. Stewart, H. B. (H. Bruce) II. Title.

QA871 .T47 2001
531′.11—dc21

2001045407

British Library Cataloguing in Publication Data

A catalogue record for this book is available from the British Library

ISBN 0–471–87645–3 (Cloth)
ISBN 0–471–87684–4 (Paper)

Typeset in 10/12pt Times by Kolam Information Services Pvt Ltd, Pondicherry, India.

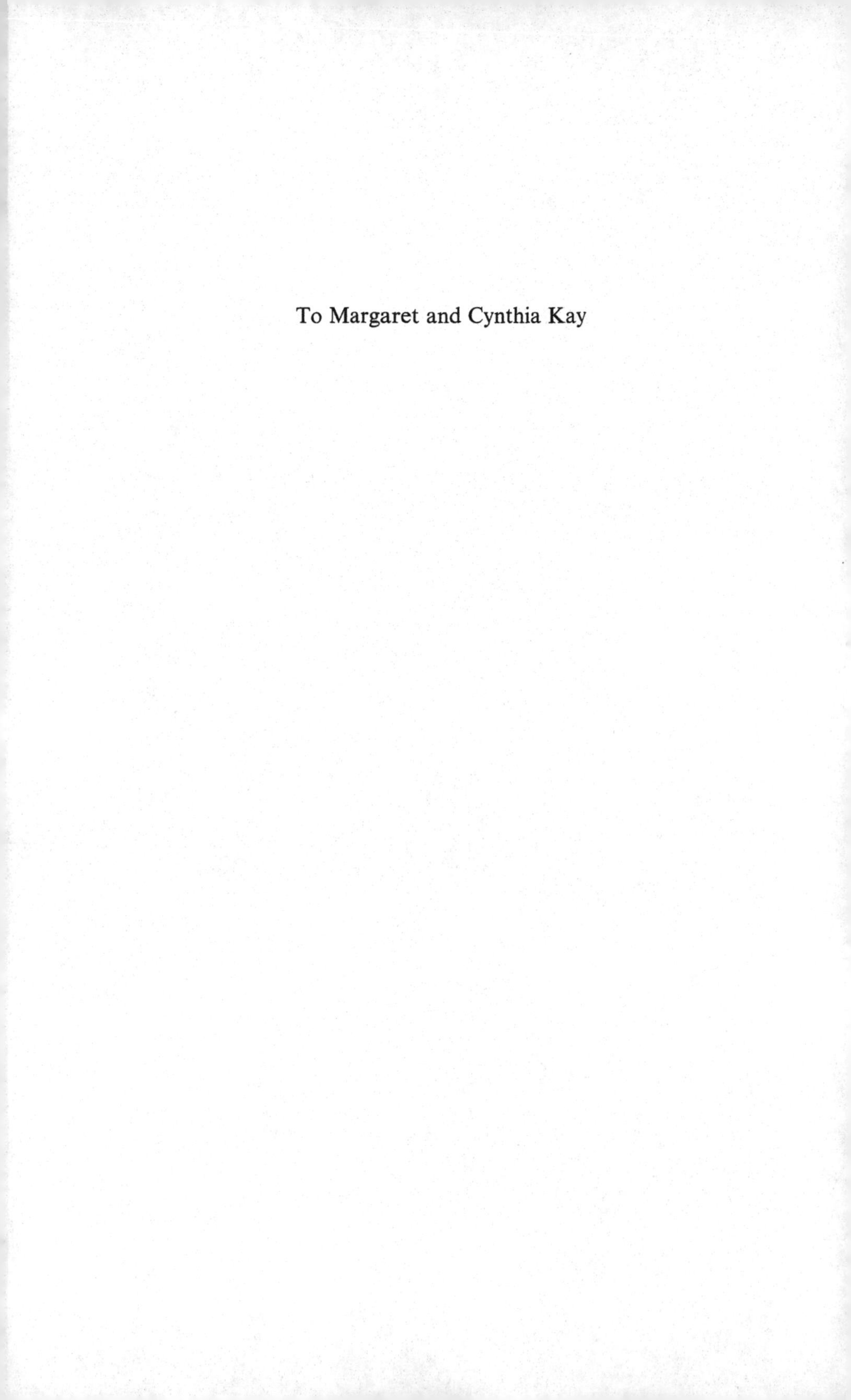

To Margaret and Cynthia Kay

Contents

Preface

Nonlinear dynamics and chaos have developed rapidly since the appearance of our first edition in 1986. New theoretical and numerical techniques have been matched by an explosion of new applications. This second edition extends and updates our phenomenological presentation of the subject, which had a major impact on this development. Major features of the new edition are a comprehensive illustrated glossary of terms, a classification of generic bifurcations, and a completely new chapter describing recent research on driven oscillators and their fractal basin boundaries. All the original text has been revised, extended and updated, and 67 new figures have been included.

The illustrated glossary of terms, definitions and formulae covers the modern geometrical concepts of nonlinear dynamics and chaos. It is arranged alphabetically and can also be read as a connected introduction to the subject by following a system of *go to* instructions. This guided tour of nonlinear dynamics and chaos starts with the entry *dynamical system* and takes the reader logically and systematically through all entries, ending with *archetypal maps*. The glossary is illustrated with nine new figures, including an illustration of horseshoe dynamics in a homoclinic tangle, two examples of indeterminate bifurcation and an outline of symbolic dynamics which makes transparent the complexities of chaotic motion.

The phenomenological classification of attractor bifurcations gives a structured guide to 18 generic bifurcations. The description of each bifurcation includes formulae, precursors, other names, examples of applications, and sources of illustrations. The four safe bifurcations, which generate no dynamic jump or enlargement, include three local supercritical forms (of the Hopf, Neimark and flip) and one global bifurcation (band merging). The six explosive bifurcations (flow, map, flow intermittency, map intermittency, regular saddle and chaotic saddle) generate a reversible enlargement of the current attractor. The eight dangerous bifurcations generate a non-reversible jump to a remote attractor, which is often indeterminate. They include the two local saddle-node events (static fold and cyclic fold), three local subcritical forms (of the Hopf, Neimark and flip) and three global bifurcations (saddle connection, regular-saddle catastrophe and chaotic-saddle catastrophe). The potential indeterminacy of the dangerous bifurcations is discussed in some detail.

The new Chapter 16, with 40 figures, is an extensive summary of the major advances that have been made in our understanding of the escape of a driven oscillator from a potential well. This is a universal problem in the physical sciences—from activation energies of molecular dynamics to the gravitational collapse of massive stars—and the results described in the chapter have offered naval architects a new paradigm for understanding and testing the stability of ships against capsize in beam seas. Chapter 16 illustrates, in a particular application, how the invariant manifolds of unstable saddle solutions structure the phase space and control the attractors, their basins and their bifurcations. Features of some new and significant indeterminate bifurcations are illustrated.

Chapter 16 includes an introduction to knots and links in periodic orbits, and shows how they can be used to determine necessary bifurcational precedences. Transients and basins of attraction are examined, and the concept of rapid basin erosion at a sharp Dover cliff is introduced. Related fractal boundaries in control space allow quick estimates of, for example, the wave height needed to capsize a model ship in a wave-tank experiment. The homoclinic tangles and heteroclinic connections that govern this fractal basin erosion are carefully described and illustrated. Melnikov predictions and Birkhoff signature changes are outlined, and particular attention is given to the heteroclinic chains and accessible orbits that play a most significant role.

The chapter concludes with an examination of the tangled saddle-node bifurcation. This robust and widely observed phenomenon gives an indeterminate jump to resonance or escape when the saddle of a saddle-node collision is located on a fractal basin boundary. The associated striation of the residual basin is examined, and its characteristics deduced from consideration of tunnelling in a one-dimensional mapping. At lower values of the forcing frequency, the tangled saddle node is replaced by the indeterminate boundary crisis of a chaotic attractor.

In Chapter 3 we have inserted a figure emphasizing the parallel linear analyses of flows and maps for two-dimensional systems. A new Section 3.5 analyses the attractors of a spinning satellite or spacecraft. This is an interesting and important application in its own right, which also serves as an introduction to competing solutions in a spherical phase space.

In Chapter 5 we have inserted a new Section 5.8 emphasizing the ABC (attractors, basins, catastrophes) of nonlinear dynamics by means of a new figure. This also shows clearly the concept of a finite residual basin which is lost instantaneously at a saddle-node bifurcation.

At the end of Chapter 6 a new schematic figure shows the template for the stretching and folding action in a twin-well chaotic attractor, which can be compared directly with its sequential Poincaré sections. The illustrated three-layer folding is similar to Smale's horseshoe. A new section is devoted to chaotic attractors in systems of mixed type, containing both nonlinear damping and nonlinear restoring force, as arises for example in self-oscillatory electric circuits. Two new figures show a chaotic attractor in Poincaré section, and the invariant manifold structure, which offers a glimpse of an important topological invariant of chaotic attractors related to the Poincaré index of unstable periodic points. A second new section on tangled basins gives an early introduction to *final state sensitivity*, which can be of great concern to engineers. An example is given of a fractal basin boundary in an attractor-basin phase portrait of the forced twin-well Duffing oscillator.

The illustrations of flow bifurcations in Chapter 7 have been substantially upgraded and extended. One of the new figures illustrates a basic theorem (Thompson and Hunt 1973) concerning the bifurcation structures of a mechanical system with one degree of freedom. Another illustrates the potential energy transformations in archetypal bifurcations. The effect of symmetry-breaking imperfections in pitchfork bifurcations is introduced, with a figure showing the two-thirds power-law cusps of *imperfection sensitivity*. A three-dimensional equilibrium surface emphasizes the link with the codimension 2 cusp of catastrophe theory. A new section explains and illustrates the basin changes that are associated with the local bifurcations of attractors; these changes govern the classification of bifurcations into safe and dangerous types.

A similar new section at the end of Chapter 8 discusses the analogous basin changes at local bifurcations of limit cycles. The cyclic fold, and the sub- and supercritical forms of the flip, Hopf and Neimark bifurcations are all discussed in considerable detail.

The revised Chapter 9 ends with a new discussion of the Hénon map, which is a convenient *archetype* for chaotic attractor dynamics in two-dimensional maps. A new figure illustrates a chaotic attractor and a fractal basin boundary in this map, the fractal layers of the attractor being strongly compressed. The remarkable qualitative similarity with the attractor-basin phase portrait of the twin-well Duffing oscillator emphasizes the usefulness of studying the Hénon map.

An important addition to Chapter 10 updates the status of the shadowing lemma of Anosov and Bowen. This addresses the question that arises when a dynamical system exhibits sensitive dependence on initial conditions: do approximate numerical solutions reflect the true behaviour of the actual system? In the presence of the right hyperbolic structure, the shadowing lemma guarantees that any reasonable approximation to an orbit is closely shadowed by some true orbit of the same dynamical system. Recent applications of these ideas to numerical solutions of prototypical systems are partly reassuring, but also in some cases cautionary.

The discussion of the rigorous topological nature of the Lorenz attractor in Chapter 11 has been updated. In Chapter 12 a new figure shows the Poincaré section of the twin-well Duffing model of the vibrated buckled beam. This shows clearly the saddle cycle which dynamically separates the two potential wells, and parts of its inset and outset exhibiting transverse homoclinic points. The chapter ends with a new section devoted to spatial chaos and localized buckling. This describes the new insights and phenomena that have emerged recently from an application of the theory of homoclinic orbits to the localization of structural post-buckling patterns.

This application is made possible by the use of a static–dynamic analogy, in which an independent spatial coordinate of a long elastic structure is identified as time in an equivalent dynamical system; a localized deformation of the structure is then identified as a homoclinic orbit. An archetypal problem is the analogy between a stretched and twisted rod and a spinning top, with applications to marine pipelines and supercoiled DNA; an *anisotropic* rod is equivalent to a non-symmetric (non-integrable) top. A new figure illustrates the infinite number of localized homoclinic post-buckling paths that can be found in buckling problems of this type.

Chapter 13 has extensive new material on the transition to chaos via quasi-periodic motion, and some significant codimension 2 bifurcations. A new figure shows a control-space diagram for the velocity-forced Van der Pol equations. This allows a direct comparison with a similar one for the circle map, the subharmonic numbers in the tongues of the Van der Pol diagram corresponding to the denominators of the map. This illustrates the usefulness of this map as an archetype. A codimension 2 point in the Van der Pol diagram illustrates the interaction of two global bifurcations. Here two chaotic attractor explosion arcs meet, giving a chaotic analogue of the cusp catastrophe. Knowledge of such codimension 2 bifurcations is of great practical use to the dynamicist who needs to construct a portrait of a dynamical system with two controls. It may even be possible to forecast a change in a codimension 1 global bifurcation, as when a chaotic attractor explosion turns into a chaotic blue sky catastrophe.

To accommodate the new material in a book of reasonable size and price, we have removed the original Chapters 16 and 17 (on particle accelerators and experimental observations of chaos). Interested readers might care to consult the first edition of the book. To update the bibliography, we have carefully selected 150 new works that appeared since the first edition. Web addresses of some online resources are also given.

J. M. T. Thompson
University College London, UK

H. B. Stewart
The Ross Institute, New York, USA

Preface to the First Edition

The aim of this book is to present the new geometrical ideas that are revolutionizing dynamical systems theory in a readable and richly illustrated form for engineers and scientists, analysts and experimentalists of all disciplines, who are concerned to model and understand the time evolution of real systems. Most real-world problems confronting the analyst, being neither linear nor even nearly linear, fall outside the domain of traditional closed-form analysis, and must be tackled in the first instance on a computer. But numerical simulations, like physical experiments, typically produce unwieldy masses of data, and for both of these, phase-space concepts must now be recognized as an essential guide to the structuring of the investigation and the interpretation of the results.

At the present time we are indeed witnessing a spectacular blossoming of nonlinear dynamics, made possible on the one hand by great theoretical strides in Poincaré's qualitative topological approach and on the other by the wide availability of powerful digital and analogue computers. This has been stimulated and sustained by important and exciting new applications that have multiplied throughout the biological, ecological, social and economic sciences, far beyond the still vibrant traditional fields of mechanics, physics, and chemistry. Indeed, wherever the time evolution of a natural or man-made system needs to be modelled and explored, the subtle and versatile techniques of dynamical systems theory are being invoked. It is to this whole field that the present book is addressed, the term 'dynamics' being used as a convenient abbreviation of dynamical systems theory, without implying any restriction to its original meaning of a branch of mechanics.

A significant element of this major thrust, which nicely epitomizes its most advanced theoretical, computational, and experimental features, is the discovery and delineation of chaotic motions in remarkably simple deterministic models from a galaxy of disciplines ranging from population dynamics and meteorology to lasers and particle accelerators. The recent discovery of quite large regimes of chaos in the long-familiar, sinusoidally driven Duffing and Van der Pol oscillators emphasizes just how much was missed by the classical analysis of ordinary differential equations.

Continuous and discrete dynamical systems, described respectively by ordinary differential equations and finite difference equations (iterated maps), are considered in the book, and attention is focused predominantly on the typical dissipative systems, familiar from the damped, energy-absorbing systems of macroscopic physics. Very little analytical knowledge is required of the reader, except perhaps a little familiarity with simple differential equations, and we aim to take him first on a gentle ramble through the foothills of the nonlinear terrain, before ascending to some of the more rarified peaks of instabilities, bifurcations, and chaos. The topological principles governing the dynamical trajectories in phase space are introduced coherently and systematically in examples of gradually increasing complexity accompanied by over 200 diagrams, most of which come directly from computer solutions of basic archetypal equations. The numerous illustrations are drawn from all areas of dynamical systems theory, including applications in electrical, mechanical, and structural engineering, meteorology, physics, chemistry, biology, and ecology. Detailed studies include the chaotic motions of impacting systems; the self-excited aeroelastic instabilities of slender structures; the wave-induced sub-harmonic resonances of offshore oil production facilities; fluid turbulence and the chaotic motions of large-scale convection in the atmosphere; the instabilities of beams in particle accelerators and storage rings; oscillatory phenomena in chemical kinetics; dynamics of the heartbeat and nerve impulse; and the surprising complexities of logistic growth in insect and animal populations.

The book emphasizes the qualitative description of long-term recurrent motions of dissipative systems governed by genuinely nonlinear equations, with no assumptions of near-linearity. General concepts of the geometric theory are illustrated using computer simulations of specific ordinary differential and difference equations. For the experimentalist, these can be viewed as ideal experiments, and show what could be achieved in a perfectly controlled noise-free laboratory: perhaps more importantly, in for example the case of chaotic motions, they show what *cannot* be achieved even under such ideal conditions. The nonlinear phenomena thoroughly discussed include the multiple attractors observable in a single system; chaotic long-term behaviour and its underlying order and structure; and discontinuous jump and hysteresis phenomena. A systematic bifurcation theory of equilibria, periodic, and chaotic motions is advanced and illustrated. The methods presented constitute guidelines for the detailed experimental and computational study of the widest variety of dynamical systems.

Before outlining in detail the contents of the book, it is perhaps informative at this stage to examine in a little detail just how the geometrical ideas can help the systems analyst.

With the advent of high-speed computers, it might have seemed that dynamical systems theory would simply fade away. But quite the reverse is true, and it is indeed one of the fastest growing disciplines of applicable mathematics. The reason is that its broad yet precise geometrical ideas are vitally needed to guide the analyst through the bewildering variety of complex behaviour that he is likely to encounter.

Time evolutions must normally be modelled by *nonlinear* equations for which closed-form analytical solutions are unobtainable. They are however readily integrated numerically by routine computer algorithms, so that the response from given starting conditions is easily established. But the starting conditions of a real system are never known precisely, and may be totally unknown: what initial conditions should be assumed for a complex model of the atmosphere, or an oilrig at sea in a developing storm? So since the motions of a nonlinear system can depend crucially on these conditions, a mammoth task begins to emerge. How can we hope to explore the responses from all possible starts?

Clearly we need some overview of what can happen in the evolution of a system, and how this is influenced by the starting conditions. Here we have our first major guidance from dynamical systems theory in the concept of an *attractor*. Typical, dissipative dynamical systems exhibit a start-up transient, after which the motion settles down towards some form of long-term recurrent behaviour. Motions from adjacent starts tend to converge towards *stable* attracting solutions.

The simplest is a stationary equilibrium point at which all motion has ceased. The archetypal example of this, studied experimentally by Newton, is the pendulum. From whatever starting values of position and velocity, this returns to its vertical hanging state, damped by air resistance and other forms of energy dissipation. In the two-dimensional abstract *phase space* whose coordinates are the displacement and velocity, all possible motions appear as non-crossing spirals converging asymptotically towards the resting state at the origin. This focus in the *phase portrait* of all possible motions is called a *point attractor*.

Secondly, we have the *periodic attractor*. If a thin steel strip is driven into resonance by an electromagnet carrying an alternating current, the strip will normally settle into a steady vibration at the frequency of the forcing. After a small knock, transients will slowly decay, and the stable fundamental oscillation will be re-established. But given a bigger disturbance, an entirely new stable steady state may be observed. Different starts of a given system may thus lead to alternative final states, with multiple attractors coexisting in the phase portait of a single dynamical system in a state of *competition*.

A third, recently discovered, attractor whose unexpected features have generated an explosion of interest is the strange or *chaotic attractor* that captures the solution of a perfectly deterministic and well defined equation into a state of steady but perpetual chaos. But although a single time history possesses an aspect of true randomness and a noisy broadband power spectrum, a comprehensive phase-space investigation reveals an underlying order in the ensemble of all trajectories whose elucidation is a major topic of research amongst mathematicians and physicists.

Clearly analysts and experimentalists should be vitally aware that such apparently random non-periodic outputs may be the correct answer, and should not be attributed to bad technique and assigned to the wastepaper basket, as has undoubtedly happened in the past. They should familiarize themselves with the techniques presented here for positively identifying a genuine chaotic attractor.

It is the final settling in the multi-dimensional phase space of a large dissipative system to attractors of low dimension that validates our presentation of general concepts via specific equations of low dimension. Two-dimensional planar spaces serve to illustrate point and cyclic attractors, while three phase dimensions are necessary to allow chaotic flows.

Multiplicities of all three attractors can coexist in the phase portrait of a given system, so that in simulations and experiments a bewildering variety of final time histories may be generated by different starts. So in a complex system with perhaps hundreds of different position and velocity coordinates giving a high n-dimensional phase space, the exploration of the full dynamical behaviour would apparently require an n-dimensional matrix of trial starts, a daunting task.

This description of *multiple attractors* is not in the least fanciful or rare. Multiplicities of competing steady states *including chaos* are constantly being discovered for the *simplest* nonlinear systems, witness the recent identification of chaos in Duffing's equation describing just the forced vibrations of our steel strip or pendulum.

Luckily numerous techniques are now available for sorting out the competing attractive basins. Running time backwards on a computer helps to locate unstable steady states, repellors or saddles, that serve to organize the *separatrix* surfaces between the catchment regimes, while Poincaré's stroboscopic sampling techniques can assist in unravelling periodic and chaotic attractors. Once all this has been done, a major test of an analyst's model is that the number, type, and catchment basins of his attractors should correlate well with those of the real system.

Now even in the most exact of the physical sciences, the coefficients of any model are never known with absolute precision. So having previously discussed the *physical stability* of final states of a given system, we must explore the stability of the modelling process itself. Would the analyst get roughly the same answers if his equations had been slightly different? This question of the robustness of the model is formalized in the concept of the *structural stability* of a phase portrait.

Many systems are in a slowly evolving environment, so their coefficients and parameters undergo gradual change. Then if the evolving system is in a steady state of equilibrium, periodic oscillation or chaos, the prediction of any sudden change is of crucial importance. Such *bifurcation* of behaviour will occur when the phase portrait undergoes a qualitative change of topological form at a point of *structural instability*. An important distinction that we emphasize in the book is between those *catastrophic* bifurcations at which there is a finite rapid dynamic jump to a new steady state, and *subtle* bifurcations in which the change in response manifests itself in the smooth growth of a new local attractor after the bifurcation point.

The succinct description of the small number of typical, structurally stable bifurcations that can arise in the real world is a major feature of this book. These descriptions relate to the low-dimensional *centre manifold*, embedded in the total phase space, on which the significant transformations occur. It is indeed the important centre manifold theory that gives solid theoretical support to the goal of understanding the loss of stability of a large complex system with a phase space of perhaps 1000 dimensions in terms of low-dimensional transformations.

Concluding this brief outline of the guidance offered by the geometric theory, we would emphasize that in many areas of science and technology a large effort has traditionally been made to model a physical system or process. Yet once the mathematical model has been constructed, only a few rather cursory computer time simulations are sometimes made. Lulled into a false sense of security by his familiarity with the *unique* response of a linear system, the busy analyst or experimentalist shouts 'Eureka, this is the solution' once a simulation settles onto an equilibrium or steady cycle, without bothering to explore patiently the outcome from different starting conditions. To avoid potentially dangerous errors and disasters, industrial designers must be prepared to devote a greater percentage of their effort into exploring the full range of dynamic responses of their systems.

Our aim then is to present a readable introduction to the geometric theory. After a brief historical introduction, Chapter 1 surveys the basic ideas of attractors, chaotic motions, and bifurcations using as a concrete example a periodically forced mechanical oscillator. In Chapter 2 the basics of second-order oscillators are reviewed, emphasizing the fundamental differences in behaviour associated with damping and nonlinearity. Chapters 3 and 4 set forth the phase-space geometry of the two elementary types of final behaviour in autonomous (unforced) systems, final equilibrium or periodic motion. A variety of examples from mechanics, ecology, biology, and chemistry are described to illustrate the geometry of real systems in the phase plane. In Chapter 5 we consider periodic motions in forced oscillators for which the full three-dimensional phase space of position, velocity, and time is necessary for a complete understanding of behaviour. Here the Poincaré mapping technique is introduced, allowing the three-dimensional flow of trajectories to be condensed into a two-dimensional mapping of return points: and in Chapter 6 the first detailed examination of chaotic attractors (in the forced Van der Pol and Duffing oscillators) makes clear the usefulness and correct application of mapping techniques. Part I concludes with a study of important stability concepts which lead naturally to a classification of the most important bifurcations for equilibria and limit cycles.

In Part II attention is shifted to iterated mappings as dynamical systems in their own right, representing evolution by discrete steps in time. Simple maps offer an excellent vehicle for further study of local bifurcation and stability in Chapter 8. Similarly, maps also provide convenient tools for the study of chaotic behaviour in Chapter 9, allowing a thorough analysis of bifurcations of prototypical chaotic attractors, as a foundation for better understanding chaos in continuous flows.

Part III explores the basic topological structures of geometric dynamics. Readers already familiar with the basics of nonlinear dynamics might scan the figures in Parts I and II and begin reading here. Chapter 10 surveys the basic sets—attractors, saddles and repellors—emphasizing the mathematical notion of hyperbolic structure. This hyperbolic structure assures the existence of well defined invariant manifolds, whose importance is illustrated in Chapter 11 using the Lorenz model of thermo-fluid convection as an example. Here the usefulness of invariant manifolds in understanding the structure of chaotic attractors and their major bifurcations is illustrated in three-dimensional phase space. Chapter 12 continues the structural investigation of chaotic attractors with a singularly important example, Rössler's folded band; and in Chapter 13, a systematic overview of dynamic bifurcations is set forth, including continuous and catastrophic types of bifurcation, and global as well as local structure of invariant manifolds. As always, the abstract ideas are illustrated with examples of specific differential equations, here mainly forced oscillators.

Finally Part IV looks at several applications at some length, to give just a flavour of how geometric methods are applied to the understanding of practical problems. Chapter 14 considers subharmonic resonant motions of an engineering marine structure, while Chapter 15 examines an oscillating system subject to impacts, such as a moored vessel. Chapter 16 gives a brief taste of the nonlinear dynamics of energy-conserving systems, exemplified by the large particle accelerators used by high-energy physicists. The book is concluded by Chapter 17 written by Harry Swinney, which surveys some recent physical experiments aimed at elucidating fundamentals of nonlinear dynamics as they occur in the real world: we are grateful to Professor Swinney and to North-Holland Publishing Company for permission to reprint this from the article that originally appeared in *Physica D*. (These last two chapters are replaced by new material in the second edition.)

Reference is made throughout the book to standard works and to the current research literature, and a critically selected bibliography of over 400 entries is included.

Michael Thompson
University College London, UK

Bruce Stewart
Brookhaven National Laboratory, Upton, New York, USA

Acknowledgements from the First Edition

We are particularly grateful to Ralph Abraham and Otto Rössler for sharing their insights with us. We also thank Ian Stewart, David Rand, Robert MacKay, Garrett Birkhoff, Charles Tresser, Joe Ford, Rob Shaw, Jim Crutchfield, Jerry Gollub, John Guckenheimer, Alan Newell, Erica Jen, Paul Manneville, Per Bak, Sue Coppersmith, and Hao Bai-Lin for stimulating discussions.

The gracious help at Brookhaven of Madeleine Windsor, and Judy Liu with library facilities, Bill Bottinger and Lily Liu with computer micrographics, Gordon Smith with graphics software, and Diane McCarron, JoAnn Langan and Muriel Kolomick with secretarial services is deeply appreciated. In London we are greatly indebted to our two assistants Ramin Ghaffari and Claudio Franciosi for their work on the preliminary drafts of several chapters: especial thanks must also be extended to Lawrence Virgin, Dennis Leung and Steve Bishop for lively discussions and help with the illustrations, to Richard Thompson for computer software, and to Alexis Lansbury for her enthusiastic reading of the whole manuscript.

It is a pleasure to acknowledge the support of the Applied Mathematical Sciences programme of the US Department of Energy, and various grants from the Science and Engineering Research Council of Great Britain.

Figures 1.5 and 1.10 are reproduced with permission from the paper 'Steady motions exhibited by Duffing's equation: a picture book of regular and chaotic motions' by Yoshisuke Ueda, in *New Approaches to Nonlinear Problems in Dynamics*, edited by Philip J. Holmes (Copyright 1980 by SIAM, all rights reserved). Figure 3.18 is reproduced from *Theoretical Ecology: Principles and Applications*, edited by Robert M. May with the permission of the publishers, Blackwell Scientific Publications Ltd. Figure 4.8 is reproduced from 'The square prism as an aeroelastic nonlinear oscillator' by G. V. Parkinson and J. D. Smith, *Quart. J. Mech. and Appl. Math.*, **17**, pp. 225–239 (1964), and reproduced by permission of Oxford University Press. Figures 5.9, 5.10 and 5.11 were kindly contributed by Chihiro Hayashi, and are reproduced with permission of the McGraw-Hill Book Company from *Nonlinear Oscillations in Physical Systems*, by C. Hayashi, McGraw-Hill (1964). Figure 9.8 due to J. P. Crutchfield is reproduced with his kind permission. Figure 10.4 is reproduced from the paper 'Dynamical instabilities and the transition to chaotic Taylor vortex flow' by P. R. Fenstermacher, H. L. Swinney and J. P. Gollub, *J. Fluid Mech.*, **94**, pp. 103–128 (1979), with the permission of Cambridge University Press. Figure 11.7 is taken from 'Deterministic nonperiodic flow' by E. N. Lorenz, *J. Atmos. Sci.*, **20**, pp. 130–141 (1963) with the permission of the American Meteorological Society. Figure 14.1 was supplied by Three Quays Marine Services, and is included with their kind permission. Material from Thompson (1984a) and Thompson, Bokaian and Ghaffari (1983) is reproduced courtesy of Butterworth and Academic Press, respectively.

Chapter 17 (not in the second edition) is an up-dated version of the original article 'Observations of order and chaos in nonlinear systems' by H. L. Swinney, *Physica*, **7D**, pp. 3–15 (1983). We are most grateful to Harry Swinney for allowing us to include this article, and acknowledge his generous help in up-dating some of the material, which is published here with the permission of the North Holland Publishing Company.

1

Introduction

The theory of any function begins naturally with its qualitative aspect, and thus the problem which first presents itself is the following: *Construct the curves defined by differential equations.*

This qualitative study, once completed, will be of the greatest utility for the numerical calculation of the function.

Furthermore, this qualitative study will be in itself of primary interest. Many important questions in Analysis and Mechanics in fact reduce to just this.

Henri Poincaré

1.1 HISTORICAL BACKGROUND

Nonlinear dynamics, in the guise of planetary motions, has some claim to be the most ancient of scientific problems. Among its few rivals in longevity is geometry; it therefore seems surprising that geometric methods in nonlinear dynamics were not pursued in earnest until the present century. The founder of geometric dynamics is universally acknowledged to be Henri Poincaré (1854–1912), who alone among his contemporaries saw the usefulness of studying topological structure *in the phase space* of dynamical trajectories. The theoretical foundations laid by Poincaré were strengthened by G. D. Birkhoff (1884–1944); but, apart from a few instances such as the stability analysis of Liapunov, Poincaré's ideas seemed to have little impact on applied dynamics for almost half a century. The reason may be in part that Poincaré and Birkhoff, motivated by problems in celestial mechanics, concentrated on energy-conserving Hamiltonian systems for which Liouville's theorem holds. Dissipative systems, on the other hand, have the property that an evolving ensemble of states occupies a region of phase space whose volume decreases with time. Over the long term, this volume contraction has a strong tendency to simplify the topological structure of trajectories in phase space. This can often mean that a complex dynamical system with even an infinite-dimensional phase space—governed for example by partial differential equations—can settle to final behaviour in a subspace of only a few dimensions.

Recent experimental observations of such low-dimensional behaviour, for example in Couette flow, chemical reaction kinetics, and solid-state plasmas, suggest that a better understanding of *a priori* low-dimensional mathematical models of dynamics would be a helpful guide to behaviour in more complex dissipative systems. Such low-dimensional models are furnished by initial value problems for ordinary differential equations. One important example is the damped, periodically forced nonlinear oscillator with displacement x:

$$\ddot{x} + k\dot{x} + x^3 = B\cos t \tag{1.1}$$

where a dot denotes differentiation with respect to time t. The behaviour of solutions to this nonlinear equation was extensively studied by Duffing (1918). Another second-order oscillator of great interest is the Rayleigh–Van der Pol equation:

$$\ddot{y} + \alpha(y^2 - 1)\dot{y} + \omega_0^2 y = A\sin\omega t \tag{1.2}$$

which describes self-excited relaxation oscillations for $\alpha \gg 1$ and $A = 0$. This equation, introduced by Lord Rayleigh (1896), was studied both theoretically and experimentally using electric circuits by Van der Pol (1927). From his experiments, Van der Pol was well aware that relaxation oscillations are readily susceptible to entrainment with the forcing frequency ω, and that the same forced oscillator (with fixed numerical values of α, ω_0, A, and ω) may become entrained to different subharmonic motions, depending only on how the oscillator is started. This phenomenon of multiple final behaviours in the same system is an important and common feature of recurrent nonlinear dynamics.

Detailed mathematical analysis of the forced Van der Pol equation by Cartwright and Littlewood (1945) and by Levinson (1949) revealed another important aspect of this phenomenon: in cases where two different final motions of (1.2) are possible, the start-up transient may hesitate between the two for an arbitrarily long time before settling to one or the other. A geometric phase-space picture of this complicated transient behaviour was provided by Smale (1963), who showed, that, despite an aspect of true randomness, these hesitating transients are governed by a relatively simple stretching and folding of ensembles in phase space. Furthermore, inspired by Andronov's ideas on structural stability in dynamics, Smale showed that his qualitative picture—the Smale horseshoe—persists under typical small perturbations of equations (1.2). Thus the paradoxical combination of randomness and structure, now called chaos, is likely to occur commonly in nonlinear dynamics.

Meanwhile, applied dynamicists began to discover other examples of chaotic behaviour in simple model systems with dissipation. Motivated by the meteorological problem of weather prediction, Lorenz (1963) studied a severely simplified model of Rayleigh–Bénard convection in fluids, which provided the first specific example of chaotic dynamics persisting for all time—a chaotic attractor. In Japan, pioneering work by Hayashi (1964, 1975) with nonlinear electric circuits provided the first detailed topological portraits of forced oscillators such as (1.1) using the technique of Poincaré mapping. This led to studies by Ueda (1980a) of steady-state chaotic behaviour of Duffing's equation (1.1).

1.2 CHAOTIC DYNAMICS IN DUFFING'S OSCILLATOR

Let us now be more concrete and consider the behaviour of the specific equation

$$\ddot{x} + 0.05\dot{x} + x^3 = 7.5\cos t \qquad (1.3)$$

In mechanical engineering, such an equation might model for example the motion of a sinusoidally forced structure undergoing large elastic deflections. Often in the past, studies of such equations concentrated on narrow parameter regimes close to a fundamental or subharmonic resonance, but here the magnitudes of damping and forcing amplitudes have been chosen to put us in one of the quite typical regimes of chaotic response identified by Ueda. An example of a time series solution x versus t obtained by digital integration of (1.3) is presented in Figure 1.1. This *chaos* has a ragged appearance, which persists for as long as time integrations are carried out. Although its *recurrent* nature is evidenced by the fact that certain patterns in the waveform repeat themselves at irregular intervals, there is never exact repetition, and the motion is truly *non-periodic*. The reason for this is that, when two identical systems are started in nearly identical conditions, the two motions diverge from each other at an exponential rate. Of course, if the starting conditions were precisely the same, then the *deterministic* nature of the equation guarantees that the motions are identical for all time. But since some uncertainty in the starting condition is inevitable with real physical systems, the divergence of nominally identical motions cannot be avoided in the chaotic regime. This is illustrated in Figure 1.2 by starting two numerical integrations from adjacent states, one with $x(0) = 3$, $\dot{x}(0) = 4$ (as in Figure 1.1) and the other slightly perturbed. This starting condition was chosen to minimize start-up transients, since it lies

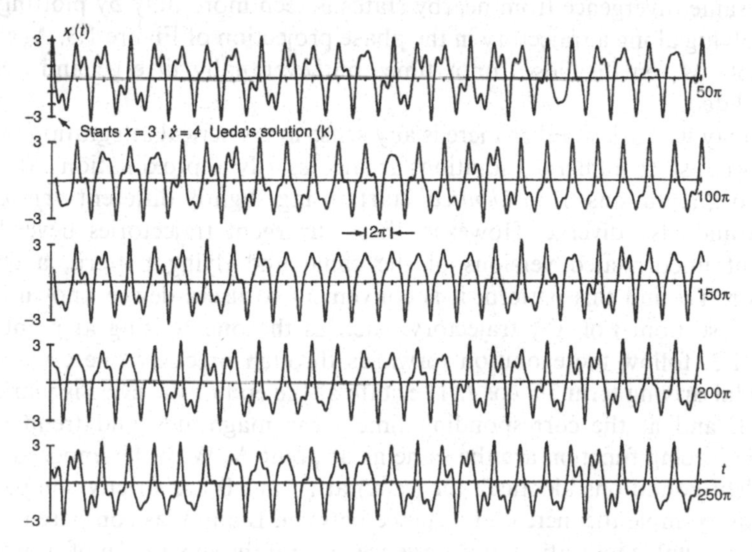

Figure 1.1 Time series of a steady-state chaotic response: $\ddot{x} + k\dot{x} + x^3 = B\cos t$ with $k = 0.05$, $B = 7.5$

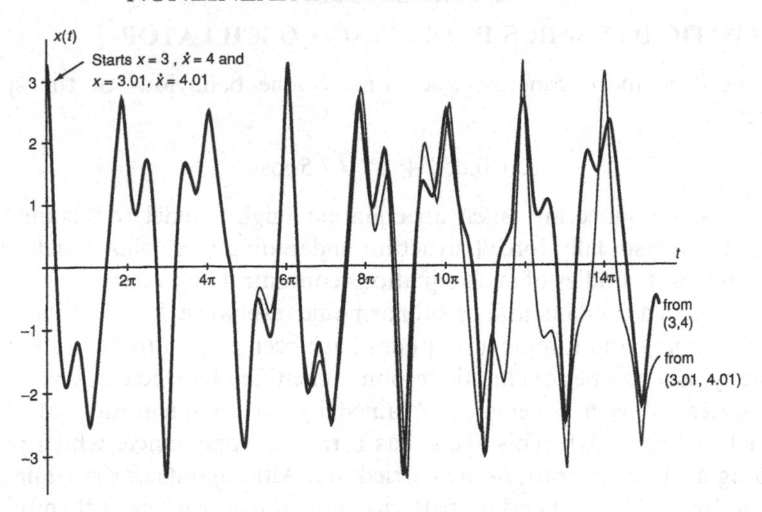

Figure 1.2 Divergence from adjacent starts: waveforms for Ueda's (k) solution, $k = 0.05$, $B = 7.5$

within the region of phase space occupied after long integration by a previously computed solution. The two adjacent starts appear to remain close to each other for a time and then rapidly become uncorrelated. On the average, their separation increases by a fixed *multiple* for any given interval of elapsed time. Because of the *exponential divergence* it is impossible to impose long-term *correlation* of the two motions by reducing the initial perturbation, since each order of magnitude improvement in initial agreement is eradicated in a fixed increment of time.

This same divergence from nearby states is seen more fully by plotting $x(t)$, $\dot{x}(t)$ evolving along a trajectory in the phase projection of Figure 1.3. Again the two motions remain close for a time, but eventually diverge and become uncorrelated.

One may well ask whether there is any sense in numerical integration of such trajectories, since numerical solutions are necessarily approximations. It is true that two integrations from *identical* starts using slightly different *approximations* would also diverge. However, these divergent trajectories nevertheless represent uncorrelated versions of the same underlying pattern, a chaotic attractor. To find this pattern, it is convenient to use a device known as the Poincaré section. For any trajectory, such as the one starting at point A in Figure 1.3, follow the evolution forwards through precisely one cycle of the sinusoidal driving term 7.5 cos t. The state at the end of the cycle is marked as point B, and at the corresponding time t the magnitude and trend of the periodic forcing function are the same as at point A. With the imposed, time-dependent conditions identical, the $(x, \dot{x})$ at A and B are strictly comparable. Thus for example the increased distance between B and b as compared with A and a is a reliable indication of divergence, even if the separation of trajectories fluctuates during the intervening period.

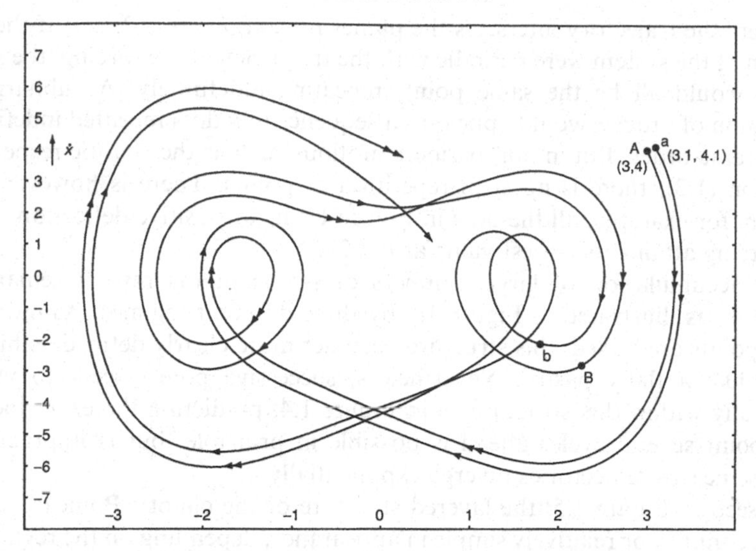

Figure 1.3 Divergence from adjacent starts: phase projections for Ueda's (k) solution

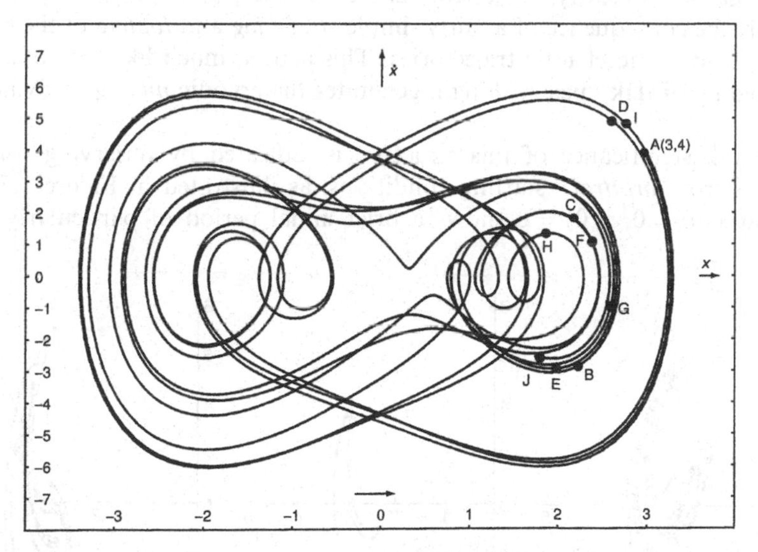

Figure 1.4 Phase projection for Ueda's (k) solution with a sequence of points in a Poincaré section

By continuing to mark the trajectory *stroboscopically* at times that are integer multiples of the forcing period 2π, a sequence of strictly comparable points A, B, C, D, etc., is accumulated in Figure 1.4. Proceeding now to simplify the picture, we erase the continuous trajectory and show only the strobed points in the *Poincaré section*. If the trajectory were traced out in the full three-dimensional phase space with coordinates x, $\dot{x}$ and t, these strobe points would lie where the trajectory intersects the planes $t = 2\pi i$, $i = 0, 1, 2, \ldots$. If the final motion of the system were periodic with the frequency of the forcing, the strobe points would all be the same point, repeating indefinitely. A subharmonic oscillation of order n would appear as a sequence of n dots repeated indefinitely in the same order. But in non-periodic motions such as the chaotic response in equation (1.3), there is no such repetition of points. There is however some pattern; for example, all the dots in Figure 1.4 lie at positive deflection, where the forcing attains its largest value at $t = 2\pi i$.

The accumulation of large numbers of strobe points reveals remarkable structure, as illustrated in Figure 1.5 by three different regimes. As more and more points are added, this structure becomes more clearly defined, exhibiting layers like a flaky pastry. Nevertheless, successive points seem to wander erratically within this structure as in Figure 1.4; predicting the exact location of a point several cycles ahead is possible in principle, but is impracticable because nearby trajectories diverge exponentially.

As seen in Figure 1.5, the layered structure of the chaotic Poincaré section can be complex or relatively simple in appearance, depending on the regime and the degree of resolution. Another high-resolution dot signature from Ueda's studies, shown in Figure 1.6, shows the possible complexity of forms, and its *fractal* nature is clearly suggested. As we shall see, this complex structure is actually the consequence of a fairly simple *stretching* and *folding* of the ensemble of steady-state chaotic trajectories. This action, much like the rolling out and folding of flaky pastry dough, generates the ergodic *mixing* of dynamical states.

The full significance of this structure is indicated by observing *transient* motions from *arbitrary* starting conditions, as illustrated in Figure 1.7. One start at $x(0) = 0$, $\dot{x}(0) = 0$ shows a brief initial period of particularly spiky

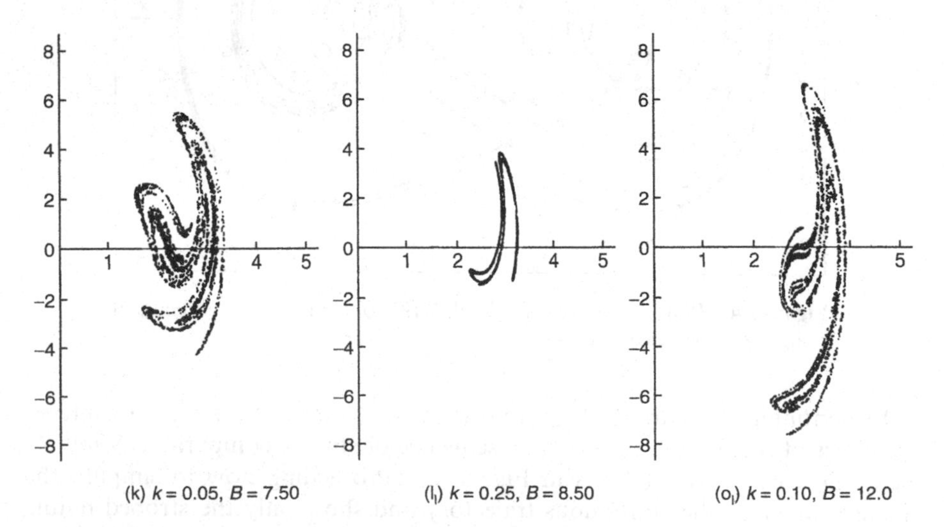

Figure 1.5 Poincaré sections in the plane $(x, \dot{x})$ for three of Ueda's chaotic attractors. Reproduced, with permission of SIAM, from Ueda (1980a)

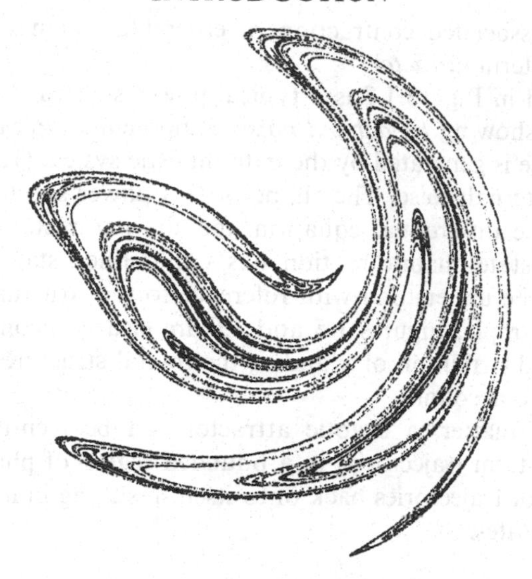

Figure 1.6 High-resolution Poincaré section due to Ueda of the unique chaotic attractor of equation (1.3)

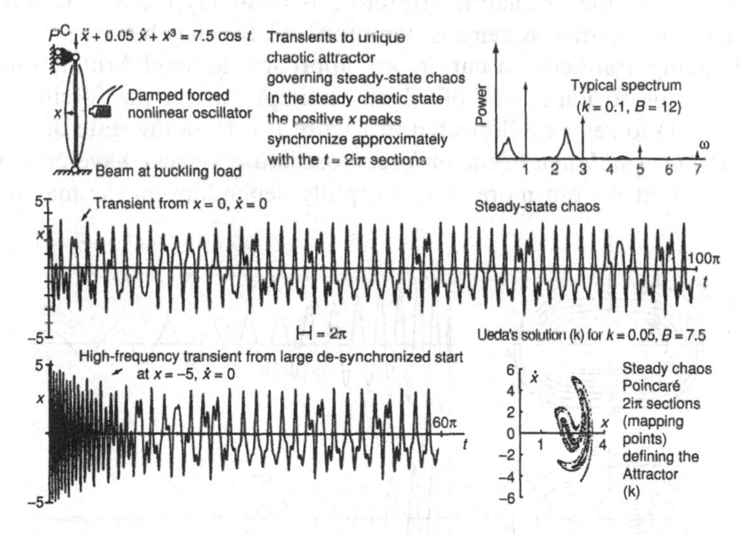

Figure 1.7 Transients settling to steady-state chaos, and a steady-state power spectrum including sharp peaks and broadband components. P^C is the critical buckling load of the beam

oscillation, after which the time series resembles Figure 1.1; another start at $x(0) = -5$, $\dot{x}(0) = 0$ shows a longer period of transient behaviour. In either case the sequence of strobe points in a Poincaré section would show points initially far away from the structure in Figure 1.5 (on the left). The succession of points would however soon approach this structure, and eventually fill in the *same structure*. This settling to well-defined final behaviour is the result of *dissipation* and associated contraction of ensemble volumes in phase space, which justify the term *attractor*.

Also illustrated in Figure 1.7 is a typical power spectrum of a steady-state chaotic motion, showing *broadband noise* components typical of chaotic dynamics. This noise is generated by the deterministic system (1.3) itself, without any source of external noise. The shape of this power spectrum is of course determined by the differential equation and the parameter regime selected. Thus the steady-state chaotic motion has well-defined statistical properties; but these should be understood with reference to the structure of the chaotic attractor. For example, Figures 1.5 and 1.6 are clearly inconsistent with any simple Gaussian distribution of x and $\dot{x}$; the fractal structure of the attractor must be taken into account.

As a practical matter, a chaotic attractor can be identified as a stable structure of long-term trajectories in a bounded region of phase space, which folds the bundle of trajectories back onto itself, resulting in mixing and divergence of nearby states.

1.3 ATTRACTORS AND BIFURCATIONS

Although the notion of chaotic attractors is relatively recent, the settling of transients in dissipative systems is common and familiar behaviour. The simplest decaying transients occur in an approach to equilibrium; numerous examples come to mind, one of which is simply to set the forcing term in equation (1.1) to zero, as illustrated in Figure 1.8. Here the state of rest is the unique *stable equilibrium* of the unforced oscillator; typical waveforms exhibit oscillations that die out more or less rapidly depending on the magnitude of

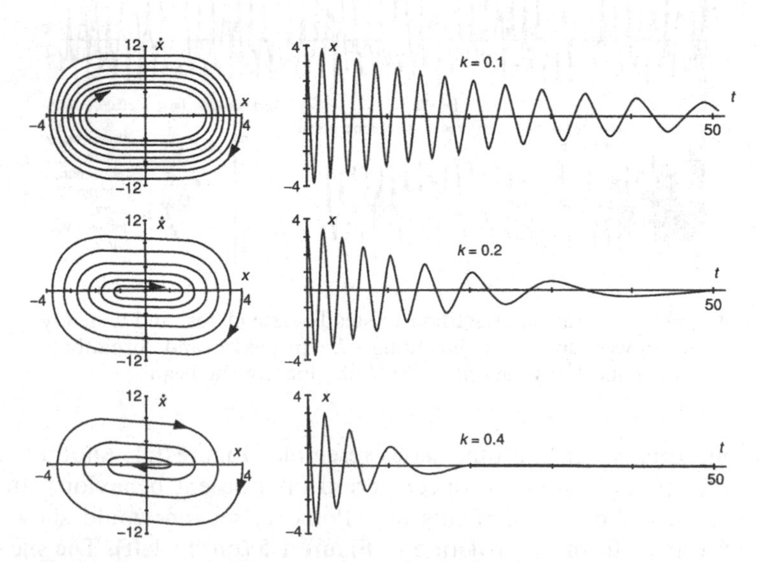

Figure 1.8 Long-term behaviour of Duffing's equation without forcing ($\ddot{x} + k\dot{x} + x^3 = 0$): point attractor

damping. Furthermore, trajectories traced in the $(x, \dot{x})$ phase plane spiral in towards the origin where $x = 0$, $\dot{x} = 0$ defines the state of rest. Thus the stable equilibrium is a *point attractor* for all trajectories in phase space.

Likewise stable periodic motions also attract nearby trajectories in phase space. Figure 1.9 shows five stable periodic motions, each of which is a possible long-term motion of the forced Duffing oscillator with $k = 0.08$ and $B = 0.2$. Not shown in the figure are transients settling to the attractors; various desynchronized starts of the oscillator from different initial states might settle to different final motions. The process of settling to a given attractor could be observed using the stroboscopic Poincaré section. For example, the first attractor is a small-amplitude motion at the fundamental frequency, with one Poincaré section point A; a transient settling to this final behaviour would show a sequence of strobed points converging on A. This stroboscopic technique will demonstrate geometrically how, as with equilibria, stable periodic motions attract nearby trajectories, justifying the name *attracting limit cycle*. The approach of transients towards the limit cycle can also be seen in the three-dimensional $(x, \dot{x}, t)$ phase space, to be illustrated in a subsequent chapter. These three types of attractors—*equilibrium* (point), *periodic* (cycle), and *chaotic* (mixing)—represent the most commonly observable long-term motions in dissipative systems. Thus we identify all stable final motions with *attracting sets* in phase space.

For any dynamical system, such as the forced Duffing oscillator with given coefficients, a qualitative description of dynamic behaviour begins with the geometric phase-space identification of all possible attractors. In nonlinear systems, multiple attractors are common, and indeed there may be periodic

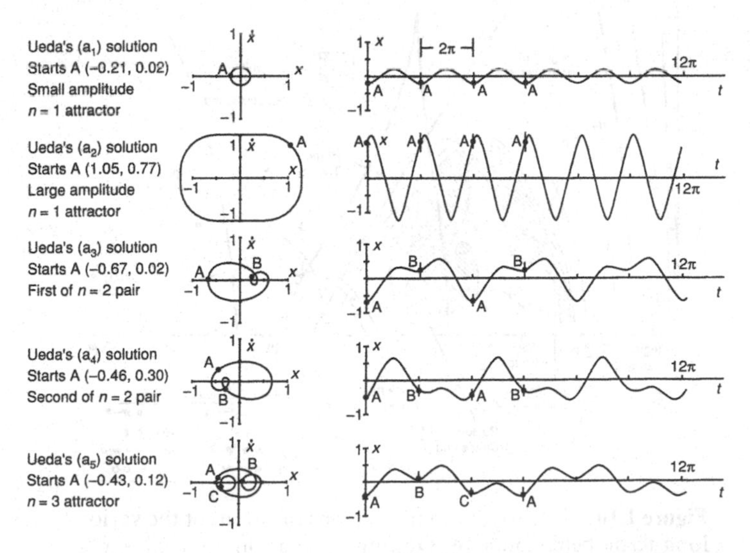

Figure 1.9 Long-term behaviour of Duffing's equation with forcing $(\ddot{x} + k\dot{x} + x^3 = B\cos t,\ k = 0.08,\ B = 0.2)$: five coexisting periodic attractors

and chaotic attractors coexisting, for example. Once the attractors and the type of each is known, one would like to associate each attractor with the ensemble of all starting conditions that settle to it; this point set is the *catchment region* or *basin of attraction*. Together, the basins of all attractors should constitute the entire phase space. These basins can be identified by massive trials of various starting conditions, but as we shall see, Poincaré's *invariant manifold* theory provides more efficient tools for this task.

Having mapped out the locations in phase space of attractors and basins, we arrive at the *AB (attractor-basin) phase portrait* of the dynamical system, as expounded in Abraham and Shaw (1992). Further investigation of the system would then involve constructing similar phase portraits for other values of the coefficients. In fact, Ueda has given a detailed and comprehensive diagram of Duffing's forced oscillator (1.1), which is the basis for Figure 1.10. Here the damping k and forcing magnitude B are treated as parameters, and the $k-B$ plane is divided into regions where various attractors (final motions) exist. To complete this diagram we would like to see phase portraits associated with each point in the parameter space; Ueda (1980a) gives a generous sample of these.

Some of the regions in Figure 1.10 have already been illustrated; for example, region (a) corresponds to the five periodic attractors illustrated in Figure 1.9, while the chaotic steady state of Figure 1.1 to 1.4 lies in region (k). Note that the regions of multiple attractors and of chaotic attractors are quite extensive

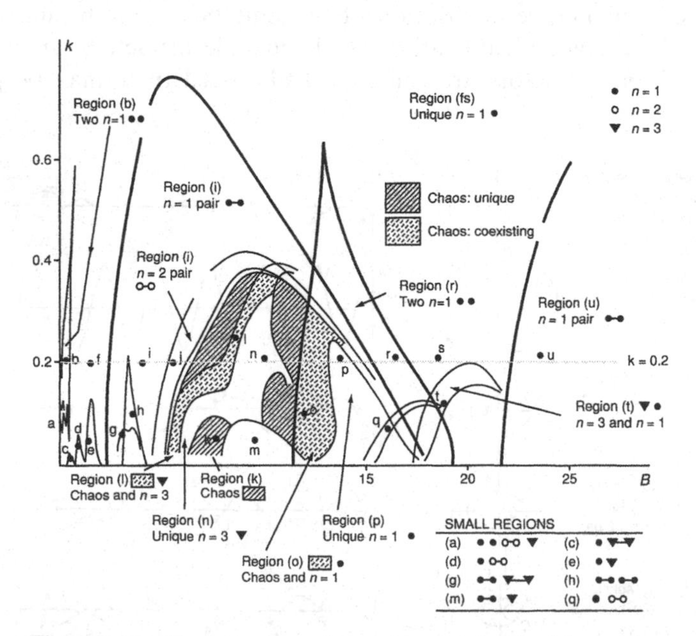

Figure 1.10 Twenty-one major regions (a) to (u) of the various long-term behaviours of Duffing's equation ($\ddot{x}+k\dot{x}+x^3 = B\cos t$) as mapped by Ueda, as a function of damping magnitude k and forcing amplitude B. Reproduced, with permission of SIAM, from Ueda (1980a)

and by no means untypical or pathological. The numbers n in the figure refer to the subharmonic ratio of periodic solutions, that is, the period of the response divided by the period of the forcing term. Thus an $n = 3$ subharmonic repeats exactly after three forcing cycles and has three strobe points, as in the bottom of Figure 1.9.

The regions in parameter space are delimited by various arcs. To interpret the meaning of these arcs, it is helpful to think of the parameters k and B as *controls*, like a throttle or rheostat used to adjust the operating regime of a real dynamical system such as an airplane, a motor, or a simulation device. We may then imagine this dynamical system running at high speed while the controls are slowly adjusted; we gradually change the controls, and let the system settle to final behaviour in each new regime. As the control settings cross one of the arcs in Figure 1.10, we observe the system settling to a qualitatively different behaviour: the motion may change from periodic to chaotic, or the previously stable motion may become unstable, in which case the system settles to a different attractor; or the change may be more subtle, as when the subharmonic number of a stable periodic motion changes. In any case, there has been a *qualitative* change in the long-term behaviour, associated with a change in (or disappearance of) an attractor. In theory there is a mathematically precise value of the control settings, lying exactly on an arc in Figure 1.10, at which the qualitative change, or *bifurcation*, occurs. A complete understanding of such qualitative changes means knowing the topological changes in structure of the phase portrait at the bifurcation threshold. Methods for constructing such *control-phase portraits* will be developed and applied in the remainder of this book.

Part I

Basic Concepts of Nonlinear Dynamics

2

An Overview of Nonlinear Phenomena

In Part I we aim to give a general outline of nonlinear dynamics, which is an essential prerequisite to our more advanced studies including our goal of understanding chaotic motions. This chapter provides a quick overview of the nonlinear dynamics field, before we begin our more detailed presentation.

2.1 UNDAMPED, UNFORCED LINEAR OSCILLATOR

We start our overview by looking at the undamped, unforced linear oscillator of Figure 2.1. The equation chosen for this first illustration has the stiffness constant $4\pi^2$, which makes the periodic time equal to unity. The solution of such an equation is simply a sine wave, the constant amplitude and phase of which are determined by the starting values of x and $\dot{x}$. So, once started, we have a constant sine wave that persists for all time, and there is no transient or

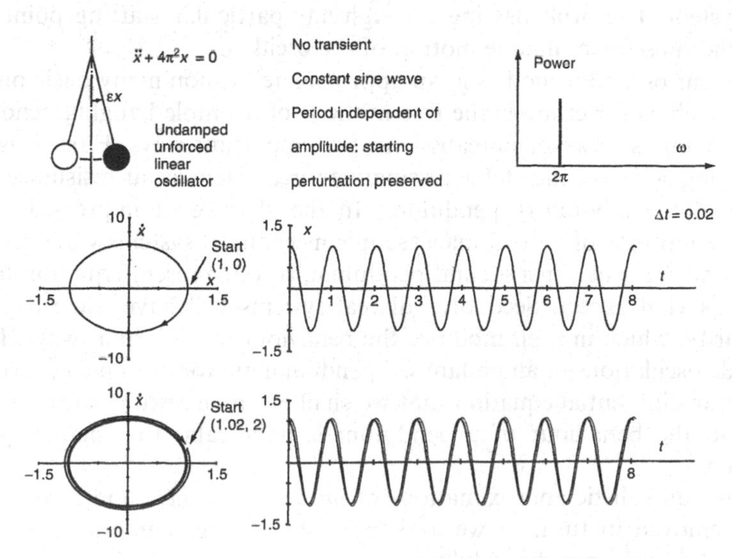

Figure 2.1 Undamped, unforced behaviour of a linear oscillator

decay of any kind. The periodic time, unity in the present example, is a constant independent of the starting conditions, the amplitude of the motion, and the time.

A typical plot of x against the time t is shown, resulting from the starting condition $(x, \dot{x})$ equal to $(1, 0)$ at the time $t = 0$. Along with the other, mainly nonlinear, problems considered in this chapter, this solution was obtained by numerical time integration using a fourth-order Runge–Kutta routine on a desktop Hewlett-Packard computer with the step size indicated, here $\Delta t = 0.02$.

If we plot not x against t but $\dot{x}$ against x, we have the phase portrait shown on the left. Starting as before at $(1, 0)$ we now have the closed ellipse shown, the representative point moving continuously round and round this closed orbit as the time goes to infinity. The power spectrum of this response, shown in the top right-hand diagram, is simply a spike (or delta function) at the circular frequency of 2π radians per second.

We must finally ask the question: what would happen if we changed the starting condition by a small amount? The answer is illustrated in the lower diagram, where we show both the *fundamental* reference motion starting at $(1, 0)$ and a perturbed motion starting at $(1.02, 2)$. We see that we have two sine waves running in step with just a small difference in amplitude and phase resulting from the slightly different starting values of x and $\dot{x}$. They continue to run nicely in step for all time because the period of oscillation of the two motions is the same (and equal to unity, as we have seen). So a starting perturbation is preserved, and the fundamental motion is *neutrally stable* in a dynamical sense.

In the left-hand phase space, the two motions appear as neatly nesting ellipses. All possible motions of this linear oscillator are indeed represented by a complete family of nesting ellipses, which represent the full phase portrait of the system. The orbit passing through any particular starting point $(x, \dot{x})$ defines the subsequent unique motion of the oscillator.

This linear oscillator models in an approximate fashion many basic physical systems, such as for example the free motions of a simple hanging pendulum. The modelling is however unrealistic in two important ways. First, it ignores the damping action of inevitable dissipative forces, such as air resistance in the example of the laboratory pendulum. In the absence of impressed driving forces, the motions of all real macroscopic mechanical systems will eventually decay, as with a free experimental pendulum, so our present equation fails to model this vital aspect. Secondly, all real systems will have some degree of nonlinearity, which in itself modifies the behaviour in important ways. Large-amplitude oscillations of an undamped pendulum are for example governed by a nonlinear differential equation that we shall examine next: a linear approximation to the behaviour of a pendulum is only valid for small angles of oscillation.

The two unrealistic approximations of *linearized stiffness* and *zero damping* will be removed in turn, so we look next at the large-amplitude, nonlinear motions of an undamped pendulum.

2.2 UNDAMPED, UNFORCED NONLINEAR OSCILLATOR

The undamped, unforced nonlinear system of Figure 2.2 represents the *exact* equation of motion of a simple pendulum undergoing arbitrarily large oscillations. This equation in terms of the angle x is easily derived using Newton's law of motion for the bob by resolving perpendicular to the light string to eliminate the unknown tension: alternatively it can be derived by Lagrangian or Hamiltonian energy methods. The length of the pendulum, relative to the gravitational constant, has been chosen to make the coefficient equal to $4\pi^2$. So for small oscillations we could *linearize* the equation by approximating $\sin x$ to x, and retrieve the linear oscillator of our earlier discussion, with periodic time equal to unity.

The solution of this nonlinear differential equation can be obtained after some algebra in terms of elliptic integrals: alternatively the equation can be easily integrated numerically on a digital computer as we have done here. Depending on the starting conditions of $(x, \dot{x})$ we now find a steady undamped oscillation corresponding to the motion of our idealized undamped pendulum. A *given motion* from a given start thus exhibits no transient or decay, just a steady waveform of constant amplitude and constant period. The waveform is not however sinusoidal, and could in fact be decomposed by Fourier analysis into a fundamental harmonic plus odd higher harmonics: this gives rise to the power spectrum shown with a large spike at a certain circular frequency ω_F and smaller spikes at 3, 5, 7, ... times this value.

The central waveform shows the steady oscillation starting at (3.054, 0) corresponding to the pendulum starting from rest with $\dot{x} = 0$ at a value of $x = 3.054 \times 180/\pi = 175°$. To visualize this physically we must suppose that the heavy pendulum bob is supported not by a string, which could become slack, but by a light rigid rod pivoted to the fixed support. Because this

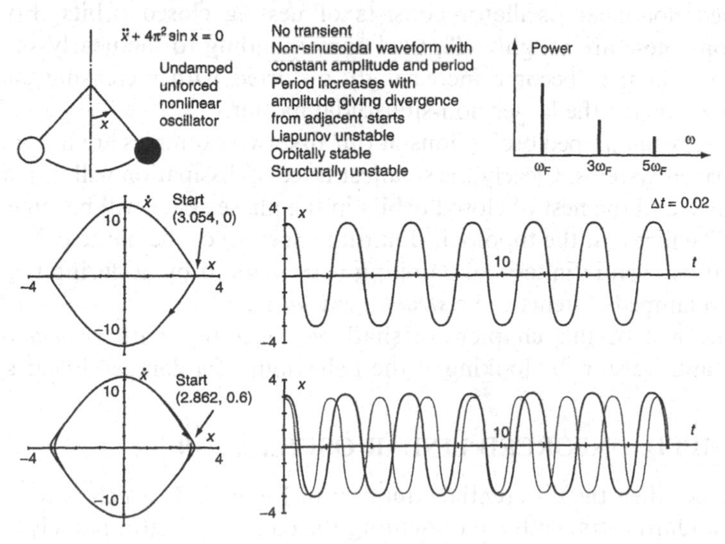

Figure 2.2 Undamped, unforced behaviour of a nonlinear oscillator

rigid-link pendulum would be in (unstable) equilibrium at $x = 180^\circ$, the motion begins very slowly and the waveform is very flat and quite noticeably non-sinusoidal. The corresponding $\dot{x}(x)$ phase picture is shown to the left-hand side: the closed trajectory is quite clearly not elliptical, and has a high curvature on the x axis corresponding to the proximity of an unstable equilibrium state.

Now the periodic time of a given motion is constant as we have just seen, but the period of different motions *increases* with the amplitude. It is clear for example that a start very close to $x = 180^\circ$ will give a motion with a very large period, since at the end of each big swing the pendulum will almost come to rest in the inverted position: indeed the periodic time goes to infinity as the amplitude approaches π. Notice that the periodic time of our displayed waveform is about 3, compared with the periodic time of unity for the small-amplitude linearized motions.

This variation of period with amplitude gives rise to a new phenomenon when we consider a perturbed motion. The lower diagram shows the fundamental motion just considered together with a perturbed motion starting from slightly different initial conditions. Because these new conditions give rise to a motion with a slightly different amplitude, the perturbed waveform has a slightly different period. So we have a *beat* phenomenon and the two motions drift in and out of phase with one another. This means that, although the two waveforms will eventually resynchronize, there is an initial *divergence* from adjacent starts. This makes the fundamental oscillatory motion unstable in the strict sense of Liapunov. In the left-hand phase diagram however, in which the *time* discrepancies of the two motions are not visible, the two closed *orbits* are seen to lie everywhere close to one another: in recognition of this fact the fundamental motion is said to be *orbitally stable*.

For the motions under consideration, the phase portrait of the present undamped nonlinear oscillator consists of nesting closed orbits. For small oscillations these are roughly elliptical corresponding to the nearly sinusoidal waveform, but they become increasingly distorted with increasing curvature near the x axis for the larger non-sinusoidal motions.

The steady undamped oscillations of our first two examples are not typical of real undriven systems. Clearly the smallest trace of dissipation will give damped waveforms, and the nest of closed orbits in the phase space will become *inward spirals*. The fact that the topological nature (closure) of the phase orbits can be destroyed by even infinitesimal damping is recognized by declaring the pathological undamped systems to be *structurally unstable*.

For the rest of this chapter we shall be concerned with *typical* damped systems, and we start by looking at the behaviour of a damped linear system.

2.3 DAMPED, UNFORCED LINEAR OSCILLATOR

We consider then the differential equation of Figure 2.3, which is written in a rather standard form, with ζ representing the damping factor, namely the ratio of the actual damping to the critical damping at which oscillatory behaviour

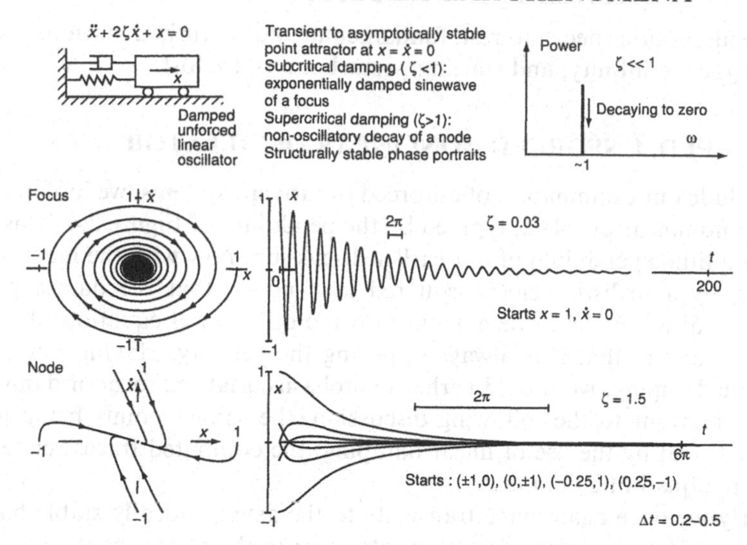

Figure 2.3 Damped, unforced behaviour of a linear oscillator

ceases. We can think of this equation as representing the motion of a mass constrained by a linear elastic spring in parallel with a dashpot full of oil, which is assumed to provide a force opposing the instantaneous velocity.

An analytical solution of this linear differential equation is readily written down: for light damping with $\zeta < 1$ we have an exponentially damped sine wave, while for heavy damping with $\zeta > 1$ we have a non-oscillatory exponential decay.

A typical lightly damped waveform is shown in the middle picture, starting at $x = 1, \dot{x} = 0$. The decaying wave has a constant period, defined for example by successive crossings of the time axis, which is nevertheless slightly dependent on the value of ζ. With the light damping shown, the period is essentially unchanged from the period, 2π, of the corresponding undamped system obtained by setting $\zeta = 0$. For light damping the power spectrum will be roughly a single spike decaying to zero along with the wave amplitude.

The corresponding phase portrait on the left is now a spiral, heading inwards towards the asymptotically stable equilibrium state at the origin (0, 0). The full linear phase portrait, termed a focus, is a set of intertwining, non-crossing spirals. Every motion here represents a transient to the asymptotically stable equilibrium state of rest at the origin, which for obvious reasons is called a *point attractor*. The whole phase portrait is now *structurally stable* since for finite damping the spiralling form cannot be topologically changed by *any* infinitesimal changes to the system.

The pictures for heavier supercritical damping shown below give the waveform and phase trajectories for six alternative starts. The system moves back to its stable state of rest in a direct non-oscillatory fashion, and the whole phase portrait is called a *node*. Once again, we have a structurally stable point attractor at (0, 0) capturing all motions of the system.

Since all motions decay to rest, fundamental and perturbed motions coalesce as time goes to infinity, and starting perturbations are lost.

2.4 DAMPED, UNFORCED NONLINEAR OSCILLATOR

To conclude our examination of unforced (undriven) systems, we look now at a damped nonlinear problem, typified by the pendulum of Figure 2.4. This is the large-amplitude pendulum of our earlier discussion, now with the modelling of air drag by a realistic velocity-squared law: notice that the damping force proportional to $\dot{x}^2$ has to be entered into the differential equation of motion as $\dot{x}|\dot{x}|$ to ensure that it is always opposing the velocity. Having put on this quadratic damping, we should perhaps emphasize that the form of damping is largely irrelevant to the following discussion, the salient points being just as well illustrated by the use of linear damping: the computed traces relate however to the quadratic damping.

Clearly we once again have transients to the asymptotically stable hanging equilibrium state representing a point attractor in the phase space.

The central waveform damps and becomes increasingly sinusoidal as x becomes small, while the power spectrum is a decaying set of spikes as shown. The phase portrait is a spiral, becoming increasingly elliptical as the trajectories approach the central attractor. A little linear damping would be needed to make this portrait structurally stable near the origin.

As with the undamped pendulum suffering large-amplitude oscillations, adjacent starts still exhibit a temporary beating character with an associated initial divergence due to the variation of the period with amplitude. But initial perturbations are eventually lost as all motions coalesce in the unique hanging state.

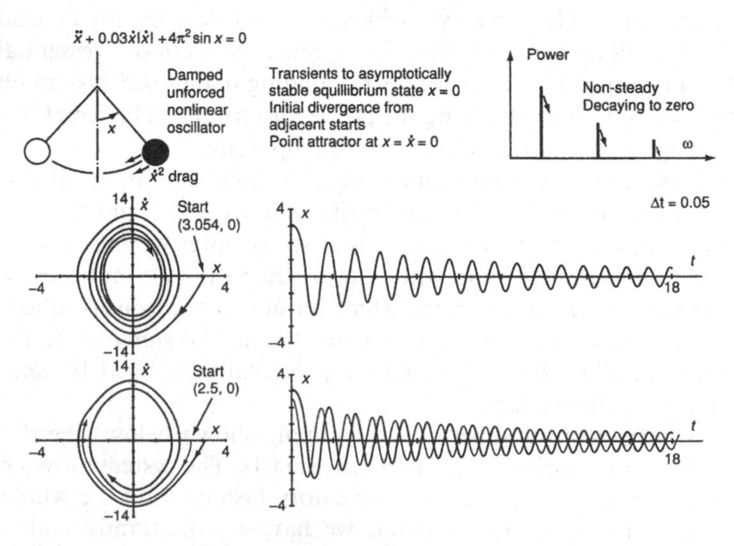

Figure 2.4 Damped, unforced behaviour of a nonlinear oscillator

This local phase portrait is a set of intertwining spirals with all motions captured by the central attractor. The full phase portrait of a pendulum including high-velocity motions passing through the inverted state is most nicely seen in a cylindrical phase space, and will be presented later (Figure 3.15).

2.5 FORCED LINEAR OSCILLATOR

We have so far looked only at autonomous unforced systems with zero on the right-hand side of the equation, but we turn now to sinusoidally driven non-autonomous oscillators. Damping, we have seen, is an essential ingredient of good modelling, so we shall start by looking at the damped, forced linear oscillator of Figure 2.5. This would be an adequate mathematical model of a pin-ended steel beam driven to small-amplitude lateral oscillations by an electromagnet carrying a sinusoidal alternating current. Here physical damping would arise from air resistance and internal material dissipation. The numerical coefficients have been chosen to provide a sharp frequency contrast between the transient and the steady-state solution, and the damping ratio of the unforced left-hand side is 0.1 (see equation 3.11).

This is a classical resonance problem of engineering texts, and the well-known analytical solution is easily written down. It is the algebraic sum of the so-called particular integral (PI) and the complementary function (CF). The CF is just the solution obtained by setting the left-hand side of the equation to zero: that is to say it is the exponentially damped sinusoidal solution of the unforced autonomous system. It has the usual two arbitrary constants of amplitude and phase obtained by applying the starting conditions to the *whole* solution. With the present choice of constants the CF is a high-frequency sine wave with quite a heavy rate of damping.

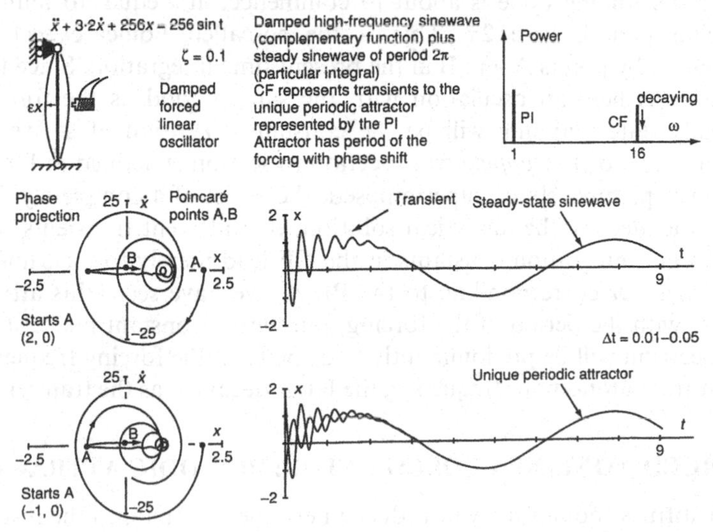

Figure 2.5 Damped, forced behaviour of a linear oscillator

The PI is a particular (known) solution of the whole equation, being in fact a steady undamped sine wave with the same frequency as the forcing term with which it has a fixed phase difference. The amplitude of the PI depends crucially on the ratio of the forcing frequency to the natural frequency of the autonomous left-hand side, being large when this ratio is close to one so that we have a condition of resonance. The conventional engineering resonance response curves simply plot the magnitude of the PI against this frequency ratio, giving for light damping a sharp peak at unity.

We should emphasize here, however, that from the qualitative dynamics point of view it is irrelevant whether the system is 'at resonance' or not. With the particular coefficients chosen, our illustration is well away from the resonant condition, but the discussion of the system's behaviour is essentially unrelated to this fact.

Since the analytical solution is just the algebraic sum of the CF and the PI, it is clear that the former damped sine wave represents a decaying transient, which leaves the PI as the unique final steady state: this is the reason for the engineer's consuming interest in the amplitude of the PI. A waveform starting at (2, 0) is shown in the central figure and we see clearly the high-frequency transient leading rapidly to the steady sinusoidal state described by the PI.

Now a forced system such as this has a three-dimensional phase space defined by the coordinates $(x, \dot{x}, t)$, the essence of phase spaces being that they are full of *non-crossing* trajectories. It is sometimes convenient, however, just to plot the *phase projection* $(x, \dot{x})$ and accept the fact that trajectories will appear to cross in this projection. The phase projection corresponding to the drawn waveform is thus shown to the left-hand side. The high-frequency transient appears as decaying circles, and the final steady state as a very long, thin ellipse pointing along the x axis.

It is also helpful in the phase projection to make a dot, or small circle, whenever the forcing cycle is about to commence, at t equal to multiples of the forcing period, here 2π. This is the so-called Poincaré section and is represented by points A and B in the present time integration. Since the final steady state is here an oscillation with the same period as the forcing, the final steady-state mapping will be the constant repetition of a fixed point, here quite close to B. *Mapping* from section to section is defined in Figure 5.1.

The lower pictures show, superimposed, the effect of a completely different start. As dictated by the analytical solution, the different transients resulting from different integration constants in the CF lead merely to the same *unique periodic attractor* corresponding to the PI. As we have seen, this attractor is sinusoidal with the period of the forcing, but with a constant phase shift. The power spectrum will be predominantly two spikes at the forcing frequency and at the natural autonomous frequency, the latter decaying as the transient is lost.

2.6 FORCED NONLINEAR OSCILLATOR: PERIODIC ATTRACTORS

Just as a stiffness nonlinearity introduced new phenomena into the response of an unforced oscillator, so a nonlinearity generates new features in a driven system. So we look now at the damped, forced nonlinear oscillator illustrated in Figure 2.6. This is the sinusoidally (here cosinusoidally) forced Duffing equation with a linear and a cubic stiffness. This could be used to model the moderately large bending deflections of an electromagnetically driven steel beam held pinned to fixed supports as shown. These fixed supports induce a membrane tension at finite deflections, which gives a hardening nonlinear stiffness modelled for moderately large deflections by the cubic term.

For such a driven nonlinear oscillator, closed-form analytical solutions are not available and recourse *must* inevitably be made to numerical time integrations. Just as with the preceding linear system, transients are observed, but after these have decayed we now find that there are two alternative stable steady states denoted here by A and B. The first plot of x against t shows these two steady oscillatory states, the starting points to eliminate transients having been found by previous trial computations. We see that the large-amplitude motion A and the small-amplitude motion B both have the same period as the forcing term and are therefore fundamental *harmonics* as opposed to subharmonics: they are noticeably out of phase with one another. The corresponding steady-state phase projections are shown in the left-hand phase diagram, each closed orbit having one Poincaré mapping denoted by a circle because the motions have the period of the forcing: these mapping points show where the system is whenever the time is a multiple of 2π.

These two steady-state solutions, A and B, can be seen on the resonance response diagram at the top right. This is a plot of the response amplitude against the ratio of the forcing frequency to the natural frequency of the autonomous system: this ratio is 1.6 for the parameters here adopted. Now in a linear resonance problem we have a vertical resonant peak, but the positive

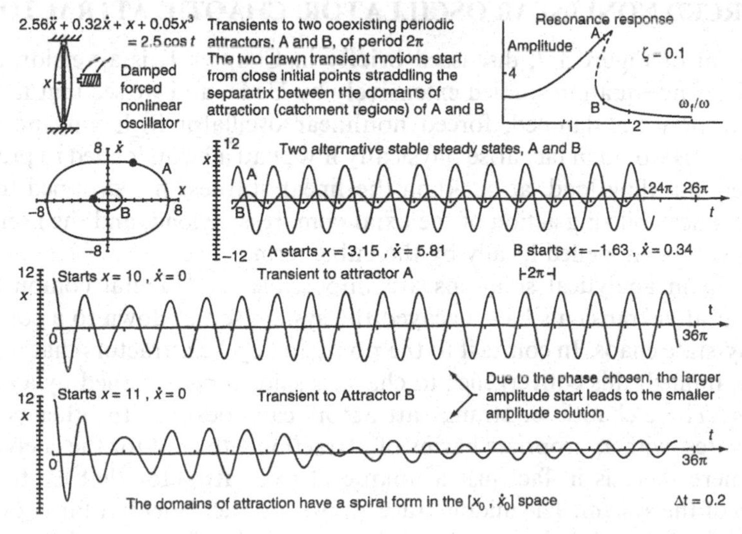

Figure 2.6 Damped, forced behaviour of a nonlinear oscillator: transients to periodic attractors

cubic stiffness of our Duffing's equation curves the peak to the right, giving a domain of frequency ratio with three steady states. The steady state of intermediate amplitude is unstable and so is not observed in a normal time integration, leaving us with the two alternative stable solutions A and B.

Now which of these two coexisting periodic attractors is picked up in a given time integration depends on the starting conditions, and two transient motions are illustrated in the lower diagrams. Starting with $(x, \dot{x})$ equal to (10, 0) gives a transient leading to attractor A while starting at (11, 0) gives a transient leading to attractor B. Notice that due to the phase chosen, the *larger*-amplitude start leads to the *smaller*-amplitude solution. The more obvious converse could equally apply, the final motion adopted being as much governed by phase as by amplitude.

Clearly in the space of the starting values of $(x, \dot{x})$ at $t = 0$ there will be *basins of attraction* such that motions originating in the basin of A lead after the decay of transients to solution A, while motions starting in the catchment region of B lead to the periodic attractor B. Between the basins of attraction (catchment regions) will be a separatrix curve, and it is clear that our two rather close starts straddle this separator. The basins of attraction tend to have a complex spiral form, which accounts for the sensitivity to both phase *and* amplitude previously mentioned.

This multiplicity of alternative stable attracting solutions (often more than the present two) dependent on the starting conditions, which is not encountered in the linear resonance problem with its unique periodic attractor, is typical of nonlinear driven oscillators.

We come at last to our final equation of this chapter giving rise, as the reader might expect, to a chaotic solution governed by a strange attractor.

2.7 FORCED NONLINEAR OSCILLATOR: CHAOTIC ATTRACTOR

The system of Figure 1.7, discussed briefly in Chapter 1, is a version of the driven Duffing equation studied extensively by Ueda, and we see that it differs from our previous damped, forced nonlinear oscillator in having no linear stiffness. This would in fact arise physically if we had a beam loaded to precisely its (Euler) buckling load: at buckling the linear stiffness has dropped to zero due to the destabilizing action of the axial compressive load, and the nonlinear stiffness can be modelled locally by the cubic term.

Once again analytical solutions are impossible, and digital computations show that after transients have decayed the system settles down to a condition of steady-state chaos. In contrast to the point and cyclic attractors that we have so far examined, this convergence to chaos is said to be governed by a *chaotic attractor*. These chaotic or strange attractors can coexist with other periodic steady states, with appropriate basins of attraction, etc., but for the coefficients chosen here there is in fact just a unique chaotic attractor that captures all motions of the system. The middle trace shows a rather brief but fairly obvious transient from (0, 0) lasting visibly for only about five forcing cycles of period 2π. The steady-state chaos covering the remaining 45 forcing cycles has a fairly regular though non-periodic appearance, and we notice that the positive x peaks synchronize approximately with the start of a forcing cycle for which t is a multiple of 2π.

The steadiness of this final chaotic state is reflected in a stationary power spectrum and a *typical* spectrum of chaos is shown in the upper right picture. This is due to Ueda, and is for a slightly different set of coefficients, with 0.1 and 12 replacing the 0.05 and 7.5 of our equation. We see spikes at the forcing frequency and odd multiples of this frequency (typical of a non-sinusoidal periodic wave with the period of the forcing) plus, however, regions of 'white noise' extended broadband peaks.

The bottom trace shows a more dramatic transient, generated by starting at large amplitude at an inconvenient phase. The high frequency is a natural consequence of the large x, because the effective stiffness increases as x^2. However, even after this start, the recognizable pattern of the steady-state chaos soon emerges.

We recollect that the phase space of this driven oscillator is three-dimensional, spanned by $(x, \dot{x}, t)$, and the Poincaré mapping is generated by the successive intersections of a trajectory with the $t = 2i\pi$ sections, where $i = 0, 1, 2, \ldots$. The steady-state chaotic mapping is shown in the last picture. Here the dots build up to form a complete shape with a fractal structure, similar to that of a Cantor set (Figure 11.9). All the points lie in the positive x regime, corresponding to our earlier observation that in the final state the positive x peaks synchronize with the beginning of the forcing cycle.

Transients would appear as rather scattered dots outside this attractor, but as we have seen the mapping points are very quickly attracted into this set. The Poincaré section, often itself referred to as the attractor of the chaotic motion, is really just a cross-section of the full attracting structure, which is a fixed geometric form in the full three-dimensional phase space to which all trajectories are finally attracted. It is the continuous stretching and folding of the sheets of this attractor that produces the turbulent mixing motions characteristic of chaotic dynamics, as we shall see in Chapter 6.

3

Point Attractors in Autonomous Systems

We begin in this chapter our detailed examination of nonlinear dynamics, treating here the point attractors that represent the simplest form of final-time recurrent behaviour. We shall look first at the general *linear* oscillator, and then at the contrasting nonlinear behaviour of a pendulum suffering large-amplitude free motions. Ecological systems are used to illustrate the modelling of non-mechanical systems, and we finally look at a spinning satellite.

3.1 THE LINEAR OSCILLATOR

Wave and phase representations

Consider a mass m, restrained by an elastic spring of stiffness s and by a dashpot damper providing a linear viscous force opposing the velocity, whose equation of motion can be written as

$$m\ddot{x} + r\dot{x} + sx = 0 \tag{3.1}$$

where a dot denotes differentiation with respect to the time t. This damped linear oscillator, shown in Figure 3.1, is the simplest of the dissipative dynamical systems with which we are mainly concerned. Plotting the displacement x against the time t, we have for light damping the familiar damped oscillatory motion typical of a pendulum vibrating with small amplitude in air. In the phase space of x against $\dot{x}$ we have a stable focus replacing the familiar circles or ellipses of an undamped system as illustrated in Figure 3.2. A three-dimensional graph of this asymptotically stable behaviour is shown in Figure 3.3, in the space of x, $\dot{x}$, and t.

Characteristic equation

Dividing the equation of motion (3.1) by the mass puts it in the convenient standard form,

$$\ddot{x} + b\dot{x} + cx = 0 \tag{3.2}$$

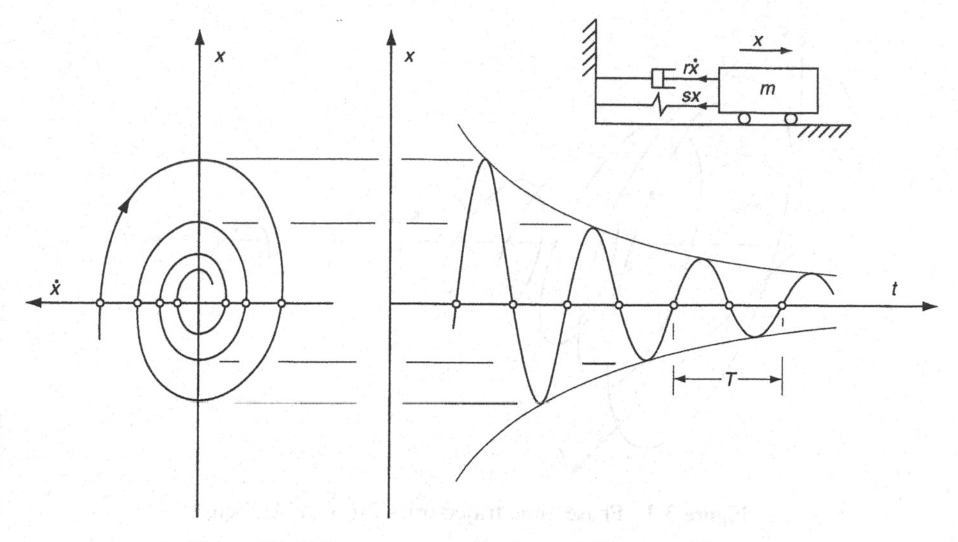

Figure 3.1 The response of a damped linear oscillator

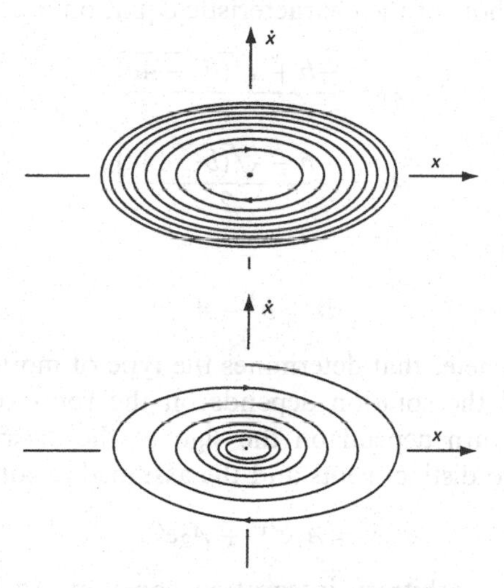

Figure 3.2 Undamped and damped phase trajectories of a linear oscillator

Assuming the solution $x = Ae^{\lambda t}$, and substituting this relation into (3.2) gives

$$(\lambda^2 + b\lambda + c)Ae^{\lambda t} = 0 \tag{3.3}$$

For a non-trivial solution we must have

$$\lambda^2 + b\lambda + c = 0 \tag{3.4}$$

The nature of the solution will now depend upon whether the roots of this *characteristic equation* are real or complex.

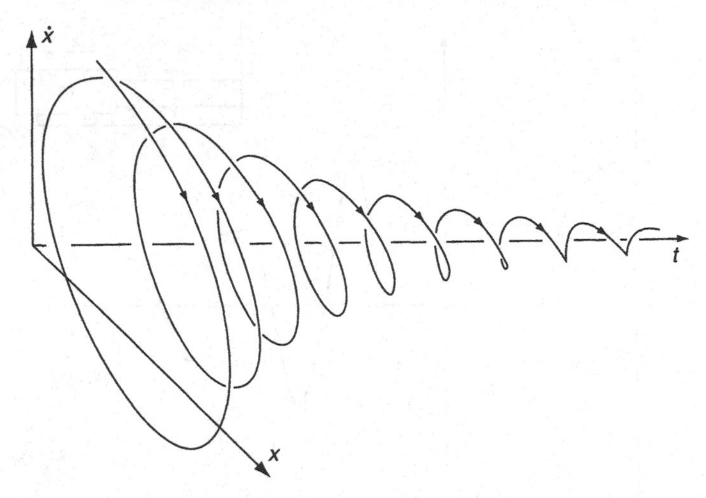

Figure 3.3 Phase–time trajectories for a stable focus

The two general roots of the characteristic equation are

$$\lambda_1 = \frac{-b + \sqrt{(b^2 - 4c)}}{2}$$
$$\lambda_2 = \frac{-b - \sqrt{(b^2 - 4c)}}{2} \tag{3.5}$$

and the discriminant

$$D = b^2 - 4c \tag{3.6}$$

is therefore the parameter that determines the type of motion.

Thus the form of the solution depends on the roots of the characteristic equation, which in turn depend on the sign of the discriminant D. If D is positive, we have two distinct roots and the assumed exponential solution,

$$x = A_1 e^{\lambda_1 t} + A_2 e^{\lambda_2 t} \tag{3.7}$$

where A_1 and A_2 are arbitrary integration constants to be found from the starting conditions of the motion.

If D is negative, we have two complex conjugate roots $\lambda_{1,2} = R \pm I\mathrm{i}$, giving us solutions of the form

$$x = e^{Rt} \sin It \tag{3.8}$$

The oscillator therefore becomes unstable if any root acquires a *positive real part*.

The response of the oscillator, typified by the nature of the equilibrium point ($x = 0$, $\dot{x} = 0$), is summarized in Figure 3.4, which shows sketches of the phase portraits next to the corresponding (R, I) Argand diagrams. The stable equilibrium points of main concern to us here lie in the quadrant $b > 0$, $c > 0$.

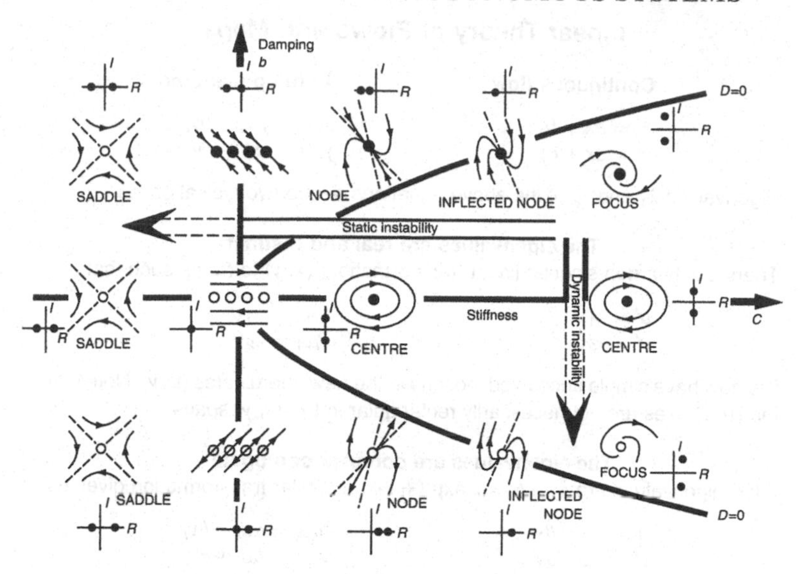

Figure 3.4 The phase portraits and root structure of a linear oscillator

Parallel linear analyses of flows of differential equations and iterations of maps are summarized in Figure 3.5.

Focus and node

In this quadrant with $D < 0$, the exponents are complex with negative real part, resulting in oscillations of exponentially decreasing amplitude, decaying more rapidly for larger b. A small typical image on the phase plane is labelled 'Focus' in Figure 3.4. In this domain of $D < 0$ the equilibrium point at the origin is called a *stable spiral* or a *stable focus*.

If we decrease the stiffness following the horizontal arrow of Figure 3.4, the roots become real as we cross the parabola of critical damping given by $D = 0$. For the special case of $D = 0$, the roots are coincident and real, with $\lambda_1 = \lambda_2$, and the phase portrait is called a *stable inflected node*, illustrated in Figure 3.6.

In the case of $D > 0$, the solutions are given by equation (3.7), where λ_1 and λ_2 are real and negative. The solution exhibits no oscillation, and the time axis is cut at most once; the phase portrait is termed a *stable node*, and is illustrated in Figure 3.7.

Static and dynamic instabilities

Now a load on an elastic structure can induce *buckling* in which the effective stiffness of the system changes from positive to negative. This *static instability*, characterized by the appearance of an adjacent position of equilibrium, is represented by the horizontal arrow on Figure 3.4. If, on the other hand, a wind blows across a flexible elastic structure it can induce dynamic galloping in

Linear Theory of Flows and Maps

Continuous flow	**Iterated mapping**
$x' = ax + by$	$x_{i+1} = ax_i + by_i$
$y' = cx + dy$	$y_{i+1} = cx_i + dy_i$

Eigenvalues λ_1 and λ_2 of the above equations define two typical cases.

The eigenvalues are real and distinct

There exists a non-singular linear transformation, (x, y) to (u, v), such that

$u' = \lambda_1 u$	$u_{i+1} = \lambda_1 u_i$
$v' = \lambda_2 v$	$v_{i+1} = \lambda_2 v_i$

We now have simple uncoupled motions in the new coordinates (u, v). Note that the (u, v) axes are not necessarily rectangular in the (x, y) space.

The eigenvalues are complex conjugate

If the eigenvalues are $\alpha \pm \mathrm{i}\,\beta = \rho \exp(\pm \mathrm{i}\,\phi)$ a similar transformation gives us

$u' = \alpha u - \beta v$	$u_{i+1} = \alpha u_i - \beta v_i$
$v' = \beta u + \alpha v$	$v_{i+1} = \beta u_i + \alpha v_i$

Writing the new coordinates in polar form, $u = r \cos \theta$, $v = r \sin \theta$, gives

$r' = \alpha r$	$r_{i+1} = \rho r_i$
$\theta' = \beta$	$\theta_{i+1} = \theta_i + \phi$

Notice that the rate of growth of r is governed by α and ρ respectively: for *asymptotic stability* a flow requires $\alpha < 0$, while a map requires $\rho < 1$.

The constant rate of rotation in (u, v) is governed by β and ϕ respectively. When the (u, v) axes are oblique in (x, y), rotation in (x, y) is not constant.

Figure 3.5 The first column summarizes the linear analysis, about an equilibrium state, of a system described by two first-order differential equations. For comparison, the second column shows the corresponding analysis of the iterated mapping that we study in Section 8.4

which the effective damping becomes negative, as indicated by the vertical arrow. In this *dynamic instability* a stable focus is transformed into an unstable focus representing growing oscillatory motion. Following either arrow, the deflections of our linear system will become infinite at the point of instability, but the behaviour of a real system will usually be controlled by nonlinear effects that we shall discuss later.

Before leaving Figure 3.4 we should notice that the *undamped* conservative system with a centre of elliptical paths is really a critical marginal case between stable and unstable zones.

Damping factor

Damping is usually referred to in terms of a dimensionless parameter called the damping factor ζ, which is the ratio of the damping constant r to the critical value r_c that makes $D = 0$.

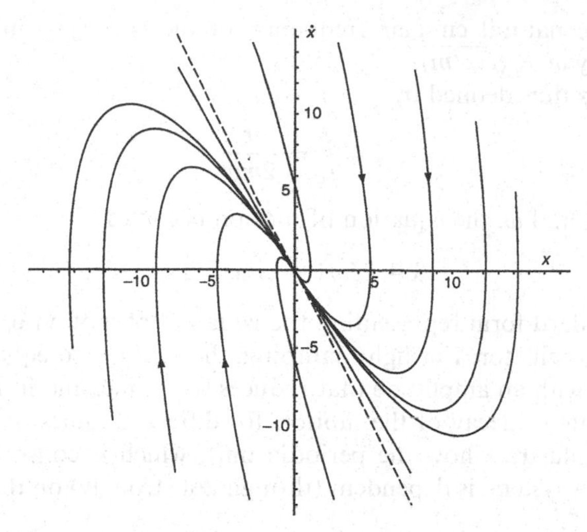

Figure 3.6 Example of a stable inflected node: $b = 4$, $c = 4$

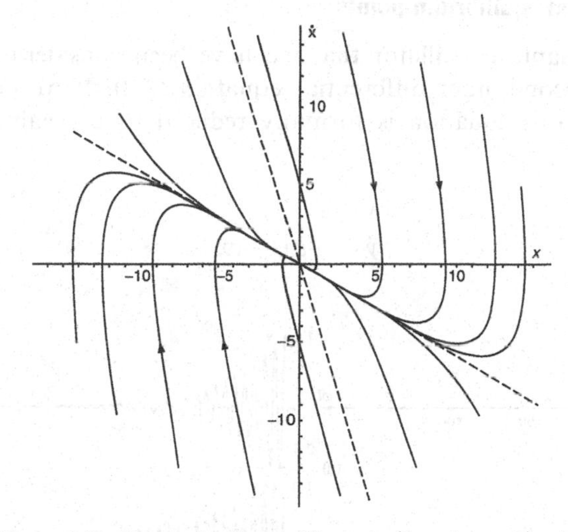

Figure 3.7 Example of a stable node ($b = 4$, $c = 2$) with the two eigenvectors shown as dashed lines. Notice the fast flow almost parallel to the 'fast' eigenvector (the steep dashed line) followed by a slow tangential approach to the 'slow' eigenvector (the shallow dashed line)

So from equation (3.6) we have

$$b_c = \sqrt{(4c)} \tag{3.9}$$

giving

$$r_c = 2\sqrt{(ms)} = 2m\omega \tag{3.10}$$

where ω is the natural circular frequency of the corresponding undamped system given by $\omega = \sqrt{(s/m)}$.

The ratio ζ is thus defined by

$$\zeta = \frac{r}{r_c} = \frac{r}{2m\omega} \tag{3.11}$$

In terms of ζ and ω, the equation of motion becomes

$$\ddot{x} + 2\zeta\omega\dot{x} + \omega^2 x = 0 \tag{3.12}$$

which is a standard form representing the equation of motion of an unforced, damped linear oscillator. For light damping, the solution to equation (3.12) is oscillatory but with an amplitude that reduces with time, and in Figure 3.8 we illustrate the rate of decay of the motion for different values of ζ. Finally, in Figure 3.9 we illustrate how the periodic time, which is constant during the decay of a given system, is dependent (though not strongly) on the value of the damping ratio, ζ.

Three-dimensional equilibrium points

The linear mechanical oscillator that we have been considering is governed by the single second-order differential equation (3.2). If we define the new variable $y = \dot{x}$, this equation is formally reduced to the pair of first-order equations

$$\begin{aligned} \dot{x} &= y \\ \dot{y} &= -by - cx \end{aligned} \tag{3.13}$$

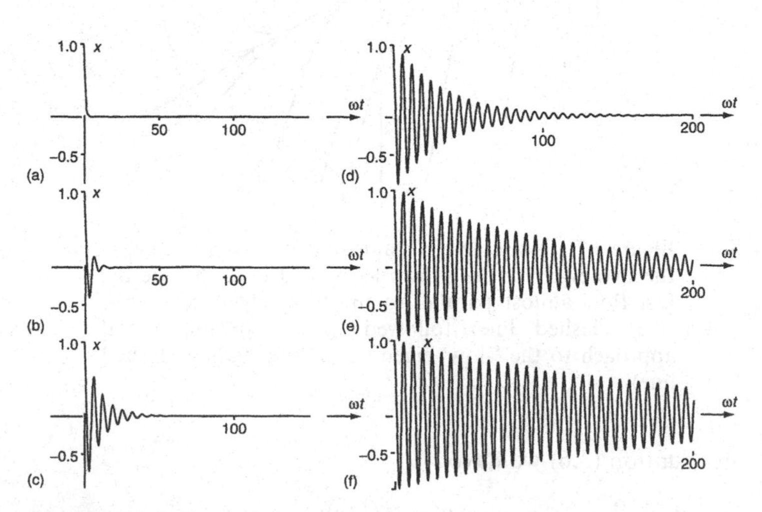

Figure 3.8 A diagram showing the rate of decay of a linear oscillator at various damping ratios, $\ddot{x} + 2\zeta\omega\dot{x} + \omega^2 x = 0 \rightarrow x = \exp(-\zeta\omega t) \cos(\omega t\sqrt{1-\zeta^2})$: (a) $\zeta = 0.9$, (b) $\zeta = 0.3$, (c) $\zeta = 0.1$, (d) $\zeta = 0.03$, (e) $\zeta = 0.01$, (f) $\zeta = 0.005$. In (f) $a_i/a_{i+1} \approx e^{2\pi\zeta}$

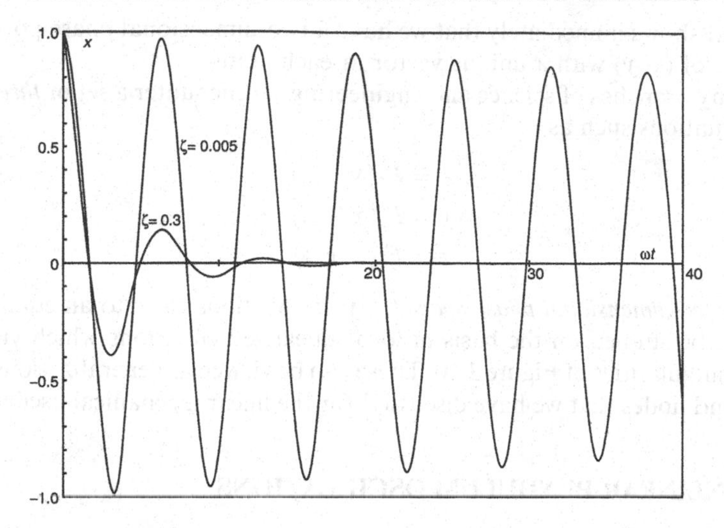

Figure 3.9 A reference diagram showing the change of period due to damping of a linear oscillator

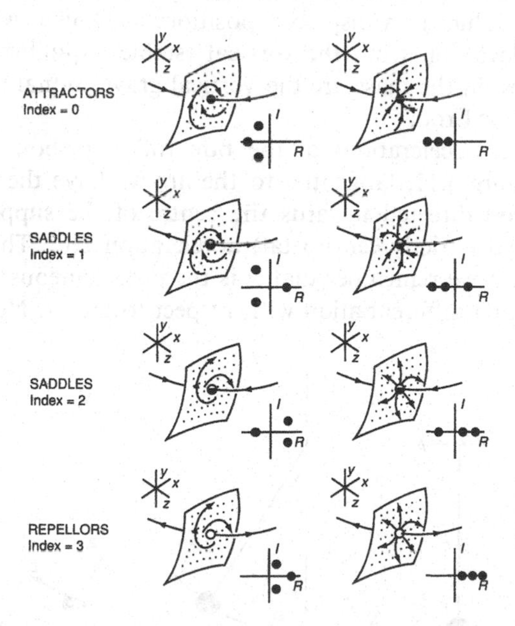

Figure 3.10 Elementary equilibrium points in three-dimensional phase space: Argand diagrams show Liapunov characteristic exponents $\lambda = R + I\mathrm{i}$ from $x = \exp(\lambda t)$; index = number of positive R = dimension of the outset

This form shows immediately that we have a two-dimensional phase portrait in the space of (x, y) with a unique vector at each point.

In many branches of science and engineering we encounter a set of *three* first-order equations such as

$$\begin{aligned} \dot{x} &= F_x(x, y, z) \\ \dot{y} &= F_y(x, y, z) \\ \dot{z} &= F_z(x, y, z) \end{aligned} \tag{3.14}$$

with a *three-dimensional phase space* (x, y, z). Motions close to an equilibrium state can be studied on the basis of local linearized equations which yield the various singularities of Figure 3.10. These can be viewed as generalizations of the spirals and nodes that we have discussed for the linear mechanical oscillator.

3.2 NONLINEAR PENDULUM OSCILLATIONS

Equation of motion

The equation of motion of a pendulum will now be formulated, to allow us to look at the free vibrations of a typical nonlinear physical system. Consider the pendulum of Figure 3.11; this comprises a light rod of length L, pivoted at point A, and carrying a bob of mass m, which is free to swing in the plane of the paper. In order to write the exact nonlinear large-angle equations of motion, we consider the pendulum in a displaced position, as shown, where the angle x designates the deviation from the vertical (stable) equilibrium position. As shown, the forces on the mass are the vertical gravitational force mg and the tension T in the light rod.

Considering the accelerations of the bob rotating about a fixed axis, as illustrated in Figure 3.11, tangential to the arc we have the acceleration $L\ddot{x}$, and along the rod directed towards the centre of the support we have the normal centrifugal (strictly centripetal) acceleration $L\dot{x}^2$. These are the well-known formulae if we remember that $\dot{x}$ is the instantaneous angular velocity, with a dot denoting differentiation with respect to time t. Now as we are not

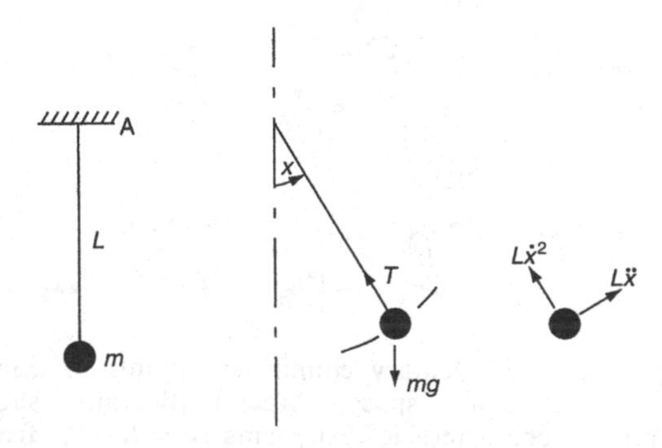

Figure 3.11 A rigid pendulum and the forces acting upon it

interested in the tension in the rod, and are for the time being neglecting all friction and dissipation, the equation of motion of the pendulum can be written by applying Newton's second law, in the direction perpendicular to the rod, to give

$$mL\ddot{x} + mg\sin x = 0 \tag{3.15}$$

which can be rearranged as

$$\ddot{x} + (g/L)\sin x = 0 \tag{3.16}$$

Setting $K = g/L$, we can write equation (3.16) in the final form:

$$\ddot{x} + K\sin x = 0 \tag{3.17}$$

which is the differential equation governing the free oscillation of the pendulum, having made no approximations relating to the angle of swing. Expanding $\sin x$ as a Taylor or power series

$$\sin x = x - \frac{x^3}{3!} + \frac{x^5}{5!} - \ldots \tag{3.18}$$

and substituting this into equation (3.17), we have

$$\ddot{x} + K\left(x - \frac{x^3}{3!} + \frac{x^5}{5!} - \ldots\right) = 0 \tag{3.19}$$

The fact that the differential equation is nonlinear means that simple closed-form analytical solutions are not readily available. However, for small angles x, the expansion of $\sin x$ will of course be dominated by the leading term, so that for small vibrations we can make the linear approximation of replacing $\sin x$ by x, to give us simply

$$\ddot{x} + Kx = 0 \tag{3.20}$$

This is a linear equation of motion, which only strictly holds in the limit as x tends to zero. The solution of this linear equation is well known; it is just a sine wave, and can be written as

$$x = A\sin(\omega t + p) \tag{3.21}$$

where the two arbitrary constants A and p are determined from the starting values $(x_0, \dot{x}_0)$ at $t = 0$. The natural circular frequency ω does not depend on the amplitude of oscillation, a general characteristic of such linear systems. The circular frequency ω is in fact equal to $\sqrt{K}$, and the periodic time T is equal to

$$T = 2\pi/\omega \tag{3.22}$$

Undamped waveforms

Returning to the complete nonlinear equation (3.17), some numerical time integrations have been performed, and the plots of the angle of deflection x against time t are shown in Figure 3.12. The response of the pendulum depends, of course, on the starting conditions $(x_0, \dot{x}_0)$. Here, in all cases, we have started

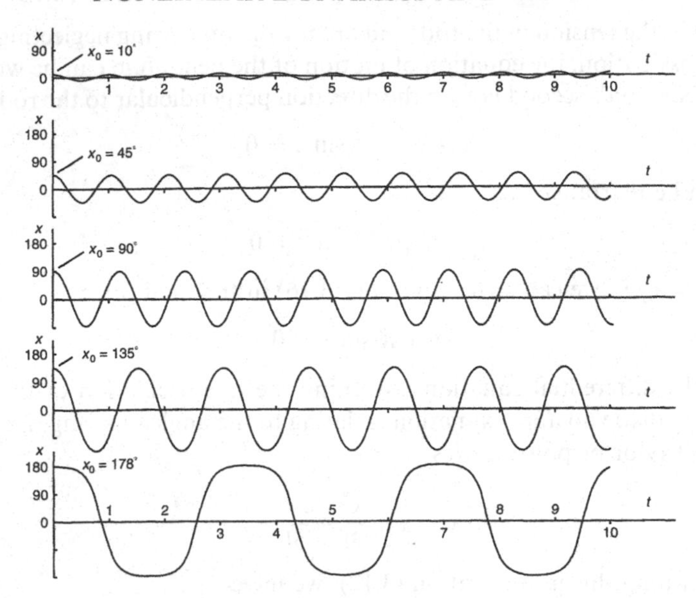

Figure 3.12 Undamped waveforms of a rigid pendulum with different starting conditions: $\ddot{x} + 4\pi^2 \sin x = 0$, $T_{\text{Lin}} = 1$, $\dot{x}_0 = 0$; fourth-order Runge–Kutta, $\Delta t = 0.02$

the pendulum with zero initial velocity (that is, $\dot{x}_0 = 0$) but with a series of different initial values of deflection x.

The top waveform has an initial deflection of $x_0 = 10°$, and it can be seen that the pendulum exhibits small undamped vibrations about the hanging position ($x = 0$). The dimensions of the pendulum have here been chosen such that the periodic time of the linearized equation is $T = 1$. For this small angle of swing, it is seen that the linear approximation is quite good, and that the pendulum is indeed oscillating in a roughly sinusoidal fashion with the period given quite accurately by 1. It should be noted that, as the model has no damping, the free vibrations will in fact continue indefinitely.

The second trace in Figure 3.12 shows a similar time integration, starting now with the initial value of $x_0 = 45°$. There is now some appreciable non-linearity, due to the quite large angles of deflection, although the waveform still appears quite sinusoidal to the eye. We notice, however, that the period of oscillation has increased and is no longer close to 1.

The next waveform corresponds to a steady oscillation with a starting value of $x_0 = 90°$. The vibration is presumably non-sinusoidal, although indeed it still looks superficially like a sine wave, but with a periodic time now quite noticeably greater than 1. This is as expected, since the pendulum has an unstable equilibrium state, in which it is completely inverted at $x_0 = 180°$, and thus as the deflections get nearer and nearer to this state, the rate of oscillation will obviously decrease.

The remaining waveforms correspond to the starting values of $x_0 = 135°$ and $178°$ respectively. This means that in the bottom diagram, the pendulum, which is made of a rigid rod, is starting to oscillate from an almost vertical position. As can be seen from the diagram, it accelerates from here quite slowly and finally undergoes large-amplitude oscillations, returning of course to 178° periodically, due to the undamped nature of the motions. The waveform is very non-sinusoidal, being flat whenever x takes its extreme values.

Planar and cylindrical phase portraits

In dealing with nonlinear dynamics it is very important to consider the motions of a system in phase space. In the case of the linearized pendulum, the phase portrait in the $(x, \dot{x})$ plane consists of the family of ellipses given by

$$x^2 + \left(\frac{\dot{x}}{\omega}\right)^2 = A^2 \tag{3.23}$$

or of a family of circles, if one uses $(x, \dot{x}/\omega)$ instead of $(x, \dot{x})$ as coordinate axes. All these circles or ellipses are traversed in the same time, $T = 2\pi/\omega$. These facts are familiar from the theory of linear oscillations.

Considering the nonlinear system of equation (3.17), the phase trajectories can be obtained most simply using the principle of conservation of energy. The total potential energy V of the model is simply the gravitational potential of the bob of mass m (ignoring the mass of the rod), and measuring its height from the support A, we have

$$V = -mgL\cos x \tag{3.24}$$

The kinetic energy T is simply $\frac{1}{2}m$ times the square of the tip velocity (ignoring again the mass of the rod), so

$$T = \tfrac{1}{2}mL^2\dot{x}^2 \tag{3.25}$$

By the conservation of total energy

$$T + V = C \tag{3.26}$$

where C is a constant of a particular motion. This yields for the pendulum

$$\tfrac{1}{2}mL^2\dot{x}^2 - mgL\cos x = C \tag{3.27}$$

The value of C for a particular motion can be established by using the initial conditions $(x_0, \dot{x}_0)$ at $t = 0$; equation (3.27) then gives the relation between x and $\dot{x}$ for the motion corresponding to these initial conditions. Thus by choosing different values for C, we can obtain this relation for any possible motion.

The phase plane is defined as before with x and $\dot{x}$ as its coordinates, and plotting the one-parameter family of curves generated by equation (3.27) for different values of C gives us the full phase portrait. The phase trajectories in Figure 3.13 can, in fact, be viewed simply as contours of constant total energy. The direction of the arrows on these curves can readily be obtained by observation since, with $\dot{x}$ positive, x must be increasing. We should note that these

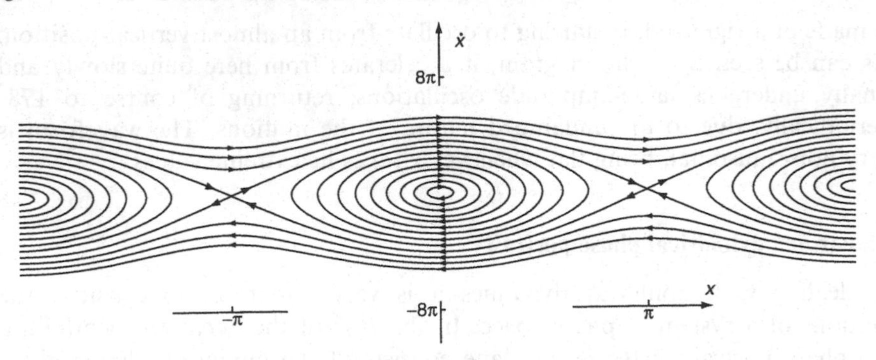

Figure 3.13 The phase-space trajectories of a rigid pendulum: $\ddot{x} + 4\pi^2 \sin x = 0$, $\Delta t = 0.02$, $x_0 = 0$, $\dot{x}_0 =$ multiples of $0.1 \times 4\pi$; $\dot{x}_0 = 4\pi$ gives separatrix

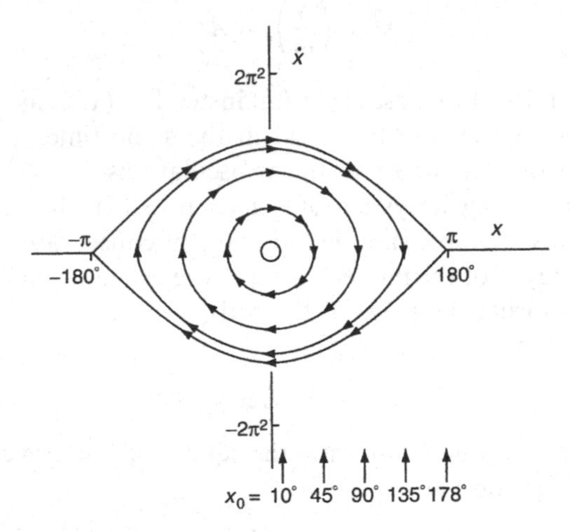

Figure 3.14 Phase trajectories corresponding to the waveforms on page 36

curves were in fact obtained by time integration and not by using the discussed conservation of energy.

The phase trajectories corresponding to the waveforms that were studied earlier are shown in Figure 3.14. Consider first the small-amplitude oscillation starting at $x_0 = 10°$; this appears in this phase diagram as an approximate circle, due to the particular choice of scales for the x and $\dot{x}$ axes. As the pendulum moves in its steady oscillations about the vertical hanging position the trajectory in phase space sweeps round this small circle. The next motion for $x_0 = 45°$ is again roughly circular and again a closed orbit. It is typical of undamped systems of this type that all phase trajectories around equilibria are closed and cyclic. As the starting condition x_0 is increased, the trajectories in the phase space are observed to become less circular, with the final trajectory for $x_0 = 178°$ being almost singular; the motion corresponding to this trajectory almost comes to rest on the x axis.

Referring back to Figure 3.13, we can see that, despite the non-appearance of the time variable in the phase diagram, it is still possible to deduce several physical features of the pendulum's possible motions. Considering first the possible states of physical equilibrium, the obvious stable equilibrium state is when the pendulum hangs without swinging (i.e. with $x = 0, \dot{x} = 0$), which corresponds to the origin. The phase path in this case degenerates to a single point. The second equilibrium position is where the pendulum is balanced vertically on end. This is an unstable equilibrium state for which $x = 180°$ and $\dot{x} = 0$. It appears as a *saddle point* in the figure, associated with the crossing of total energy contours.

The wavy lines at the top and bottom of Figure 3.13, on which $\dot{x}$ is of constant sign and x continuously increases or decreases, correspond to the whirling motions of the pendulum. The phase path that corresponds to the motion 'originating' at the unstable equilibrium state at $x = 180°$ is called a *separatrix*. It separates motions that represent oscillations about the hanging stable state from those motions involving a continuous whirling rotation of the pendulum. A separatrix is often asymptotic to a saddle point or to a saddle cycle as $t \rightarrow \infty$. This particular separatrix is asymptotic to a saddle point as $t \rightarrow +\infty$, and also as $t \rightarrow -\infty$.

Because of the periodicity of the system, with x a cyclic coordinate in the sense that the state of the pendulum is unchanged by the addition of 360° to x, it is an advantage topologically to introduce a cylindrical phase space. This is illustrated in Figure 3.15, where $\dot{x}$ is measured along the axis and x around the circumference. The periodic solutions circle the stable equilibrium position on the surface of the cylinder, without encircling the cylinder, whereas the large whirling motions of the rotating pendulum correspond to curves that enclose the cylinder. This cylindrical phase space shows that the separatrix is

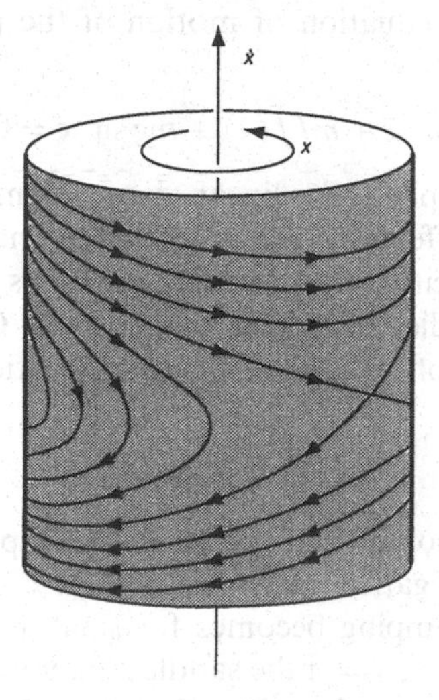

Figure 3.15 Cylindrical phase diagram for the rigid pendulum

asymptotic to the *same* saddle point both as $t \to +\infty$ and as $t \to -\infty$. Such a doubly asymptotic trajectory is called a *homoclinic saddle connection.*

Clearly, the linear representation fails completely for large angles x. In fact, the nonlinear differential equation (3.17) has an additional equilibrium position $x = 180°, \dot{x} = 0$, whereas the 'linearized' differential equation (3.20) has only a single critical point at $x = 0, \dot{x} = 0$. Thus, we would expect to obtain a better representation of the behaviour of the solutions of equation (3.17) by keeping not only the first term in the Taylor series but also as many terms of higher order as possible.

The free motion followed in practice by a given undamped pendulum depends, of course, on the starting conditions, that is, on the initial values of the displacement and its time derivative, which serve to define a unique trajectory. The continuous whirling motions occurring at large positive and negative values of $\dot{x}$ represent the rapid complete rotations of the pendulum that could be induced by a high-velocity start.

Linear and nonlinear viscous damping

Hitherto, we have assumed that no damping forces of any kind were acting on the pendulum, with the result that the total mechanical energy was conserved. This assumption naturally is not satisfied in practice: it is known that damping results in a steady decrease of the amplitude of oscillation. The kind of damping terms included in the mathematical analysis will depend on judgements concerning the physical modelling of the pendulum.

The most commonly assumed damping laws are linear viscous damping, which provides a term proportional to velocity $\dot{x}$; quadratic viscous damping, written as $c\dot{x}|\dot{x}|$; and damping due to Coulomb dry friction. These *damping laws* not only occur in applications from mechanics but also play an important role in other fields. The equation of motion of the pendulum could here be taken as

$$mL\ddot{x} + mLf(\dot{x}) + mg\sin x = 0 \tag{3.28}$$

where $f(\dot{x})$ would be simply $c\dot{x}$ for linear viscous damping.

For the x-linearized differential equation, where $\sin x$ has been replaced by x, the solution for the velocity-proportional damping is familiar from the theory of linear oscillators, studied earlier in this chapter. Consider, then, the nonlinear equation of motion of the pendulum, equation (3.28), which can be rearranged in the form

$$\ddot{x} + c\dot{x} + K\sin x = 0 \tag{3.29}$$

The singularities corresponding to the equilibrium points are still at $x = n\pi$, where n is a positive or negative integer or zero. At $x = 0$ the previous neutrally stable centre for zero damping becomes for small c an asymptotically stable spiral or focus. However, at $x = \pi$ the saddle remains unchanged in appearance on the introduction of viscous damping. Every typical motion now tends to the stable hanging equilibrium position.

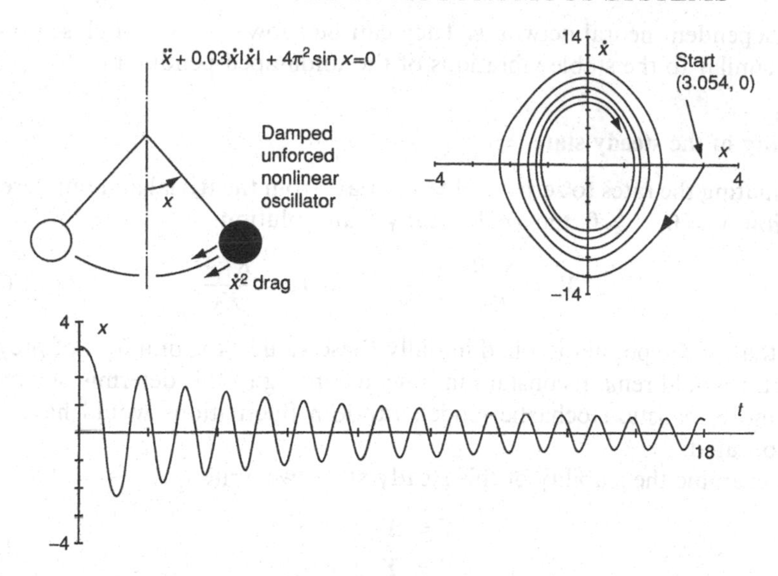

Figure 3.16 Motions of a pendulum with quadratic damping

Consider next the case of a pendulum immersed in a medium that exerts a force proportional to the square of its velocity and in a direction opposite to the velocity. The differential equation for the pendulum in this case can be written as

$$\ddot{x} + c\dot{x}|\dot{x}| + K \sin x = 0$$

The behaviour of the above equation is shown in Figure 3.16 and more fully in Figure 2.4. Qualitatively the motions do not differ from those of equation (3.29). The singularity at $x = 0$ has the general appearance of a focus, while at $x = \pi$ we have a saddle point.

3.3 EVOLVING ECOLOGICAL SYSTEMS

Lotka–Volterra prey–predator equations

The dynamical behaviour governing the growth, decay, and general evolution of interacting biological species is perhaps most easily modelled by the Lotka–Volterra prey–predator equations. These can be written as

$$\begin{aligned} \dot{X} &= K_1AX - K_2XY \\ \dot{Y} &= K_2XY - K_3BY \end{aligned} \tag{3.30}$$

where X is the population of the prey, say rabbits, Y is the population of the predator, say foxes, and A, B, K_1, K_2, and K_3 are positive constants. These nonlinear coupled evolution equations form an interesting and instructive introduction to population dynamics. They have been used for many years to model many basic biological phenomena such as biological clocks and time-dependent neural networks. They can be shown to exhibit closed oscillations similar to the stable vibrations of the undamped pendulum.

Stability of the steady state

By equating the rates to zero we obtain, apart from the trivial and uninteresting solution $X = 0$, $Y = 0$, the single steady-state solution

$$X_s = \frac{K_3 B}{K_2} \quad \text{and} \quad Y_s = \frac{K_1 A}{K_2} \tag{3.31}$$

such that, if the population had initially these values, the numbers of prey and predator would remain constant in time according to this deterministic model. In a more realistic stochastic model, random fluctuations would have to be incorporated.

To examine the stability of this steady state, we write

$$\begin{aligned} X &= X_s + x \\ Y &= Y_s + y \end{aligned} \tag{3.32}$$

assuming that the increments x and y are small quantities so that after substitution into equation (3.30) their products can be ignored.

We thus obtain the *linearized* variational equations

$$\begin{aligned} \dot{x} &= -K_3 B y \\ \dot{y} &= +K_1 A x \end{aligned} \tag{3.33}$$

which describe small population changes around the steady state. Eliminating y between these two equations gives

$$\ddot{x} + K_1 K_3 A B x = 0 \tag{3.34}$$

which is the familiar equation of a simple harmonic oscillator of circular frequency

$$\omega = (K_1 K_3 A B)^{1/2} \tag{3.35}$$

and periodic time $T = 2\pi/\omega$.

Phase trajectories

The phase trajectories in (X, Y) space are thus concentric ellipses for small deviations from (X_s, Y_s), and the steady state is thus neutrally stable. For larger finite oscillations about (X_s, Y_s), the phase trajectories are no longer elliptical, but they remain *closed* curves with, however, a continuous change in the periodic time. The large- and small-amplitude behaviour is thus entirely analogous to that of the *undamped* pendulum.

We should note carefully here that a linear prediction of elliptical centres does not in general guarantee centres in the corresponding nonlinear system. Exclusively nonlinear damping in a mechanical system could, for example, give asymptotically stable 'foci', even though the linearization, with no damping, predicted centres.

Because of the constantly varying periodic time, the dynamical motions along the phase trajectories are themselves only *orbitally stable*, and random stochastic disturbances will induce a constant drifting between orbits.

Three phase trajectories are shown in Figure 3.17, for which the constants have been set equal to unity, i.e.

$$K_1 A = K_3 B = K_2 = 1$$

giving the steady state

$$X_s = Y_s = 1$$

The linear theory ellipses become, in this case, circles.

The phase space (X, Y) is in reality filled with an infinity of such nesting phase trajectories, each with its own periodic time, and the trajectory through any starting point summarizes the dynamical evolution of the ecosystem. Thus if the system starts away from the steady state, S, it will exhibit undamped oscillations about (X_s, Y_s), the amplitude of the oscillations being governed by the original departure from S.

Structural stability

One acknowledged deficiency of the Lotka–Volterra equations is thus that they have the structural instability of the *undamped* conservative mechanical

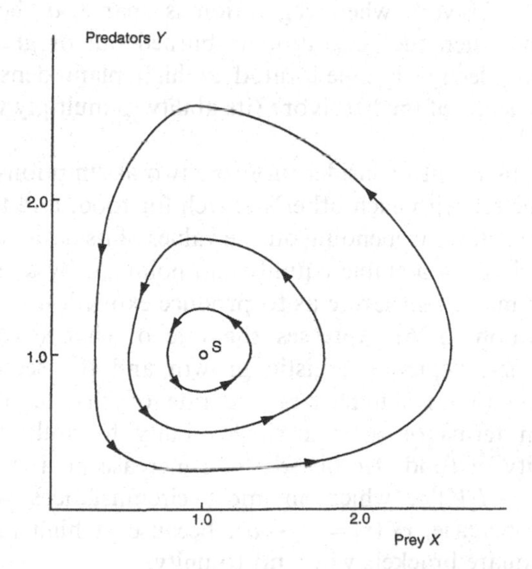

Figure 3.17 Closed phase trajectories for the predator–prey model, showing analogy with an undamped pendulum ($K_1 A = K_2 = K_3 B = 1$): for more details see Thompson (1982)

system, the phase trajectories of which can be topologically changed by the introduction of infinitesimal viscous damping.

A more realistic set of equations could be expected to yield not an infinity of neutrally stable trajectories but structurally stable *focuses* and *limit cycles*. An attracting limit cycle would, for example, give rise to more coherent cyclic behaviour with a well-defined periodic time T. Despite this lack of structural stability, the usefulness of the Lotka–Volterra equations in predicting prey–predator oscillations in a remarkably simple manner is widely acknowledged.

Plant–herbivore evolution

We now look at the dynamic behaviour of a structurally stable ecological model, the plant–herbivore system. This system explores the interaction between plants and grazing animals.

Of the many models available, one will suffice to illustrate the behaviour of this type of system:

$$\frac{\mathrm{d}V}{\mathrm{d}t} = r_1 V(1 - V/K) - c_1 H[1 - \exp(-d_1 V)] \tag{3.36}$$

and

$$\frac{\mathrm{d}H}{\mathrm{d}t} = H\{-a + c_2[1 - \exp(-d_2 V)]\} \tag{3.37}$$

where V is the standing crop of plants, H the size of herbivore population, r_1 the intrinsic rate of increase of plants, K the maximum ungrazed plant density, c_1 the maximum rate of food intake per herbivore, d_1 the grazing (searching) efficiency of the herbivore when vegetation is sparse, a the rate at which herbivores decline when the vegetation is burned out or grazed flat, c_2 the rate at which this decline is ameliorated at high plant density, and d_2 the demographic efficiency of the herbivore (its ability to multiply when vegetation is sparse).

The modelling takes into consideration the two assumptions that the herbivores do not interfere with each other's search for food, and that the animals range free of persecution. Depending on the values of its parameters, the system may be characterized by a stable equilibrium point, or by a stable limit cycle whose amplitude may be so severe as to produce extinction.

The first equation (3.36) expresses the rate of change of vegetation by two terms, the first depicting logistic growth and the second the rate of grazing. Equation (3.37) summarizes the rate of change of the herbivore population H in terms of their intrinsic ability to multiply, as modified by the availability of food. Herbivores can increase at a maximum rate of $\{-a + c_2[1 - \exp(-d_2 V)]\}$, which in most circumstances will equal their intrinsic rate of increase, r_2 ($r_2 = c_2 - a$), because at high plant density the term inside the square brackets will tend to unity.

Figure 3.18 shows the growth of a population of herbivores, and the resultant changes in plant density, as the two variables spiral towards their mutual equilibrium point. For this illustration, the parameter values are $r_1 = 0.8$,

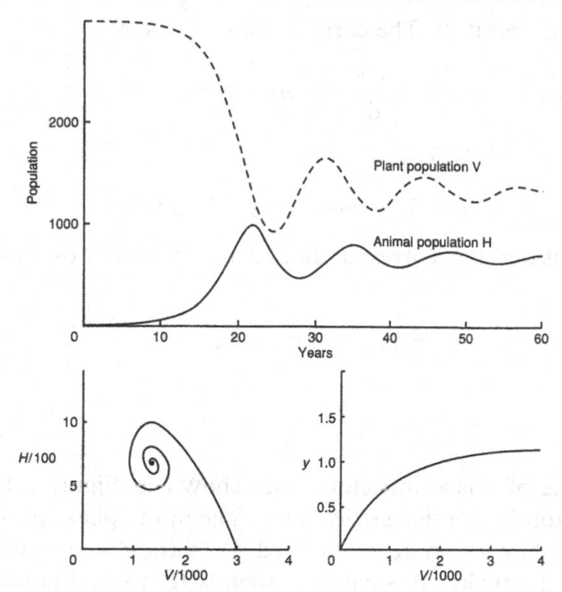

Figure 3.18 Time evolution of the vegetation V and herbivore population H during a herbivore eruption. The intake y of vegetation per herbivore is also displayed against the density of vegetation. Reproduced by permission of Blackwell Scientific Publications Ltd

$K = 3000$, $c_1 = 1.2$, $d_1 = 0.001$, $a = 1.1$, $c_2 = 1.5$ and $d_2 = 0.001$. This example might represent white-tailed deer colonizing a mosaic of grassland and forest. Wildlife managers will recognize the growth curve as a deer eruption.

Further information on nonlinear behaviour in ecological models, including a study of chaotic fluctuations in Canadian lynx populations based on skin records of the Hudson's Bay Company, can be found in a recent survey paper by Schaffer (1985). Important applications of nonlinear dynamics relate to the prediction of epidemics of human diseases, such as the measles cycle observed in New York.

3.4 COMPETING POINT ATTRACTORS

In this section we will consider the competition between more than one stable attractor. The easiest way to devise a system with two stable equilibria is to consider the motion of a mass in a potential field exhibiting two minima.

Nonlinear potential

A suitable total potential energy function is

$$V = -\tfrac{1}{2}ax^2 + \tfrac{1}{4}bx^4 \tag{3.38}$$

where a and b are positive. The corresponding force is

$$\frac{\mathrm{d}V}{\mathrm{d}x} = -ax + bx^3 \tag{3.39}$$

with equilibria (zero force) at

$$x = 0 \quad \text{and} \quad x = \pm\sqrt{(a/b)}$$

So we can consider the corresponding damped oscillator described by the equation

$$m\ddot{x} + c\dot{x} - ax + bx^3 = 0 \tag{3.40}$$

Phase portraits

Two typical sets of phase trajectories are shown in Figure 3.19 beneath the corresponding total potential energy curve. The upper phase portrait shows the pathological undamped system obtained by setting $c = 0$, while the lower portrait shows the typical dissipative system with a small amount of positive damping.

In the undamped system, motions that cross the potential barrier repeatedly are divided in phase space from motions that remain on one side by separatrices, which are saddle connections, doubly asymptotic to the saddle fixed point.

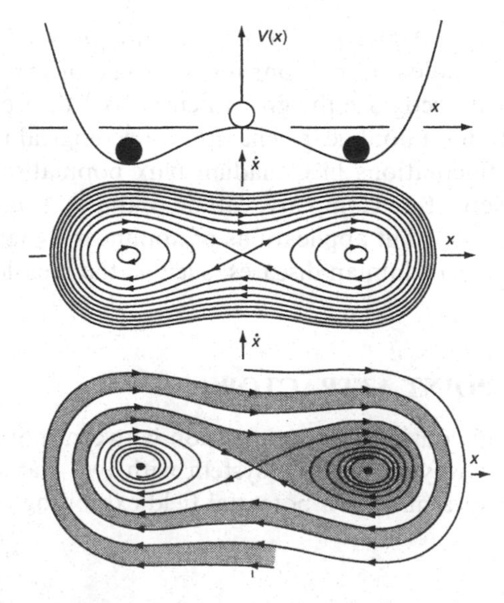

Figure 3.19 Undamped and damped phase trajectories for a nonlinear oscillator with two stable equilibrium states, representing competing point attractors

Similar to the pendulum, on the introduction of light damping, the two neutrally stable centres become asymptotically stable foci. The saddle remains locally unchanged, but the saddle connection is broken. The two stable equilibrium states are now competing *point attractors*, and in this symmetric problem half of the phase space (shown dotted) leads to one attractor, while the other half (left blank) leads to the other. The dotted region is thus the *domain* or *basin of attraction* of its enclosed attractor.

We see that for large-amplitude motions the catchment regions are interlocking spirals. This is as we would expect physically. Motions starting at $\dot{x}_0 = 0$ with larger and larger values of x_0, for example, will clearly undergo a large number of big oscillations, right across both the minima, before settling into one of them; and the one finally chosen will clearly alternate, as x_0 is varied.

The separatrix between these two basins is an *inset* (stable manifold), a pair of trajectories asymptotic to the saddle point as time goes to infinity.

3.5 ATTRACTORS OF A SPINNING SATELLITE

A second illustration of competing fixed-point attractors arises in the spherical phase space of a spinning satellite or spacecraft. The detailed analysis is not essential here, but we sketch it briefly for those who are already familiar with rigid-body mechanics. The motions of a free rigid body in space are governed by Euler's equations of motion, which can be written in the form of three first-order differential equations,

$$\dot{m}_1 = \frac{I_2 - I_3}{I_2 I_3} m_2 m_3$$
$$\dot{m}_2 = \frac{I_3 \quad I_1}{I_1 I_3} m_1 m_3$$
$$\dot{m}_3 = \frac{I_1 - I_2}{I_1 I_2} m_1 m_2$$

Here the I_i $(i = 1, 2, 3)$ are the principal moments of inertia of the body, and we suppose them to be unequal with $I_1 > I_2 > I_3$. The m_i are the body components of the angular momentum defined by $m_i = I_i \omega_i$, where ω_i is the instantaneous angular velocity about the ith principal axis. The starting conditions of a motion can be taken as (m_1, m_2, m_3). In the absence of any applied torques, there are two (independent) conserved quantities during any motion. The first is the Hamiltonian H, equal in magnitude to the total rotational energy, and the second is the magnitude R of the angular momentum. These are

$$H = \frac{1}{2}\left(\frac{m_1^2}{I_1} + \frac{m_2^2}{I_2} + \frac{m_3^2}{I_3}\right)$$
$$R^2 = m_1^2 + m_2^2 + m_3^2$$

Notice that the angular momentum *vector* remains fixed in space with constant components, but the body components m_i are not themselves conserved.

Associating the m_i with rectangular coordinates, we see that the phase trajectories (orbits) will be given by the intersection of the ellipsoids defined by $H = \text{constant}$ with the spheres of constant angular momentum. Focusing on the sphere of radius R, it is clear from the geometry that the flow will exhibit two fixed-point saddles at $(0, \pm R, 0)$, and four fixed-point centres at $(\pm R, 0, 0)$ and $(0, 0, \pm R)$ as illustrated in Figure 3.20. The saddles are connected by four trajectories that tend to one saddle as time goes to plus infinity, and to the other as time goes to minus infinity. These trajectories are said to be *heteroclinic* to the saddle fixed points.

At centre A on the m_1 axis, the energy on the sphere takes its minimum value of $H = R^2/2I_1$; while at centre C, on the m_3 axis, it takes its maximum value $H = R^2/2I_3$. Both of these centres are stable, so in the absence of any dissipation a rigid body is stable when spinning freely about either its axis of maximum inertia, I_1, or its axis of minimum inertia, I_3. Meanwhile the saddle, B, on the m_2 axis implies that the body is unstable when spinning about its principal axis of intermediate inertia, I_2. These facts can be demonstrated by spinning a book into the air about each of its three axes. Two motions will be clearly stable, while the third will lead to appreciable tumbling during the throw.

The four centres are only marginally stable (not asymptotically stable) in the sense of Liapunov, and as with any conservative Hamiltonian system, we expect

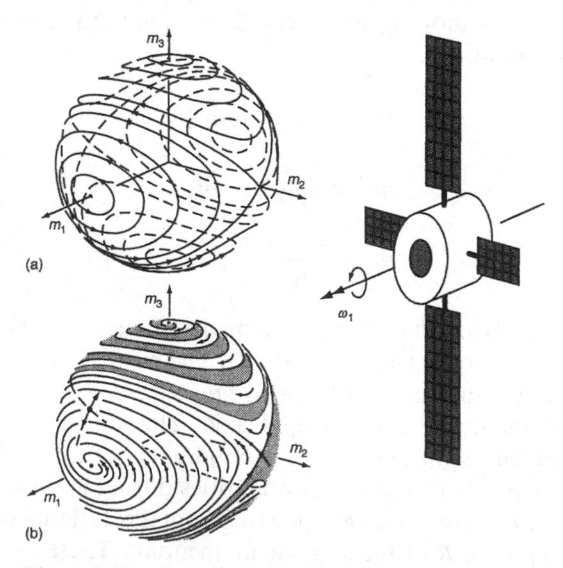

Figure 3.20 Spherical phase portraits of a spinning satellite: (a) without dissipation, (b) with internal dissipation. A typical satellite is also shown; with internal dissipation it is asymptotically stable spinning about its axis of maximum inertia. Part (a) is reproduced, with permission, from S. Wiggins, 1998, *Global Bifurcations and Chaos*, Springer Verlag, New York, page 177

the phase portrait to be structurally unstable against the addition of a little dissipation. So we now assume there is some internal dissipation in the system as would arise, for example, if the body were slightly viscoelastic instead of rigid. During a tumbling motion, angular momentum is still conserved, but the kinetic energy will decrease slowly as the continuously distorting body warms up. Only at the fixed points, where the force field in the body is not time-dependent, will the kinetic energy remain constant.

Heuristically, we can assess the behaviour by assuming that the change in response is dominated by the slow reduction of the kinetic energy, H. We now have the geometry of a sphere, radius R, intersecting a slowly shrinking ellipsoid, and the result is shown in Figure 3.20(b). We see that A has become an asymptotically stable attractor, B has become an asymptotically unstable repellor, while fixed point C remains an unstable saddle. The heteroclinic connections between saddles are broken, and the insets (stable manifolds) entering the saddles as time goes to plus infinity become basin boundaries. These boundaries separate the basin of attraction of A from that of its counterpart, A′, located diametrically opposite: in the picture the basin of attraction of A′ is shown with a tint. Notice that A and A′ differ only in the sense of their rotation about the I_1 axis.

The final stable motion from any typical start thus corresponds to either A or A′ in either of which the body is spinning (like a galaxy) about its principal axis of maximum rotational inertia, I_1. This state has the minimum energy for a given angular momentum. In the satellite literature this result is known as the *major axis rule*.

4

Limit Cycles in Autonomous Systems

After looking at point attractors, we move now to a discussion of cyclic attractors, looking in this chapter at limit cycles exhibited by autonomous unforced dynamical systems.

The single limit cycle and its stability characteristic are discussed first, with an example drawn from a neural model of brain activity. The generation of a trace of limit cycles by a Hopf bifurcation is next illustrated with reference to chemical oscillations. We finally consider multiple coexisting limit cycles, arising for example in the wind-induced galloping of bluff elastic bodies.

4.1 THE SINGLE ATTRACTOR

Asymptotically stable equilibrium states are not the only attractors that can arise in a two-dimensional dissipative phase space. A second type of attractor is the stable *limit cycle*, namely a steady closed oscillation that attracts all adjacent motions. To get a single stable limit cycle it is necessary to ensure that the origin (0, 0) is unstable so that trajectories of small amplitude move outwards, while ensuring at the same time that trajectories of large amplitude move inwards.

We consider, then, the oscillator

$$m\ddot{x} - c\dot{x} + d\dot{x}^3 + kx = 0$$

and typical trajectories are shown in Figure 4.1. For very small amplitudes we can linearize the above differential equation, by dropping the $d\dot{x}^3$ term, and we then have an unstable focus, the negative linear damping giving trajectories spiralling outwards away from the central point *repellor*. For large amplitudes the nonlinear term dominates, ensuring that all motions of the system tend towards a stable steady-state oscillation, the heavily drawn limit cycle. This is the only attractor of this phase portrait, and the whole phase space is its basin of attraction.

Stability of a limit cycle

Three typical attracting limit cycles in a three-dimensional phase space are shown schematically in Figure 4.2. To discuss the stability of such a cycle, it

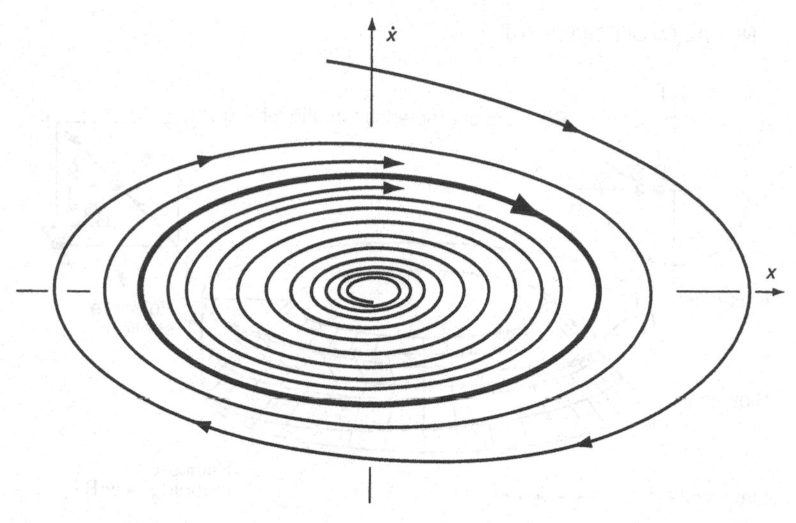

Figure 4.1 Phase trajectories of a nonlinear oscillator exhibiting a single stable limit cycle

is simplest to consider the intersection of adjacent trajectories with a Poincaré section as shown, giving rise to a mapping A → B, etc. Linearizing the problem for small deviations from the central limit cycle, a stability discussion then hinges on an inspection of the Poincaré characteristic multipliers as indicated.

For the untwisted nodal cycle of the top diagram, the two characteristic multipliers are both real and positive, corresponding to inwards flows on the fast and slow insets (eigenvectors).

For the spiral cycle, the adjacent trajectories approach the limit cycle in an oscillatory fashion, as shown in the central diagram. The characteristic multipliers are now complex as indicated in the (I, R) Argand diagram.

A particularly interesting cycle is the twisted nodal one of the lower diagram, corresponding to a pair of real but negative multipliers. The fast and slow insets are now Möbius bands as drawn, resulting in a mapping A → B that alternates across the central fixed point on successive returns.

In the Argand diagram, the stability boundary for the Poincaré characteristic multipliers is the unit circle, the transit of a root outwards through this circle signalling the loss of stability of the central limit cycle.

4.2 LIMIT CYCLE IN A NEURAL SYSTEM

We choose now to illustrate the dynamical behaviour of a mathematical model of the human brain, as an example of a system exhibiting cyclic attractors. Here the relationships between the apparently simple activity of the individual neurons and the high levels of organization associated with thought and consciousness are only just beginning to be explored.

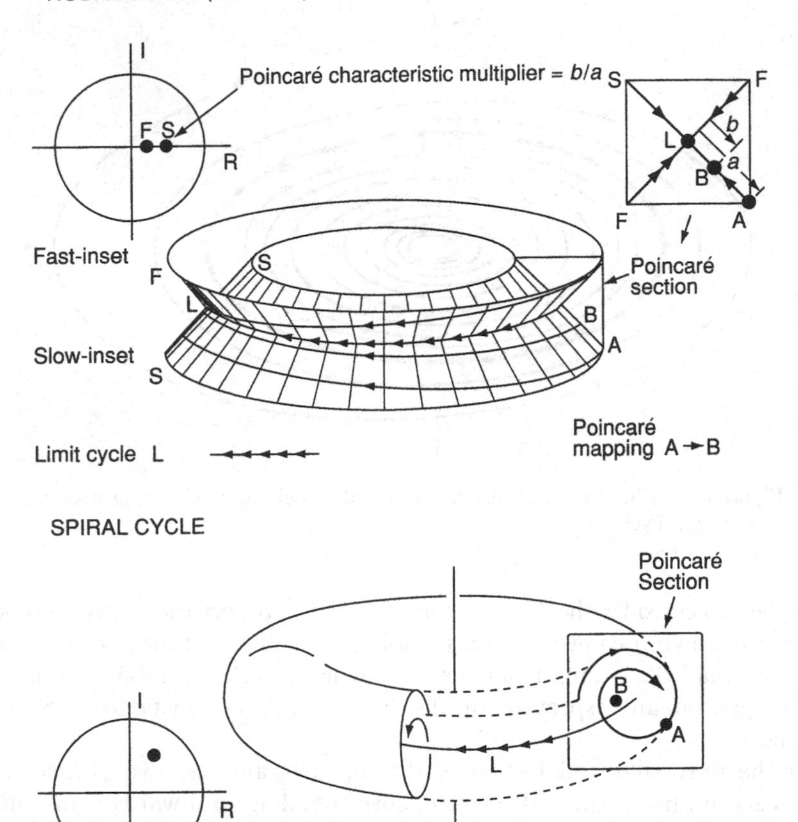

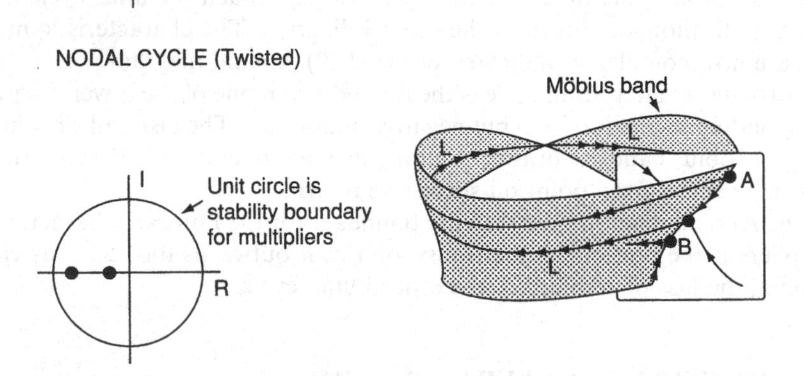

Figure 4.2 A schematic picture of attracting limit cycles in three-dimensional phase space. There is a full discussion in Section 10.4

Dynamical equations of populations of excitatory and inhibitory neurons have been developed to model the neural tissue of the brain. For spatially localized populations, coupled nonlinear differential equations are obtained, and have been studied using phase-plane methods and numerical analysis. *Folds* in the steady-state solutions are found to generate multiple hysteresis phenomena, while *limit cycles*, modelling brain rhythms, are observed in which the frequency of oscillation is found to be a monotonic function of a stimulus intensity.

The brain and central nervous system

The brain can be highly idealized as a network of *neurons* connected in a random manner by *synapses*. When a neuron *fires*, the stimulus is transmitted through the synaptic connections to adjacent neurons, which may then be induced to fire after the synaptic *delay*.

The neuron population can be divided into *excitatory* neurons, which give out a positive stimulus when they fire, and *inhibitory* neurons, which give out a negative stimulus. A neuron will fire when the sum of the received stimuli exceeds a certain *threshold* value: and having once fired it remains inactive for a certain *refractory* period, even if it receives a stimulus above its threshold. Such a discrete *neural net* can be readily modelled on a digital computer, and waves of firing activity have been observed in computer simulations.

Mechanics of excitation and inhibition

The dynamical 'continuum' model illustrated here introduces as two fundamental variables $E(t)$, the proportion of excitatory cells firing per unit time, and $I(t)$, the proportion of inhibitory cells firing per unit time. We assume that E and I at time $(t+\tau)$ after a *delay* τ will be equal to the proportion of cells that are *sensitive* and also receive at least *threshold* excitation.

Non-sensitive cells are those that, having recently fired, cannot fire again for their *refractory* period. Thus, if the absolute refractory period is r, the proportion of sensitive excitatory cells can be approximated as

$$E_s = 1 - r_e E$$

with a similar expression for I_s. Notice that the refractory period for E is r_e, which might be different from r_i.

Now the expected proportions of the subpopulations receiving at least threshold excitation per unit time will be a mathematical *function* of E and I, which for the excitatory cells is

$$\mathcal{S}_e(x) = \mathcal{S}_e[c_e E - g_e I + P(t)]$$

and for the inhibitory cells is

$$\mathcal{S}_i(x) = \mathcal{S}_i[c_i E - g_i I + Q(t)]$$

Here the coefficients are constants representing the average number of synapses per cell, and $P(t)$ and $Q(t)$ are *external* excitations.

The *response functions* $\mathcal{S}(x)$ will depend on the probability distribution of neural thresholds. It is argued that they will have the *sigmoidal* shape of an integral sign, rising monotonically with x from zero and becoming asymptotic to a value equal to or near to unity as x tends to infinity. In the analytical work they are taken as (Wilson and Cowan 1972)

$$\mathcal{S}(x) = \frac{1}{1+\exp[-a(x-\theta)]} - \frac{1}{1+\exp(a\theta)}$$

with different values of the constants a and θ for the two types of neurons.

Now if the probability of a cell being sensitive is independent of the probability that it is currently excited above its threshold, we can multiply our probabilities to get, with some time *coarse-graining* assumptions,

$$E(t+\tau_e) = (k_e - r_e E)\mathcal{S}_e[c_e E - g_e I + P(t)]$$
$$I(t+\tau_i) = (k_i - r_i I)\mathcal{S}_i[c_i E - g_i I + Q(t)]$$

Here k_e and k_i replace unity in our earlier expressions for E_s and I_s: they are in fact very close to unity, being defined as

$$k = \mathcal{S}(\infty)$$

They are part of a small adjustment to make $E = I = 0$ a stable *resting* state under zero external excitation.

If we now write the Taylor approximation

$$E(t+\tau_e) = E(t) + \frac{dE}{dt}\tau_e$$

and likewise for I, we have our final differential equations

$$\tau_e \frac{dE}{dt} = -E + (k_e - r_e E)\mathcal{S}_e[c_e E - g_e I + P(t)]$$
$$\tau_i \frac{dI}{dt} = -I + (k_i - r_i I)\mathcal{S}_i[c_i E - g_i I + Q(t)]$$

It can be shown that this present model can exhibit a *damped* oscillatory response to *impulsive* external stimulation, as indeed could be demanded of a satisfactory model.

Moreover, with Q equal to zero and P equal to a certain constant value, the model can, with an appropriately chosen set of coefficients, exhibit a stable limit cycle as shown in the two-dimensional (E, I) phase space of Figure 4.3. These limit cycles arising from a realistic neural model provide a concrete physiological base for the study of electroencephalogram (EEG) rhythms, such as the important alpha rhythms.

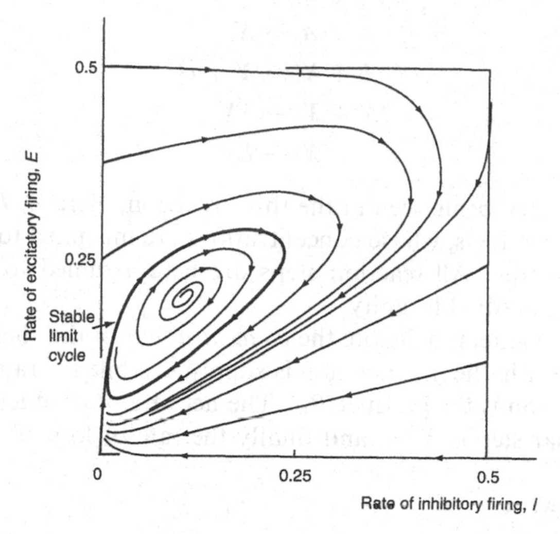

Figure 4.3 A stable limit cycle in the (E, I) phase space of a neural system. The limit cycle represents a steady oscillation in the two firing rates

4.3 BIFURCATIONS OF A CHEMICAL OSCILLATOR

It is well established that the kinetics of chemical reactions are governed by nonlinear differential equations. Reacting chemical systems are thus capable of exhibiting a wealth of both temporal and spatial bifurcation phenomena.

The Belousov–Zhabotinsky reaction, which involves the complex cerium-catalysed bromination and oxidation of malonic acid by a sulphuric acid solution of bromate, is a well-studied system confirming this conclusion. With no spatial variations in a stirred flow reactor, it can exhibit complex oscillations and even chaotic phase motions; see Epstein and Pojman (1998).

Not surprisingly it can also undergo spatial evolutions. For example, a shaken homogeneous chemical mixture, if left in a shallow dish, can organize itself into spiral patterns. Because such a dish is essentially a closed system, this self-organization is here only temporary and eventually the system reverts to a homogeneous state: the chemical 'organism' dies! A permanently organized state can, however, be maintained if appropriate chemicals are fed continuously into and out of such a system. (See Thompson, 1982 for details.)

The Brusselator model chemical reaction

We shall focus our attention here on a trimolecular model system, the so-called *Brusselator*, which is one of the simplest to exhibit these phenomena. The *real* Belousov–Zhabotinsky reaction is by contrast an exceedingly complex reaction, which is even now not fully understood. This model considers the hypothetical reactions (Nicolis and Prigogine 1977)

$$A \rightarrow X$$
$$B + X \rightarrow Y + D$$
$$2X + Y \rightarrow 3X$$
$$X \rightarrow E$$

the trimolecular step being seen in the third reaction. Here A, B, D, and E are initial and final products, whose concentrations are imagined to be imposed as constants throughout. All reaction steps are here assumed to be irreversible with rate constants equal to unity.

Using the same letters to denote the *concentrations* of the chemicals, the rate of production of X in the first reaction is simply A, while the rate of loss of X in the second equation is the product BX. The net rate of production of X in the third trimolecular step is X^2Y, and finally the rate of loss of X in the fourth reaction is X.

Thus we can write

$$\dot{X} = A - (B+1)X + X^2Y$$

and similarly

$$\dot{Y} = BX - X^2Y$$

These are the coupled nonlinear rate equations that must be solved for the time evolution of X and Y with A and B prescribed constants.

Setting the time derivatives equal to zero gives us the primary solution of the thermodynamic branch

$$X = A, \quad Y = B/A$$

and we write

$$X = A + x, \quad Y = (B/A) + y$$

where x and y are now changes in concentration from the primary values. The evolution equations are now

$$\dot{x} = x(B-1) + y(A^2) + [x^2(B/A) + 2xyA + x^2y]$$
$$\dot{y} = x(-B) + y(-A^2) - [x^2(B/A) + 2xyA + x^2y]$$

and for a linear stability analysis we need retain only the terms in x and y and ignore higher-order terms in x^2, xy, and x^2y, given in square brackets. The linear equations can therefore be written in the standard form

$$\dot{x} = c_{11}x + c_{12}y$$
$$\dot{y} = c_{21}x + c_{22}y$$

where

$$c_{11} = B - 1, \quad c_{12} = A^2$$
$$c_{21} = -B, \quad c_{22} = -A^2$$

and the characteristic equation becomes

$$\lambda^2 - \lambda(c_{11} + c_{22}) + c_{11}c_{22} - c_{12}c_{21} = 0$$

where the required coefficients are

$$c_{11}c_{22} - c_{12}c_{21} = A^2$$

and

$$-(c_{11} + c_{22}) = 1 + A^2 - B$$

Since A is necessarily positive, we can never have a static instability at which the first coefficient would have to vanish, but we see that the vanishing of the second coefficient predicts a dynamic instability at

$$B^{\mathrm{c}} = 1 + A^2$$

Setting $A = 1$ and regarding the concentration B as a control parameter, we see that we have a stable focus for $B < B^{\mathrm{c}}$ and an unstable focus for $B > B^{\mathrm{c}}$ where

$$B^{\mathrm{c}} = 2$$

We have in fact at $B = B^{\mathrm{c}}$ a dynamic Hopf bifurcation at which a stable supercritical limit cycle is generated, and some computed phase portraits from the full nonlinear rate equations are shown in the following three figures. Figure 4.4 shows a damped stable chemical oscillation for $B = 1.5$ with for reference the stable limit cycle generated at $B = 3$. At this subcritical value of $B = 1.5$, disturbances are damped out, and the chemical concentrations always return to those of the thermodynamic branch, $x = y = 0$. Figure 4.5 shows the system moving to a stable supercritical limit cycle for $B = 3$, and here a permanent chemical oscillation is established with the well-defined periodic time of this cycle. Figure 4.6 shows the nest of limit cycles generated at $B^{\mathrm{c}} = 2$ for a series of supercritical values of B.

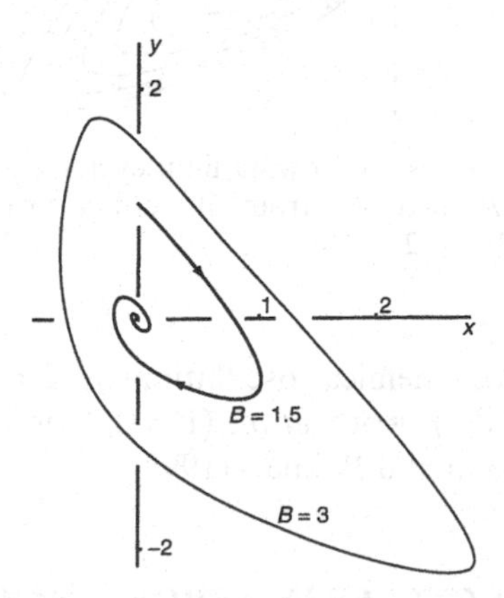

Figure 4.4 Damped oscillatory motion of a chemical reaction to a stable focus at a subcritical value of B and $A = 1$

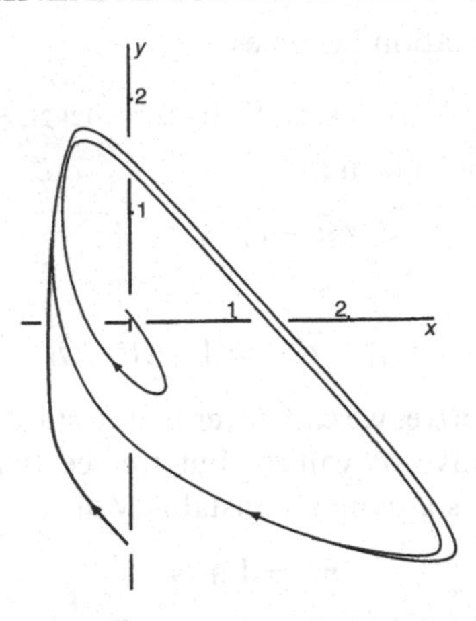

Figure 4.5 Motions to a stable limit cycle of sustained chemical oscillation at a supercritical value of B ($B = 3$) and $A = 1$

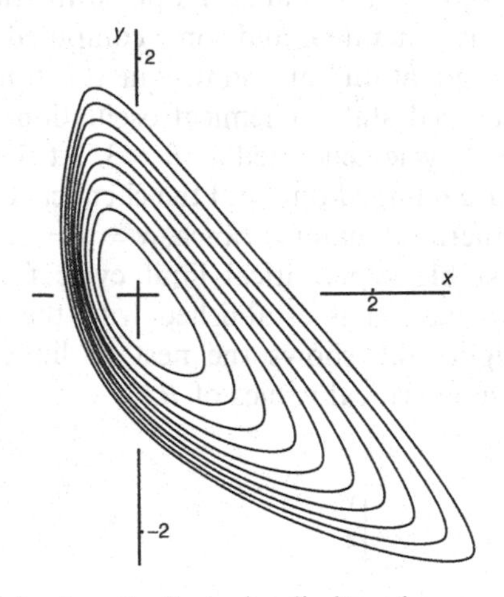

Figure 4.6 A nest of growing limit cycles generated for $A = 1$ as B increases from its critical value of 2: $B = 2.1, 2.2, \ldots, 2.9, 3.0$

Further information on chemical oscillations and models can be found in Mankin and Hudson (1984), Roux *et al.* (1983), Tomita (1982), Hudson and Rössler (1984), and Epstein and Pojman (1998).

4.4 MULTIPLE LIMIT CYCLES IN AEROELASTIC GALLOPING

Just as two point attractors can be in a state of competition, so can two attracting stable limit cycles.

Now, when a steady wind blows across a flexible elastic structure, it can induce and maintain large-amplitude oscillations such as those that destroyed the Tacoma Narrows suspension bridge. Three distinct mechanisms of aero-elastic excitation are unimodal *galloping*, *vortex resonance*, and bimodal *flutter*, and it would seem that the Tacoma bridge was destroyed by a combination of more than one of these phenomena. We shall, however, focus our attention on simple galloping, which is observed in its purest form in the dangerous vibrations of ice-coated power cables.

Consider, for example, the wind of velocity V blowing past the square prism of Figure 4.7, which is constrained by a spring and a dashpot to move vertically at right angles to the wind. Now when the prism moves downwards with velocity $\dot{y}$ the velocity of the air *relative* to the body is given by V_R in the triangle of velocities drawn. This relative wind will give rise to a vertical component of force as shown. Under a quasi-static assumption the force coefficient C is simply dependent on the angle α, which in turn depends on $\dot{y}$. Two typical variations of C with α are shown in the lower diagram.

The equation of motion of this aeroelastic oscillator is thus

$$m\ddot{y} + r\dot{y} + ky = \tfrac{1}{2}\rho V^2 a[A_1(\dot{y}/V) - A_3(\dot{y}/V)^3 + A_5(\dot{y}/V)^5 - A_7(\dot{y}/V)^7]$$

where the left-hand side corresponds to the elastic structure itself, while the right-hand side represents the aerodynamic forces. Here ρ is the air density, V is the wind velocity (which can be viewed as a controlled loading parameter), and a is the frontal area of the prism. The positive constants A_1, A_3, A_5, and A_7 are usually determined empirically from calibration tests on a tilted stationary prism.

Following Parkinson and Smith (1964), we shall look at the nonlinear behaviour of a square prism, taking the coefficients that give the best fit to the experimental points, as shown in Figure 4.8(a).

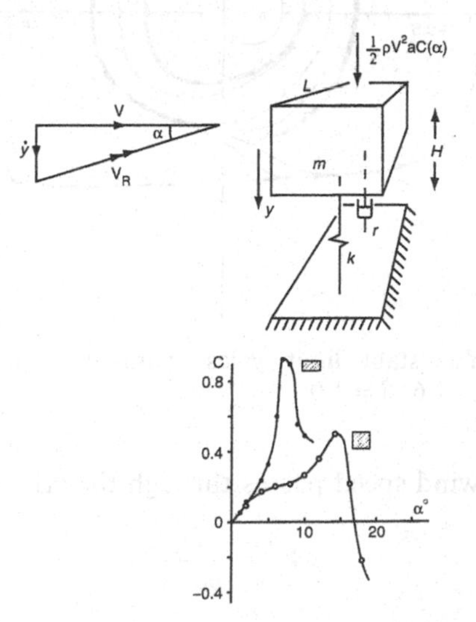

Figure 4.7 The model and aerodynamic characteristic for the galloping of a square prism in a steady wind: $a = HL$ and $V^c = 2r/\rho a A_1$

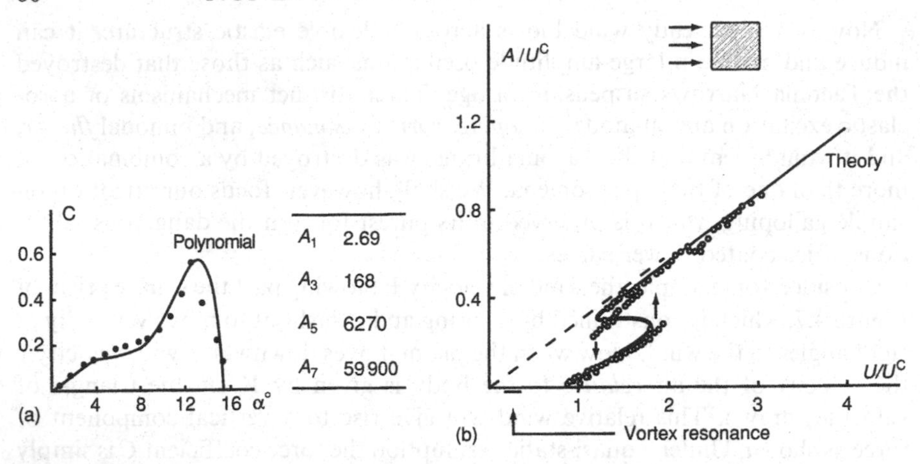

Figure 4.8 A polynomial fit to an experimentally determined $C(\alpha)$ characteristic for a square cross-section and the corresponding theoretical and experimental results of Parkinson and Smith (1964) on a plot of vibration amplitude against wind velocity. Reproduced, with permission of Oxford University Press, from Parkinson and Smith (1964)

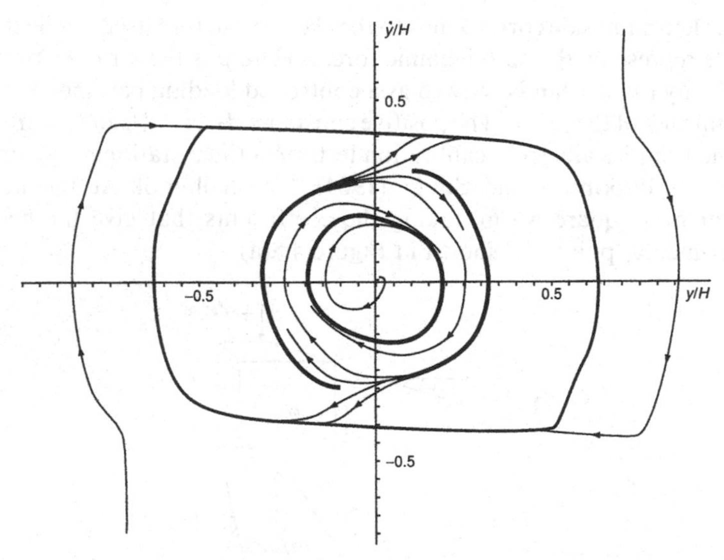

Figure 4.9 Two stable limit cycles separated by an unstable limit cycle: $U/U^{c} = 1.6$, $\beta = 1.0$

As the controlled wind speed passes through the critical velocity

$$V^{c} = \frac{2r}{\rho a A_1}$$

the negative linear aerodynamic damping exceeds the positive structural damping, and a limit cycle forms at a supercritical *Hopf bifurcation*. This is the simplest dynamic bifurcation, in which under the variation of a single control parameter a stable focus bifurcates into an unstable focus surrounded by a growing stable limit cycle.

At higher wind velocities, there exists a nest of three limit cycles, two stable cycles separated by one unstable cycle, and a sample phase portrait is shown in Figure 4.9. Here an unstable central focus is surrounded by a stable attracting limit cycle. This is in turn surrounded by an unstable limit cycle, which is finally surrounded by an outer stable limit cycle. We see that here the unstable limit cycle plays the role of a separatrix, dividing the phase space into two basins of attraction. Motions starting inside the separating unstable limit cycle all tend towards the inner cyclic attractor, while all motions starting outside the separatrix tend towards the outer cyclic attractor.

4.5 TOPOLOGY OF TWO-DIMENSIONAL PHASE SPACE

Qualitatively, dynamical systems with a plane phase space can only have the two types of final behaviour described so far, equilibria and periodic cycles. This was demonstrated at the turn of the century by Poincaré and Bendixson; the proofs rely on the simple topological fact that any closed curve in a planc divides the plane into an interior region and an exterior region. In addition, unless the dynamical system is energy-conserving, cycles will be limit cycles, either attracting nearby trajectories, or separating basins of other attractors. The Poincaré–Bendixson theory is described for example by Rosen (1970), Lefschetz (1957), and Coddington and Levinson (1955).

Another important two-dimensional phase space is the torus, which arises naturally in connection with forced oscillations to be discussed in the next two chapters. A remarkable topological theorem of Peixoto (1962) shows that in toroidal phase space (and in other two-dimensional manifolds as well) only one additional type of final behaviour is possible, namely quasi-periodic motion produced by two incommensurate frequencies. This type of motion is structurally unstable however, since an arbitrary small perturbation of the governing differential equation can cause frequency locking at some integer ratio of frequencies, although admittedly the two integers in the ratio could be large. Quasi-periodic motion is discussed more fully in Chapters 10 and 13.

Because of these topological theorems, it is clear that, for a continuous dynamical system described by differential equations to have *chaotic* final behaviour, the dimension of the phase space must be 3 or more. Notice however that the topological theorems do not apply to discrete dynamical systems described by iterated maps, which exhibit chaos even in one dimension as we shall see.

The most important examples of nonlinear dynamics in three-dimensional phase spaces are the periodically forced oscillators, which we turn to next.

5

Periodic Attractors in Driven Oscillators

As we have seen, an unforced mechanical oscillator with a single generalized coordinate x has a two-dimensional phase space associated with x and $\dot{x}$. A mechanical oscillator with two degrees of freedom, x_1 and x_2, has a four-dimensional phase space generated by $(x_i, \dot{x}_j)$ where $i, j = 1, 2$. This raises obvious problems of visualization.

An intermediate case that we shall consider in this chapter is a *forced* or *driven* mechanical oscillator, described for example by the *non-autonomous* differential equation

$$m\ddot{x} + f(x, \dot{x}) = F_0 \sin \omega_{\mathrm{f}} t$$

since here the phase space is three-dimensional spanned by x, $\dot{x}$, and t.

This space is full of non-crossing trajectories, which now tend to spiral around the time axis looking like a stranded cable. It may of course sometimes be convenient just to plot the *phase projection* in the subspace of x and $\dot{x}$, but here the trajectories will constantly cross one another.

We notice that this non-autonomous second-order differential equation can be converted into three autonomous first-order equations if we set $\dot{x} = y$ and employ the standard trick of regarding t as one of the variables with the third equation simply $\dot{t} = 1$. For numerical integration there is a straightforward extension of the Euler finite difference method to the present problem.

5.1 THE POINCARÉ MAP

A standard technique in dealing with the three-dimensional phase space $(x, \dot{x}, t)$ of our periodically driven oscillator is to inspect the projection $(x, \dot{x})$ whenever t is a multiple of $T = 2\pi/\omega_{\mathrm{f}}$ as shown in Figure 5.1. Here T is the periodic time of the forcing.

Clearly a similar trajectory bundle emerges from each $t = mT$ plane $(m = 0, 1, 2, 3, \ldots)$ so that photographs of every interval would be identical. This is not to say, however, that a *particular trajectory* repeats itself with period T: we show one in the diagram that repeats every $2T$. The repetition after T of the trajectory bundle is conveniently treated in topological dynamics by imagining the whole bundle of one interval twisted back on itself to form a solid

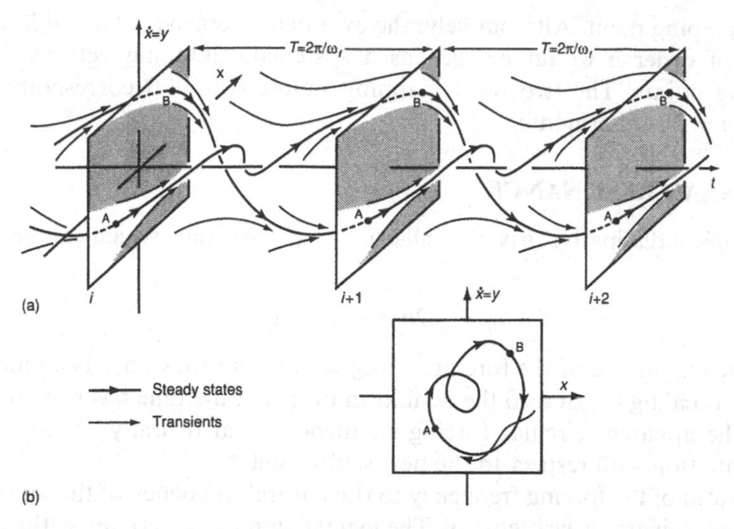

Figure 5.1 (a) Three-dimensional phase portrait of a forced mechanical oscillator ($m\ddot{x}+f(x,\dot{x})=F_0\sin\omega_f t$) and (b) its two-dimensional phase projection. The Poincaré mapping is

$$\begin{bmatrix} x(t) \\ y(t) \end{bmatrix} \to \begin{bmatrix} x(t+T) \\ y(t+T) \end{bmatrix}$$

$x_{i+1}=G(x_i,y_i)$, $y_{i+1}=H(x_i,y_i)$. A → B → A represents an $n=2$ fixed point (subharmonic)

torus. Most of the fibres, or trajectories, represent *transient motions*, but within the bundle of a dissipative system will usually exist some steady-state attracting trajectories, or *attractors*. In a nonlinear problem of any complexity, there will usually be a multiplicity of competing attractors, and a number of repellors and saddles.

A steady-state trajectory cannot normally repeat itself exactly *within* a T interval, since this would require the system to repeat an identical motion under different conditions of forcing. The most common result is for a steady trajectory to repeat itself with the period T of the forcing: the corresponding steady-state oscillation is then termed a *fundamental solution*. If, alternatively, a trajectory repeats itself after n intervals, so that it has period nT, it is called a *subharmonic of order n*. The figure shows two stable steady-state $n=2$ trajectories, differing only by a phase shift of T. Just like the more general transient motion, this steady-state subharmonic may cross itself any number of times in the $(x,\dot{x})$ phase projection: in our illustration we have shown just one such crossing.

The result of inspecting the phase projection only at the specific times, $t=mT$, is to see a sequence of dots, representing the so-called Poincaré mapping. This is often called the *stroboscopic technique*, for obvious reasons. Transient motions will appear as rather scattered dots, and the emergence of a stable fundamental solution would be seen as the eventual repetition of just one fixed mapping point. Alternatively, the eventual emergence of a stable subharmonic of order n would be seen as a systematic jumping between n fixed mapping points. The two fixed mapping points A and B correspond to the drawn $n = 2$ subharmonic.

5.2 LINEAR RESONANCE

The simplest dissipative driven oscillator is the linear one, which can be written as

$$\eta^2 \ddot{x} + 2\eta\zeta\dot{x} + x = \sin\tau$$

Here the magnitude of the forcing, along with the stiffness, has been incorporated as a scaling factor into the definition of x, and the time has been scaled to make the apparent circular forcing frequency equal to unity. A dot denotes differentiation with respect to the new scaled time τ.

The ratio of the forcing frequency to the natural frequency of the undamped, undriven oscillator is written as η. The usual damping ratio (namely the ratio of the actual damping to the critical damping) of the undriven oscillator is written as ζ. We notice that this equation for a linear driven oscillator is the special case of $\alpha = 1$ for the *bilinear oscillator* that we shall be discussing later.

The complementary function of this linear differential equation, which represents the damped transient behaviour, can be written as

$$x_c = \exp(-\zeta\tau/\eta)[A\sin(\omega_d\tau + p)]$$

where

$$\omega_d = \frac{1}{\eta}\sqrt{(1-\zeta^2)}$$

Here A and p are the arbitrary integration constants to be found from the starting conditions.

The particular integral of this equation, which represents the final steady-state solution, can be written as

$$x_p = \frac{1}{[(1-\eta^2)^2 + (2\eta\zeta)^2]^{1/2}}\sin(\tau - \psi)$$

where

$$\tan\psi = \frac{2\eta\zeta}{1-\eta^2}$$

The complete general solution is then just the sum of these two contributions,

$$x(\tau) = x_c + x_p$$

The complementary function is just the solution of the unforced damped oscillator, dependent on the starting conditions through A and p, at the fixed autonomous frequency ω_d. Once the complementary function has decayed, we are left with the single stable steady-state solution represented by the particular integral which has the frequency of the forcing term. So there are no subharmonics involved, and the unique attracting fundamental steady state captures all motions of the system: its domain of attraction is the whole phase space.

There are just the two independent frequencies involved, and in Figure 5.2 we show a typical trajectory in the phase projection for the damping $\zeta = 0.1$ and the frequency ratio $\eta = 1/16$. The motion has been started at the origin at $\tau = 0$, and shows two Poincaré points, A at the origin and B. In this one forcing period, the high-frequency transient has essentially damped out, leaving us with the circular steady-state fundamental solution: this has a single fixed Poincaré mapping close to B, almost at the top of the circle.

A second typical trajectory in the phase projection is shown for $\zeta = 0.8$ and $\eta = 6$ in Figure 5.3. This is again started at the origin, and we see the circular particular integral, with a Poincaré point at approximately its lower extremity, being swept first to the right and then to the left by the low-frequency transient motion. Under the heavy damping prescribed, the fundamental circular steady state centred on the origin has almost been reached after one full transient vibration. The transitional Poincaré points A to F are marked, and with $\eta \gg 1$ the final fixed mapping point of the steady state is now at about the bottom of the final circle.

The well-known resonance response diagram relates purely to the steady-state particular integral, and is shown in Figure 5.4. Here the top graph shows the amplitude, y of x as a function of the frequency ratio, at a fixed damping, and the two lower diagrams show the corresponding movement of the fixed Poincaré mapping point $(x_P, \dot{x}_P)$.

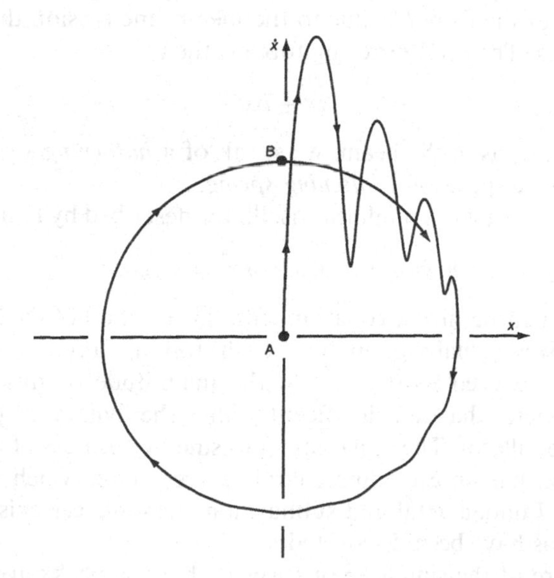

Figure 5.2 Phase projection of a driven linear oscillator, showing damping of a high-frequency transient, leading to a circular steady state: $\zeta = 0.1$, $\alpha = 1$, $\eta = 1/16$

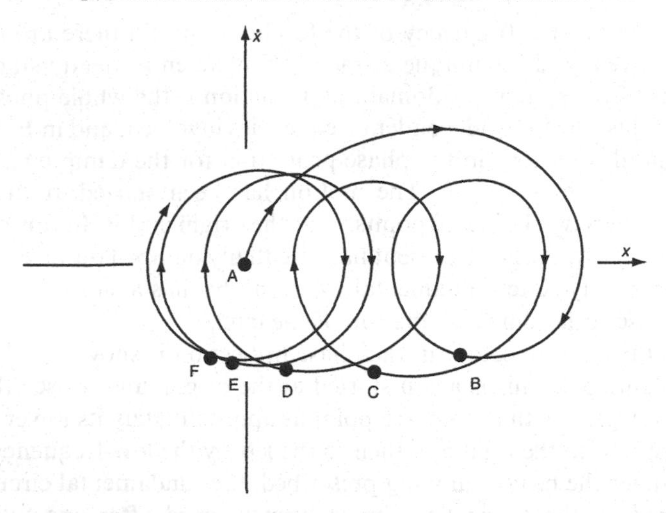

Figure 5.3 Phase projection of a driven linear oscillator, showing a low-frequency transient sweeping the circular particular integral to the right and then to the left: $\zeta = 0.8$, $\alpha = 1$, $\eta = 6$

5.3 NONLINEAR RESONANCE

Nonlinear effects arise in mechanical and structural systems when deflections become large. In the resonance of a beam clamped between fixed supports, the linear stiffness due to bending action is for example augmented by a nonlinear restoring force of the form bx^3 due to the membrane tension that builds up at large amplitudes. The total restoring force is then

$$ax + bx^3$$

and for b positive, as in the beam, we speak of a *hardening spring*: conversely with b negative we speak of a *softening spring*.

Consider then the driven nonlinear oscillator described by Duffing's equation

$$\eta^2\ddot{x} + 2\eta\zeta\dot{x} + x + \alpha x^3 = F_0 \cos\tau$$

which is just our earlier linear equation with the addition of the hardening term αx^3. Notice, however, that the amplitude of the forcing cannot now be scaled out, and has therefore been written as F_0: the magnitude of this is now a new operative parameter that can significantly alter the dynamical phenomena exhibited by the oscillator. The replacement of sine by cosine is of no significance.

This equation has an exceedingly complex response, which is even yet not fully explored. Fundamental and subharmonic resonances exist, and recently regimes of chaos have been discovered.

A new feature of the simple resonances is that the peaks are curved, as we shall see shortly: they curve to the right for a hardening spring, since the inherent frequency of the system then increases with amplitude, but to the left for a softening spring.

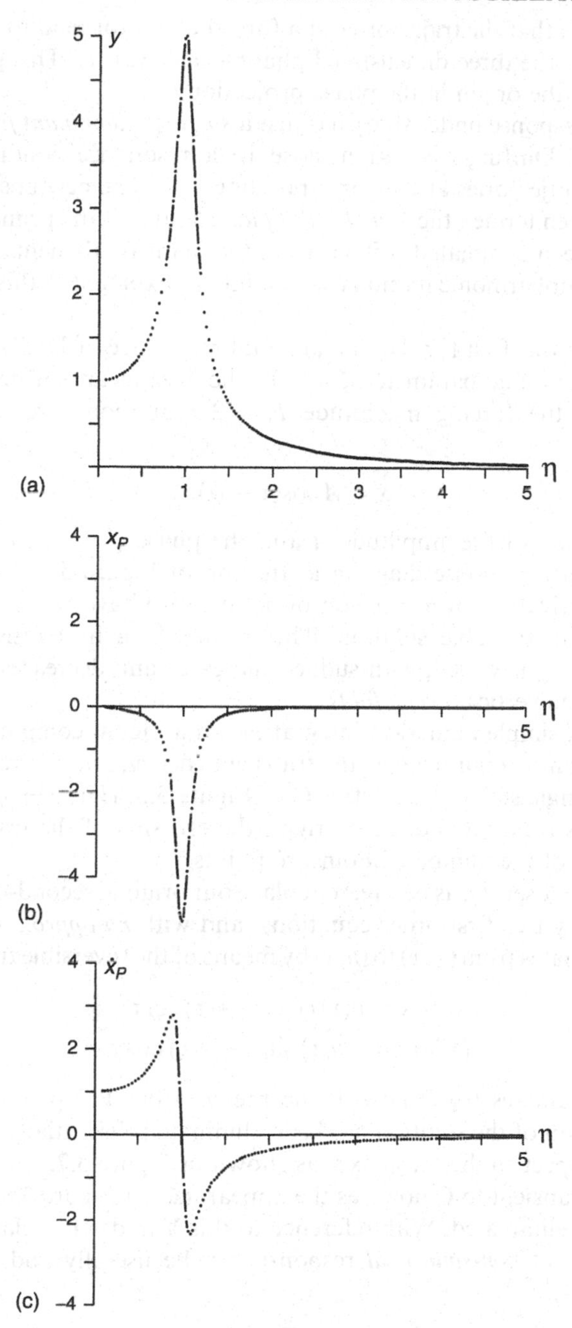

Figure 5.4 (a) Resonance response diagram of a driven linear oscillator, showing the amplitude of the steady state as a function of frequency ratio ($\alpha = 1$, $\zeta = 0.1$). Diagrams (b) and (c) show movement of steady-state Poincaré mapping points

We have seen that the trajectories of a forced oscillator tend to spiral around the time axis in the three-dimensional phase space $(x, \dot{x}, \tau)$. This gives rise to a rotation about the origin in the phase projection $(x, \dot{x})$.

Now if the response under study has just a *single predominant frequency* (as it might have in Duffing's equation close to a resonance condition) we can 'unscrew' the trajectories at the corresponding rate. The new unscrewed phase projection is then termed the *Van der Pol plane*, and in it the primary dominant rotation has been eliminated. Either a predominant fundamental motion or a predominant subharmonic motion can usefully be explored in this way, but not both at once.

We illustrate this first for the fundamental resonance of Duffing's equation. Taking the damping parameter $\zeta = 0.1$, the nonlinear stiffness parameter $\alpha = 0.05$, and the forcing magnitude $F_0 = 2.5$, motions are approximately given by

$$x = A\cos(\tau - \psi)$$

and the variation of the amplitude A and the phase ψ with η is given by the curved resonance response diagram at the top of Figure 5.5. Notice that the curves are multivalued over a region of η where we have two stable solutions separated by an unstable solution. This results in a hysteresis loop as the forcing frequency is varied, with sudden increases and decreases in amplitude and phase at the vertical *cyclic folds*.

For $\eta = 2$, a simple numerical integration on a digital computer reveals the phase projection of Figure 5.6 as the transients decay, and the response settles down to the single stable steady state C of Figure 5.5. Here $y = \eta\dot{x}$, and we see the continuous rotation about the origin, the crossing of the trajectories, and the movement of the numbered Poincaré points.

Writing $\eta\dot{x} = y$ serves, as before, to replace our original second-order differential equation by two first-order equations, and with *no approximation* we can change coordinates from (x, y) to (u, v) by means of the reversible transformation

$$x = u(\tau)\cos\tau - v(\tau)\sin\tau$$
$$(1/\eta)y = -u(\tau)\sin\tau - v(\tau)\cos\tau$$

The new coordinates (u, v) now define the Van der Pol plane in which the primary rotation of the solution has been eliminated. Essentially the (u, v) axes rotate with respect to the (x, y) axes as shown in Figure 5.7.

The *exact* transient to C now has the appearance of Figure 5.8, with most of the crossovers eliminated. With reference to this Van der Pol plane, the amplitude of the exact *non-sinusoidal* response can be usefully and appropriately *defined* as

$$A = \sqrt{(u^2 + v^2)}$$

and the *phase* can be *defined* as

$$\psi = -\tan^{-1}(v/u)$$

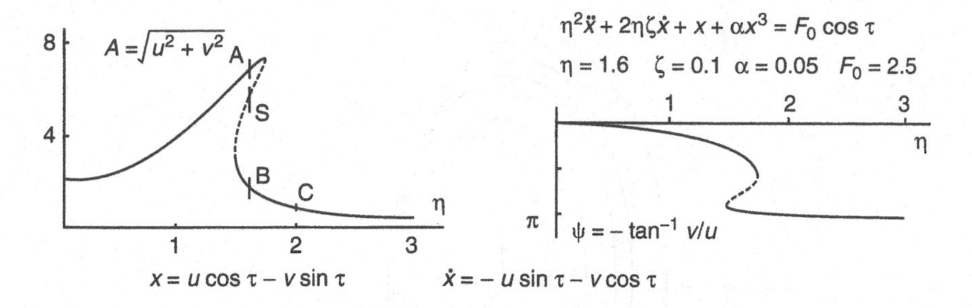

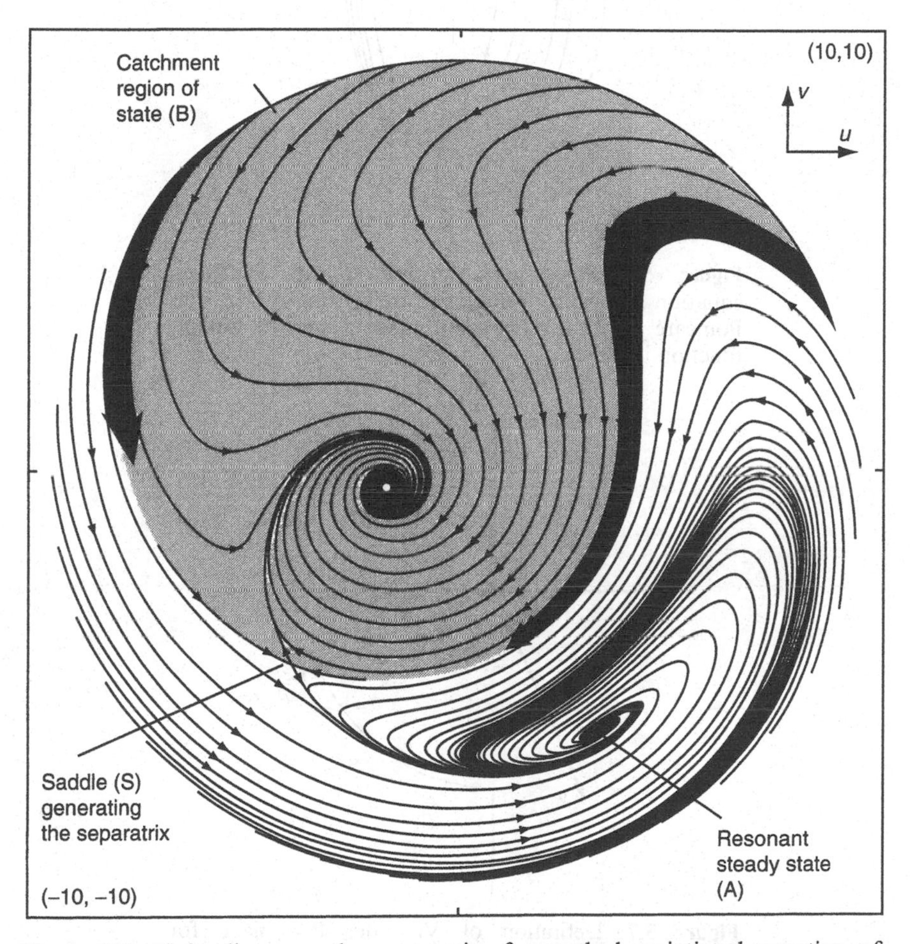

Figure 5.5 Main diagram: phase portrait of smoothed variational equation of Duffing's equation in the Van der Pol plane. The two small graphs show the nonlinear resonance response and the phase diagram; they reveal a hysteresis regime in the fundamental resonance

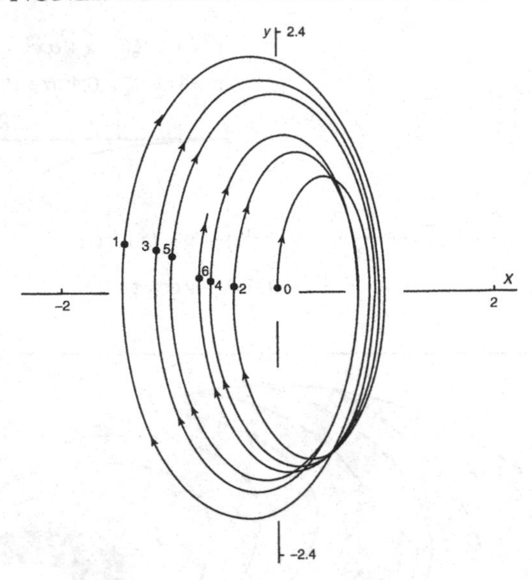

Figure 5.6 Phase-space projection of Duffing's equation showing transients to C for $\eta = 2$: the Poincaré points $\tau = 2n\pi$ are numbered and the forcing function is $F_0 \cos \tau$

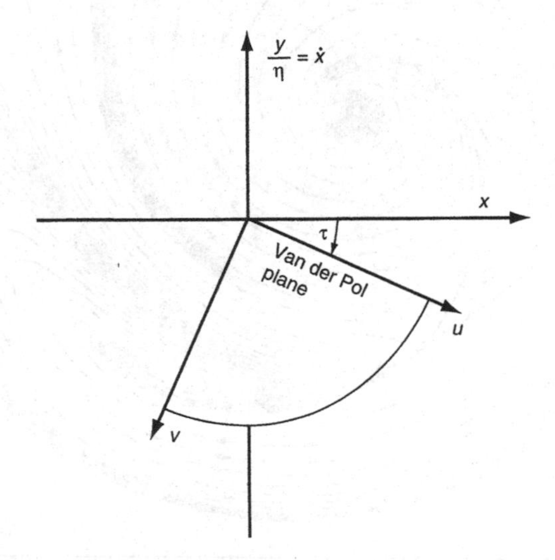

Figure 5.7 Definition of Van der Pol plane for examination of a fundamental solution

So far we have made no approximations, but simply introduced a rotating system of coordinates at the forcing frequency to define a Van der Pol plane in which rotations at this frequency have been eliminated. This plane will only be useful, in the sense that the trajectories in it are easily seen and understood,

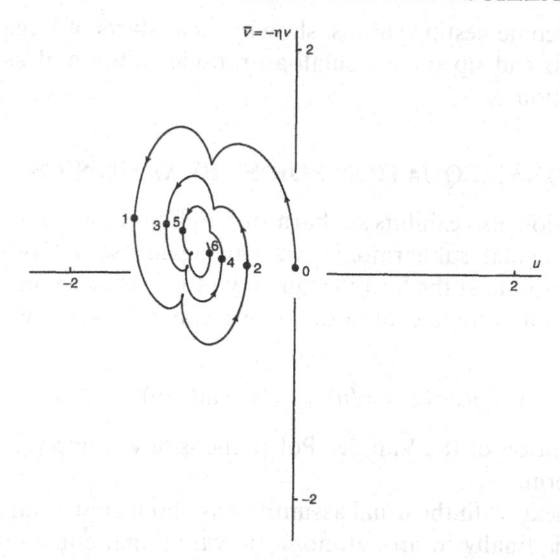

Figure 5.8 Exact trajectories of Duffing's equation in Van der Pol plane, showing transients to C for $\eta = 2$

provided that all dominant motions of the system are at a frequency close to the forcing frequency.

5.4 THE SMOOTHED VARIATIONAL EQUATION

For our nonlinear oscillator described by Duffing's equation in the range of parameters under discussion, we can finally make an approximation by *averaging* the exact equations in u, v, and τ following the methods of Krylov and Bogoliubov as given, for example, by Hale (1969). This averaging procedure, in which higher harmonics are suitably ignored, eliminates the explicit appearance of τ and leaves us with an autonomous system described by u and v. For the resulting autonomous variational equation, the Van der Pol (u, v) plane is the true two-dimensional phase space, so no crossovers can occur and the high-frequency cusping behaviour of Figure 5.8 is smoothed out.

A phase portrait for $\eta = 1.6$ is shown at the bottom of Figure 5.5, with the previously given parameters. Here we see transient motions heading towards the two steady-state sinks corresponding to A and B on the top of the same figure. The flow in this diagram can be interpreted in polar coordinates as the continuous adjustment by the system of its amplitude A and its phase ψ, the averaging procedure being often called the method of *slowly varying amplitude and phase*.

The catchment regions for the two competing stable steady-state vibrations A and B, represented in this Van der Pol plane by two sinks, are bounded by the separatrix passing through the unstable saddle solution: one of the domains of attraction is shown dotted, the other being left blank. For large amplitudes these clearly become nesting spirals, showing that starts of large amplitude are just as likely to end up on the small-amplitude solution B as on the large-amplitude solution A.

5.5 VARIATIONAL EQUATION FOR SUBHARMONICS

Duffing's equation also exhibits subharmonic solutions, and if we are interested in just one particular subharmonic, we can again use a Van der Pol plane, defined now to rotate at the lower frequency of the subharmonic under investigation. For a subharmonic of order n we can for example write x in the form

$$x = u_n(\tau)\cos(\tau/n) + v_n(\tau)\sin(\tau/n) + k\cos\tau$$

where the definition of the Van der Pol plane is now complicated by the final fundamental term.

Averaging, next, with the usual assumptions about slowly varying amplitude and phase leads finally to an autonomous variational equation in u_n and v_n. Transients to the steady-state subharmonics can now be seen in the two-dimensional (u_n, v_n) phase space of this variational equation.

A typical picture due to Hayashi for Duffing's equation with a hardening cubic spring, linear damping but no linear stiffness is shown in Figure 5.9 for the $n = 3$ subharmonic. This contains seven equilibrium solutions, those numbered

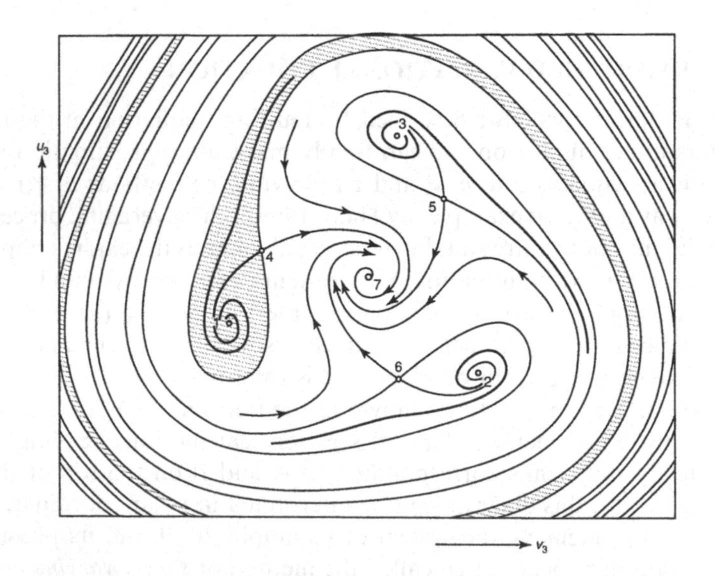

Figure 5.9 Phase portrait of smoothed variational equation of Duffing's equation in Van der Pol plane corresponding to subharmonics of order $n = 3$: (1, 2, 3) stable subharmonics; (4, 5, 6) unstable subharmonics; (7) fundamental harmonic, stable. Reproduced, with permission of McGraw-Hill Book Company, from Hayashi (1964)

1, 2 and 3 representing the stable steady-state subharmonic of order $n = 3$, with the domain of attraction shown shaded. The central sink, numbered 7, with $u_3 = v_3 = 0$, represents a condition of no subharmonic oscillation, corresponding to the possibility of a harmonic fundamental response. The domains of attraction are bounded by the separatrix trajectories associated with the three unstable saddle solutions numbered 4, 5 and 6. We see that the domains (or basins) of attraction of the $n = 3$ subharmonic spiral outwards as thinning tails.

We should always remember that these variational techniques only work well in specific parameter regimes, and give, naturally, no information about the possible occurrence of motions other than those that are specifically incorporated in the analytical procedure.

5.6 BASINS OF ATTRACTION BY MAPPING TECHNIQUES

The preceding approximate method has two drawbacks. First, if the initial conditions are far removed from the steady state, the assumption of slowly varying amplitude and phase will break down, and inaccuracies will arise. Secondly, as we have indicated, it cannot deal with a situation in which a number of different types of response coexist. An exact method for determining domains of attraction is based on the Poincaré mapping that we have introduced earlier.

To elaborate this, let us consider first the rather general forced oscillator

$$\ddot{x} + f(x, \dot{x})\dot{x} + g(x) = E(t)$$

where $E(t)$ is periodic with period L. Once again we set $\dot{x} = y$ to obtain two first-order equations

$$\begin{aligned} \dot{x} &= y \\ \dot{y} &= -f(x, y)y - g(x) + E(t) \end{aligned}$$

These are simply a particular case of the more general equations that we now consider:

$$\begin{aligned} \dot{x} &= X(x, y, t) \\ \dot{y} &= Y(x, y, t) \end{aligned}$$

where X and Y are both periodic in t, with the same period, L. Suppose $x(t)$ and $y(t)$ are solutions of the latter, starting at a point P_0 in the (x, y) phase projection. So P_0 has the coordinates $x(0)$, $y(0)$. We study the Poincaré mapping, looking at the points P_m at the time $t = mL$, where $m = 0, 1, \ldots$. We call the *transformation* $P_0 \rightarrow P_1$ the mapping, T, and write $P_1 = TP_0$. In an obvious extension of this notation we write *iterates* of the mapping as

$$\begin{aligned} P_2 &= TP_1 = T^2P_0 \\ P_3 &= TP_2 = T^3P_0 \qquad \text{etc.} \end{aligned}$$

We can also write the inverse mapping, $P_0 = T^{-1}P_1 = T^{-2}P_2$, etc.

If a solution $x(t)$, $y(t)$ is a fundamental harmonic with the period L of the forcing, then the point P_0 is a *fixed point* of the mapping T. If, alternatively, we have a subharmonic of order n (= 2,3, . . .) (sometimes called a subharmonic of order $1/n$) with a *minimum* period of nL, the steady-state mapping points $P_0, P_1, \ldots, P_{n-1}$ are called periodic points. They are in fact all fixed points of the nth iterate, T^n, of the mapping T. We shall see that a study of just the mapping T of the (x, y) plane onto itself serves to determine the domains of attraction of our original continuous differential equation.

As an example, for Duffing's equation with zero linear stiffness, Hayashi has located the fixed points 1, 2 and 3 of Figure 5.10, using a specially designed computer. Here we see that, with the parameters chosen, the equation has just three fundamental solutions, as in our approximate study of Figure 5.5. Also shown in his figure are two curves that are *invariant* under the mapping. Thus along the invariant curve denoted by a fine line, successive iterates of the mapping step in the direction of the arrows and approach either the stable solution 2 or the coexisting stable solution 3. Alternatively, successive images of the mapping step along the heavy invariant curve and approach the unstable saddle point 1.

The location of this heavy invariant curve is the key to the delineation of the domains of attraction, since it divides the whole (x, y) plane into its two catchment regions. Any initial point in one of the domains steps, under the repeated iterations of the mapping T, to the particular stable fixed point (2 or 3) that is located in the interior of that domain. The heavy invariant curve itself can be located, once the unstable fixed point 1 has been found by running time backwards from close to this point, using the inverse mapping T^{-1}.

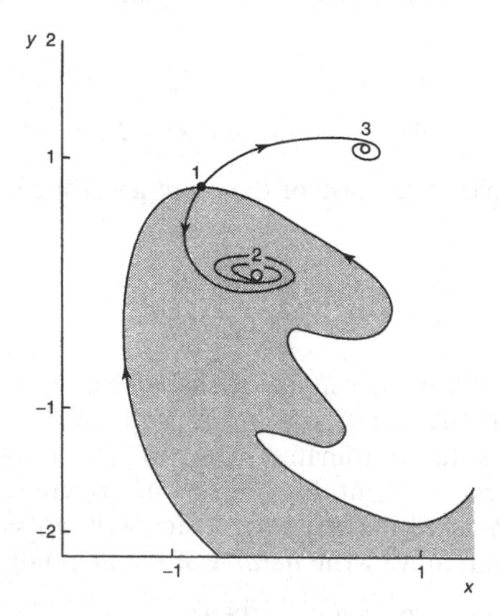

Figure 5.10 Fixed points and invariant curves for harmonic oscillations of Duffing's equation due to Hayashi. Reproduced, with permission of McGraw-Hill Book Company, from Hayashi (1964)

A more complex example, again for Duffing's equation but with a different set of system parameters, has also been studied in depth by Hayashi, and his results for the domains of attraction are shown in Figure 5.11. Here we have *three coexisting harmonic solutions*, these being fixed points of the mapping T, denoted by points 1, 2 and 3: of these, number 1 is *unstable*, while 2 and 3 are

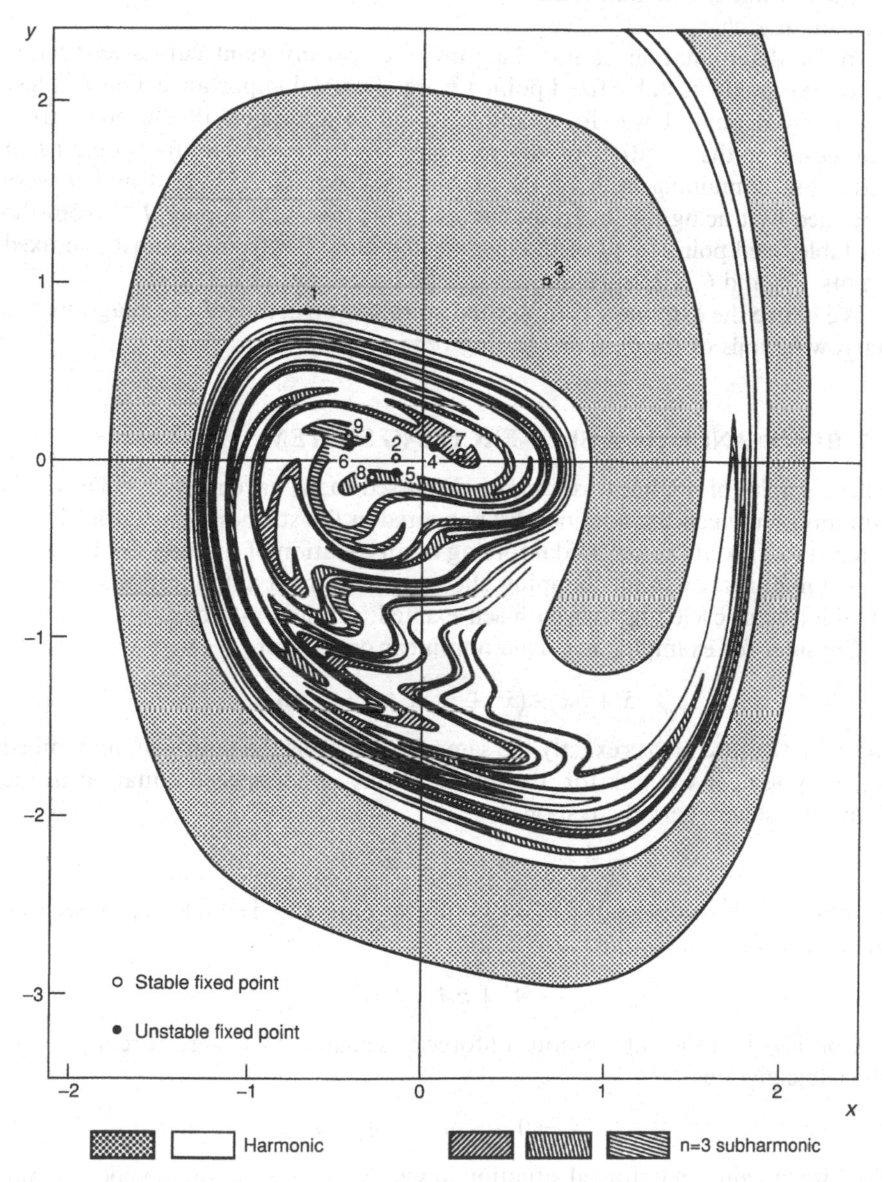

Figure 5.11 Domains of attraction for three harmonic and two coexisting subharmonics of order $n = 3$ for Duffing's equation, as determined by Hayashi. Reproduced, with permission of McGraw-Hill Book Company, from Hayashi (1964)

stable. We notice that, as before, the unstable solution 1 lies on the domain boundary. In addition we have six fixed points, numbered 4 to 9, of the mapping T^3: these represent *two coexisting subharmonics of order 3*, there being of course three mapping points per solution. The subharmonic solution represented by points 4, 5 and 6, which lie on the domain boundary, is clearly *unstable*, while the subharmonic represented by points 7, 8 and 9, lying within a domain, is *stable*.

In the determination of this diagram, the two invariant curves associated with the directly unstable fixed point 1 were of crucial importance. One of these curves from point 1 was for example found to approach all the other fixed points under the iteration of the mapping T. In particular, the domains of attraction containing each of the stable fixed points 2, 3, 7, 8 and 9 were obtained by tracing the invariant curves of the inverse mapping T^{-1} from the unstable fixed point 1 and of the inverse mapping T^{-3} from the unstable fixed points 4, 5 and 6. These techniques will be used again in Part III.

We notice the extremely complicated spiralling structure of this diagram, the narrowing tails of the domains having been omitted.

5.7 RESONANCE OF A SELF-EXCITING SYSTEM

The examples of nonlinear resonance discussed so far have related to Duffing's equation, in which the nonlinearity appeared in the stiffness term. It is also of interest to look at the sinusoidal forcing of an equation of the Van der Pol type: here a nonlinearity in the damping allows the unforced autonomous system to exhibit a limit cycle together with self-excited transient motions.

Consider for example the driven nonlinear osciallator

$$\ddot{x} + b\dot{x} - (x^2 + \dot{x}^2)\dot{x} + x = f \sin t$$

in which the forcing has exactly the same frequency as the undriven, undamped linear system. If we look for a solution of this rather special equation in the form of a steady periodic response

$$x = A \cos t$$

we find that this indeed satisfies the nonlinear differential equation provided we have the *cubic relationship*

$$-A^3 + bA + f = 0$$

Consider first the autonomous unforced oscillator. With zero forcing, $f = 0$, the cubic gives us

$$A = 0 \qquad \text{or} \qquad A^2 = b$$

So if we imagine an unforced situation in which the linear damping coefficient b decreases from a positive value to a negative value, due for example to the fluid loading of a structure, we see that the undeflected solution $x = \dot{x} = 0$ becomes unstable at $b = 0$. This loss of stability is associated with an unstable Hopf bifurcation, at which is generated a trace of unstable limit cycles whose amplitudes are given by $A^2 = b$. So at $b = 0.5$, for example, we have a stable origin, $x = \dot{x} = 0$, surrounded by an unstable sinusoidal limit cycle of amplitude $A = 1/\sqrt{2}$. The situation is then similar to that shown earlier in Figure 4.1 with time reversed. For large starts outside the unstable limit cycle, the amplitude will now grow without limit, and we have a self-excited fluttering motion.

Suppose we now force the system with, for example, a value of $f = 0.1$. The cubic relationship then gives us three values of A, corresponding to three co-existing steady-state sinusoidal oscillations. Notice that $A = 0$ with $x = \dot{x} = 0$ is no longer a solution. Of these three solutions, we shall see that the small-amplitude motion P (with $A = -0.222$) is stable, while the large-amplitude motions S (with $A = -0.570$) and C (with $A = 0.791$) are unstable. Here the negative signs are really giving us information about the phase relative to the forcing.

To study the transient motions between these steady states, we can follow our earlier examples and use the method of slowly varying amplitude and phase to obtain an autonomous variational equation in the Van der Pol plane. This gives us the phase portrait of Figure 5.12, which summarizes in polar coordinates the amplitude of the response and its phase relative to the forcing term. In this picture, the three steady-state sinusoidal solutions appear as fixed equilibrium points, the central sink being the one stable solution. The circular basin of attraction is clearly seen, starts outside this leading to divergent motions with amplitudes tending to infinity.

If we were now to vary the system by decreasing b, at constant $f = 0.1$, the sink would move towards the left-hand saddle S and at a critical value of b we would have a saddle-node coalescence. Thereafter the right-hand source C

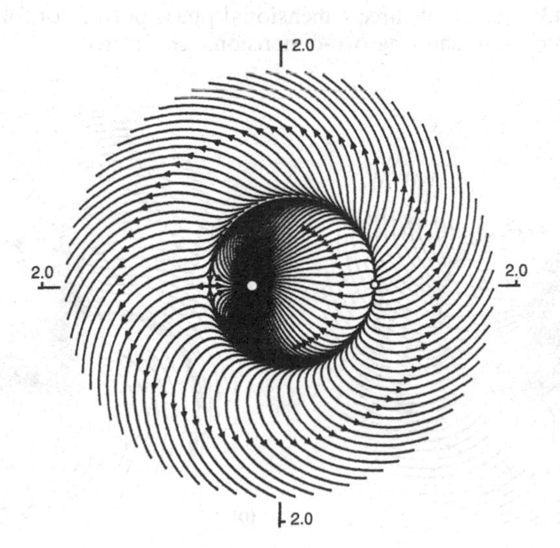

Figure 5.12 Phase portrait in Van der Pol plane of smoothed variational equation of a forced self-exciting nonlinear oscillator ($b = 0.5, f = 0.1$)

would be the only remaining steady state. We see that at this saddle-node bifurcation the finite catchement area of P is instantaneously lost, all trajectories henceforth flowing from C out to infinity (see Figure 5.14).

On the basis of this smoothed variational diagram we can sketch the complete $(x, \dot{x}, t)$ phase space of the nonlinear driven oscillator as shown in Figure 5.13. Here we have a two-dimensional separatrix, outside of which are the unbounded transients tending to infinity. On the separatrix are the two unstable steady-state oscillations: one corresponds to a saddle, being stable for motion on the separatrix, but unstable off it; the other corresponds to a totally

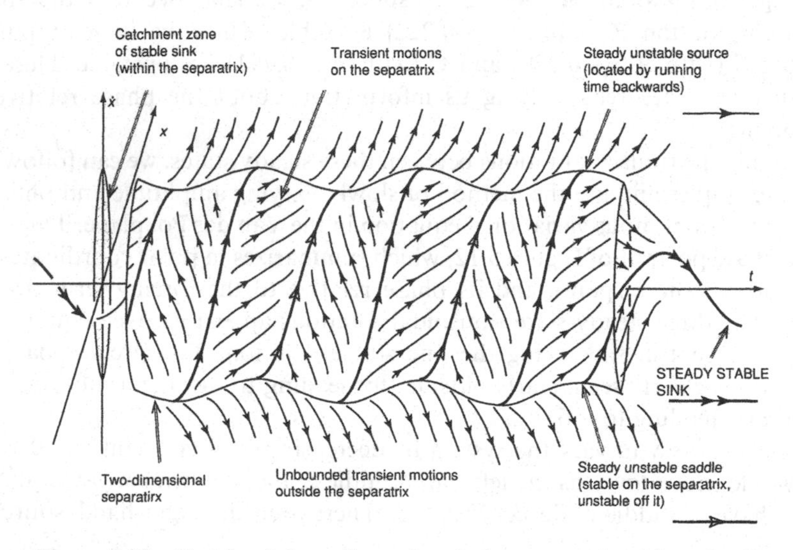

Figure 5.13 Sketch of three-dimensional phase portrait of forced self-exciting oscillator, showing two-dimensional separatrix

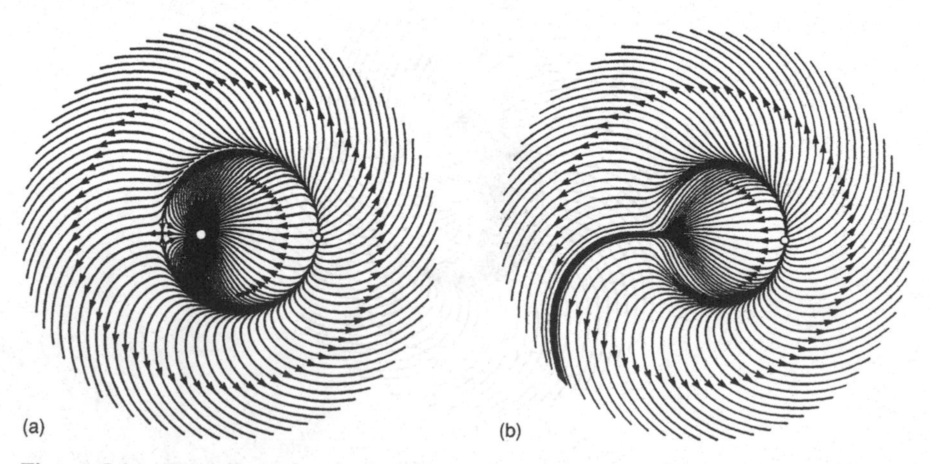

Figure 5.14 Two-dimensional phase portraits showing the effect of a saddle-node bifurcation on an attractor and its basin: (a) before the bifurcation, (b) after the bifurcation

unstable source, which can therefore be conveniently located numerically by running time backwards. Transient motions on the separatrix flow between these two oscillations as shown. Within the separatrix is the basin of attraction of the small-amplitude stable sink.

5.8 THE ABC OF NONLINEAR DYNAMICS

The portraits of the previous section give a nice introduction to the ABC of nonlinear dynamics. Here A stands for attractor, B stands for basin, and C stands for catastrophe (conveniently used here to mean bifurcation). To emphasize this, we show two portraits side by side in Figure 5.14. The first is the earlier Figure 5.12, for $b = 0.5$, $f = 0.1$; the second is for $b = 0.375$, $f = 0.1$. Between these two parameter sets, a finite residual basin of the central attractor is instantaneously lost at a saddle-node bifurcation.

6

Chaotic Attractors in Forced Oscillators

Having introduced the three-dimensional phase space of periodically forced oscillators in the previous chapter, we are now prepared to examine the topological structure of chaotic attractors in forced oscillators. The most important examples are the Van der Pol equation with nonlinear damping, and oscillators of the Duffing type having a nonlinear restoring force.

6.1 RELAXATION OSCILLATIONS AND HEARTBEAT

The second-order oscillator equation

$$\ddot{x} + F(\dot{x}) + \omega^2 x = 0 \tag{6.1}$$

with a nonlinear damping function F was introduced by Lord Rayleigh (1896, §68a) as a model of a vibrating clarinet reed or a violin string. The function $F(\dot{x})$ might for example be polynomial as in the model of wind-induced galloping of Figures 4.7 to 4.9. A nonlinear damping function that exerts a force always opposed to the direction of velocity will yield qualitative behaviour like a linearly damped system, as for example in Figure 3.16. However, the behaviour of solutions of (6.1) is qualitatively different from a damped linear oscillator if F sometimes acts with the same direction as the velocity $\dot{x}$, indicating the presence of an energy source. In qualitative studies it is usual to follow Rayleigh and consider a polynomial F with linear and cubic terms only. The final behaviour of the equation (6.1) may then be a limit cycle in the autonomous case, as we have seen.

Upon differentiating equation (6.1) with respect to time, and substituting v for $\dot{x}$ above, we obtain

$$\ddot{v} + F'(v)\dot{v} + \omega^2 v = 0 \tag{6.2}$$

Choosing $F(v) = \alpha(v^3/3 - v)$ leads to the Van der Pol equation

$$\ddot{v} + \alpha(v^2 - 1)\dot{v} + \omega^2 v = 0 \tag{6.3}$$

This equation was extensively studied by Van der Pol both theoretically and in analogue simulation using vacuum-tube circuits, where the function F

corresponds to the nonlinear characteristic of a triode tube. Van der Pol observed that limit cycle oscillations of equation (6.3) are nearly sinusoidal functions of time when α is small compared with ω, but approach a square wave when α becomes large, as illustrated in Figure 6.1. This latter, highly nonlinear oscillation was called a *relaxation oscillation* by Van der Pol, because each half-cycle corresponds to a build-up of charge on a capacitance C with relaxation time $\tau = RC$. Thus the frequency of the relaxation oscillation is determined not by the restoring force reflected in ω^2 but by a relaxation time.

This feature of relaxation oscillations makes them particularly susceptible to locking at some external driving frequency, even when the external frequency differs widely from the natural frequency of the unforced relaxation oscillations. As Van der Pol noted, 'it is a well-known fact that outer circumstances may much easier influence a resistance than a mass or elasticity'. Relaxation oscillators are ideally suited to control systems in which an input stimulus should produce a response of fixed amplitude but adaptable frequency or repetition rate. An example is the beating of the heart: it is known that each contraction of the ventricle is stimulated by a nerve impulse generated upon contraction of the auricle. Van der Pol and Van der Mark (1928) constructed an electrical circuit composed of coupled relaxation oscillators, which they proposed as a qualitative model for the beating heart. They reported that, by adjusting the coupling from one oscillator to the other, convincing simulations of both normal heartbeat and of certain disorders were observed.

A closer connection between the heartbeat and the Van der Pol equations was discovered by FitzHugh (1961) and pursued by Nagumo *et al.* (1962). FitzHugh suggested a variant of the Van der Pol equation as a simplification of the

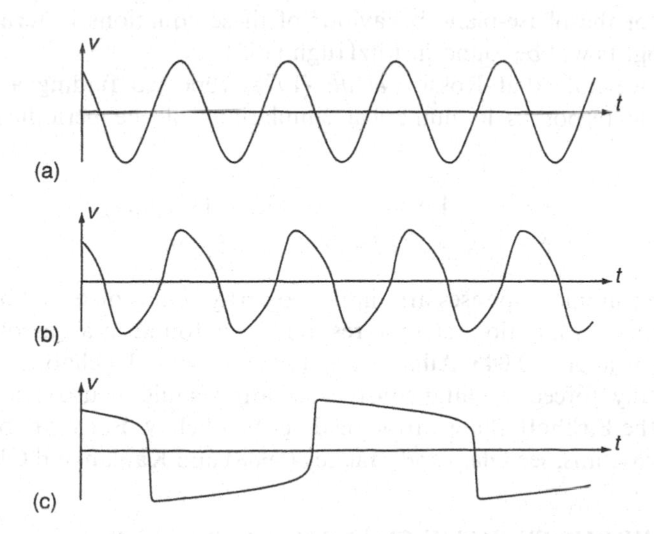

Figure 6.1 Solution versus time of the unforced Van der Pol equation (6.3) for $\omega = 1$ and (a) $\alpha = 0.1$, (b) $\alpha = 1$, (c) $\alpha = 10$. The solutions progress from nearly sinusoidal to highly nonlinear relaxation oscillations

successful model of nerve axon response developed by Hodgkin and Huxley (1952) consisting of a system of four first-order ordinary differential equations. It is interesting to note that although the Hodgkin–Huxley equations are in part derivable from Maxwell's electromagnetic field equations applied to a cylindrical model of the nerve axon (e.g. Scott 1975), the crucial step taken by Hodgkin and Huxley was the inference of electrodynamic characteristics of the nerve membrane from detailed experimental studies of excised animal axons, and not from electrochemical first principles. The FitzHugh–Nagumo equations are thus an inspiring example of a qualitative model derived from observation of dynamic behaviour, rather than from fundamental physical laws. This approach will undoubtedly be of particular importance in the growing fields of nonlinear dynamics of biological, ecological, and social phenomena. Another example of this phenomenological approach to nonlinear model-making can be found in Abraham *et al.* (1985).

The equations proposed by FitzHugh are based on the two first-order equations

$$\begin{aligned} \dot{v} &= \alpha w - F(v) \\ \dot{w} &= -v/\alpha \end{aligned}$$

which are equivalent to (6.3) with $\omega = 1$. FitzHugh added additional terms to obtain

$$\begin{aligned} \dot{v} &= \alpha(w + z) - F(v) \\ \dot{w} &= -(v - a + bw)/\alpha \end{aligned} \tag{6.4}$$

where a and b are fixed parameters and z is stimulus intensity. A detailed discussion of the phase-plane behaviour of these equations in terms familiar to physiologists will be found in FitzHugh (1961).

Finally, we note that Rössler *et al.* (1978) reported finding a variety of subharmonic responses in numerical simulations of the periodically forced system

$$\begin{aligned} \dot{x} &= (y - 1.5) + (x - x^3/3) - A\sin(0.2t) \\ \dot{y} &= -[x - 0.467 + 0.8(y - 1.5)]/9 \end{aligned} \tag{6.5}$$

These subharmonic responses are similar to arrhythmias observed in malfunctioning hearts. In addition, chaotic response was found in a narrow range of parameters near $A = 0.045$. Aihara *et al.* (1986) observed a chaotic attractor in a sinusoidally forced, spontaneously oscillating squid neuron; its structure resembles the Birkhoff–Shaw attractor described below. For more on chaos in biological rhythms, see Glass and Mackey (1988) and Kaplan and Glass (1995).

6.2 THE BIRKHOFF–SHAW CHAOTIC ATTRACTOR

Van der Pol's studies have inspired a number of investigations of the periodically forced equations

$$\ddot{v} + \alpha(v^2 - 1)\dot{v} + \omega^2 v = A \sin \omega_0 t \tag{6.6}$$

The periodically forced Van der Pol equation is particularly interesting because it represents a system with autonomous oscillations influenced by a second, external cyclic force. Since cyclic behaviour is one of the most important qualitative phenomenon in non-equilibrium dynamics, the coupling of one cycle to another has special significance in any systematic study of dynamics.

As might be expected, the behaviour of solutions of (6.6) for small α (i.e. weak nonlinearity) can be studied by analytical methods. The method of averaging, by which an autonomous system is obtained from averaging (6.6) over one forcing cycle, leads to a fairly complete understanding of the behaviour for small α. The milestones in this line of research are Cartwright (1948) and the phase portrait diagrams on pages 71 and 72 of Guckenheimer and Holmes (1983), which suggest the possibility of mildly chaotic behaviour in a small range of control parameters α, ω, A and ω_0.

The more interesting case of large α, corresponding to strong nonlinearity in equation (6.6), has proved more difficult to analyse. Cartwright and Littlewood (1945) and Levinson (1949) studied this problem and showed that steady-state subharmonic responses with two different multiples of the forcing frequency ω_0 can occur simultaneously for the same values of the control parameters. In the Van der Pol equations, the existence of these competing attractors led Levinson to conclude that transient response—before settling to one of the two periodic attractors—may be very complex. In more recent terminology, this would be called *transient chaos*. A definition of this term will be postponed until Chapter 12 when we discuss the Smale horseshoe. In fact, Smale was led to his horseshoe construction as a means of understanding Levinson's analysis of the forced Van der Pol equation. For additional information, see pages 74–82 of Guckenheimer and Holmes (1983). Again we emphasize that this line of research has not found chaotic *attractors*; with probability 1, trajectories finally settle to one of two competing periodic attractors.

In spite of this venerable history of efforts to understand the behaviour of periodically forced Van der Pol equations, it was only recently that *steady-state* chaotic behaviour was observed over a substantial range of control parameter values, as reported by Shaw (1981). Shaw discovered in analogue simulation that the equations

$$\begin{aligned} \dot{x} &= 0.7y + 10x(0.1 - y^2) \\ \dot{y} &= -x + 0.25 \sin(1.57t) \end{aligned} \tag{6.7}$$

have a chaotic attractor with a particularly interesting topological structure. Note that if the periodic forcing term is removed from the second line of (6.7), the equations can be combined, taking $v = y$, $\dot{v} = -x$, to give the Van der Pol equation (6.6) with $A = 0$. However, the forcing term on the right of equation (6.6) acts on the acceleration, and would appear in the *first* line of (6.7); Shaw has moved the driving term to the second line, forcing the velocity instead. This would be an unusual type of forcing in a mechanical system, but might well be realized in an electrical or chemical system.

The waveform of a typical steady-state response obtained by numerically integrating equations (6.7) is shown in Figure 6.2 over 50 cycles of the sinusoidal forcing function. The forcing term always repeats exactly, but the response cycles, while showing similarities, never repeat exactly; this continues to be true for as long as we care to observe the response. Figure 6.3 plots another

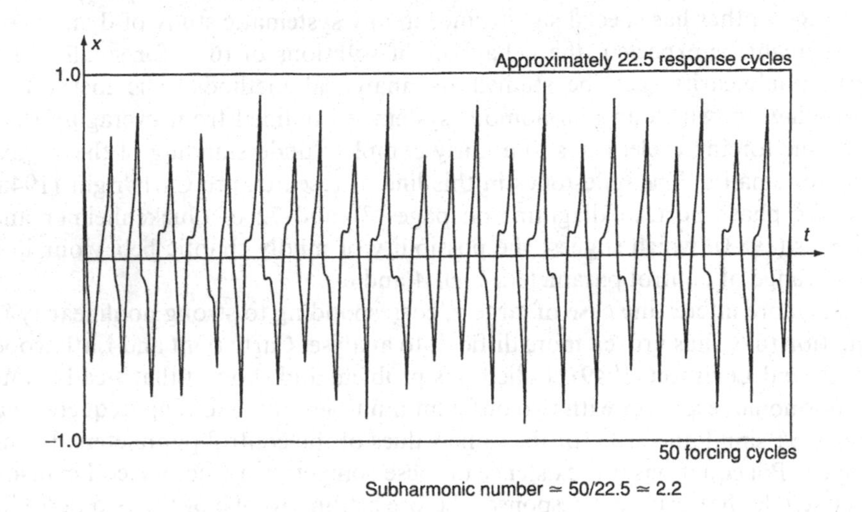

Figure 6.2 Steady-state numerical solution of Shaw's forced Van der Pol equations (6.7) versus time

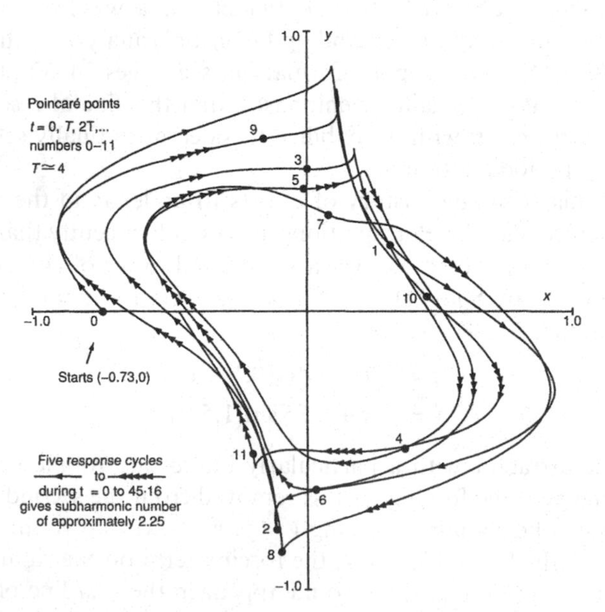

Figure 6.3 A steady-state chaotic trajectory of equations (6.7) in the (x, y) plane, with numbered Poincaré section points

steady-state trajectory of the same equations, this time in the (x, y) plane over five response cycles. Again the erratic nature of the response is evident. The orderly aspect of this steady-state chaotic response is not visible in the (x, y) plane, which is not a complete phase space. Viewed in the plane, the trajectory cuts across its own path repeatedly, because a coincidence of (x, y) coordinates can occur at different angles of the sinusoidal driving function. Poincaré's idea for making sense of such a response is based on the observation that only those points in a trajectory that correspond to the same angle of the periodic forcing term can be meaningfully compared. Examples are shown in Figure 6.3 as numbered points at multiples of the period $T = 2\pi/1.57$ of the forcing cycle. Thus we must turn to the three-dimensional phase space of forced oscillators, in which the third coordinate is the angle in the forcing cycle.

The three-dimensional attracting structure

Final steady-state trajectories of (6.7) in the three-dimensional (x, y, t) phase space are illustrated in Figure 6.4. The (x, y) planes are perpendicular to the line of view, while time is measured on an axis receding into the background. The timescale is measured by the angle $\phi = 1.57t$ of the sinusoidal driving term, progressing from 0° at the front edge of this three-dimensional attracting structure to 360° at the back edge. Since the equations at any angle ϕ_0 are repeated exactly at $\phi_0 + 2\pi$, we expect that once transients have died away the structure formed by *all* final trajectories will repeat itself identically at each full cycle of the driving term.

Before discussing the structure in Figure 6.4, let us describe exactly how this picture was constructed. A single trajectory of equations (6.7) starting from an arbitrary initial value was computed over many forcing cycles until transients died away. Thereafter, this single trajectory was computed over 1000 forcing cycles, and its position in the (x, y) plane was recorded each time the driving angle returned to an exact multiple of 2π. The result is the loop with wings visible at the front edge of the attractor in Figure 6.4, i.e. the Poincaré section of the chaotic attractor at driving angle $\phi = 0$. All 1000 points in this Poincaré section were then advanced by integration in 10° steps through one complete forcing cycle, generating a sequence of 36 successive Poincaré sections. The outlines of these sections are drawn in Figure 6.4; a number of continuous trajectories running front to back and transverse to the sections have also been superimposed.

Alternative ways of viewing the three-dimensional phase space of forced oscillators can be helpful; Figure 6.5 illustrates three equivalent ways of viewing such a phase space, using a periodic trajectory as a simple example. The middle view shows the (x, y, t) space as the Cartesian product of the (x, y) plane with the time axis. If the oscillator is driven by any periodic function of time, then the bundle of all trajectories in phase space consists of identical units $0 \leq \phi < 2\pi, 2\pi \leq \phi < 4\pi$, and so on. It suffices to consider one such unit: when a trajectory reaches the time where $\phi = 2\pi$, translate its x and y coordinates back to the plane $\phi = 0$ and continue. Topologically, the planes $\phi = 2\pi$

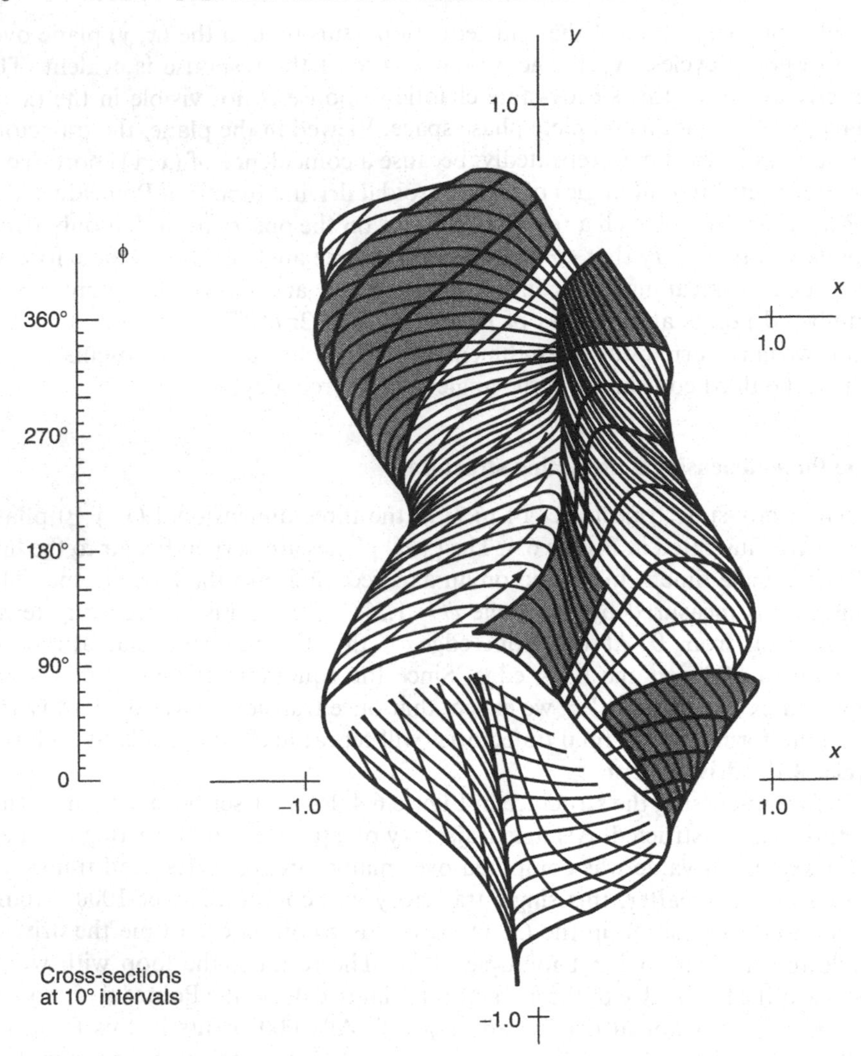

Figure 6.4 Chaotic attractor of the velocity-forced Van der Pol system (6.7) in the three-dimensional (x, y, t) phase space: cross-sections at 10° intervals

and $\phi = 0$ are *identified*. This identification is conveniently expressed in the geometry of the upper view, where the time axis bends around in a circle and the planes $\phi = 2\pi$ and $\phi = 0$ are glued together. The continuous trajectory lies on a locating torus in Figure 6.5. Time is measured by the coordinate ϕ, which is now clearly an angle repeating after each forcing cycle.

The bottom view in Figure 6.5 is a completely unwrapped view of the surface of the torus at top, or of the cylinder in the middle. The radial direction in the (x, y) plane is suppressed. To follow trajectories in the unwrapped view, two identifications must be made: points at $\phi = 2\pi$ are identified with points at $\phi = 0$, and points at $\theta = 2\pi$ are identified with those at $\theta = 0$. The advantage of

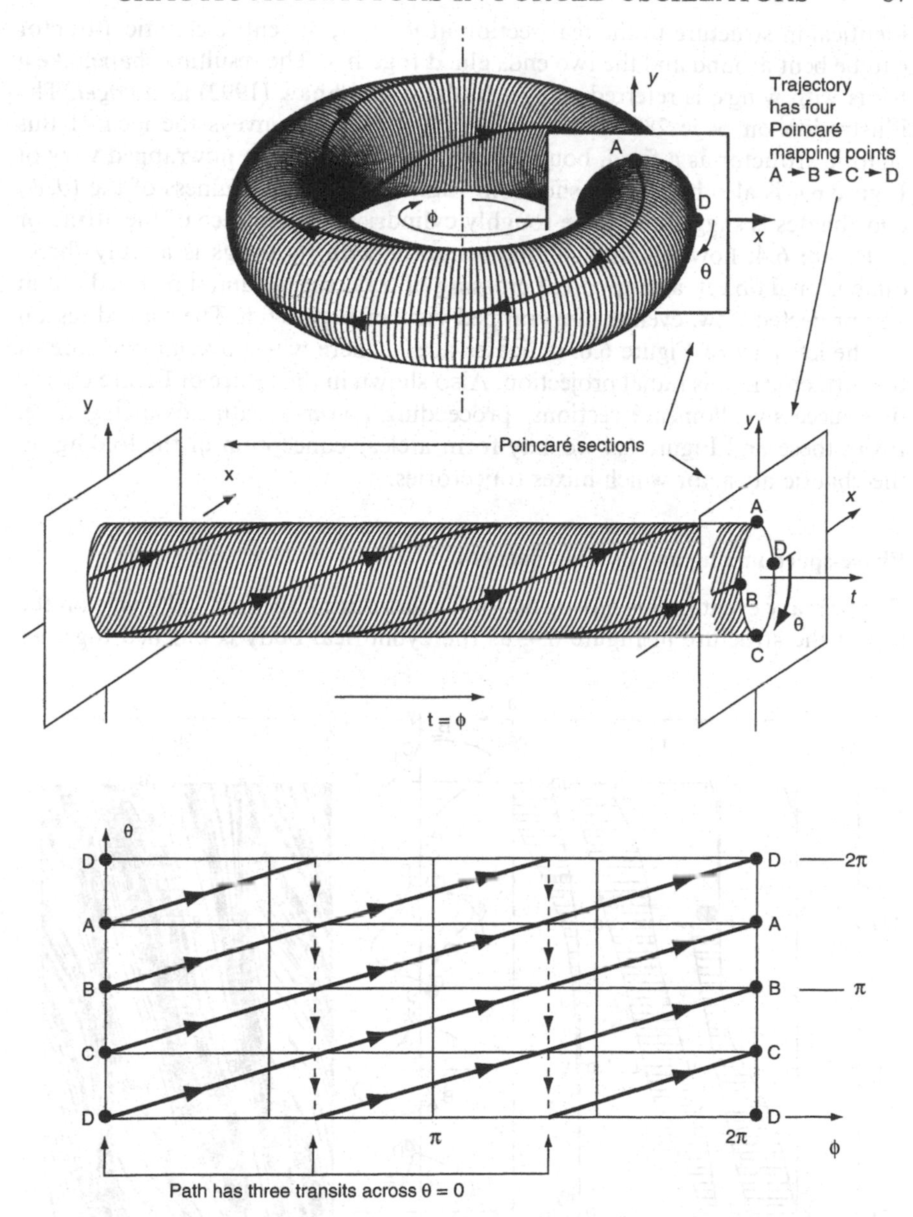

Figure 6.5 Transformations of forced oscillator phase space: toroidal or ring model, cylindrical and unwrapped

the bottom view is that, if trajectories do indeed lie on the two-dimensional surface of a torus, they are displayed in the unwrapped view with all parts of the surface visible, and no apparent self-crossing.

Returning to the chaotic trajectories of Figure 6.4, we may use the same viewing transformations. Because the front Poincaré section at $\phi = 0$ is identical in structure to the rear section at $\phi = 2\pi$, the entire chaotic attractor can be bent around and the two ends glued together. The resulting shape, like a torus with wings, is referred to by Abraham and Shaw (1992) as a *bagel*. The illustration on page 280 in their book most clearly conveys the idea of this chaotic attractor as a fixed, bounded asymptotic form. An unwrapped view of Figure 6.4 is also helpful, as shown in Figure 6.6. The usefulness of the (θ, ϕ) coordinates is suggested by the roughly cylindrical appearance of the attractor in Figure 6.4; however, the chaotic attractor with its wings is a truly three-dimensional object, and apparent crossing of trajectories cannot be ruled out in any projected view, even the unwrapped view of Figure 6.6. The shaded region on the left part of Figure 6.6. indicates areas where wings overlap the core of the attractor in this radial projection. Also shown in the centre of Figure 6.6 are five successive Poincaré sections, proceeding *upwards* with advancing time. From these and Figure 6.4 we may form a clear conception of the folding on the chaotic attractor which mixes trajectories.

Phase-space mixing and its consequences

The primary event in this mixing action occurs during the first half-cycle on the top of the structure in Figure 6.4, as the cylindrical body is pinched together

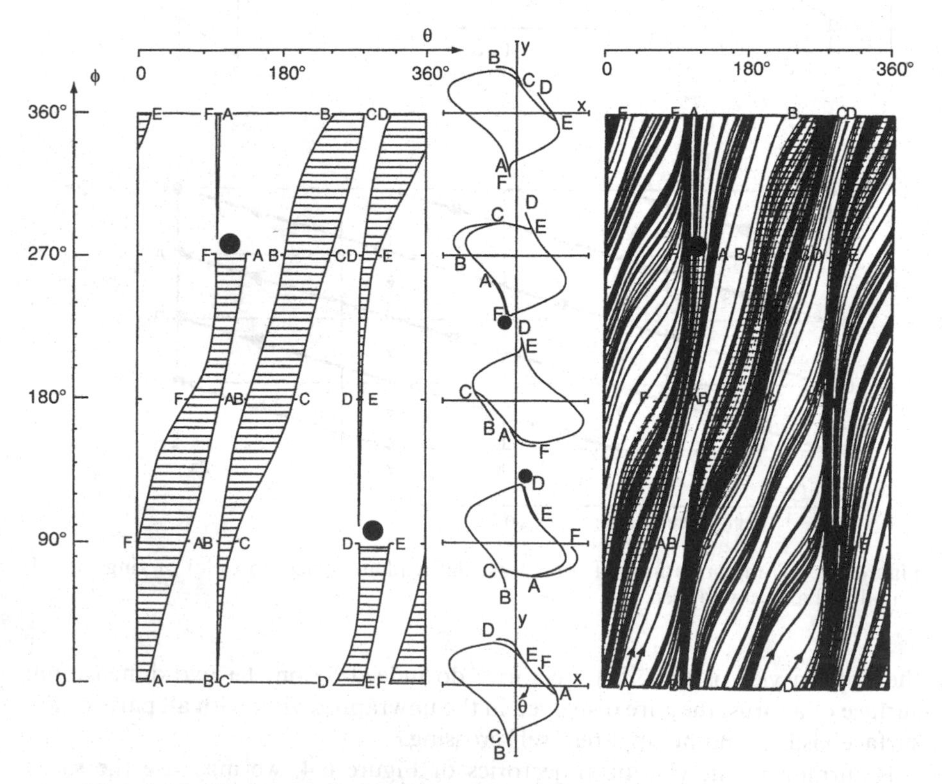

Figure 6.6 Unwrapped projections of the Birkhoff–Shaw chaotic attractor

and pulled upwards. In Figure 6.6 this is visible in the second and third Poincaré sections from the bottom; the point D is near the apex of the newly formed beak. This same pinching action also compresses the existing wing—identified by D and E in the bottom of Figure 6.6—virtually onto the core of the attractor. Although the uniqueness of solutions of equations (6.7) implies that the wing will never precisely touch the core, the distance separating them is no longer visible after $\phi = 90^\circ$. A solid dot in Figure 6.6 marks the apparently completed folding of the wing; immediately thereafter, the new beak begins to appear.

The evolution of this beak can be followed partially in Figure 6.4, and more fully in Figure 6.6 by making the proper topological identifications: top to bottom and left edge ($\theta = 0^\circ$) to right edge ($\theta = 360^\circ$). The new beak grows to form a wing, which then moves clockwise with the core trajectories. As seen in Figure 6.4, this new wing at $\phi = 360^\circ$ is identified with its match at $\phi = 0^\circ$, which then disappears around the bottom and out of view. But from Figure 6.6 we see that this wing, now associated with F and A, is folded (seemingly) onto the core at $\phi = 270^\circ$, that is one and a half forcing cycles after its formation. At this point, yet another new beak is formed, resulting in a new wing, and so on ad infinitum.

The repeated formation of beaks which stretch to form wings and fold back onto the core causes mixing of trajectories. This is emphasized in Figure 6.7, in which trajectories seeded around the core at $\phi = 0^\circ$ are numbered for identification and followed to $\phi = 360^\circ$. From the trajectory numbers at the beak, we see that 1, 2 and 3 have been spread apart by stretching of a beak into a wing. Folding when the pinched beak emerges brings trajectories 10 and 20 together, for example. Additional folding takes place when a wing is compressed back onto the core.

Imagining two trajectories that begin extremely close together, we see they must be stretched apart by repeated encounters with the spiralling wing–beak structure. Their separation is multiplied over each forcing cycle, leading to *sensitive dependence* on initial conditions. Once the trajectories become macroscopically separated, they soon fall on opposite sides of a folding and are uncorrelated thereafter. Exponential stretching can also occur in (divergent) linear systems; the essential discovery of chaos is the simplicity of *nonlinear folding* which keeps separating trajectories in a *bounded* region of phase space, and makes the chaotic attractor possible.

As a result of the infinitely repeated stretching and folding of the bundle of trajectories, the chaotic attractor must have a *fractal* structure. Even though wings appear to merge with the core after folding, this cannot be the case, since complete merging would mean that pairs of trajectories meet to form one. By the uniqueness of trajectories forwards and backwards in time, this is impossible. Hence one folding produces two layers, which are folded again to make four layers, and so on. In principle, the attractor contains an infinite succession of layers within layers, which would be resolved by finer and finer magnification. An abstract set with such a structure was first described by Georg Cantor at the end of the nineteenth century; the *Cantor set* will be discussed

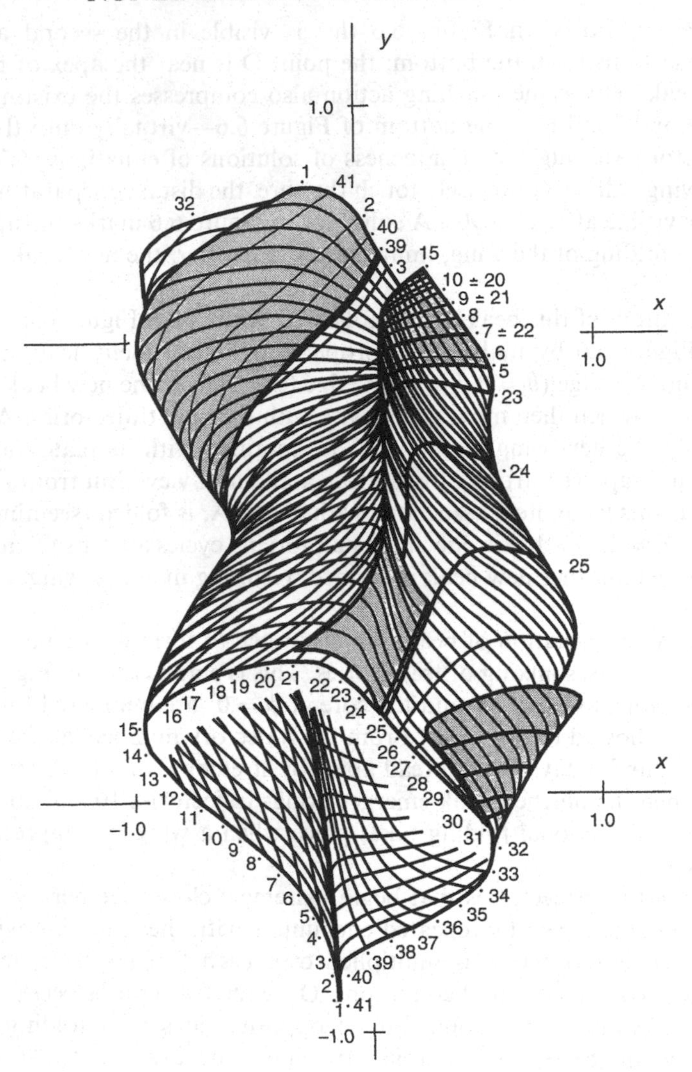

Figure 6.7 Trajectories of the forced Van der Pol equations (6.7) are mixed by stretching and folding on the attractor

in Chapter 11. G. D. Birkhoff realized in the 1920s that Cantor's structure might arise naturally in a dynamical system, and described in detail a repeated folding much like that of the attractor in Figure 6.4; see Birkhoff (1932). For this reason we refer to it as the Birkhoff–Shaw attractor.

Infinitely layered structures such as the Cantor set have been studied in a systematic way by Mandelbrot (1983), who finds them in many problems of practical interest. Mandelbrot has coined the term *fractal* for such objects, both because of their infinitely *broken* structure, and because a standard mathematical prescription for determining spatial dimension of arbitrary points sets yields a *non-integer* value for fractals. This same prescription for computing dimension, and related rules formulated as numerical algorithms, has been applied to chaotic attractors, and their dimension has so far been found to be greater than 2; see for example Farmer *et al.* (1983) and Grassberger and Procaccia (1983).

As a practical matter, the fractal structure of a chaotic attractor may be of greater or lesser significance, depending on the attractor. In Figure 1.6 the layers are quite apparent, while in the Birkhoff–Shaw attractor they are quickly compressed to invisibility. One factor that can influence the visibility of fractal structure is the amount of dissipation, as we shall see below in connection with forced Duffing oscillators.

The first encounters with chaotic attractors in the 1960s happened at a time when sets like the Cantor set had rarely been studied in the applied sciences, and were felt by some to be a perverse invention of mathematicians. Perhaps because of this, chaotic attractors were first referred to as 'strange attractors'. As more examples are studied and become familiar, the fractal structure is appreciated as a natural consequence of a basic stretching and folding action infinitely repeated. We therefore feel that the adjective strange is unnecessary, and prefer the term 'chaotic attractor'. A *chaotic attractor* may be defined as any bounded attractor that stretches and folds the bundle of final steady-state trajectories, producing sensitive dependence on initial conditions and longterm unpredictability. From what is known to date, it appears that a typical attractor in low-dimensional phase space is likely to be a chaotic attractor.

Poincaré mapping and chaotic attractors

A *Poincaré mapping* is the action that takes all points in a Poincaré section to their image points by following trajectories until they first return to the Poincaré section. The Poincaré mapping of the Birkhoff–Shaw bagel attractor is clearly indicated by the numbered transits in Figure 6.7. Technically, this mapping can be specified by connecting points at $\phi = 0$ to their images at $\phi = 360^\circ$ with straight lines. Such a picture might at first appear confusing, but it would economically summarize the mixing action over one forcing cycle.

Any surface transverse to the bundle of trajectories may be used to define a *Poincaré section*, the intersection of the bundle with the surface. In periodically forced systems, a surface of section can always be defined at any fixed angle of the forcing function. For forced oscillators, we use this definition exclusively in this book. If any other prescription for a surface of section is adopted, it is essential to verify that the proposed surface is never inflected or tangent to any trajectory. This will be easier if a complete geometric picture of the attractor can be constructed before defining the surface of section.

A picture of one Poincaré section of an attractor, while often suggestive, does not specify the mixing action of the attractor unambiguously. For this one must know the Poincaré mapping, i.e. the correspondence of section points with their first return images. Often the Poincaré mapping can be deduced by examining a sequence of Poincaré sections between a reference section and the first return to the reference section, as in the centre of Figure 6.6. This approach will be used in the remainder of this chapter. The logical extension of this sequence-of-section technique is an animated view of very closely spaced sections, equivalent to watching a thin transverse slice moving continuously down the time axis in Figure 6.4. Examples of such animations can be seen in the beautiful computer-generated movie of Crutchfield (1984).

Other chaotic regimes of forced Van der Pol systems

Chaotic attractors with the same structure as Figure 6.7 occur for a wide range of forcing amplitudes and frequencies. The angular forcing frequency 1.57 in equations (6.7) is approximately twice the frequency of the corresponding unforced periodic oscillations, and the chaotic attractor is near a region of subharmonic entrainment. An amplitude–frequency control-space diagram for the velocity-forced Van der Pol equation (similar to the picture in Figure 1.10 constructed by Ueda for the amplitude–damping control space of the forced Duffing oscillator) will be seen in Chapter 13. There we shall illustrate the generic bifurcations of chaotic attractors using examples from equations (6.7) with different control values, in the style of the attractor gallery in Ueda (1980a). Conceptually, this implies a *control-phase space* (Abraham 1985) formed by attaching a copy of the phase space to each point in the control space; the appropriate phase portrait is entered in each copy of phase space. Ueda's control-phase space is thus five-dimensional, and cannot be directly visualized, but in Part II of this book we will discuss a discrete dynamical system – the logistic map – whose control-phase space is two-dimensional and so can be seen in its entirety. Note that for the forced Van der Pol equation (6.6) or (6.7), a complete control space would require three dimensions – amplitude, frequency, and nonlinearity (corresponding to the parameter α).

For the Van der Pol equation with acceleration forcing (6.6) there are two detailed pictures of parts of control-phase space. The diagram for weak nonlinearity in Guckenheimer and Holmes (1983) has already been mentioned. For strong nonlinearity, Hayashi (1964) gives a diagram covering a wide range of frequency and amplitude at a fixed value of $\alpha = 4$. This control space is dominated by regions of periodic limit cycle responses at various subharmonic numbers. For small forcing amplitudes, Hayashi identifies regions in the amplitude – frequency control plane in which the final response is *drift*, i.e. the Van der Pol oscillator does not lock onto the forcing frequency. Here the net motion in steady state is compounded of two frequencies, the forcing frequency and the natural frequency of the unforced oscillator. Topologically, this *quasi-periodic* motion wanders over the entire surface of a torus in the $(x, \dot{x}, \phi)$ phase space. A single trajectory eventually fills the surface of the torus when the ratio of the two frequencies is not a ratio of integers. In quasi-periodic motion, nearby trajectories do not spread apart exponentially, and the dynamics remain predictable over long times.

It should be remembered that Hayashi's diagram was published at a time when the understanding of chaotic dynamics was sketchy indeed; it may be that there are subregions of Hayashi's drift regions in control space where in fact the attractor folds like the Birkhoff – Shaw attractor, perhaps with wings that are small compared to the toroidal (or cylindrical) core.

It has recently been noted by Abraham and Simó (1986) that almost all studies of forced Van der Pol oscillators to date have focused on systems with symmetries. And yet these symmetries are easily broken, for example by simply adding a DC bias to the AC forcing function. Abraham and Simó began a valuable systematic study of the effects of asymmetries on the unforced Van der Pol equation; and in a preliminary exploration of forced asymmetric systems they found chaotic attractors in the acceleration-forced Van der Pol system, which had only yielded transient chaos under symmetric forcing. The potential practical importance of asymmetric systems is illustrated by the example of FitzHugh's equations (6.4).

In conclusion, recent studies have shown that, despite the tendency of forced Van der Pol systems to lock on to the forcing frequency, there are many ways of adjusting controls to produce steady-state chaotic response. Systematic study of the entire range of controls has barely begun, but in view of the many applications in which relaxation oscillations can arise, continued study appears well justified.

6.3 SYSTEMS WITH NONLINEAR RESTORING FORCE

As mentioned above, two classes of nonlinearity that generalize the damped, linear oscillator are systems with nonlinear friction and oscillators with nonlinear restoring force. In turning now to the latter category, it is convenient to consider various potential functions $V(x)$ leading to

$$\ddot{x} + k\dot{x} + \operatorname{grad} V = A \sin \omega t \tag{6.8}$$

For the linear oscillator this potential function is of course the harmonic potential $V(x) = \frac{1}{2}ax^2$ proportional to the displacement squared. The forced oscillator with harmonic potential gives only periodic motion in the steady state, as we have seen, but various *anharmonic* potentials – corresponding to nonlinear restoring force – can lead to chaotic steady-state response. Here we consider *qualitatively* different shapes of anharmonic potential. We shall, for example, be concerned with whether the restoring force grad V increases more or less rapidly than a linear function of displacement (stiffening or softening spring, respectively), so that the potential rises more or less steeply than a quadratic. But we shall not be concerned with how much more or less rapidly; that is, we assume that the quantitative difference between cubic and quintic restoring force leads to mainly quantitative, not qualitative, differences in the observed dynamics and steady-state response.

Softening spring

The qualitative effect of a softening spring can be studied in the potential $V(x) = \frac{1}{2}x^2 - x^4$, illustrated in Figure 6.8. Here the potential shape is like the

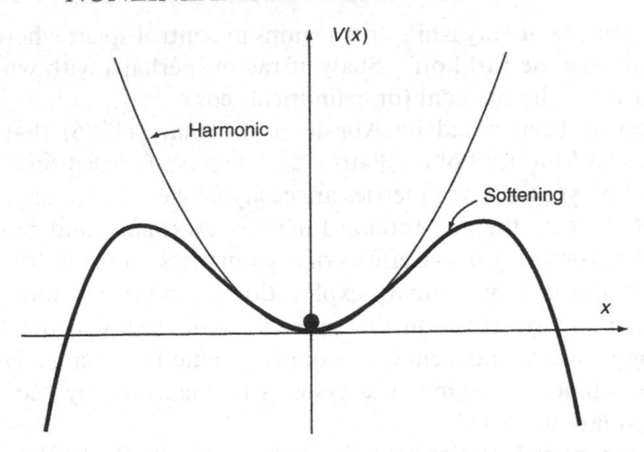

Figure 6.8 An anharmonic potential with softening spring effect

harmonic potential for small displacement, but at larger displacement the restoring force is weakened. At the two smooth maxima of the potential, the restoring force becomes zero, and for larger displacements a trajectory can diverge to infinity, but we will only consider trajectories that remain within the potential well. In particular, oscillations that visit the portion of the well just inside the maxima experience the softening spring effect most strongly.

A chaotic steady-state response under these conditions was discovered by Huberman and Crutchfield (1979); Figure 6.9 shows an example of a steady-state trajectory in the phase plane, plus eight Poincaré sections of the chaotic attractor. The equation used to calculate the pictures is

$$\ddot{x} + 0.4\dot{x} + x - 4x^3 = 0.1185 \sin 0.555t \tag{6.9}$$

and in the phase plane, $y = \dot{x}$. This chaotic response occurs in a rather small region of control space, perhaps because the softening spring effect in this potential is noticeable only in a narrow range of response amplitudes.

This chaotic response coexists with a smaller-amplitude periodic limit cycle at the same control parameter values. Thus the situation is similar to Figure 5.5 with the larger-amplitude periodic motion now replaced by a chaotic one in which the peak amplitude varies from one cycle to the next. The fluctuations in peak amplitude are not large, but their size is unpredictable over long times.

The Poincaré sections in the lower part of Figure 6.9 show that, in spite of this randomness, there is again a simple underlying structure. In fact, this is topologically the simplest chaotic attractor. The eight sections shown are taken at eight equally spaced angles through one complete cycle of the sinusoidal driving function, and sections progress clockwise in the plane. Most of the folding action occurs as the attractor passes the x axis, and is more easily seen in the three closely spaced Poincaré sections in Figure 6.10, viewed close up. The earliest section is on top; the middle section is the same as the one near the

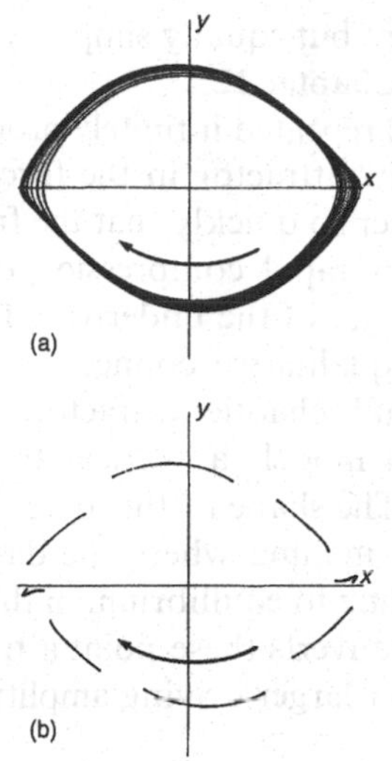

Figure 6.9 Forced oscillations in an anharmonic potential with softening spring effect (6.9): (a) a single trajectory and (b) eight Poincaré sections through one cycle

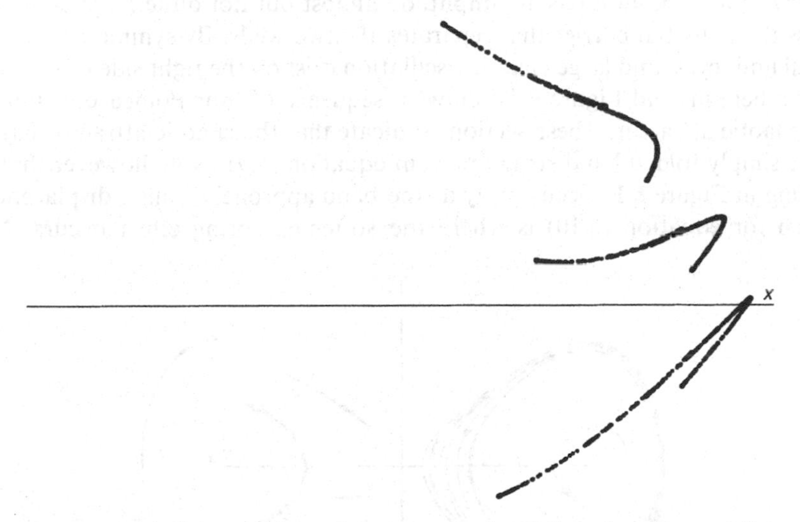

Figure 6.10 Close-up view of three successive Poincaré sections of the attractor of the forced oscillator equation (6.9), showing (top to bottom) the folding action

positive x axis in Figure 6.9; by the time of the bottom section, folding is well advanced. Imagining the three-dimensional bundle of steady-state trajectories in the $(x, \dot{x}, t)$ phase space, we see it forms an apparent ribbon or band, which is smoothly and simply folded onto itself. This *folded band* is the simplest structure for a chaotic attractor; its existence was established by Rössler in simulation of a very different but equally simple system of ordinary differential equations to be studied in Chapter 12.

The simple folding action repeated infinitely produces a fractal structure; but as with the Birkhoff – Shaw attractor in the forced Van der Pol system, the folding brings layers together so quickly that the fractal structure is not visible without magnification. This rapid compression of layers allows us to more readily recognize the simplicity of the underlying folding action.

As confirmation of the qualitative connection between a softening spring effect and the folded band chaotic attractor, we consider the potential $V(x) = -\frac{1}{2}x^2 + \frac{1}{4}x^4$, which models a vertical Euler support column loaded beyond its buckling point. The shape of this potential was illustrated in Figure 3.19, showing two smooth minima where the damped, unforced system has point attractors corresponding to equilibrium in the buckled state. Once again, a small external vibration converts these point attractors to limit cycle attractors in the forced system. For larger driving amplitude and certain frequencies, such as in

$$\ddot{x} + 0.25\dot{x} - x + x^3 = 0.191 \sin t \tag{6.10}$$

we find a small-amplitude limit cycle competing with a larger-amplitude chaotic motion, as illustrated in Figure 6.11 in the left half of the phase plane. The steady chaotic motion has an amplitude almost but not quite large enough to cross the potential barrier that separates the two wells. By symmetry the same small limit cycle and large chaotic oscillation exist on the right side of the phase plane; here instead Figure 6.11 shows a sequence of four Poincaré sections of the chaotic attractor. These sections indicate that the chaotic attractor has the same simply folded band structure as in equation (6.9). Note however that the folding in Figure 6.11 occurs only as the band approaches small displacement, which for equation (6.10) is where the softening spring effect occurs. Near

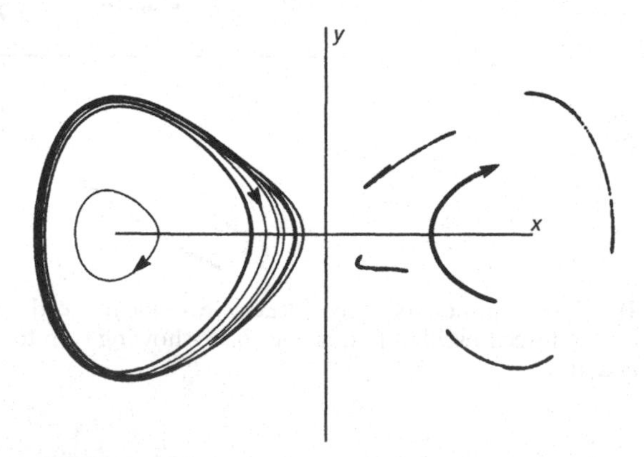

Figure 6.11 Small-amplitude periodic attractor coexisting with a larger-amplitude chaotic steady state in a two-well potential oscillator (6.10) representing a vibrating buckled beam

maximum displacement, the buckled beam exhibits a hardening spring effect, and there is no folding as the attractor in Figure 6.11 passes maximum displacement. Thus we may associate the simple folding of Figures 6.9, 6.10 and 6.11 with the softening spring effect, and expect that, at least for certain forcing amplitudes and frequencies, other softening spring oscillators may be found to have chaotic attractors like the folded band. This turns out to be the case in practical problems, such as the capsize of a vessel in regular seas (Thompson 1997); we shall return to this in Chapter 16.

Stiffening spring

The opposite of a softening spring is a stiffening spring, in which the restoring force rises more rapidly than a linear function of the displacement, and the potential rises more rapidly than the harmonic potential $V(x) = \frac{1}{2}ax^2$. We have already seen an example in the forced Duffing oscillator with cubic restoring force (and quartic potential), whose response has been extensively studied by Ueda.

Figure 1.10 shows clearly when chaos can be expected in the forced Duffing oscillator, and is one of the few such thorough control-space diagrams published to date. The engineer used to dealing with practical problems might be concerned that this case of zero linear stiffness, corresponding to loading a column precisely to the buckling point, is somehow atypical of stiffening springs generally. Some reassurance is offered in Figure 6.12, which shows a Poincaré section of the chaotic attractor computed from

$$\ddot{x} + 0.05\dot{x} + ax + x^3 = 7.5 \sin t \tag{6.11}$$

where $a = -0.2$ on the left, $a = 0$ in the middle corresponding to Ueda's region (k), and $a = +0.2$ on the right. The effect on the chaotic attractor of introducing a small linear stiffness is difficult to detect. In fact, even for larger amounts

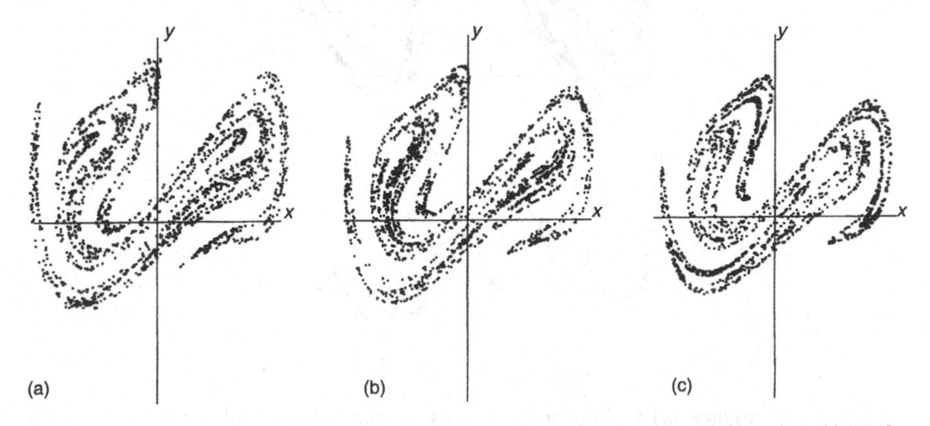

Figure 6.12 Poincaré sections of the forced Duffing-type oscillator equation (6.11) for different values of the linear stiffness: (a) $a = -0.2$, (b) $a = 0$, (c) $a = +0.2$

of linear stiffness the general appearance of Figure 1.10 would change mainly by distortions and displacements rather than by disappearance of the various response regions. For this reason we associate the chaotic responses mapped out by Ueda with a stiffening spring effect.

Within each of the control-space regions of chaotic response in Figure 1.10 there may be considerable variation in the structure of the attracting set. This is illustrated in Figure 6.13, which shows Poincaré sections of chaotic attractors in region (1) of Ueda's diagram. The form with the simplest appearance occurs with the largest value of damping. The overall folding action in three dimensions is similar in all three cases, but with higher damping the layers formed by folding are more rapidly compressed together. Thus the succession of layers remains more visible with less damping. This illustrates a more general principle that the folding and mixing action of a chaotic oscillator can be seen in simplest form with maximum dissipation.

The folding action in region (1) with large damping turns out to be similar to the folded band that we associated with the softening spring. We have observed this similarity in computer-generated film animation of a continuously evolving Poincaré section, similar to Crutchfield (1984). Unfortunately, the results are difficult to convey on the printed page; the simple folding is compounded with continuous distortions, which contribute nothing to the mixing action but make a confusing picture without continuous animation. We simply note that, in the flattened S-shape appearance of the highly damped case in Figure 6.13, two successive folds are visible. The most recent fold produced the more open bend of the S, while the fold during the previous forcing cycle produced the bend that now appears more tightly compressed.

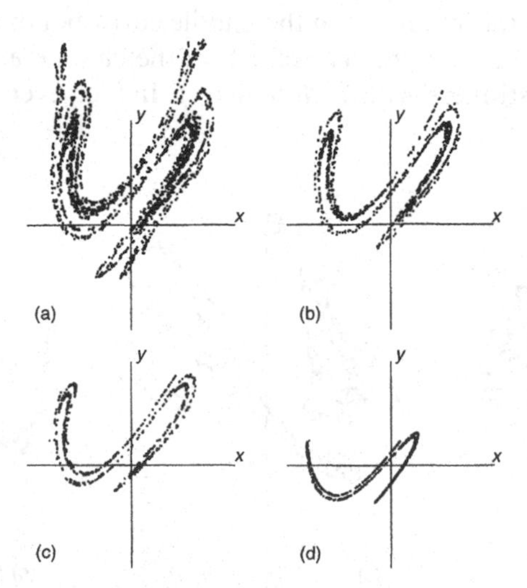

Figure 6.13 Poincaré sections of the forced Duffing oscillator in the region (1) of Ueda's response diagram, with increasing damping from (a) to (d)

Multiple potential wells

Another important and qualitatively different anharmonic oscillation occurs when the potential has more than one well. One example is the Euler column loaded past the buckling point, with $V(x) = -\frac{1}{2}ax^2 + bx^4$ and two potential wells. In addition to the chaotic response associated earlier with the softening spring effect within one potential well, another type of chaotic response can be found in which the oscillator jumps back and forth between the two wells in an erratic manner. This behaviour has been studied by Moon and Holmes (1979), Greenspan and Holmes (1982), and Holmes and Whitley (1983), and is described in Guckenheimer and Holmes (1983). Related studies have been published by Arecchi *et al.* (1984) and Arecchi and Califano (1984).

An important example of multiple potential wells is the periodic potential $V(x) = -\cos x$, an infinite lattice of wells having minima at $x = 2\pi n$, $n = \ldots, -2, -1, 0, 1, 2, \ldots$. This potential models the nonlinear oscillations of a pendulum introduced in Chapter 3, and is of interest in the study of Josephson junctions and of charge-density waves in plasmas; see Huberman *et al.* (1980), MacDonald and Plischke (1983), Miracky *et al.* (1983), Yeh and Kao (1983), Bak *et al.* (1984), Yeh *et al.* (1984).

Figure 6.14 shows a chaotic trajectory of the two-well potential oscillator

$$\ddot{x} + 0.4\dot{x} - x + x^3 = 0.4 \sin t \tag{6.12}$$

in the $(x, \dot{x})$ phase plane. Comparing this single trajectory with Figure 1.4, it might not be evident that this chaotic motion corresponds to a chaotic attractor that is topologically different from the attractors found by Ueda at higher forcing amplitude and zero linear stiffness.

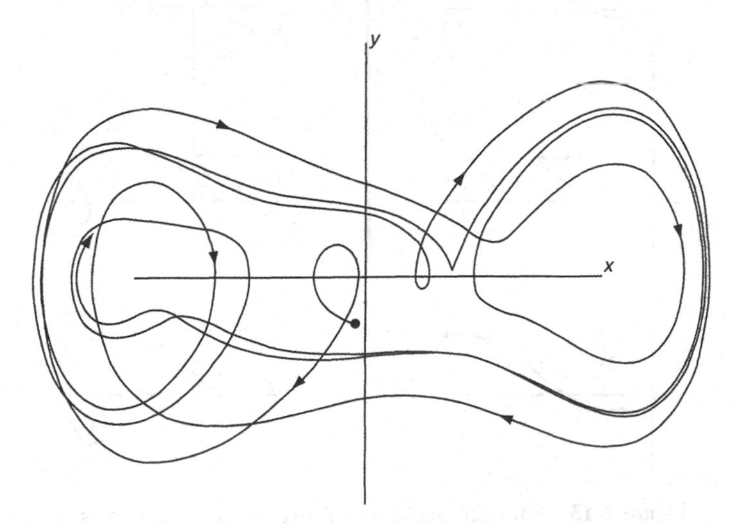

Figure 6.14 A steady-state chaotic trajectory in the phase plane of the vibrating buckled beam equation (6.12) with negative linear stiffness and moderately large forcing amplitude

Successive Poincaré sections of this attractor, and of a chaotic attractor of the equation

$$\ddot{x} + 0.5\dot{x} + \sin x = 1.1 \sin 0.5t \tag{6.13}$$

are shown progressing top to bottom in Figure 6.15; in each case the sections are for equally spaced angles through one half of a forcing cycle. Because of a symmetry in both equations (6.12) and (6.13) under the transformation $x \to -x, \dot{x} \to -\dot{x}, t \to t + \pi/\omega$, the section at the bottom of each column that is exactly one half-cycle advanced from top section is also identical to the top section rotated 180° about the origin in the $(x, \dot{x} = y)$ phase plane. The second

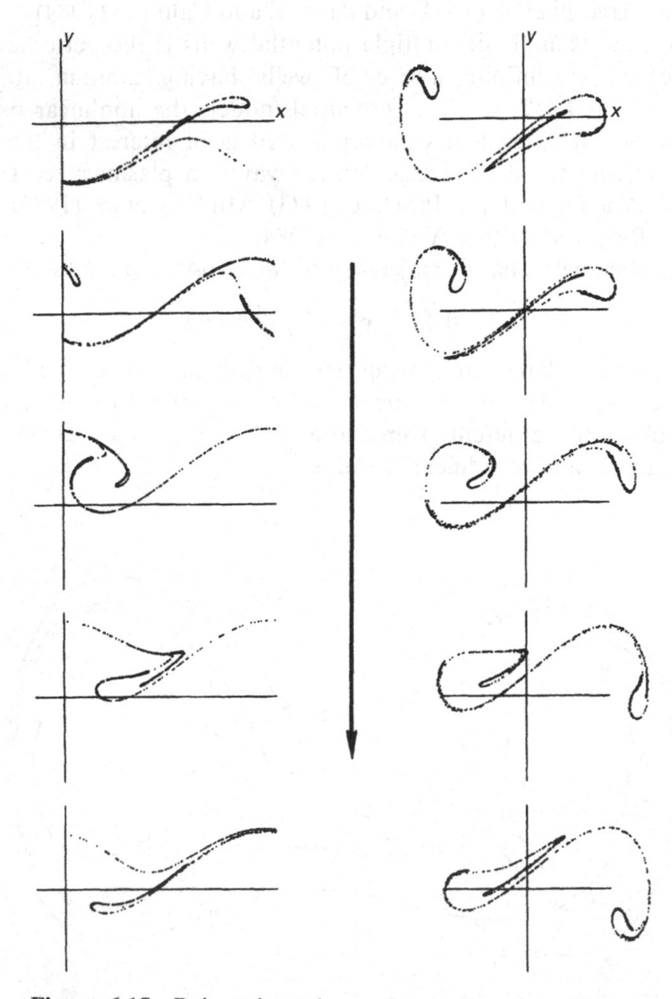

Figure 6.15 Poincaré sections of two forced oscillators, progressing top to bottom with time through one half-cycle: the left-hand column shows the periodic potential (6.13); the right-hand column shows the two-well potential (buckled beam) of (6.12)

half of the forcing cycle is therefore described by rotating each picture in a column by 180°. Note that with the periodic potential of equation (6.13) the phase space is periodic in x as well as in t. In the left column of Figure 6.15, the infinitely many identical intervals $2\pi n \leq x < 2\pi(n+1)$ are all plotted as one interval $0 \leq x \leq 2\pi$; trajectories above the x axis (i.e. $\dot{x} > 0$) may move off the picture to the right and re-enter at the left edge, while trajectories below the x axis may leave on the left and reappear on the right.

Chaotic attractors of these types were first reported by Huberman *et al.* (1980) for the infinite well, and by Moon and Holmes (1979) in the two-well case. We have chosen parameters for Figure 6.15 involving larger dissipation in both cases, thus gaining more compression of the fractal layers and a simpler picture. This helps reveal the similarity of the two-well and infinite-well attractors.

Figure 6.16 shows a schematic representation of the stretching and folding action in the twin-well attractor in Figure 6.15. The top rectangle ABCD in Figure 6.16 encloses the right half of the twin-well attractor in Figure 6.15 at the start of the forcing cycle. During the first half of the forcing cycle, the right half of the attractor in Figure 6.15 is stretched and bent. In Figure 6.16 the second rectangle abcd represents the result of stretching; the partial bending is not shown in this intermediate stage of the schematic representation.

During the second half of the forcing cycle, the folding of the right half of the attractor is completed. By symmetry, the sequence of Poincaré sections through the second half of the forcing cycle is obtained by rotating each individual Poincaré section in Figure 6.15 through 180°; notice that the bottom

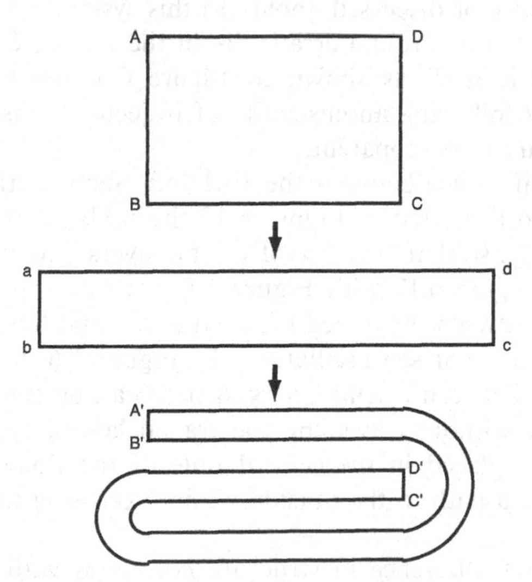

Figure 6.16 Template for the stretching and folding action in the twin-well attractor. Compare it with the sequence on the right in Figure 6.15 by following the right half of the attractor, and complete the folding on the left half, imagining each snapshot turned through 180°

section in each sequence of Figure 6.15 is identical with the top section rotated 180°. Thus the left half of either attractor in Figure 6.15 shows how the folding is completed, with a fold and tuck to generate three layers. The net effect on the original enclosing rectangle is shown schematically at the bottom of Figure 6.16. This three-layer folding was analyzed for the two-well attractor by Holmes and Whitley (1983).

This simplified representation of the folding action is similar to Smale's approach to the horseshoe (Chapter 12). Such folding models, called *templates*, can be used as a basis for classifying chaotic attractors in three-dimensional phase space topologically (e.g. Solari *et al.* 1996). They also provide a useful tool for understanding jump and escape phenomena near resonance, as we shall see in Chapter 16.

Systems of mixed type

Chaotic attractors have also been studied in forced oscillators of mixed type, containing both nonlinear damping and nonlinear restoring force. As an example, consider the equation

$$\ddot{x} - 0.2(1 - x^2)\dot{x} + x^3 = 17 \sin 4t \tag{6.14}$$

studied by Ueda *et al.* (1973); see also Ueda (1992). This equation describes a self-oscillatory electric circuit with a negative resistance element.

Figure 6.17 shows a chaotic attractor of this system in Poincaré section. In three-dimensional phase space, this attractor is a folded torus, much like the Birkhoff–Shaw attractor discussed above. In this system, an individual trajectory will appear to jump around erratically in the Poincaré section from one forcing cycle to the next, as shown in Figure 6.3 for the Birkhoff–Shaw attractor. Only by following an ensemble of trajectories, as in Figure 6.7, is the folding structure made apparent.

The principle difference between the Birkhoff–Shaw attractor and the attractor in Figure 6.17 is that in Figure 6.17 the folds on the torus are compressed more slowly, so that evidence of fractal layers remains clearly visible.

By comparing Figure 6.17 with Figure 1.6, we can appreciate the fundamental difference between the folded torus structure and the chaotic attractors in systems which are not self-oscillatory. In Figure 1.6, if a pencil point is placed in any gap between fractal layers, a path can be traced to a point far from the attractor without traversing the fractal layers. However, in Figure 6.17, if a pencil is placed in the central hole of the Poincaré section, it is impossible to trace a path to the outside without crossing through the fractal layers.

This fundamental difference in structure correlates with the number and topological type of unstable periodic orbits of saddle type located in the attractor, as discussed in Stewart (1991). Figure 6.18 shows (as open circles) the unstable subharmonics of period 2 which are located within the chaotic attractor of Figure 6.17, together with their associated invariant curves. As in Figure 5.10, successive iterates of the Poincaré mapping step along the invariant

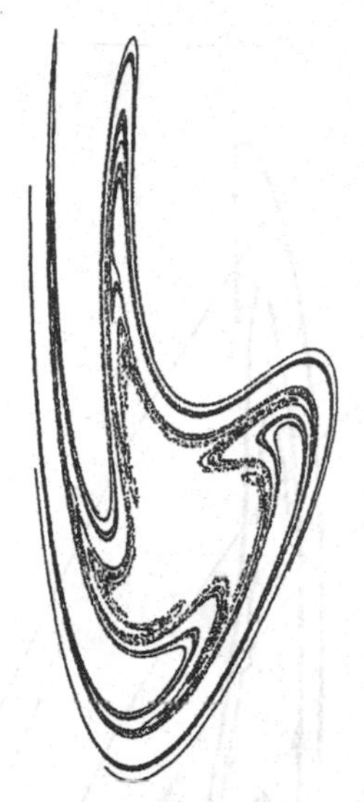

Figure 6.17 Chaotic attractor in a forced self-oscillator with nonlinear restoring force, shown in Poincaré section; the structure in three-dimensional phase space is a folded torus, with more visible fractal structure than the Birkhoff–Shaw attractor. Courtesy of Y. Ueda

curves in the direction of the arrows, either toward or away from the periodic points. A single trajectory may move by large steps along the invariant curve; tracing these invariant curves requires construction of a large number of trajectories, all starting (or ending) very close together near the periodic points.

In Figure 5.10, the invariant curves asymptotic to the unstable saddle orbit were of particular interest, as they constitute the boundary between two basins of attraction. In Figure 6.18 the invariant curves leading away from an unstable saddle orbit are of special interest, since they are contained within the chaotic attractor and indicate its structure. Here the invariant curves asymptotic to an unstable saddle orbit do not separate basins of attraction, but rather give additional information about the folding structure of phase space on and around the chaotic attractor.

Tangled basins

The sensitive dependence on initial conditions that obstructs prediction in steady-state chaos also has counterparts in transient dynamics. In severe cases a small change in initial conditions can have an unpredictable effect on which of two or more attractors is the long-term behaviour of the system. In mechanical and electrical systems it is common for one attractor to represent a desired behaviour, with other attractors representing undesirable conditions. The attractors may all be periodic, or some may be chaotic. What matters more is the general location in phase space, some locations being safer than

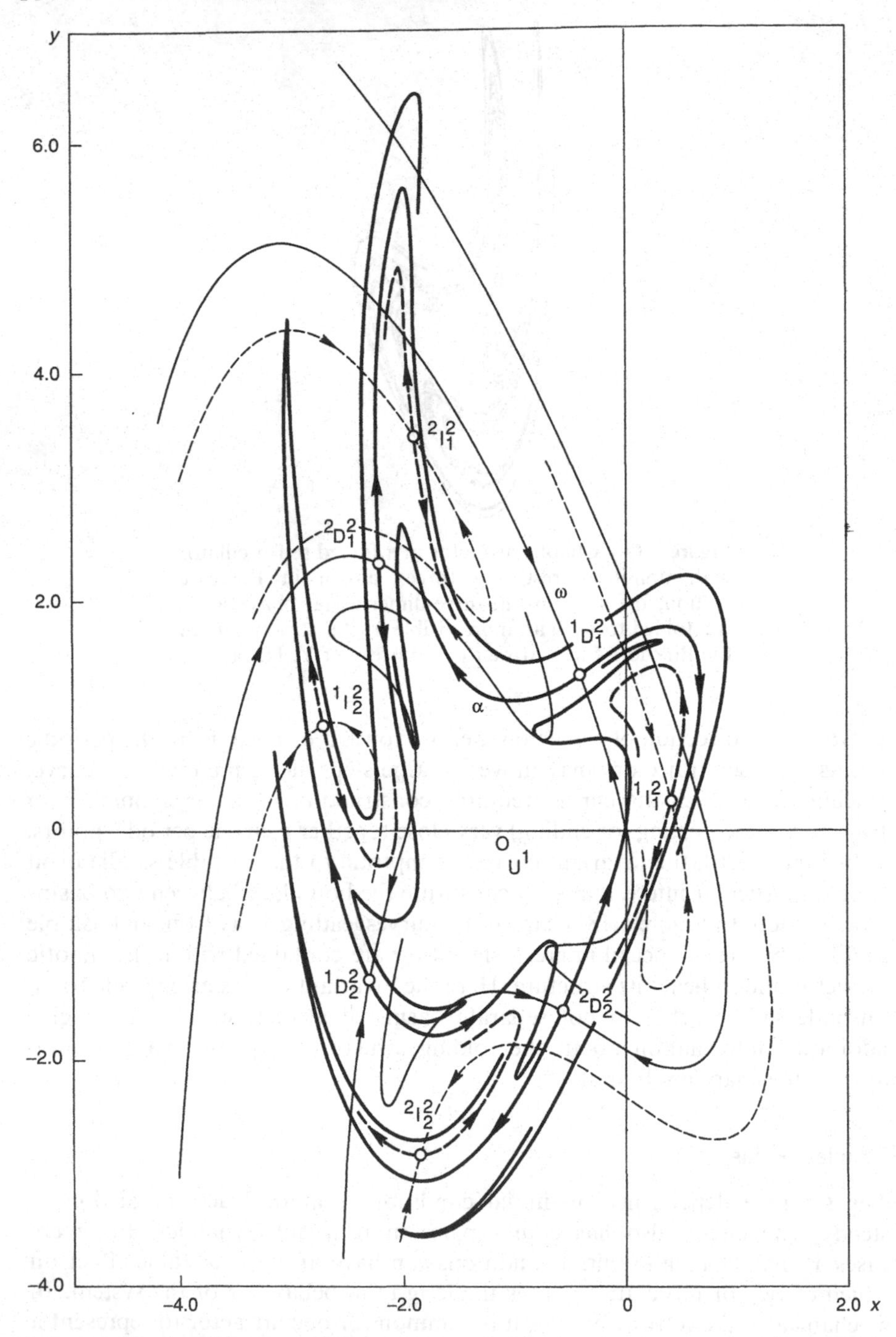

Figure 6.18 Invariant manifold structure of the chaotic attractor in Figure 6.17. Courtesy of Y. Ueda

others. This type of sensitive dependence in transients is called *final state sensitivity* (McDonald *et al.* 1985). As we shall see in Chapter 16, it may be of great concern to engineers.

In the geometry of phase space, this kind of sensitive dependence is manifested in repeated folding structure of the basins of attraction. The repeated folding of basins generates fractal structure in the basin boundaries. Figure 6.19 shows an example of a *fractal basin boundary* structure in a Poincaré section of the forced twin-well Duffing oscillator (6.10). As in Figure 6.11 there are two attractors in each potential well: a periodic motion near the bottom of the well, and a resonant response that is chaotic. The smaller-amplitude periodic oscillation is shown in Figure 6.19 as a white dot near the upper right corner of the phase portrait, and its basin is black. The chaotic attractor in Figure 6.19 (confined to the right potential well) is the same as the snapshot in Figure 6.11 nearest the origin of the coordinate axes. In Figure 6.19 its basin is white. The gray regions lead to attractors in the opposite potential well.

Fractal basin boundaries will help understand and classify attractor bifurcations in Chapter 13; their importance in engineering will be highlighted in Chapter 16.

Figure 6.19 Attractor-basin phase portrait of the twin-well Duffing oscillator, obtained by determining a steady state for each initial condition in a grid of 640×480 points in a Poincaré section. The fractal structure suggested as the boundary between the black and white basins indicates that, for many initial conditions, the fate of a trajectory is unpredictable

7

Stability and Bifurcations of Equilibria and Cycles

In this chapter we shall outline the classical definitions for the Liapunov stability of steady-state solutions (equilibrium points or periodic orbits) and for the structural stability of phase portraits. Then we will briefly examine the most common bifurcations that can occur in nonlinear dynamical systems. In the discussion of Liapunov stability the steady state is first supposed to be an equilibrium point, and some of the basic bifurcations are examined in detail. The system is assumed to be controlled by a single parameter, so we restrict our study to the so-called bifurcations of codimension 1. The typical losses of stability at folds and Hopf bifurcations are illustrated, together with two other structurally unstable bifurcations, which arise when additional conditions are imposed upon the flow. The simplest examples are selected on the basis of centre manifold theory. Finally the three structurally stable co-dimension 1 bifurcations for cycles are discussed, and the geometrical description of the invariant manifolds is emphasized.

7.1 LIAPUNOV STABILITY AND STRUCTURAL STABILITY

We have seen in the previous chapters that an equilibrium state of a linear damped oscillator can lose its stability in two quite distinct ways. In the first case the stiffness of the system changes from positive to negative, as adjacent positions of equilibrium appear; while in the other case the damping becomes negative, and growing oscillatory motion is initiated.

Both of these phenomena are characterized by the fact that beyond the critical state the system moves away from its fundamental equilibrium configuration as soon as a small perturbation is superimposed. It is useful therefore, to give a geometrical definition of stability of an equilibrium state, without energy considerations, but referring solely to the topological properties of the flow.

Consider an n-dimensional system of differential equations

$$\begin{aligned}\dot{x}_1 &= F_1(x_1, x_2, \ldots, x_n)\\ \dot{x}_2 &= F_2(x_1, x_2, \ldots, x_n)\\ &\vdots\\ \dot{x}_n &= F_n(x_1, x_2, \ldots, x_n)\end{aligned}$$

and suppose that the point $P_{\mathrm{E}} = (x_1^{\mathrm{E}}, x_2^{\mathrm{E}}, \ldots, x_n^{\mathrm{E}})$ is an equilibrium state characterized by

$$F_i(x_1^{\mathrm{E}}, x_2^{\mathrm{E}}, \ldots, x_n^{\mathrm{E}}) = 0 \quad (i = 1, 2, \ldots, n)$$

We say that this equilibrium point is stable if every nearby solution stays nearby for all future time. If the equilibrium configuration is represented by the point P^{E} in the space of the variables x_i, it is clear that a perturbation can be represented by a point P in the neighbourhood of P^{E}. We will say that P^{E} is (Liapunov) stable if, for every neighbourhood U of P^{E} in this phase space, there exists a smaller neighbourhood U_1 of P^{E} contained in U, such that every solution starting in U_1 will remain in U for all $t > 0$.

If all solutions tend to the equilibrium as t tends to infinity, then P^{E} is said to be asymptotically stable. Conversely, if it is possible to find any local perturbation that moves the system away from rest, P^{E} is called an unstable equilibrium point. These three different qualities of equilibria are illustrated in Figure 7.1.

An example of an equilibrium point that is stable but not asymptotically stable was given in Chapter 2, namely the motion of the undamped, unforced linear oscillator shown in Figure 2.1. We note that the equation of motion for this system is

$$\ddot{x} + 4\pi^2 x = 0$$

and therefore the oscillator can be thought to be in a critical state in which the damping has dropped to zero. Adding a non-zero damping to the equation will change the quality of the whole phase portrait, as we have already pointed out in Chapter 2. This qualitative change is clearly an instability phenomenon involving not a single equilibrium point but the whole flow pattern.

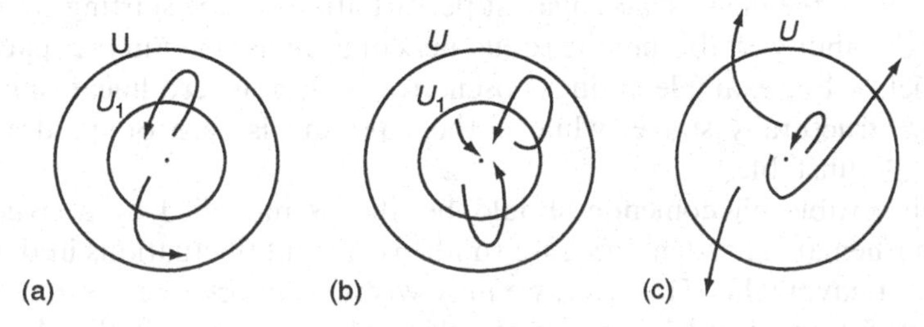

Figure 7.1 Three phase portraits illustrating the character of Liapunov stability for an equilibrium state: (a) stable, (b) asymptotically stable, (c) unstable

In much the same way as we can investigate the stability of an equilibrium by perturbing the initial conditions, it is possible to inspect the *robustness* of the phase portrait by perturbing the whole vector field via a perturbation of the differential equation. A system is said to be structurally stable if, for any sufficiently small perturbation of the defining equation, the resulting flow is topologically equivalent to the initial one.

To give a precise topological meaning to this definition we must define the *perturbation* of the vector field, and we must make clear what we mean by an *equivalent system.* It is convenient to identify a vector field $F(x)$ with a point in a given function space. In this way it is natural to say that another vector field $G(x)$ of the same space is in a neighbourhood of radius ε of F if

$$|F(x) - G(x)| < \varepsilon$$

for every x. That is, F and G are ε-close *in the C^0 topology.*

The Jacobian $JF(x)$ of F

$$JF(x) = \begin{bmatrix} F_{1,1} & \dots & F_{1,n} \\ \vdots & \vdots & \\ F_{n,1} & \dots & F_{n,n} \end{bmatrix}$$

where $F_{i,j} = \partial F_i/\partial x_j$, can be considered as a vector field in another function space, so that we can define the neighbourhoods of $JF(x)$.

A vector field G is a perturbation of size ε of F *in the C^1 topology* if it belongs to a C^0 neighbourhood of radius ε of F, and if its Jacobian $JG(x)$ belongs to a C^0 neighbourhood of radius ε of $JF(x)$. In some cases it may be useful to consider perturbations in which all derivatives up to order r remain within radius ε; this is the C^r topology of vector fields.

The resulting perturbed vector field G must now be compared with the original vector field, in order to detect if they are equivalent. Since our main concern is the behaviour embodied in the phase portrait, we define as *equivalent* two vector fields F and G if there is a continuous invertible function that transforms the phase portrait of F into the phase portrait of G.

With these definitions, the meaning of structural stability is clear: a system is *structurally stable* if sufficiently nearby vector fields always have an equivalent phase portrait. Thus while Liapunov stability is related to the robustness of a single point in the phase space, against perturbations of the starting conditions, structural stability establishes the robustness of a single point in the space of the vector fields. For example, a linear oscillator with non-zero linear damping is certainly structurally stable, while if the damping is zero or quadratic it is structurally unstable.

An observable phenomenon should be always modelled by a structurally stable mathematical system, because small errors and fluctuations in the coefficients are unavoidable. However, we may wish to impose a restriction that the perturbations envisaged be of a particular nature, restricting the choice to a subspace of the original function space. For example, in certain circumstances it may be useful to demand that the perturbations of the vector field given by

$$\ddot{x} + 4\pi^2 x = 0$$

must preserve the conservativeness of the system. In this case it is not immediately clear whether or not the previous vector field is structurally stable in the subspace of all conservative vector fields.

The mathematical study of structural stability was initiated by Andronov and Pontryagin (1937), and is central to the work of Peixoto, Smale, and Thom; in the Russian literature (e.g. Arnold 1983) a structurally stable system is called a *coarse* system. The work of Thom revolves around a profoundly geometric view of the function spaces of vector fields, which Abraham has called 'the big picture of René Thom'.

7.2 CENTRE MANIFOLD THEOREM

Consider again an n-dimensional system of first-order differential equations

$$\dot{x} = F(x)$$

where x is an n-dimensional vector and F is a real n-dimensional vector function. Suppose that $x = x^{\mathrm{E}}$ is an equilibrium point, and let us proceed to examine its stability. According to our definition, we must superimpose a disturbance ξ to x^{E}, obtaining the *perturbed* equation

$$\dot{\xi} = F(x^{\mathrm{E}} + \xi)$$

Next, $F(x^{\mathrm{E}} + \xi)$ can be expanded in a Taylor series around x^{E}, so that

$$\dot{\xi} = F(x^{\mathrm{E}}) + F_x(x^{\mathrm{E}})\xi + \frac{1}{2}F_{xx}(x^{\mathrm{E}})\xi^2 + \ldots$$

where for example $F_x(x^{\mathrm{E}})$ is the Jacobian of F evaluated at the rest state.

As we have seen, stability can be detected by examining a *small* neighbourhood of the equilibrium point, so ξ can be assumed small, and its successive powers $\xi^2, \xi^3, \ldots$ can *normally* be neglected. On the other hand, $F(x^{\mathrm{E}})$ is zero for equilibrium, and the following linear *variational* equation can be written:

$$\dot{\xi} = F_x(x^{\mathrm{E}})\xi$$

The solution $\xi(t) = \{\xi_1(t), \ldots, \xi_n(t)\}$ must tend to vanish when t goes to infinity, if the equilibrium state is to be asymptotically stable, and the condition for this is that the real parts of all the eigenvalues of $H = F_x(x^{\mathrm{E}})$ must be negative.

Hence the topological condition of the previous paragraph has been converted into an algebraic condition, since to examine the stability of a normal equilibrium point we have now merely to solve the characteristic equation, and to examine the real parts of its roots. Sometimes it is also possible to obtain the sign of the real parts without explicitly finding the roots, by means of the Routh–Hurwitz criterion, as illustrated for example by Leipholz (1970).

Suppose now that our system is dependent on a control parameter μ. Varying μ, the eigenvalues λ_j of the matrix $H(\mu)$ describe some paths in the complex plane. Suppose that at $\mu = \mu_0$ the eigenvalues are all in the negative half-plane, so that the system is asymptotically stable, and let μ increase. If the eigenvalues are assumed to cross the imaginary axis *transversely*, it is easy to see that the simplest ways in which the system can lose its stability are shown in Figure 7.2. The first case involves essentially one eigenvalue, and is therefore the simplest transition that can occur. The second, in which two eigenvalues cross the stability boundary as a pair, involves two eigenvalues, but there are strong simplifications because they are complex conjugate. In the presence of the necessary nonlinear coefficients these two transitions give rise to the only two typical bifurcations of equilibria that can be observed under the influence of a single control parameter. The first manifests itself as the *fold bifurcation*, associated with an inherent stiffness dropping to zero, while the second gives the *Hopf bifurcation*, associated with an inherent damping changing from positive to negative. In either case we should observe that the phase portrait is structurally unstable at the point of bifurcation, and it is precisely at these critical states that a linear stability analysis must be supplemented by a non-linear investigation.

The remaining three cases in Figure 7.2 can generate typically observed codimension 2 bifurcations under the influence of *two* control parameters. Their study is more complex, however, and we refer the interested reader to the excellent book of Guckenheimer and Holmes (1983).

As μ increases and the eigenvalues move in the complex plane, the original equilibrium point describes an equilibrium path in the n-dimensional phase

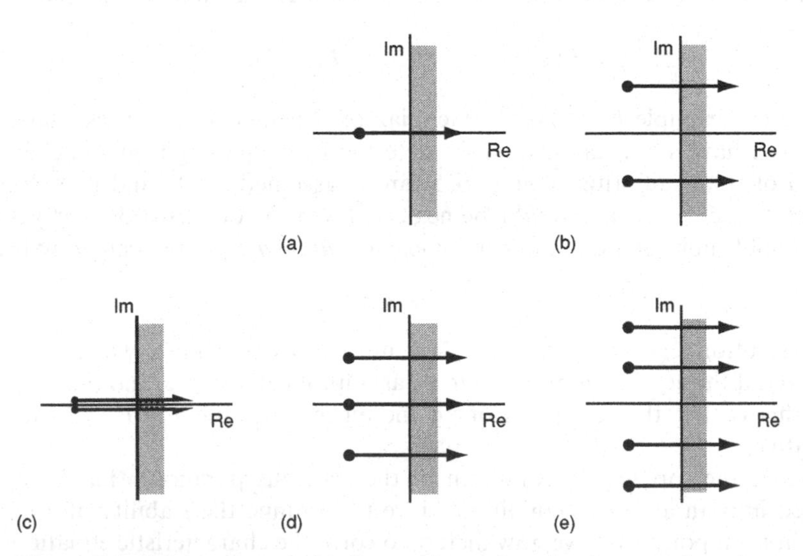

Figure 7.2 Movement of the linear eigenvalues in the complex plane corresponding to (a) the fold and (b) the Hopf bifurcation. With appropriate nonlinearities these two are structurally stable under one control. Structurally stable under two controls we have (c) two coincident zero eigenvalues, (d) a pure imaginary pair and a simple zero eigenvalue, and (e) two pure imaginary pairs of eigenvalues

space of the variable $x_1, x_2, \ldots, x_n$. We could expect that this path has some interesting geometrical properties at the points of stability loss, but it is difficult to visualize them in an n-dimensional space. On the other hand we have seen that the codimension 1 phenomena essentially involve one or two eigenvalues; therefore it is reasonable to search for a reduced system, in one or two dimensions, that preserves all the qualitative properties of the general n-dimensional system.

Such a reduction can in fact be performed in various ways, using the elimination of passive coordinates (Thompson and Hunt 1973), the Fredholm alternative (Iooss and Joseph 1980) or the Liapunov–Schmidt reduction (Chow and Hale 1982). However, it seems that the centre manifold theory represents perhaps the easiest and most general approach to the problem. More than that, it is in the spirit of the present work, relying on a geometrical grasp of the phase portrait, rather than on complex analytic computations.

Suppose that at the critical point under consideration the matrix H has r eigenvalues with negative real parts, s eigenvalues with zero real parts, and t eigenvalues with positive real parts. The eigenvectors associated with these three sets of eigenvalues span three distinct eigenspaces. The centre manifold theory allows us to view these eigenspaces as local approximations of invariant manifolds that in some sense organize the phase portrait.

Further geometrical description of these manifolds will be postponed until Chapter 10, because here we wish only to stress that the system exhibits its essential bifurcational behaviour on the *centre* manifold associated with the s critical eigenvalues of the second set. The centre manifold theory, as illustrated in Guckenheimer and Holmes (1983), reduces the flow to this manifold. It is thus clear that the fold bifurcation will be captured by a one-dimensional manifold in the phase space, while the greater complexity of the Hopf bifurcation will require a two-dimensional manifold.

7.3 LOCAL BIFURCATIONS OF EQUILIBRIUM PATHS

We can define as a *bifurcation* point every point in the parameter-control space corresponding to a structurally unstable vector field. This definition is very general, embracing static and dynamic local bifurcations, as well as global bifurcations of the phase portraits. Consider in fact a general two-dimensional system of differential equations

$$\dot{x}_1 = F_1(x_1, x_2)$$
$$\dot{x}_2 = F_2(x_1, x_2)$$

which will allow us to discuss the local bifurcations without any loss of generality, in view of the previously outlined centre manifold ideas.

The local phase portraits of this system can be classified according to the nature of the eigenvalues of the linearized system. All the possible cases are analogous to those of Figure 3.4, and it is easy to see that the structurally unstable phase portraits correspond to points of changing stability.

The fold

In a small neighbourhood of a bifurcation point the system can be considered to be restricted to its centre manifold, and a simple application of the theory shows that the vector field near a fold can be expressed as

$$\dot{x} = \mu - x^2$$

where x is now a scalar. The equilibrium equation ($\dot{x} = 0$) gives a quadratic relationship between μ and x, and the diagram in the plane (x, μ) shows a parabolic equilibrium path (Figure 7.3). One of the branches of this parabola is unstable, and marks the separation between the basin of attraction of the stable path and the basin of attraction of the point at infinity.

The fold bifurcation is often called a saddle-node bifurcation, and to understand this name it is necessary to consider it embedded in a two-dimensional system. This *lifting* can be accomplished in various ways. For example, we can simply add an acceleration to the previous vector field, arriving at the equation of a nonlinear damped oscillator

$$\ddot{x} + \dot{x} + x^2 - \mu = 0$$

or

$$\dot{x} = y$$
$$\dot{y} = \mu - x^2 - y$$

A second possible lifting of the one-dimensional equation to a two-dimensional space is achieved by considering two decoupled equations, as for example in Guckenheimer (1980b), where he considers the following flow:

$$\dot{x} = \mu - x^2$$
$$\dot{y} = -y$$

An advantage of this latter procedure is to obtain an *easier* phase portrait, but the analogy with the motion of an oscillator can no longer be pursued. In Figure 7.4(a) three phase portraits of the second lifting are shown. The first diagram is drawn at $\mu = 0.5$, so that there is a sink at $x = \sqrt{0.5}$ and a saddle at $x = -\sqrt{0.5}$. As we decrease the control parameter these two equilibria approach one another and at $\mu = 0$ they collide in a saddle node. The resulting critical phase portrait can be seen in the second diagram. Here the inset of the equilibrium point at the origin is given by the two-dimensional manifold $x \geq 0$. The outset (trajectories asymptotic to the fixed point as $t \to -\infty$) coincides with the line $y = 0, x < 0$. If μ decreases past zero the equilibrium is lost, and every trajectory goes away to infinity, as illustrated in the third diagram.

The fold bifurcation or limit point is very common in all branches of applied mathematics. The buckling of a shallow arch is the most typical example in engineering, while the equilibrium response of a massive cold star with two stable regimes should be more familiar to astronomers. Very often models exhibit a succession of folds, whose stability transitions can be predicted following Thompson (1979). Figure 7.4(b) shows the basin change at a fold.

The great importance of the fold lies in the fact that it is a *structurally stable bifurcation*. Suppose that we start with an n-dimensional vector field F_μ with an equilibrium, one of whose eigenvalues is real and changes sign at $\mu = \mu_c$. We know that in the appropriate centre manifold the phase portraits near $\mu = \mu_c$ are topologically equivalent to Figure 7.4(a). Now consider a perturbed vector field G_μ, which is sufficiently close to F_μ when μ is near μ_c. Then there is again a centre manifold, in which the phase portraits of G_μ near μ_c^* (close to μ_c) are again topologically equivalent to Figure 7.4(a).

Here we may imagine Figure 7.4(a) expanded to form a movie, with one value of μ per frame; structural stability of a *vector field* relates to the robustness of one frame of this movie, while structural stability of a *bifurcation* relates to the robustness of the movie as a whole.

Hopf bifurcation

Consider an unforced, damped nonlinear oscillator, whose motion can be described by

$$\ddot{x} + b\dot{x} + \omega^2 x + f(x, \dot{x}) = 0$$

where $f(x, \dot{x})$ contains the nonlinear terms. This second-order equation can be written as a system of two first-order equations

$$\dot{x} = y$$
$$\dot{y} = -by - \omega^2 x - f(x, y)$$

It is easy to verify that, as b changes from positive to negative, the eigenvalues of the linearized system cross the imaginary axis as a pair at

$$\lambda, \bar{\lambda} = \pm\sqrt{(-\omega^2)}$$

Hence the so-called Hopf bifurcation can be associated with the vanishing of the linear damping in the motion of an oscillator.

If $f(x, \dot{x}) = 0$ the system is linear and the post-critical motion is typified by oscillations whose amplitude grows to infinity. The addition of nonlinear terms can destroy this feature. Consider for example the equation that we encountered in the previous chapter,

$$\ddot{x} + \alpha(x^2 - 1)\dot{x} + \omega^2 x = 0 \quad (\alpha > 0)$$

which can represent the motion of a nonlinear oscillator after a Hopf bifurcation. If x is small it is possible to neglect x^2 in the term in parentheses, and the resulting motion shows oscillations of growing amplitude. When x increases, however, the nonlinear damping term $x^2\dot{x}$ becomes dominant, and the motion must remain bounded. The system will thus exhibit a post-critical limit cycle in its phase portrait.

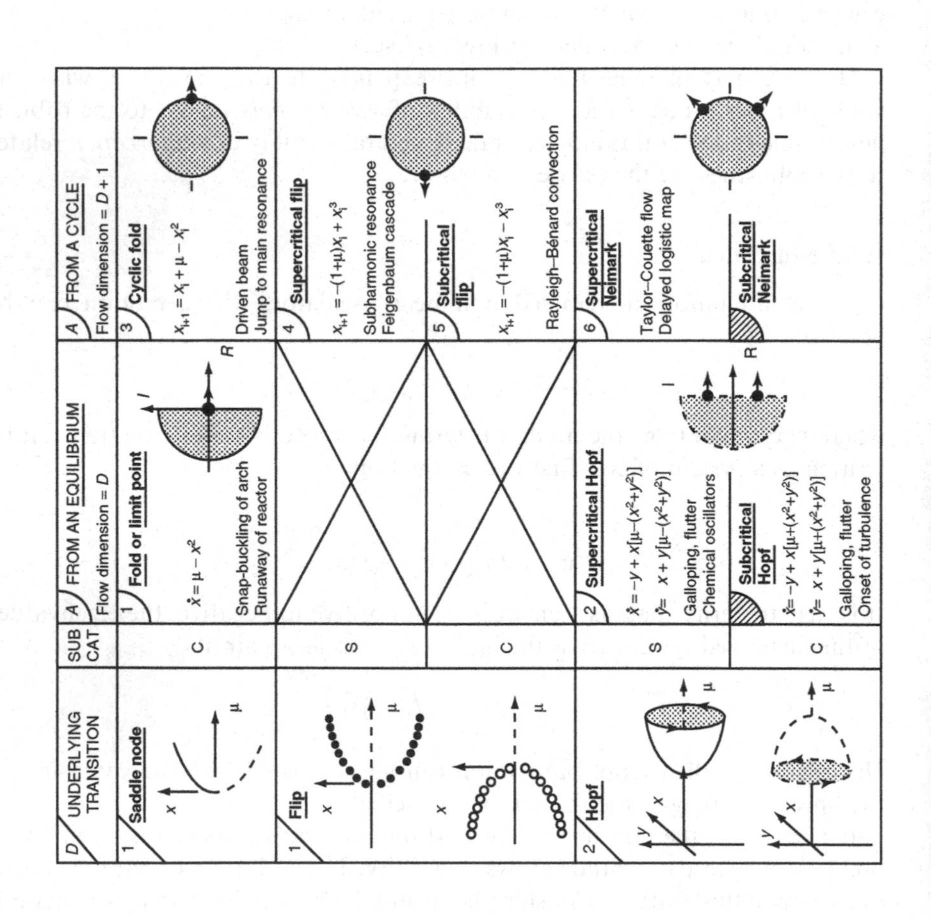
D
UNDERLYING TRANSITION
SUB CAT
A
FROM AN EQUILIBRIUM
Flow dimension = D
A
FROM A CYCLE
Flow dimension = D + 1
1
Saddle node
x
μ
C
1
Fold or limit point
$\dot{x} = \mu - x^2$
Snap-buckling of arch
Runaway of reactor
I
R
3
Cyclic fold
$x_{i+1} = x_i + \mu - x_i^2$
Driven beam
Jump to main resonance
1
Flip
x
μ
S
C
4
Supercritical flip
$x_{i+1} = -(1+\mu)x_i + x_i^3$
Subharmonic resonance
Feigenbaum cascade
5
Subcritical flip
$x_{i+1} = -(1+\mu)x_i - x_i^3$
Rayleigh–Bénard convection
2
Hopf
y
x
μ
S
C
2
Supercritical Hopf
$\dot{x} = -y + x[\mu-(x^2+y^2)]$
$\dot{y} = x + y[\mu-(x^2+y^2)]$
Galloping, flutter
Chemical oscillators
Subcritical Hopf
$\dot{x} = -y + x[\mu+(x^2+y^2)]$
$\dot{y} = x + y[\mu+(x^2+y^2)]$
Galloping, flutter
Onset of turbulence
I
R
6
Supercritical Neimark
Taylor–Couette flow
Delayed logistic map
Subcritical Neimark

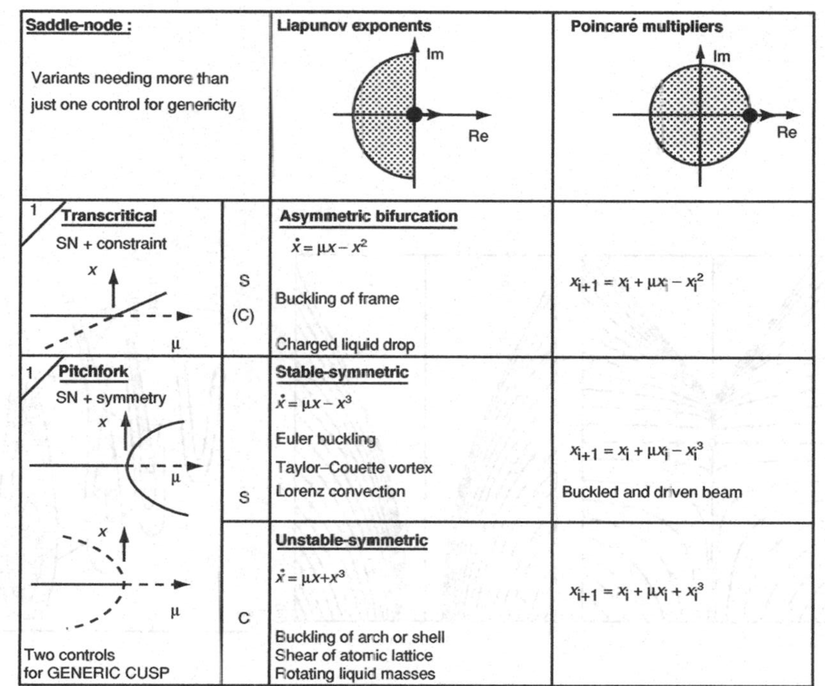

Figure 7.3 A summary of the generic local bifurcations of equilibria and cycles under one control parameter, following Guckenheimer and Holmes (1983). The cycles are discussed via their Poincaré return maps. *D* denotes the phase-space dimension of the centre manifold, SUB (S) denotes a subtle bifurcation, and CAT (C) denotes a catastrophic bifurcation. *A*-numbers refer to the pictorial sequence of Abraham and Marsden (1978). Some typical examples are quoted from Thompson (1982). For more details of the codimension-one bifurcations see the Appendix

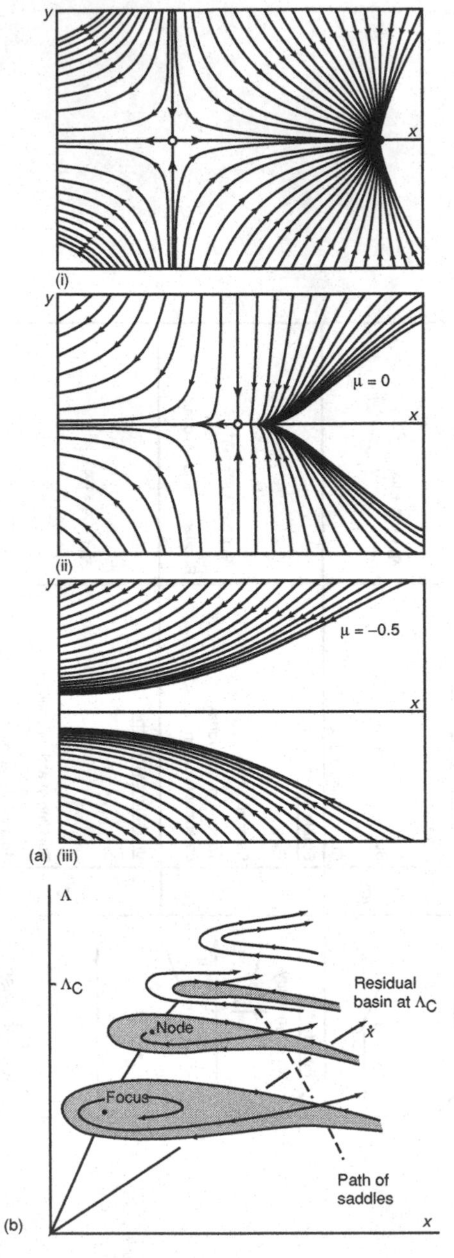

Figure 7.4 (a) Phase portraits for the saddle-node bifurcation. (b) Sketch of the portraits and basins under the variation of a control parameter Λ; note the residual basin at Λ_C. This is the *static fold* of the Appendix

In the above equation the nonlinear term is $f(x, \dot{x}) = \alpha x^2 \dot{x}$, but it is easy to see that the same conclusions can be drawn if we consider the simpler term $f(\dot{x}) = \dot{x}^3$, giving the equation

$$\ddot{x} - \mu\dot{x} + x + \dot{x}^3 = 0$$

This equation has a fixed point at the origin that exhibits a Hopf bifurcation as μ changes from negative to positive values. A sequence of phase portraits is shown in Figure 7.5(a). When $\mu < -2$ the origin is an asymptotically stable node which becomes a stable spiral as the eigenvalues become complex at $\mu = -2$. At the bifurcation at $\mu = 0$ the convergence is very slow (third picture), while at $\mu > 0$ the origin becomes an unstable spiral and the flow tends to the newborn limit cycle (fourth picture). Finally at $\mu > 2$ the origin is an unstable node, while the limit cycle continues to grow.

Guckenheimer and Holmes (1983) consider the following model for the Hopf bifurcation:

$$\dot{r} = r(\mu - r^2)$$
$$\dot{\theta} = 1$$

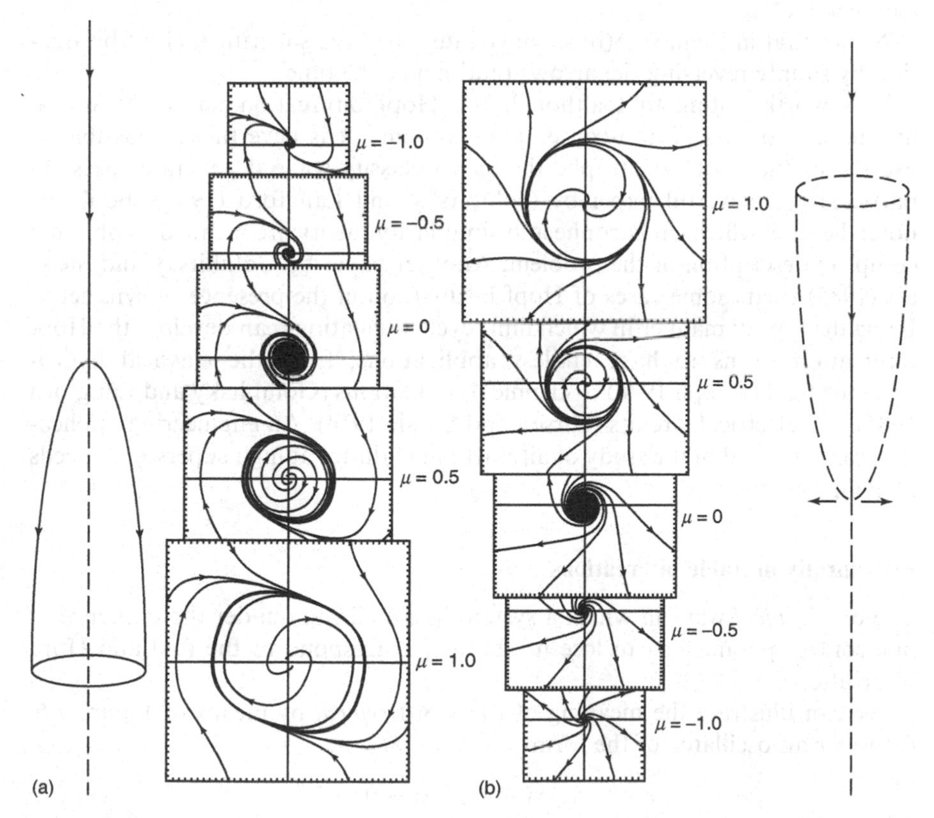

Figure 7.5 Illustrations of the Hopf bifurcation for (a) a nonlinear oscillator with cubic damping, showing the supercritical form; (b) the Guckenheimer–Holmes model, showing (by time reversal) the subcritical form

where the use of polar coordinates makes it possible to decouple the equations. At $\mu < 0$ the sole steady state is the fixed point at the origin, but at $\mu > 0$ there is also the solution

$$\theta = t + \text{const.}$$
$$\mu = r^2$$

Hence the limit cycles are circles of radius $\sqrt{\mu}$, and in Figure 7.5(b) a similar sequence of phase portraits is shown as for the previous equation. According to a theorem of Hopf (1942), equivalent phase portraits can be found in the centre manifold of an n-dimensional system with an equilibrium and a pair of complex conjugate eigenvalues crossing the imaginary axis transversely, given the appropriate nonlinearities.

Another more dangerous type of Hopf bifurcation is sketched in Figure 7.3, where the coalescence of an unstable periodic orbit with the stable fixed point marks the end of the stability range. This is called a subcritical Hopf bifurcation, or Hopf bifurcation in its catastrophic form. A nice example of a catastrophic Hopf bifurcation occurs in the Lorenz equations, as we shall see in Chapter 11.

Notice that in Figure 7.5(b) we have illustrated the subcritical Hopf bifurcation by simply reversing the arrows (and hence the time).

It is worth noting that although the Hopf bifurcation has a dynamical nature, and occurs in non-conservative systems, it is nevertheless possible to use the methods of catastrophe theory to classify some degenerate cases. In particular, a powerful paper by Golubitsky and Langford (1981) should be noted here, in which catastrophe and singularity theory are blended to obtain a complete description of the problem. Another paper by Golubitsky and Stewart (1985) treats some cases of Hopf bifurcation in the presence of symmetry. Being the typical manner in which limit cycle oscillations can develop, the Hopf bifurcation seems to have endless applications, from the classical hydrodynamic field (Joseph 1976) to chemical oscillations (Golubitsky and Langford 1981) and electrical circuits (Hirsch and Smale 1974). An engineering application can be found in the study of aircraft panel flutter at high supersonic speeds (Thompson 1982).

Structurally unstable bifurcations

The only *typical* ways in which a system in equilibrium under the influence of one control parameter can lose its stability correspond to the fold and Hopf bifurcations.

We can illustrate the meaning of the word *typical* by means of Figure 7.6. Consider an oscillator of the form

$$\ddot{x} + \dot{x} + f(x, \mu) = 0$$

where the nonlinear stiffness function is the derivative, V_x, of a total potential energy $V(x, \mu)$. With fixed positive damping, there can be no Hopf bifurcation,

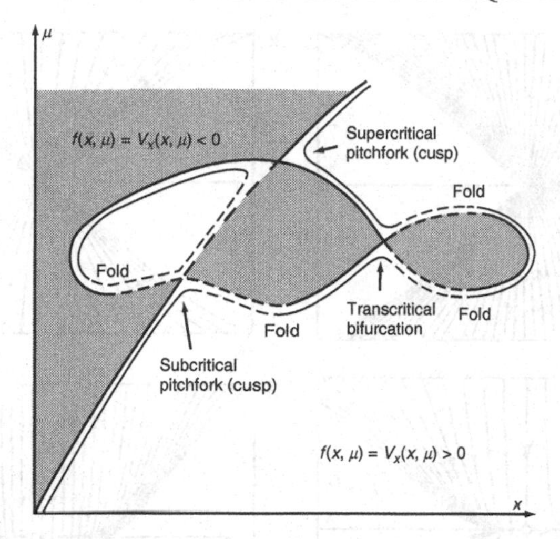

Figure 7.6 For a positively damped oscillator, the stability of the thick equilibrium paths can be deduced from the sign of the stiffness function $f(x, \mu)$ and the bifurcations can be identified. Under a typical perturbation, only the folds persist in the thin paths

and a minimum of V with respect to x is both necessary and sufficient for stability (Thompson and Hunt 1973). Now if we draw the equilibrium paths corresponding to $f(x, \mu) = V_x(x, \mu) = 0$ in the space of μ against x, they will divide the space into regions where f is positive or negative. In the schematic figure, the tint shows where f is negative. The necessary and sufficient condition for the stability of a path is now that it should have a tint on its left-hand side. In this way we can deduce the stability regions of the thickly drawn paths (solid line means stable, dotted line means unstable) and identify the bifurcations. If we now perturb the system, by perturbing the function V, the thick paths will typically be replaced by the thin paths. Folds persist under this perturbation, because they are codimension 1 events, but the pitchfork and transcritical bifurcations (discussed next) do not.

Despite this result, many physical examples exhibit transcritical bifurcations and cusps instead of the general saddle node. This latter should *not* be viewed as an entirely pathological phenomenon, because the cusp is 'perhaps the most important example of bifurcation in the classical literature' (Golubitsky and Schaeffer 1985).

Suppose in fact that the flow is *constrained* to admit the trivial state $x = \dot{x} = 0$ as an equilibrium configuration, for every value of the control parameter. It is clear that now the folding saddle node is ruled out, and perturbations must be chosen in the subspace of the vector fields that satisfy the above condition.

In this subspace the transcritical bifurcation is indeed structurally stable and its simplest form is

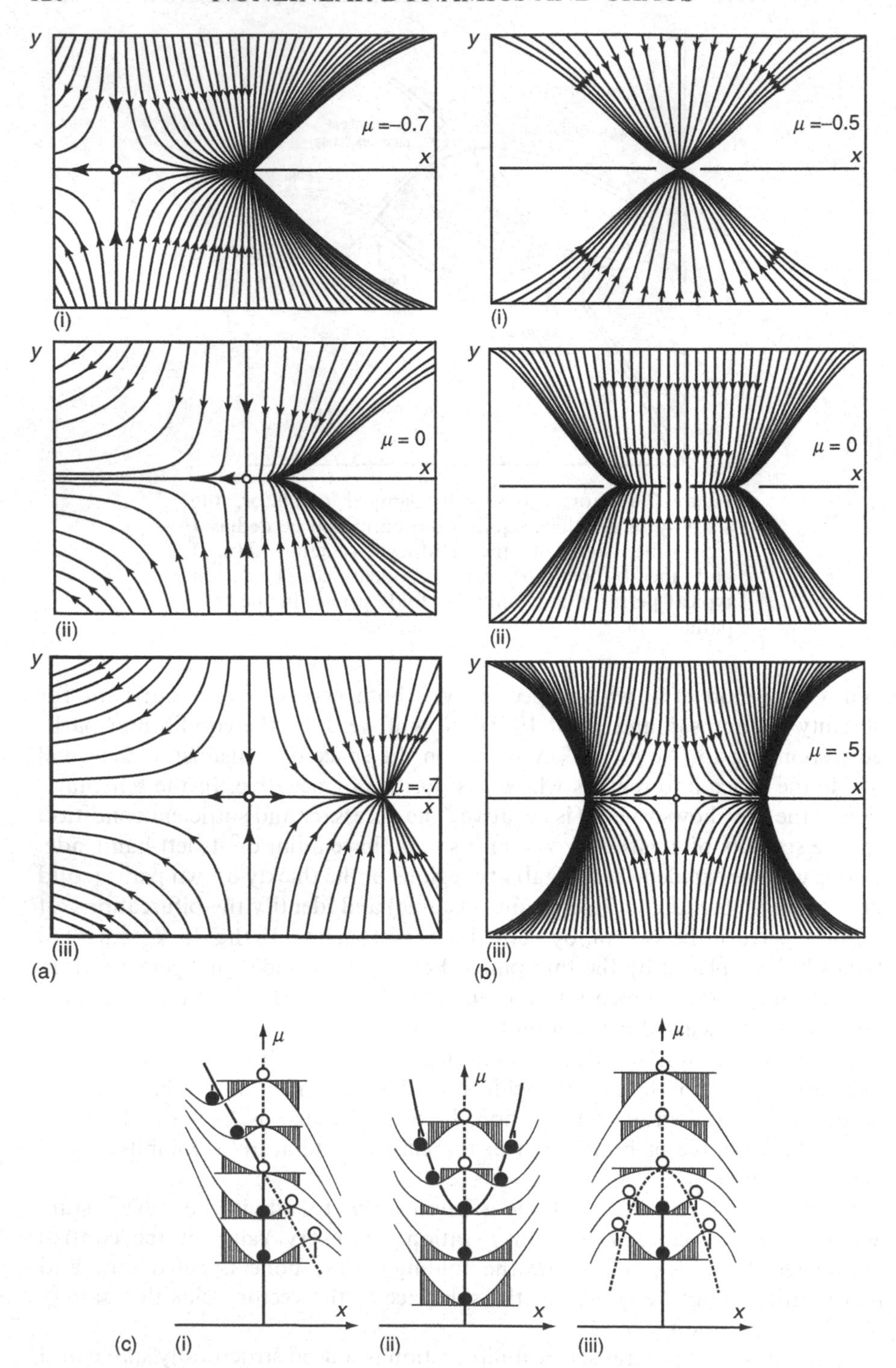
y
μ =−0.7
x
(i)
y
μ = 0
x
(ii)
y
μ = .7
x
(iii)
(a)
y
μ =−0.5
x
(i)
y
μ = 0
x
(ii)
y
μ = .5
x
(iii)
(b)
μ
x
μ
x
μ
x
(c)
(i)
(ii)
(iii)

$$\dot{x} = \mu x - x^2$$

The bifurcation diagram is shown in the second diagram of Figure 7.3. Two fixed points are present for every value of μ, and at $\mu = 0$ they coalesce in a degenerate equilibrium. The usual exchange of stability follows from a theorem of Poincaré, and for $\mu > 0$ the trivial equilibrium is unstable. Owing to the form of the diagram this bifurcation is also called an *asymmetric point of bifurcation*.

It is again possible to lift the flow, in much the same way as for the saddle node, by means of the equations

$$\dot{x} = \mu x - x^2$$
$$\dot{y} = -y$$

and the resulting phase portraits are sketched in Figure 7.7(a). In the first picture there is a sink at the origin, while the saddle is at $x = -0.7$. As we increase the parameter, this saddle moves towards the sink, and the resulting flow at $\mu = 0$ is shown in the second diagram. If $\mu > 0$ the origin is unstable, while the non-trivial equilibrium gains stability. The resulting flow is shown in part (iii) of Figure 7.7(a).

The critical phase portrait at $\mu = 0$ is the same as for the saddle node, because the governing equation is the same. This suggests that the only difference between these two bifurcations is the route through which the critical point is achieved. In fact, it is shown in Thompson and Hunt (1984) how both phenomena can be seen as different routes through the fold catastrophe.

If the system has some intrinsic symmetry, the transcritical bifurcation is no longer possible, and we must restrict our attention to the subspace of symmetric vector fields with trivial equilibrium paths. The resulting structurally stable bifurcation is the *cusp*, in which the sink at the origin becomes a saddle and gives rise (in its supercritical form) to two symmetric sinks. The archetypal model is given by

$$\dot{x} = \mu x - x^3$$

and the bifurcation diagram is illustrated in the continuation of Figure 7.3. The form of the diagram justifies the classical name of pitchfork bifurcation or symmetric point of bifurcation.

The catastrophic subcritical form of the cusp is shown in the next diagram of the same figure, where two saddles shrink towards the sink at the origin.

The usual lifting of the flow allows us to draw the three diagrams in Figure 7.7(b), for the supercritical form. In the first there is only a sink at the origin, and all the trajectories tend towards it. This convergence becomes more and

---

Figure 7.7 (a) Phase portraits for the transcritical bifurcation: (i) $\mu = -0.7$, (ii) $\mu = 0$, (iii) $\mu = 0.7$. (b) Phase portraits for the supercritical pitchfork (cusp) bifurcation: (i) $\mu = -0.5$, (ii) $\mu = 0$, (iii) $\mu = 0.5$. (c) Balls rolling on potential energy curves, illustrating (i) the transcritical bifurcation, (ii) the supercritical pitchfork bifurcation and (iii) the subcritical pitchfork bifurcation

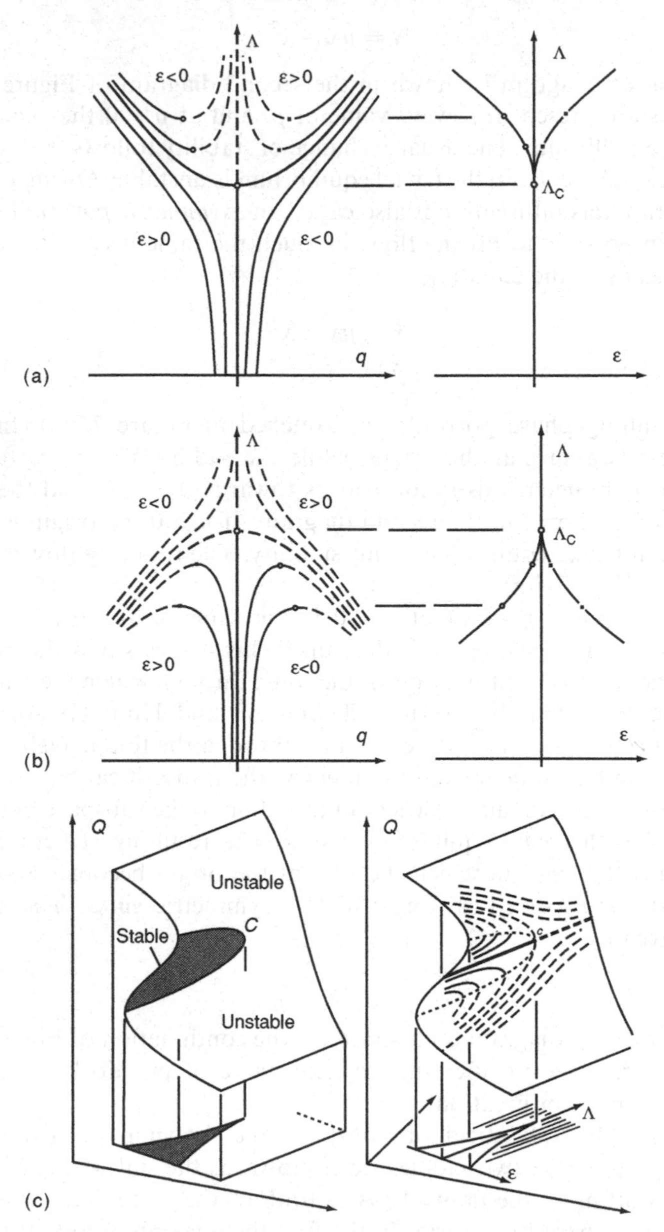

Figure 7.8 Effect of a symmetry-breaking imperfection ε on the control (Λ) versus response (q) diagram for (a) the supercritical pitchfork and (b) the subcritical pitchfork: the right-hand diagrams show the projection of the folds into two-thirds power-law cusps in the (Λ, ε) control space. (c) How the subcritical form can be displayed as an equilibrium surface with response Q plotted above the control plane, on which can be drawn the (Λ, ε) axes

more sluggish as the parameter approaches the critical value. The critical case is shown in the second diagram, where the role of the centre manifold becomes evident. In fact, every trajectory is attracted strongly to the line $y = 0$, and the subsequent dynamics is essentially restricted to this one-dimensional subspace. As we increase μ, two sinks develop from the origin, and the phase portrait is divided into two basins of attraction by the stable manifold of the saddle, as is shown in part (iii) of Figure 7.7(b).

The cusp is a codimension 2 event requiring two controls to render it structurally stable and robustly observable (Thompson 1982). This is emphasized in catastrophe theory by Figure 7.8(c). For a complete classification of higher static singularities, the reader can usefully consult the book of Gilmore (1981).

7.4 LOCAL BIFURCATIONS OF CYCLES

After the equilibrium point, the next most common form of recurrent behaviour is the limit cycle, as we have already seen in Chapter 4. It is possible to define the stability of a cycle following the previous definition of a stable equilibrium point, leading us to an investigation of the various bifurcations. The best method to examine the ways in which a periodic orbit can lose its stability is by using a Poincaré section, and by studying the resulting *discrete* dynamical system. We will discuss extensively these discrete systems in Part II; for now we want to describe the behaviour of a periodic orbit near a bifurcation point, emphasizing some aspects of the *flow phenomena* that are lost in the analysis of the return Poincaré map.

The cyclic fold

The first case to be considered is the *cyclic fold* (Figure 7.3), which is very similar to the fold bifurcation for equilibria. Two limit cycles of different stability coexist for certain values of the control parameter, and approach one another as we vary the control. At the bifurcation point they collide, and afterwards every trajectory goes out of the immediate neighbourhood. The cyclic fold is often associated with the so-called *jump phenomenon* or *hysteresis*, where for a limited parameter range two stable closed orbits coexist, separated by a repelling limit cycle.

Consider for example the Duffing equation

$$\ddot{x} + d\dot{x} + x + \alpha x^3 = f \cos \omega t$$

and fix the values of the three parameters as $\alpha = 0.05, f = 2.5$ and $d = 0.2$. As we slowly vary the remaining control parameter ω it is found that the experimentally observed amplitude of the response is a smooth function of ω, at all but two ω values. At these values a jump is observed, following the well-known hysteresis diagram of Figure 7.9. This figure was drawn following the variational method outlined by Holmes and Rand (1976), and attention was focused on a small range of the parameter space, so that a single cyclic fold can be clearly seen at F. Other similar hysteretic regions can be detected

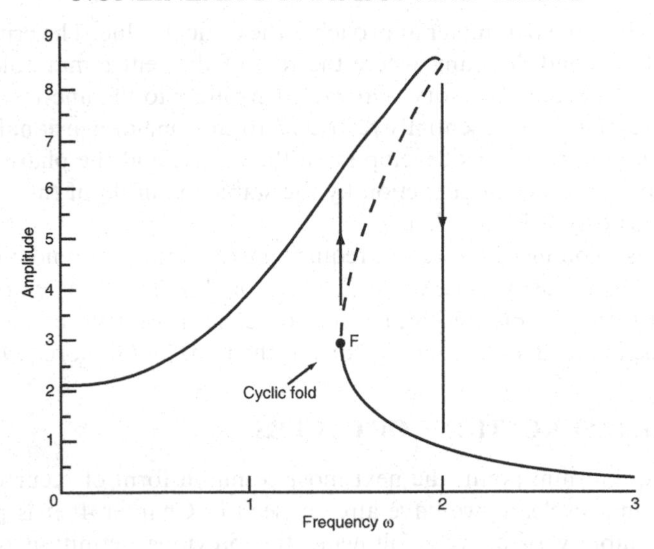

Figure 7.9 Resonance response curve for Duffing's equation

numerically for smaller ω values, and in each a pair of cyclic folds is always necessary, in order to create the pair of stable periodic orbits (Parlitz and Lauterborn 1985).

If a periodic orbit is constrained to exist for every value of the control parameter, the cyclic fold is no longer possible, and the analogues of transcritical bifurcations and cusps can arise.

The flip bifurcation

Instead of considering in more detail these and other atypical singularities, we proceed to describe the *period-doubling* flip phenomenon (Figure 7.3), which is peculiar to cycles and has no correspondence for equilibria. In this type of *flip bifurcation* a stable limit cycle loses its stability, while another closed orbit is born whose period is twice the period of the original cycle. A rather classical sequence of phase portraits leading eventually to period doubling is illustrated in Figure 7.10. A fixed point at the origin loses its stability and a limit cycle is created through a supercritical Hopf bifurcation at $\mu = \mu_1$. Every trajectory then tends to the closed orbit, as shown in the second diagram. As μ increases, the stability of the cycle descreases until, as the parameter μ is further increased, a period-doubled orbit is born as the original cycle becomes unstable. In the last diagram of Figure 7.10 the new stable orbit generated by the *supercritical* flip is seen lying on the edge of a Möbius strip. The width of this Möbius strip increases continuously from zero as μ passes the bifurcation point. This bifurcation requires at least a three-dimensional phase space, and we should observe that it can arise in both subcritical (catastrophic) and supercritical (subtle) forms.

In many physical examples a sequence of n period-doubling bifurcations is observed, in which a stable limit cycle with period $2^n T$ is finally obtained. If the

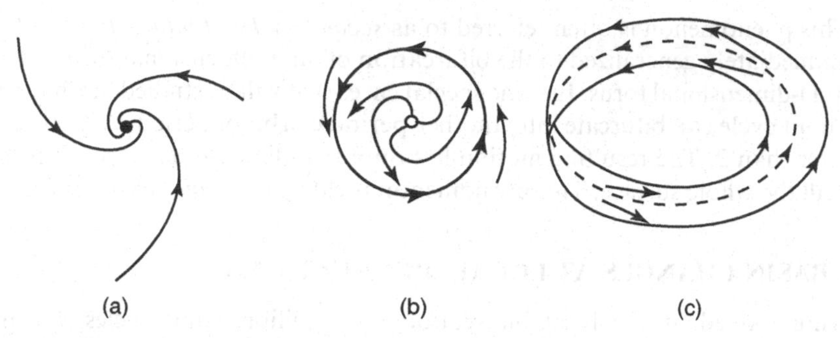

Figure 7.10 A typical scenario in which a fixed point gives rise to a limit cycle through a Hopf bifurcation, and the limit cycle subsequently gives rise to a two-periodic orbit by means of a flip bifurcation

sequence is infinite, with a finite accumulation point, the resulting motion beyond this point is chaotic. On the other hand, it is also possible to observe a *finite* number of period doublings, followed by the same number of reverse bifurcations. This latter behaviour seems to be typical of two-parameter systems. Consider for example the Duffing equation (Parlitz and Lauterborn 1985)

$$\ddot{x} + 0.2\dot{x} + x + x^3 = f \cos \omega t$$

It can be checked that for $f = 22$ a periodic orbit of period $T = 2\pi/\omega$ bifurcates into a $2T$-periodic orbit as we increase ω. Then this new orbit loses stability and a $4T$-periodic orbit is obtained. If ω is allowed to increase further, there is a first reverse bifurcation in which the $4T$-periodic orbit merges into the $2T$-periodic orbit, which regains stability. In the same manner the $2T$-periodic orbit remerges into the original T-periodic orbit. For a different, higher f value the sequence of period doublings will advance further.

Neimark or secondary Hopf bifurcation

We have seen that the Hopf bifurcation of an equilibrium point can be characterized as the separation point between an area-contracting region with positive damping and an area-expanding region with negative damping, at which the previously stable fixed point becomes unstable, and a limit cycle develops.

To detect a similar bifurcation for a periodic orbit it is necessary to consider a three-dimensional differential system

$$\dot{x} = F(x, \mu)$$

exhibiting a limit cycle at $\mu = \mu_1$. Suppose also that at this parameter value the system is locally dissipative. Let μ increase in such a way that at $\mu = \mu_2$ the behaviour is locally conservative, while at $\mu > \mu_2$ it is area-expanding in a neighbourhood of the limit cycle. The analogy with the Hopf bifurcation is evident, so that a bifurcation is expected at $\mu = \mu_2$. With *stabilizing* nonlinearities, we might expect that, as the limit cycle loses its stability, an attracting torus is born.

This phenomenon is often referred to as *secondary Hopf bifurcation*, and can be immediately generalized to the bifurcation of an n-dimensional torus into an $(n+1)$-dimensional torus. In some special cases that will be studied in Chapter 8, the limit cycle can bifurcate into another periodic orbit of period kT, where k is greater than 2. The resulting motion lies on a two-dimensional torus, but does not fill the whole surface. This phenomenon is called *resonance* or *phase locking*.

7.5 BASIN CHANGES AT LOCAL BIFURCATIONS

Having looked at the local bifurcations of equilibria and cycles, we now examine their implications for the basins of attraction. We start with the fold of Figure 7.4. Figure 7.4(b) shows the basin of attraction of the stable node in a succession of phase portraits as the control parameter Λ is varied through its critical value Λ_C. The basin boundary is formed by the inset (namely the stable manifold) of the saddle solution, comprising all the points that tend to the saddle as time goes to plus infinity. The inset shown consists of two curves, each a trajectory asymptotic to the saddle. Each of the two trajectories separates the flow of nearby trajectories.

As the control Λ approaches Λ_C, saddle and node move towards each other. The attractor and its basin boundary are moving steadily closer together. The size of the basin appears to be shrinking, but in the totally dissipative oscillator illustrated, it actually remains infinite in area for $\Lambda \leq \Lambda_C$; this is a consequence of the negative divergence of the flow, which becomes positive divergence under time reversal. The so-called residual basin at $\Lambda = \Lambda_C$ is perhaps something of a surprise and warrants further discussion.

For a particle moving with dissipation in a cubic potential well, $V = x^3$, it means that the unstable critical equilibrium state at the origin has a residual basin of attraction of infinite area; somewhat counterintuitively, there are an infinite number of trajectories that settle at the origin. In fact, as can be seen in Figure 7.4(a), where the critical condition has $\mu = 0$, the critical equilibrium state has a two-dimensional inset, and a one-dimensional outset. It is this outset (unstable manifold, defined by points that tend to the equilibrium as time goes to minus infinity) which renders the critical state unstable in the classical sense of Liapunov.

Notice that the critical phase portrait ($\Lambda = \Lambda_C$ or $\mu = 0$) is structurally unstable. The slightest change in the value of the control parameter results in a topologically different phase portrait.

For $\Lambda > \Lambda_C$ there is no local equilibrium solution; all trajectories move off towards another attractor, the one whose basin surrounded the basin of the node at subcritical values of Λ. The basin change at the fold is summarized in Figure 7.11.

Consider next the two variants of the Hopf bifurcation illustrated in Figure 7.5. In the supercritical Hopf bifurcation there is a unique attractor throughout the sweep of the control parameter μ; the whole of the drawn rectangular region of phase space lies within a single basin of attraction for all values of μ. For $\mu \leq 0$ the basin of attraction is that of the stable fixed point at the origin, while for $\mu > 0$ it is that of the stable limit cycle created at the

WITH NO HYSTERESIS AND REMOTE UNRELATED BASIN BOUNDARIES

SUBTLE (CONTINUOUS) : SAFE

Supercritical
Hopf
Flip
Neimark
(Cusp, 2μ)

Basin of attraction

μ^c μ

CATASTROPHIC (DISCONTINUOUS) : EXPLOSIVE

Explosions
Flow
Map
Intermittency
Chaotic

x

μ^c μ

WITH HYSTERESIS, BLUE-SKY JUMPS AND SHRINKING BASINS

CATASTROPHIC (DISCONTINUOUS) : DANGEROUS THRESHOLDS

Subcritical
Hopf
Flip
Neimark
(Cusp , 2μ)

Folds
Static
Cyclic

Saddle connection

Chaotic blue sky catastrophe

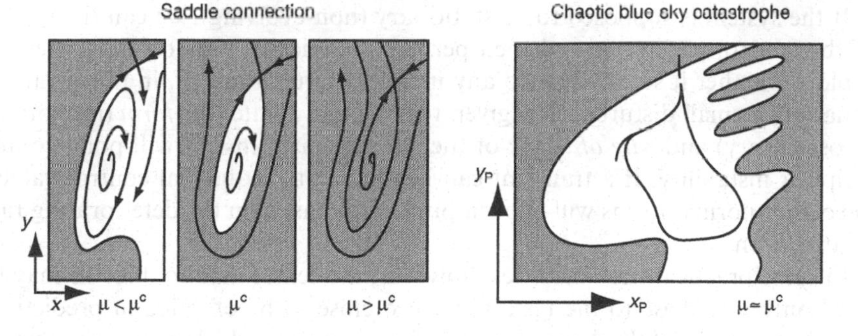

Figure 7.11 A schematic summary of the basin changes at different types of codimension 1 bifurcation (generic under one control parameter μ). The large arrows denote fast dynamic jumps to a remote, unrelated attractor. These bifurcation types are discussed in detail in Chapter 13

bifurcation. So the supercritical Hopf bifurcation has no significant involvement with the basin structure, as illustrated in Figure 7.11. The same is also true of the supercritical pitchfork and the supercritical flip bifurcation of limit cycles (Chapter 8).

A significant, indeed radical, basin metamorphosis is encountered in the subcritical Hopf bifurcation, in which we shall here suppose the control parameter μ to be decreasing. In the two-dimensional phase portraits, the unstable limit cycle before the critical condition at $\mu = 0$ separates nearby trajectories, and forms the basin boundary of the stable fixed point at the origin. So as we approach $\mu = 0$ we observe the basin of attraction of the fixed point shrinking to zero around the point. Past the critical point, there is no attractor in the locality, and all trajectories leave the rectangle towards another, remote attractor. It is this dramatic change of basin structure that generates an inevitable finite dynamic jump from the fixed point as μ is decreased through 0. This behaviour, typical of the subcritical bifurcations, is also illustrated in Figure 7.11.

Entirely analogous basin metamorphoses arise during the instabilities of limit cycles which are best discussed in terms of Poincaré maps (Chapter 8). Equivalent to the fold is the cyclic fold; equivalent to the Hopf bifurcation with subcritical and supercritical forms are the flip and Neimark bifurcations.

7.6 PREDICTION OF INCIPIENT INSTABILITY

From a practical point of view some bifurcations are more dangerous than others, either because they are unexpected or because they are followed by a discontinuous change in the system behaviour. For example, all the subcritical forms of bifurcation are sudden phenomena, so it would be useful to have a method to predict the incipient *jumps* before their catastrophic appearance. In some cases this problem can be considered solved, but in its general terms it remains open, at least for the so-called evolving systems in which the control parameter is allowed to vary slowly with time according to a given law.

If the system is supposed to be stationary (non-evolving), we can fix a value of the control parameter and then perform a series of tests on the system to explore whether it seems close to any incipient bifurcation. If, in this idealized situation, a small disturbance is given to the steady state, the *form* (monotonic or oscillatory) and *rate of decay* of the subsequent transient will point to any incipient instability. If a transient can be studied at sequential control values, the earlier normal forms will allow a prediction based on the deteriorating rate of attraction.

In some practical circumstances, however, the centre manifold forms may be valid only *very close* to the instability, too close to be of value in prediction. This is true of *lightly damped* mechanical systems, which may continue to exhibit oscillatory transients until very close to a fold or cyclic fold. The one-dimensional forms of these are of course only valid after the damping has become supercritical, and oscillation has ceased. To overcome this, recent research has considered the behaviour of lightly damped oscillators, perturbations of undamped systems (Virgin 1986a). Considering, for example, the Hamiltonian fold,

$$\ddot{x} + x^2 - \mu = 0$$

we can examine ω, the frequency of linear transient vibrations about the current stable state. It is easy to show that ω^2 varies linearly with progress along the parabolic equilibrium path, so it is ω^4 that drops linearly to zero at criticallity when plotted against μ. This is illustrated in Figure 7.12, where the transient frequency of an *evolving system* for which μ decreases linearly with time, is seen to vanish as time approaches $t^c = 1000$.

A similar situation has been explored for the cyclic fold (Thompson and Virgin 1986). In an undamped system the approach to a cyclic fold follows the left-hand diagrams of Figure 7.13, where μ might measure the driving frequency of a Duffing oscillator. In the complex plane, the eigenvalues of a

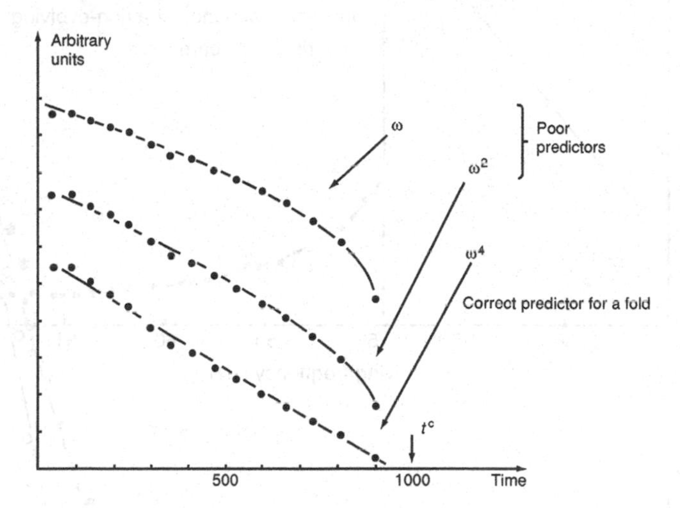

Figure 7.12 Instability predictors for an undamped fold, showing that the correct predictor is the fourth power of the frequency

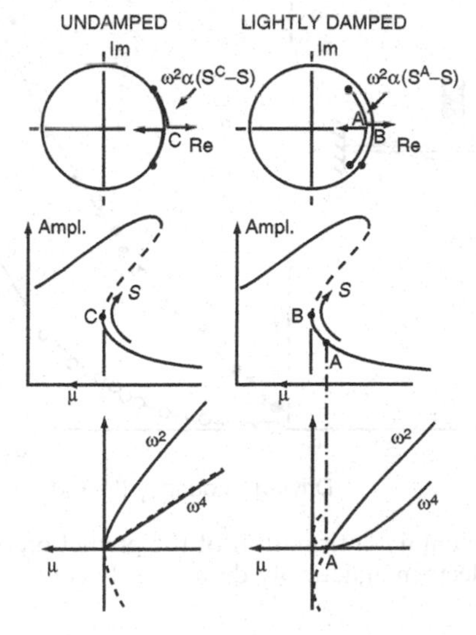

Figure 7.13 Outline of the scheme for predicting a jump to resonance at a lightly damped cyclic fold on the basis of the local transient frequency

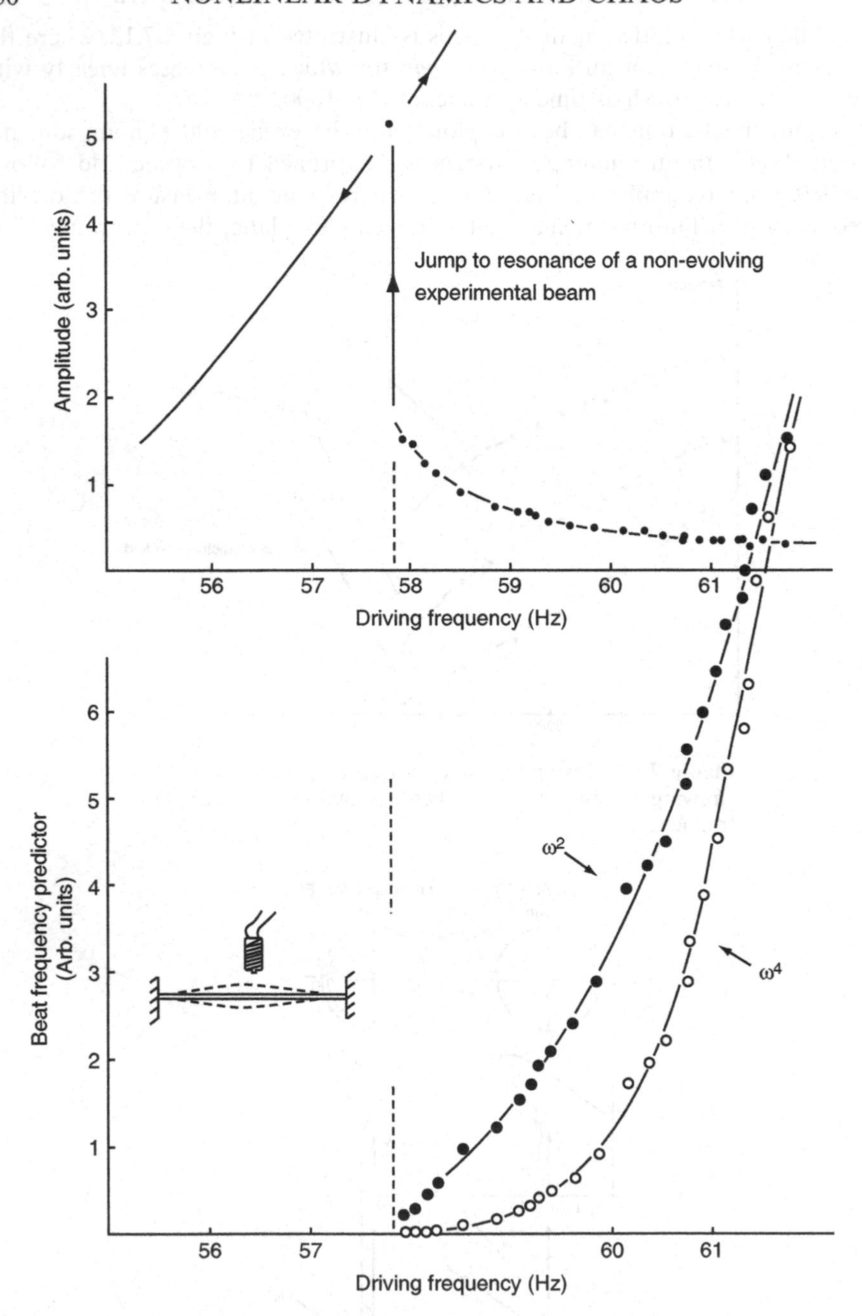

Figure 7.14 Experimental confirmation of the prediction scheme for the jump to resonance of an electromagnetically driven steel beam

Poincaré mapping move around the unit circle, with ω^2 proportional to a path parameter so that ω^4 is again the linear predictor to plot against μ. Here ω is a transient *mapping* frequency, which can be identified as the frequency of transient *beats* superimposed on the steady-state oscillation. Light damping generates the right-hand pictures, with the vanishing of ω now predicting, not the fold at B, but the critical damping at A. Since A is very close to B, this is nevertheless a satisfactory and *safe* prediction of the instability. An experimental confirmation is shown in Figure 7.14 for the jump to resonance of an electromagnetically driven beam.

The excellent programme of experimental work in nonlinear dynamics pursued by Lawrence Virgin is recorded in his recent book, Virgin (2000).

Part II

Iterated Maps as Dynamical Systems

8

Stability and Bifurcation of Maps

In this chapter we shall commence the study of mappings in their own right as dynamical systems. The main motivation for this is our understanding of the Poincaré mapping technique for flows, which condenses the behaviour of three-dimensional trajectories to a mapping of a two-dimensional surface of section to itself. Thus the stability properties of the map reflect the stability properties of the flow. We have also seen that centre manifold techniques sometimes permit further reduction, so that we shall also be concerned with mappings of a one-dimensional phase space to itself. Furthermore, such mappings sometimes arise directly as models of dynamics evolving in discrete time steps.

After a brief historical introduction, we study stability conditions and some various possible bifurcations of one-dimensional maps, starting with the fold. For two-dimensional maps an additional possible kind of structurally stable bifurcation is described, similar in many ways to the Hopf bifurcation from a fixed point for a continuous flow. An ecological model exhibiting this bifurcation is used as an illustration.

8.1 INTRODUCTION

Historical outline

Poincaré first recognized the importance of studying the dynamical behaviour of mappings, as defined by a difference equation

$$x_{i+1} = F(x_i) \tag{8.1}$$

where x is an n-dimensional vector and F is a nonlinear transformation. The index i refers to either discrete steps in time or successive returns to a surface of section. Of course the time for a flow to return to the Poincaré section is always the same in a periodically forced oscillator, but need not be in autonomous systems, as we shall see in Chapters 11 and 12.

The earliest theories of Poincaré and Birkhoff centred mainly on topological studies of abstract classes of mappings likely to be of general relevance to differential equations. This study of diffeomorphisms has continued to the present day, and we mention Birkhoff (1913), Denjoy (1932), Arnold (1965) and Smale (1967) as important examples.

More recently, a growing number of researchers have considered the behaviour of specific maps with explicit transformations given by simple algebraic or transcendental formulae. The aim here is to choose the mapping judiciously so that its dynamical behaviour will be prototypical, and then to carry out analyses and/or direct simulations on a computer. Notable examples include the logistic map discussed in May (1976) and a two-dimensional map introduced by Hénon (1976), both of which will be studied in the next chapter.

In this chapter we shall consider general but low-dimensional maps of the form (8.1), concentrating on the determination of local forms as described for example by Guckenheimer and Holmes (1983). These local forms describe the stability transitions of periodic orbits, and lead to a systematic classification of their bifurcations.

Discrete dynamical systems and Poincaré section

There are essentially two ways in which a flow can give rise to a map: in non-autonomous periodically forced oscillations and in autonomous nonlinear systems exhibiting periodic orbits.

Consider a system

$$\ddot{x} + g(x, \dot{x}) = f(t) \tag{8.2}$$

where the forcing term is periodic with period T. The phase portrait is here three-dimensional, spanned by x, $\dot{x}$, and t, and the phase projection $(x, \dot{x})$ is two-dimensional. A Poincaré *section* is obtained simply by plotting the points $(x, \dot{x})$ in the plane projection whenever t is a multiple of the period T. The Poincaré *map* transforms the points in the section according to the behaviour of the three-dimensional flow. Another important class of problem that can usefully be studied using Poincaré maps is represented by differential equations with periodically varying coefficients, as for example

$$\ddot{x} + a_1(t)\dot{x} + a_2(t)x = 0 \tag{8.3}$$

Here again, since the flow repeats with period T, a similar Poincaré map summarizes the dynamics.

Perhaps the central way in which a flow leads to a Poincaré map is through the study of periodic orbits of autonomous systems. Let us consider a nonlinear system

$$\dot{x} = f(x) \tag{8.4}$$

which exhibits a closed orbit in its n-dimensional space, and choose a surface, of dimension $(n - 1)$, such that the flow is everywhere transverse to it. Mathematically this means that we must have $f(x) \cdot n(x) \neq 0$ where $n(x)$ is the unit normal to the section at the point x. The Poincaré map is then defined on this surface, and links a point of the flow to its first return point on the surface, as illustrated in Figure 8.1.

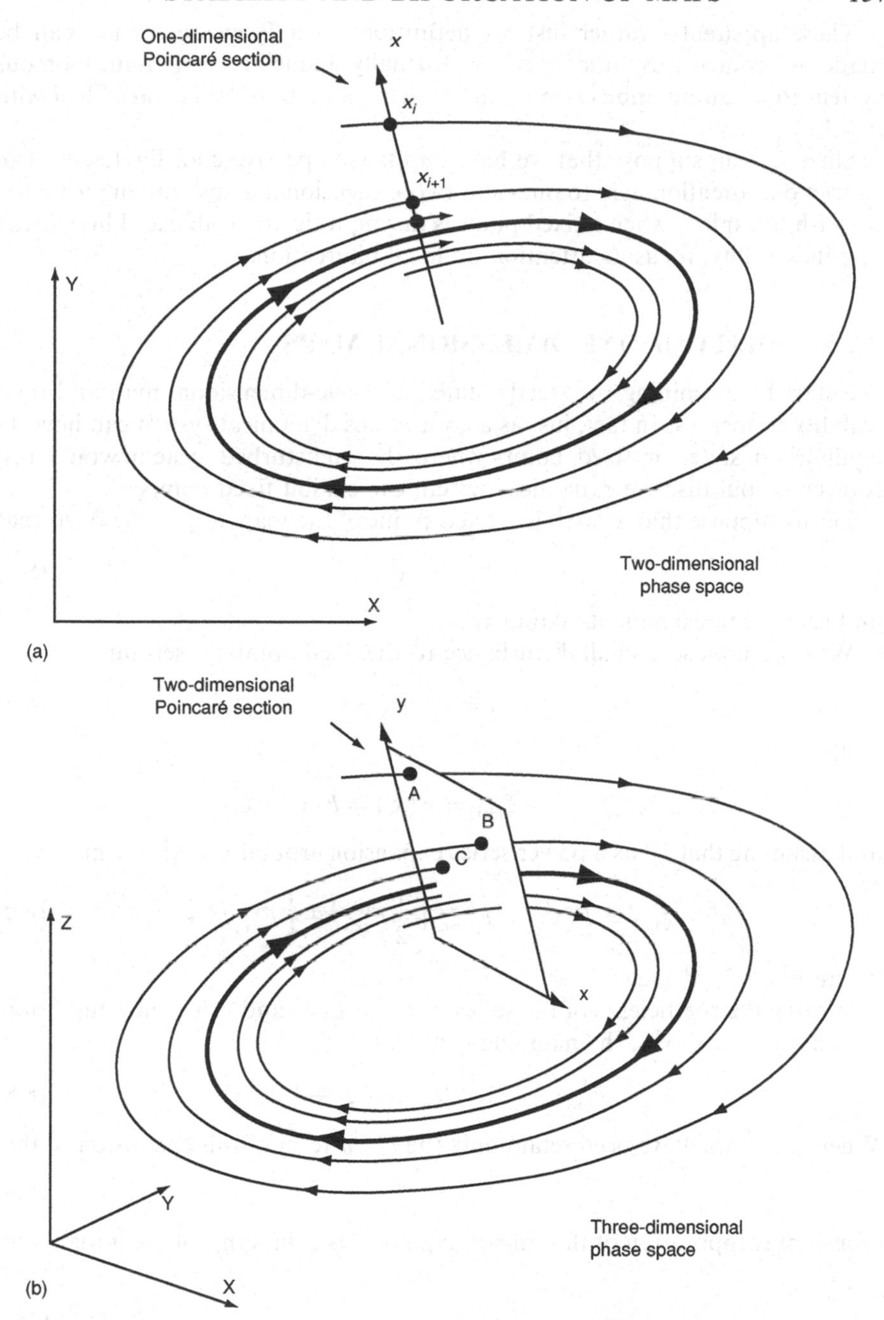

Figure 8.1 Poincaré sections: (a) one-dimensional, mapping $x_{i+1} = F(x_i)$; (b) two-dimensional, mapping $x_{i+1} = F(x_i, y_i)$, $y_{i+1} = G(x_i, y_i)$

These apparently rather distinct definitions of a Poincaré section can be made to coincide by the trick of formally reducing a non-autonomous system to an autonomous one by introducing a dummy cyclic variable θ with $\dot{\theta} = 1$.

Since we can suppose that we have suppressed passive coordinates, we can restrict our attention here to one- and two-dimensional maps, and in particular we wish to explore when a fixed point is (asymptotically) stable and how it can lose its stability, focusing attention on local bifurcations.

8.2 STABILITY OF ONE-DIMENSIONAL MAPS

We start by examining the steady states of a one-dimensional map and their stability properties. In fact, just as a continuous dynamical system can have an equilibrium state, or fixed point, where the undisturbed system would rest forever, so our discrete dynamical system can exhibit fixed points.

Let us suppose that $x = x^{\mathrm{E}}$ is a fixed point of the map $x_{i+1} = F(x_i)$, so that

$$x^{\mathrm{E}} = F(x^{\mathrm{E}}) \tag{8.5}$$

and proceed to examine its stability.

We superimpose a small disturbance to the fixed point x^{E}, setting

$$x = x^{\mathrm{E}} + \xi \tag{8.6}$$

so that

$$x_{i+1} = x^{\mathrm{E}} + \xi_{i+1} = F(x_i) = F(x^{\mathrm{E}} + \xi_i)$$

and assuming that F has a power series expansion around $x = x^{\mathrm{E}}$, we may write

$$x^{\mathrm{E}} + \xi_{i+1} = F(x^{\mathrm{E}}) + F_x^{\mathrm{E}}\xi_i + \frac{1}{2}F_{xx}^{\mathrm{E}}\xi_i^2 + \frac{1}{6}F_{xxx}^{\mathrm{E}}\xi_i^3 + \dots \tag{8.7}$$

where $F_x^{\mathrm{E}} \equiv F_x(x^{\mathrm{E}})$, etc.

Writing the coefficients of the series as $C, D, E, \dots$ and using the equilibrium condition $x^{\mathrm{E}} = F(x^{\mathrm{E}})$, the mapping becomes

$$\xi_{i+1} = C\xi_i + D\xi_i^2 + E\xi_i^3 + \dots \tag{8.8}$$

When ξ_{i+1} is small we need retain only the *linear* term of this expansion so that

$$\xi_{i+1} = C\xi_i \tag{8.9}$$

Constant reapplication of this linear map gives us ξ_i in terms of the initial point ξ_0 as

$$\xi_i = C^i\xi_0 \tag{8.10}$$

Clearly, therefore, the fixed point is linearly stable if $-1 < C < 1$.

If C lies between 0 and 1, disturbances decay monotonically, while if C is greater than 1 they grow monotonically. We shall call this monotonic growth *divergence*. If C is exactly equal to 1 we have neutral behaviour; in this critical (bifurcating) situation, however, the linear theory yields no conclusion about the stability of the original nonlinear system, and we must retain higher-order terms in the Taylor expansion (8.8).

If C lies between -1 and 0, disturbances decay in an oscillatory fashion with an alternating sign of ξ; while if C is less than -1 they grow in an oscillatory fashion. We shall call this oscillatory growth *flipping*. If C is exactly equal to -1, we have incipient flipping, and within the linear approximation we observe neutral behaviour, with $\xi_{i+2} = \xi_i$ for any starting condition. Once again a complete discussion about the stability or instability of such a critical (bifurcating) state can only be made by examining higher-order, nonlinear terms of the Taylor expansion.

In addition to these fixed points (designated by $n = 1$) characterized by

$$x_i = F(x_i) \tag{8.11}$$

we can also observe so-called *periodic orbits* of period $n = k$, defined by k points such that

$$\begin{aligned} x_{i+1} &= F(x_i) \\ x_{i+2} &= F(F(x_i)) \equiv F^2(x_i) \\ &\vdots \\ x_{i+k} &= F^k(x_i) = x_i \end{aligned} \tag{8.12}$$

it being presumed that there has been no periodicity of lower order. Notice that here F^2 is a notational shorthand, and is not the square of F. The simplest is a period 2 orbit. Writing for convenience $x_i = x, x_{i+1} = y$ and $x_{i+2} = z$, this orbit is characterized by $z = x$, it being understood that $y \neq x$. Now z is generated from x by the equation

$$z = F^2(x) \equiv T(x) \tag{8.13}$$

Clearly we can examine the stability of an $n = 2$ periodic orbit by inspecting the function T along the lines of our earlier examination of F.

The Taylor coefficients of this function can of course be obtained in terms of the Taylor coefficients of F by chain rule differentiation, as for example

$$T_x(x) = F_x(y)F_x(x) \tag{8.14}$$

8.3 BIFURCATIONS OF ONE-DIMENSIONAL MAPS

We are usually concerned with the response of a map under the operation of a control parameter P, which will appear in the function F, so that

$$y = F(x, P) \tag{8.15}$$

As the control parameter is varied, the various steady states (fixed points and periodic orbits) will trace out paths in (x, P) space, and we are interested in the bifurcations of these paths.

Consider first the $n = 1$ fixed-point solutions. These are characterized by $y = x$, so that equation (8.15) can be written as

$$x - F(x, P) = 0 \tag{8.16}$$

Now, any function of a *single* variable x can be integrated (in principle) to yield an *effective potential* V such that equation (8.16) becomes

$$V_x(x, P) = 0 \tag{8.17}$$

For incipient divergence we have $C = F_x = 1$, which means that

$$V_{xx} = 1 - C = 0 \tag{8.18}$$

This is precisely the condition for the incipient instability of the equivalent (conservative, mechanical) potential system governed by V.

For an incipient flip we have $C = F_x = -1$, so that

$$V_{xx} = 1 - C = 2$$

The equivalent potential system is therefore not now at a critical state, and there will be no singularity in the paths of the $n = 1$ fixed points. We shall see later, however, that there is indeed a bifurcation into an $n = 2$ path.

The equivalent potential system governed by V can of course be embedded in elementary catastrophe theory, so the equilibrium paths of the fixed points can be expected to undergo stability transitions at folds, cusps, and higher singularities, in the normal way (Figure 8.2). Notice, however, that of these only the fold will be structurally stable under a single control parameter.

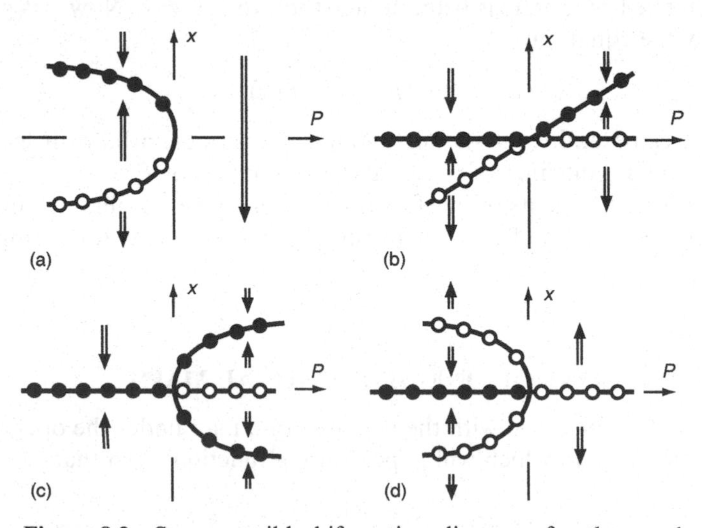

Figure 8.2 Some possible bifurcation diagrams for the $n = 1$ fixed points: (a) the fold, (b) the transcritical bifurcation, (c) the stable cusp, (d) the unstable cusp. This figure can be compared to Figures 7.6 and 7.7

To study a singularity at a value of $P = P^c$ it is now convenient to expand F as a Taylor series in both the variable x and the change in P from its critical value, $p = P - P^c$, as follows:

$$\begin{aligned} y = B + Cx + Dx^2 + Ex^3 + \ldots \\ + p(B_1 + C_1x + D_1x^2 + E_1x^3 + \ldots) \\ + p^2(B_2 + C_2x + D_2x^2 + E_2x^3 + \ldots) + \ldots \end{aligned} \tag{8.19}$$

Here, as we shall define later, x may be either an increment in the original x measured from a fixed critical point (a fold) or a sliding increment measured always from a fundamental *path*. This follows the approach of Thompson and Hunt (1973) to elastic bifurcation.

Folds and saddle-node bifurcations

To analyse a fold in the fixed points, we make use of the above Taylor expansion, measuring x from the fixed critical point itself. Since $x = p = 0$ is a fixed point, we must have $B = 0$ and we assume that we have arrived at a critical state of incipient divergence at which $C = 1$. Setting $y = x$ to locate the path of the fixed points, the y cancels with the Cx, and considering the typical case in which D and B_1 are non-zero, the local first-order solution for the path is

$$Dx^2 + B_1p = 0 \tag{8.20}$$

This is the local parabolic form of the fold, an example of which is shown in Figure 8.3. In this particular case the upper branch is unstable while the lower branch is stable. Two sequences of iterations are illustrated. In the left-hand

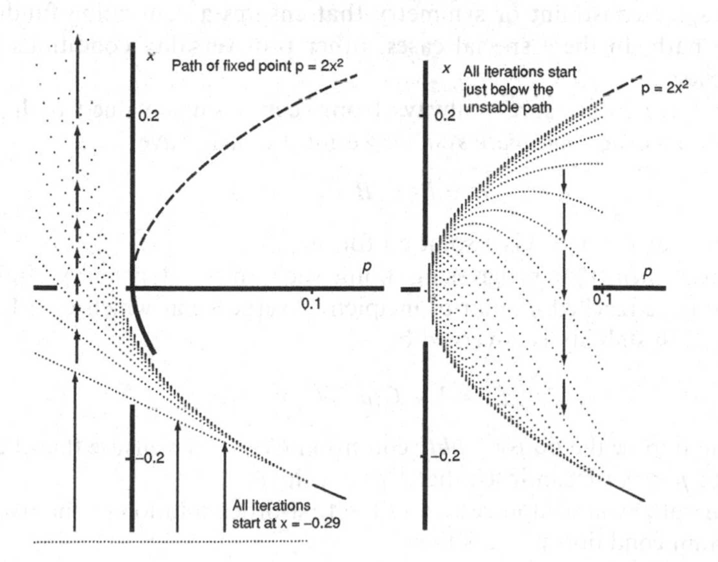

Figure 8.3 The saddle-node bifurcation as a fold ($x_{i+1} = x_i - p + 2x_i^2$) in the $n = 1$ fixed points. This illustrates the *cyclic fold* of the Appendix

diagram all the iterations, at a spectrum of p values, start at $x = -0.29$: for positive p these head monotonically towards the stable branch of the parabola, while for negative p they go to infinity, slowing perceptibly as they pass the control axis. In the right-hand diagram all the iterations start just below the unstable branch, and proceed monotonically towards the stable states. Starts just above the unstable branch would go to plus infinity, so we see that the unstable path is the separatrix between the catchment regions of the attracting branch and the attractor at infinity.

To study the stability transition we must now imagine a stability discussion being made at various points along the path, the local value of the stability coefficient corresponding to C being given by the derivative F_x, which can be expanded as

$$F_x = C + 2Dx + 3Ex^2 + \ldots + C_1 p + \ldots$$

to give the first-order solution

$$F_x = 1 + 2Dx \tag{8.21}$$

This changes from being less than 1 (stable) to being greater than 1 (unstable) as x changes sign along the path.

We note that (8.20) could also be deduced from the normal forms theorem, as given in Guckenheimer and Holmes (1983); in this case the transversality conditions of the Sotomayor theorem are entirely equivalent to our requirement that D and B_1 are non-zero. These conditions ensure that bifurcation takes place in the most typical way.

Other interesting but less typical bifurcations involving the intersection of two distinct paths arise under the operation of a single control parameter if there is some constraint or symmetry that ensures a continuing fundamental primary path. In these special cases, other transversality conditions become appropriate.

If we agree to measure x always from such a single-valued path using if necessary a *sliding coordinate system*, we must clearly have

$$B = B_1 = B_2 = \ldots = 0$$

to ensure that $y = x = 0$ is a solution for all p.

Suppose, then, that progressing along such an $n = 1$ primary equilibrium path, we have reached a point of incipient divergence at which $C = 1$. On the primary path stability is governed by

$$F_x = 1 + C_1 p + C_2 p^2 + \ldots \tag{8.22}$$

so we can impose the *transversality* condition $C_1 \neq 0$. To ensure that the path is stable for $p < 0$, we can insist that C_1 is positive.

In general D will be non-zero, and the first-order solution of the fixed-point equilibrium condition $y = x$ is then

$$Dx^2 + C_1 xp = 0 \tag{8.23}$$

We thus have the primary path $x = 0$ and the local equation of the asymmetric branch given by

$$Dx + C_1 p = 0 \tag{8.24}$$

This corresponds to the asymmetric or transcritical $n = 1$ bifurcation.

The first-order form of F_x is now

$$F_x = 1 + 2Dx + C_1 p \tag{8.25}$$

and evaluating on the *secondary* path we have

$$F_x = 1 - C_1 p \tag{8.26}$$

This confirms the familiar exchange of stabilities illustrated. A numerical example of an asymmetric point of bifurcation is shown in Figure 8.4 for the displayed equation. Iterations are shown at various fixed values of the control p, and the monotonic convergence towards the stable paths is easily observed.

If we retain also a cubic term of the Taylor expansion we have the example of Figure 8.5. This higher term gives rise to a stabilization of the secondary subcritical path at the fold C. Iterates at fixed values of p are again shown, including two non-monotonic approaches denoted by the letter D, but the *local* picture near the origin is unchanged from Figure 8.4 as a result of this higher-order contamination.

Bifurcations of distinct paths are frequently generated by symmetry, which will often ensure additionally that $D = 0$. The first-order solution of the equilibrium condition is then

$$Ex^3 + C_1 xp = 0$$

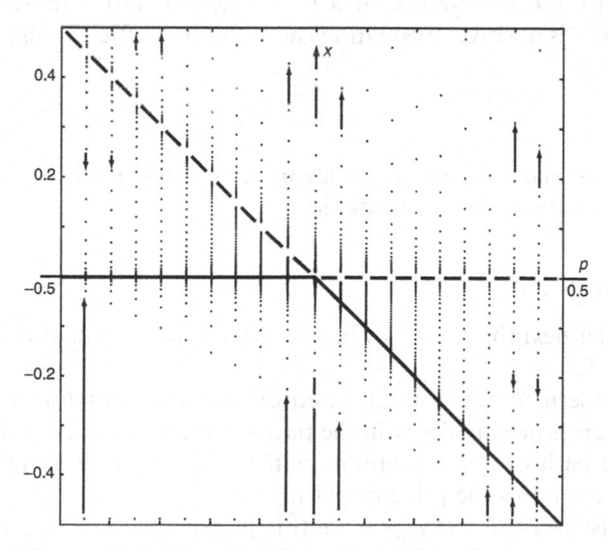

Figure 8.4 The saddle-node bifurcation as a transcritical bifurcation: $y = x(1 + p) + x^2$. For $n = 1$, $y = x$; $x = 0$ or $p = -x$

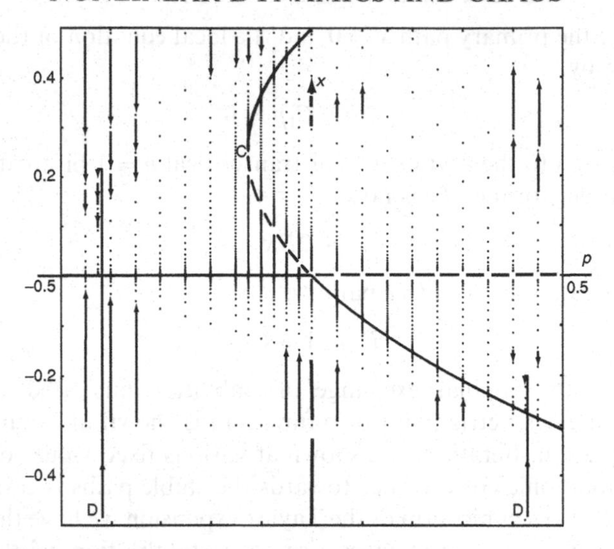

Figure 8.5 Higher-order contamination of a transcritical bifurcation: $y = x(1+p) + x^2 - 2x^3$. For $n = 1$, $y = x$; $x = 0$ or $p = -x + 2x^2$. Two non-monotonic approaches to a stable $n = 1$ solution are labelled D

giving the trivial solution plus the parabolic secondary path

$$Ex^2 + C_1 p = 0 \tag{8.27}$$

With C_1 positive, we thus have a supercritical pitchfork (stable-symmetric point of bifurcation) if E is negative, or a subcritical pitchfork (unstable point of bifurcation) if E is positive. Evaluating the stability coefficient on the secondary path we have

$$F_x = 1 - 2C_1 p \tag{8.28}$$

which confirms that a secondary solution for positive p will be stable while a solution for negative p will be unstable.

The flip bifurcation

Let us consider next the onset of what we have called a flip, at a fixed point at which $C = -1$.

As we have seen, at such a point the equivalent $n = 1$ potential system is non-critical, so there is no singularity in the paths of the $n = 1$ fixed points. A locally single-valued path of $n = 1$ solutions will therefore pass through the critical point, and we call this the primary path.

We agree to measure x always from this primary path by means of a sliding coordinate system, so that again we have

$$B = B_1 = B_2 = \ldots = 0$$

and if we want the most transversal, that is, most typical loss of stability under increasing p, we must have C_1 non-zero and negative.

Now the oscillatory response of the linearization suggests that an $n = 2$ path will bifurcate from the primary $n = 1$ path, so we must focus attention on the second iteration

$$z = F^2(x) = T(x)$$

At the point (0, 0)

$$T_x = (F_x)^2$$

so at our incipient *flip* of the map F where $F_x = -1$ we have incipient *divergence* of the map T with $T_x = 1$.

Now the divergence of the second iterate and the form of the associated $n = 2$ paths is controlled by an equivalent $n = 2$ potential system based on T, just as for the $n = 1$ case based on F, so the forms of elementary catastrophe theory must be expected. Notice however that since $y = x$ implies $z = x$, the $n = 2$ solutions will include the primary $n = 1$ path. There will also be constraints on the form of T arising from its relationship to F, and these we must now explore.

Writing out the transformation F^2 in full, we have the leading terms

$$z = C(Cx + Dx^2 + Ex^3 + C_1xp) + D(Cx + Dx^2 + C_1xp)^2 + E(Cx)^3 + C_1p(Cx)$$

For an $n = 2$ solution, characterized by $z = x$, the z cancels with the C^2x so we have the first-order $n = 2$ solution

$$Ex^3 + D^2x^3 + C_1px = 0$$

giving $x = 0$ and

$$C_1p + x^2(E + D^2) = 0 \tag{8.29}$$

We observe that the assumed existence of a continuous single-valued $n = 1$ path, which inevitably reappears analytically as an $n = 2$ path, eliminated the possibility there of an $n = 2$ fold. Meanwhile the cancelling of the x^2 terms eliminates the possibility of a transcritical bifurcation. So we have an $n = 2$ pitchfork bifurcation: with C_1 negative we have supercritical flip bifurcation if $(E + D^2)$ is positive and subcritical flip bifurcation if $(E + D^2)$ is negative.

A numerical example of a supercritical flip bifurcation is shown in Figure 8.6 where our first-order perturbation solution of (8.29) is in fact the exact solution of the $n = 2$ path. This is due to the complete symmetry and the simplicity of the map. Convergence towards the stable solution is seen, and all iterations generate a change in sign of x as indicated on the far left.

A second example of a supercritical flip bifurcation without any overall symmetry is shown in Figure 8.7. The first-order perturbation solution is now only a local approximation to the real secondary $n = 2$ path, which has no symmetry about the control axis. Once again oscillatory convergence is observed, and it is seen that for large values of p the real secondary $n = 2$ path becomes in some way unstable with iterations going to infinity.

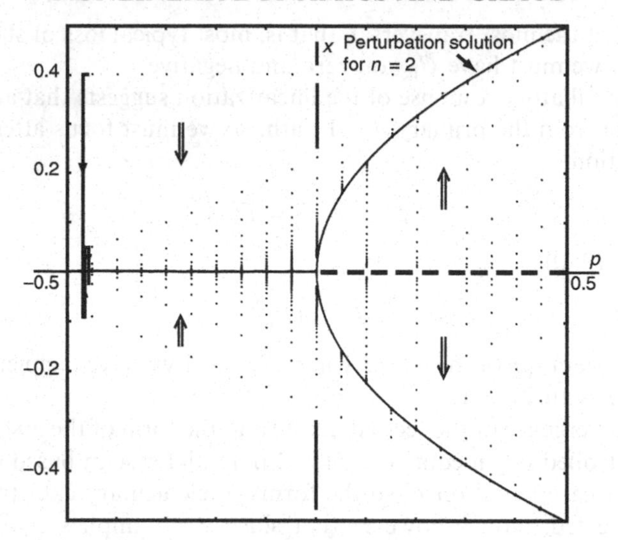

Figure 8.6 Symmetric supercritical flip bifurcation: $y = -x(1+p) + 2x^3$; first-order perturbation solution for $n = 2$ is $p = 2x^2$; all runs start at $x = 0.4$ or $x = 0.01$; all iterations alternate in x, $+-+-+-$, etc.

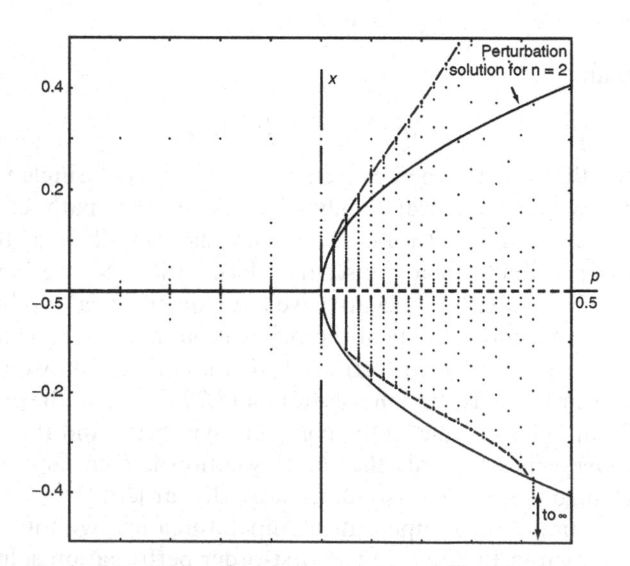

Figure 8.7 Non-symmetric supercritical flip bifurcation: $y = -x(1+p) + 2x^2 - x^3$; perturbation solution in $n = 2$ is $p = (-1 + 2^2)x^2 = 3x^2$; all runs start at $x = 0.3$ or $x = 0.01$; all iterations alternate in x, $+-+-+-$, etc.

An unstable flip bifurcation is shown in Figure 8.8 and the oscillatory motion is seen for all the indicated starting points.

The mechanism of flipping beyond a supercritical flip bifurcation is fully illustrated in Figure 8.9, where the 'bouncing' between $F(x)$ and the 45° line now generates oscillations around the origin. With F assumed to be odd, the stable secondary $n = 2$ solutions are on the $-45°$ line, and the map is seen to approach the associated square. We notice in particular the oscillatory motion away from the unstable origin where $F_x < -1$.

The stability of the supercritical path and the instability of the subcritical path are guaranteed by our effective potential ideas, and can be checked by inspecting the sign of $T_x - 1$ on the secondary $n = 2$ path. Notice carefully that writing here

$$T_x = F_x(y)F_x(x)$$

the two F_x derivatives are evaluated at the *two* distinct mapping points of the $n = 2$ orbit.

We should note here that the introduction of sliding coordinates has greatly simplified our transversality conditions (cf. Guckenheimer and Holmes 1983, p. 158), while the constant necessary to detect stability of the secondary path remains unchanged.

The bifurcation behaviour has similarities with the $n = 1$ pitchfork and in (x, p) space the diagram is superficially similar, but a glance at Figure 8.9 and 8.10 will clearly show the difference between divergence and flip instability. To visualize the iterates of the map we have 'bounced' the solution between the 45°

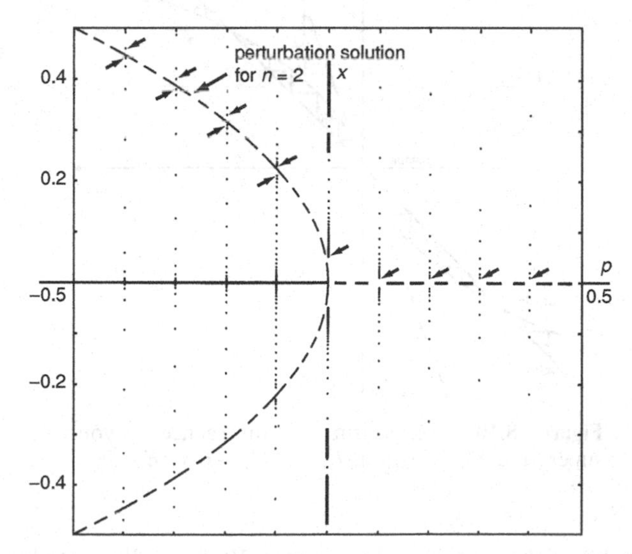

Figure 8.8 Subcritical flip bifurcation: $y = -x(1+p) - 2x^3$; perturbation solution in $n = 2$ is $p = -2x^2$; all starts are indicated by an arrow (↙); all iterations alternate in x, $+-+-+-$, etc.

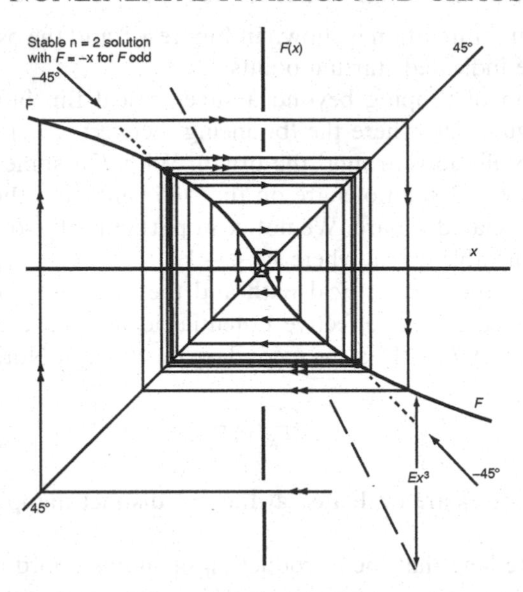

Figure 8.9 Mechanism of flipping beyond a supercritical bifurcation with $D = 0$: $n = 2$ solution with $F = -x$ for F odd, $C_1 < 0$ so $E > 0$

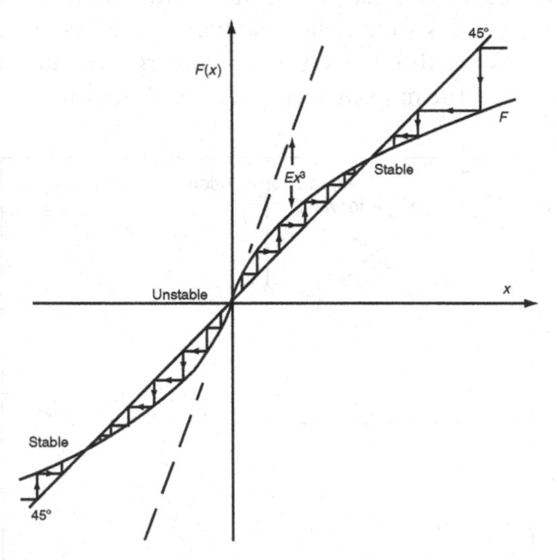

Figure 8.10 Mechanism of divergence beyond a supercritical bifurcation: $D = 0$, $C_1 > 0$ so $E < 0$

line and the plot of the map $F(x)$. In Figure 8.10 this is illustrated for a function $F(x)$ corresponding to the condition just beyond a stable-symmetric point of bifurcation. The map escapes from the unstable origin, where $F_x > 1$, towards the stable secondary solution.

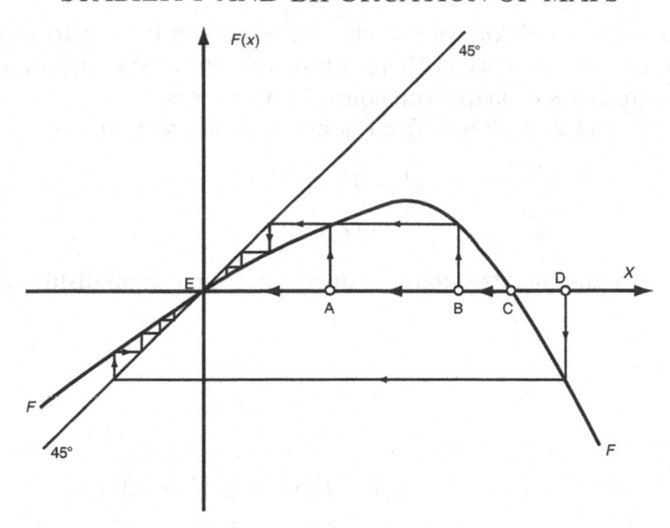

Figure 8.11 Complications that can arise when $F(x)$ is not invertible

Complications that arise when $F(x)$ passes over a maximum so that its inverse is not single-valued are shown in Figure 8.11. The map is now non-invertible and we see how iterates from A and B can merge to become indistinguishable. A sequence starting at D immediately changes the sign of x and then approaches the stable origin from below, while a sequence starting at C goes straight to the origin in a single step.

Thinking of these maps as generated by the intersection of a continuous flow with a Poincaré section, it is clear that in a two-dimensional planar phase space (see the top diagram of Figure 8.1) a limit cycle cannot be associated with oscillatory intersections, so that bifurcation to a periodic orbit of period 2 is not possible.

Meanwhile two-dimensional flow on a Möbius strip must always give oscillatory maps, and can never bifurcate into another limit cycle of the same period. In phase spaces of three or more dimensions both divergence and flip are possible, flip being here associated with a Möbius-like centre manifold.

All these one-dimensional mapping phenomena are included in Figure 7.3.

8.4 STABILITY OF TWO-DIMENSIONAL MAPS

We turn now to two-dimensional invertible maps

$$\begin{aligned} x_{i+1} &= F(x_i, y_i) \\ y_{i+1} &= G(x_i, y_i) \end{aligned} \tag{8.30}$$

These maps can be regarded as Poincaré maps of three-dimensional flows (see Figure 8.1) or as extensions of non-invertible one-dimensional maps (Ott 1981).

In both cases it is well known that chaotic behaviour is likely to appear, as we shall see later, but first we wish to study the linear stability of (8.30) and possible bifurcations of maps with control parameters.

Let $x_i = x^E$ and $y_i = y^E$ be a fixed point of (8.30), so that

$$\begin{aligned} x^E &= F(x^E, y^E) \\ y^E &= G(x^E, y^E) \end{aligned} \tag{8.31}$$

We proceed to examine its stability. Superimposing a small disturbance we set

$$\begin{aligned} x_i &= x^E + \xi_i \\ y_i &= y^E + \eta_i \end{aligned}$$

so that

$$\begin{aligned} x_{i+1} &= x^E + \xi_{i+1} = F(x^E + \xi_i, y^E + \eta_i) \\ y_{i+1} &= y^E + \eta_{i+1} = G(x^E + \xi_i, y^E + \eta_i) \end{aligned}$$

Expanding the functions F and G in Taylor series at the fixed point and utilizing (8.31), we see that

$$\begin{aligned} \xi_{i+1} &= F_x\xi_i + F_y\eta_i + \tfrac{1}{2}(F_{xx}\xi_i^2 + 2F_{xy}\xi_i\eta_i + F_{yy}\eta_i^2) + \ldots \\ \eta_{i+1} &= G_x\xi_i + G_y\eta_i + \tfrac{1}{2}(G_{xx}\xi_i^2 + 2G_{xy}\xi_i\eta_i + G_{yy}\eta_i^2) + \ldots \end{aligned} \tag{8.32}$$

where all the derivatives of F and G are evaluated at the fixed point.

For small disturbances we need retain only the linear part of the expansion (8.32), and abbreviating the first derivatives as a, b, c, d we can obtain the variational equation

$$\begin{aligned} \xi_{i+1} &= a\xi_i + b\eta_i \\ \eta_{i+1} &= c\xi_i + d\eta_i \end{aligned} \tag{8.33}$$

or, written in matrix form,

$$\boldsymbol{\zeta}_{i+1} = \mathbf{H}\boldsymbol{\zeta}_i$$

Let λ_1 and λ_2 be the eigenvalues of the matrix $\mathbf{H}$, and neglect the rather exceptional case in which $\lambda_1 = \lambda_2$. If λ_1 and λ_2 are *real* eigenvalues it is always possible (Hirsch and Smale 1974) to find an invertible transformation of coordinates, so that system (8.33) becomes (compare this with flow analysis, Figure 3.5)

$$\begin{aligned} u_{i+1} &= \lambda_1 u_i \\ v_{i+1} &= \lambda_2 v_i \end{aligned} \tag{8.34}$$

In this way the problem is decoupled into two one-dimensional maps, and the stability question is immediately solved; the system is (asymptotically) stable if $-1 < \lambda_{1,2} < 1$, but unstable if either λ_1 or λ_2 are greater than 1 in absolute value. If one of the eigenvalues, say λ_1, has modulus equal to 1 and λ_2 is less than 1 in absolute value, then the linear approximation is not sufficient to establish the stability or instability of the fixed point.

We turn now to the case in which λ_1 and λ_2 are *complex conjugate* eigenvalues, say

$$\lambda_{1,2} = \alpha \pm \mathrm{i}\beta$$

It is again possible (Hirsch and Smale 1974) to find an invertible change of variable, so that equation (8.33) becomes

$$\begin{aligned} u_{i+1} &= \alpha u_i - \beta v_i \\ v_{i+1} &= \beta u_i + \alpha v_i \end{aligned} \tag{8.35}$$

It is convenient to introduce polar coordinates r and θ so that

$$\begin{aligned} u_i &= r_i \cos\theta_i \\ v_i &= r_i \sin\theta_i \end{aligned} \tag{8.36}$$

and to express the eigenvalues λ_1 and λ_2 in exponential form

$$\begin{aligned} \lambda_1 &= \alpha + \mathrm{i}\beta = \rho \mathrm{e}^{\mathrm{i}\phi} \\ \lambda_2 &= \alpha - \mathrm{i}\beta = \rho \mathrm{e}^{-\mathrm{i}\phi} \end{aligned} \tag{8.37}$$

It is now possible to solve equation (8.35) with respect to the initial point $(u_0, v_0) = (r_0, \theta_0)$. In fact, writing (8.35) when $i = 0$ gives

$$\begin{aligned} r_1 \cos\theta_1 &= \rho r_0 \cos(\phi + \theta_0) \\ r_1 \sin\theta_1 &= \rho r_0 \sin(\phi + \theta_0) \end{aligned}$$

and when $i = 1$

$$\begin{aligned} r_2 \cos\theta_2 &= \rho^2 r_0 \cos(2\phi + \theta_0) \\ r_2 \sin\theta_2 &= \rho^2 r_0 \sin(2\phi + \theta_0) \end{aligned}$$

We see that we have, in general,

$$r_k = \rho^k r_0 \tag{8.38}$$

$$\theta_k = \theta_0 + k\phi \tag{8.39}$$

From (8.39) we can extract k, giving

$$r_k = r_0 \rho^{(\theta_k - \theta_0)/\phi} \tag{8.40}$$

We can now immediately deduce the stability of a fixed point. If $\rho < 1$ the trajectory spirals inwards and the point is stable, while if $\rho > 1$ the trajectory spirals outwards and the point is unstable. If $\rho = 1$ the equilibrium point is called a centre; the stability can be described as neutral, and the linear approximation is no longer sufficient to establish stability.

We see that the stability criterion for two-dimensional maps is best discussed in the complex plane: if the eigenvalues are both inside the unit circle the system is (asymptotically) stable; if at least one of the eigenvalues is outside the circle the system is unstable. The stability boundary is the unit circle itself.

If the eigenvalues are real there are only two points at which they can cross the stability boundary, +1 and −1. When one of the eigenvalues, say λ_1, is equal to 1 and $|\lambda_2| < 1$ the system is in a state of incipient divergence; when λ_1 is equal to −1 and $|\lambda_2| < 1$ the system is in a state of incipient flip. These two kinds of instability are essentially one-dimensional, involving only one eigenvalue.

If the eigenvalues are complex conjugate, they can cross the unit circle at an angle $\phi \neq 0, \pi$ and this is the so-called Neimark instability. In Figure 8.12 the stability transitions for equilibrium states of flows and periodic orbits or cycles (studied via their Poincaré section) are summarized in the complex plane. We see that the stability transition with $\lambda = -1$ does not have an analogue in flows, and we shall see that Neimark instability is more complex than the analogous Hopf bifurcation for equilibria. But before turning to the Neimark instability, we wish to discuss stability in terms of the invariants of the matrix **H**. The eigenvalues of this matrix are given by

$$\lambda^2 - (a+d)\lambda + (ad - bc) = 0$$

and writing

$$a + d = \text{tr}\,\mathbf{H} = \mathcal{T}$$

$$ad - bc = \det\mathbf{H} = \mathcal{D}$$

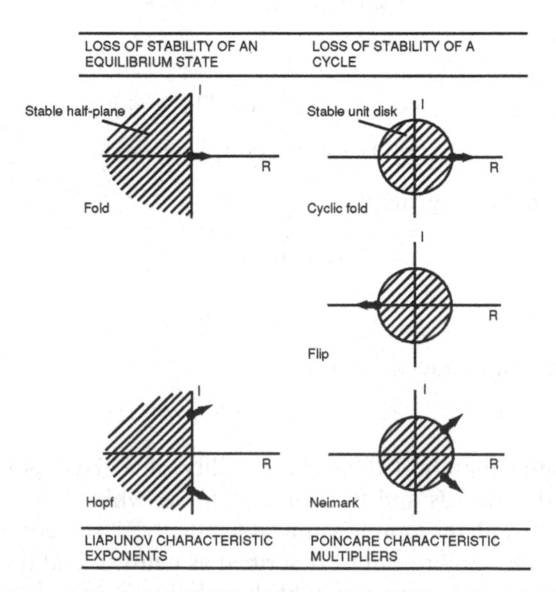

Figure 8.12 Stability transitions in the complex plane for flows and maps: the Neimark bifurcation is often called a secondary Hopf bifurcation

we can express the eigenvalues in term of these invariants, as follows

$$\lambda_{1,2} = \tfrac{1}{2}[\mathcal{T} \pm \sqrt{(\mathcal{T}^2 - 4\mathcal{D})}]$$

The stability criteria in the $(\mathcal{T}, \mathcal{D})$ plane are sketched in Figure 8.13. Divergence occurs on the straight line $\mathcal{T} - \mathcal{D} = 1$, flip on the straight line $\mathcal{T} + \mathcal{D} = -1$, and flutter (or Neimark) bifurcation on the line $\mathcal{D} = 1$. The stability boundary is the triangle LMN, the points inside this triangle being representative of (asymptotically) stable systems.

In Figures 8.14 to 8.17 various routes through the stability boundary are summarized, the points at which they cross the triangle LMN being indicated in Figure 8.13. Figure 8.14 shows the behaviour of the system at divergence, and in the top three diagrams the non-critical eigenvalue is less than zero, while in the bottom three diagrams it is greater than zero. All the diagrams show very clearly that the centre manifold is one-dimensional, roughly on the 45° line, and we could indeed focus our attention on the behaviour of the system on it. In the diagrams on the left the critical eigenvalue is slightly less than 1; hence the equilibrium point at the origin is stable, and all the starting points are first attracted by the centre manifold, and then converge to the fixed point. In the diagrams on the right the critical eigenvalue is greater than 1; hence the origin is unstable, and the iterates go first towards the centre manifold, and then diverge to infinity along it. In the middle diagrams, the neutral case is shown, and we can observe that this represents a structurally unstable portrait. In the top diagrams we see that every starting point converges in an oscillatory fashion to the centre manifold, while in the bottom diagrams the system converges monotonically to it.

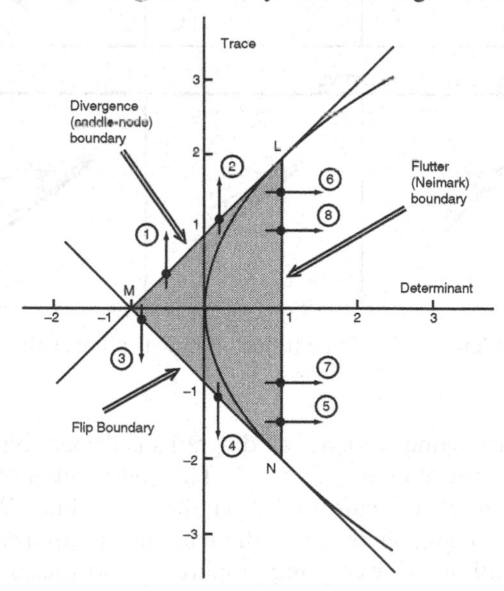

Figure 8.13 Stability criteria in the trace–determinant plane. Eigenvalues are complex to the right of the parabola

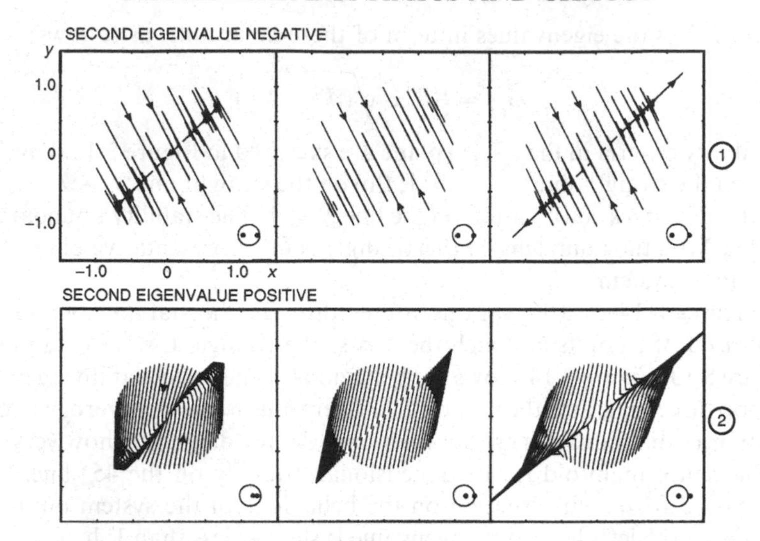

Figure 8.14 Some trajectories near a linear divergence

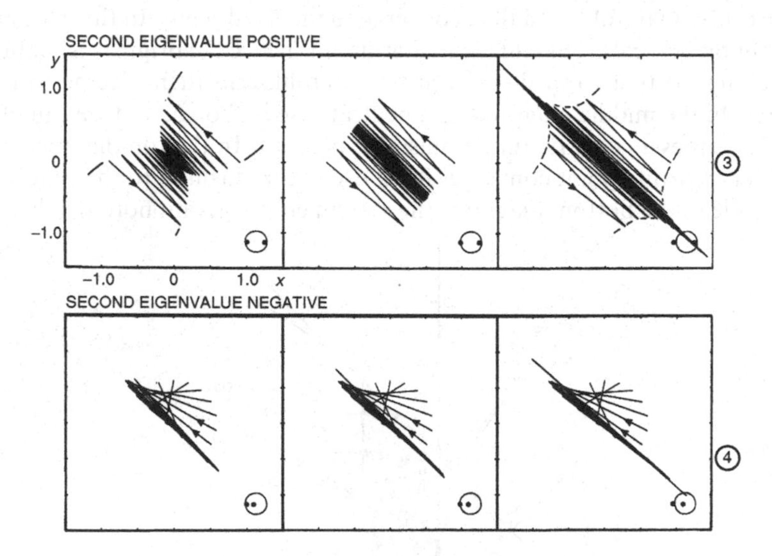

Figure 8.15 Some trajectories near a linear flip

In Figure 8.15 the same sequence of diagrams is drawn, but now the critical eigenvalue crosses the unit circle at -1. The centre manifold is again one-dimensional, but now it seems to lie on the $-45°$ line. When the critical eigenvalue is slightly greater than -1 the map slowly approaches the (stable) origin, with oscillations of decreasing amplitude, and the rate of convergence decreases as the eigenvalue tends to -1. At the critical value of -1 the map settles down on the centre manifold but continues to oscillate in a neutral fashion. If the critical eigenvalue decreases past -1 the system first converges to the centre manifold but oscillations on this manifold have increasing amplitudes that finally go to infinity.

The behaviour of the system near the flutter stability boundary is more complicated, because the centre manifold is two-dimensional, and the eigenvalues are not constrained to cross the unit circle at a particular point. The top row of Figure 8.16 shows a route through point 5 of Figure 8.13. This point is near to the flip stability boundary, which explains the oscillatory behaviour of the map. The system spirals and oscillates at the same time: in the diagram on the left the successive iterations spiral inwards towards the (stable) fixed point, in the right diagram they spiral away to inifinity, while in the middle diagram the neutral system shows an infinite oscillation of constant amplitude. The bottoms of Figure 8.16 relates to point 6 in Figure 8.13, near the divergence boundary, illustrating the simple spiralling structure of the map. These diagrams of Figure 8.16 were all traced for a *single* starting point, emphasizing the difference between the flutter instability and the flip or divergence.

The behaviour of the map at the stability boundary can change if the eigenvalues cross the unit circle at rational points where the eigenvalues are roots of unity and the response turns out to be periodic. Two cases are shown in Figure 8.17, together with their stable and unstable counterparts. In the top diagrams the eigenvalues are assumed to cross the unit circle at an angle of $\pm 120°$, so that $\lambda^3 = 1$ and the system has period 3. The name *strong resonance* refers to an associated nonlinear phenomenon. The stable and unstable companion pictures show a triangular-shaped function spiralling inwards and outwards respectively. The lower three diagrams relate to the eigenvalues crossing the stability boundary at an angle of $\pm 60°$, so that the system is periodic of period 6.

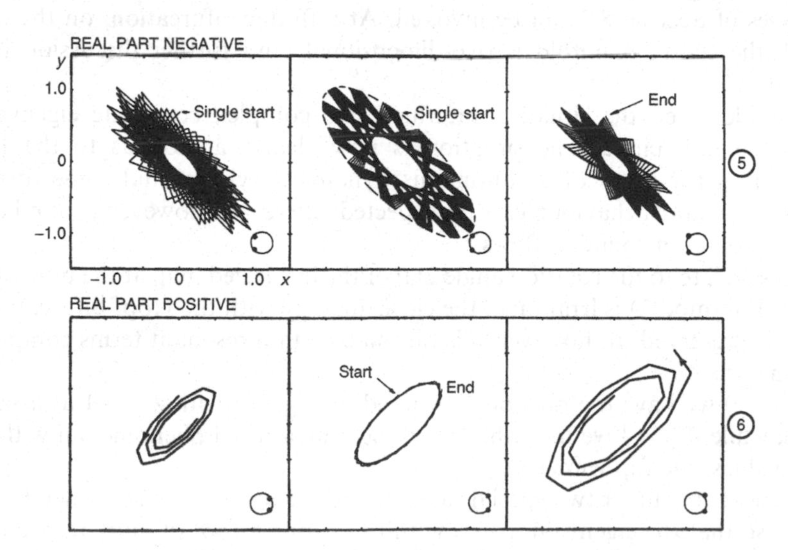

Figure 8.16 Some trajectories near a linear flutter

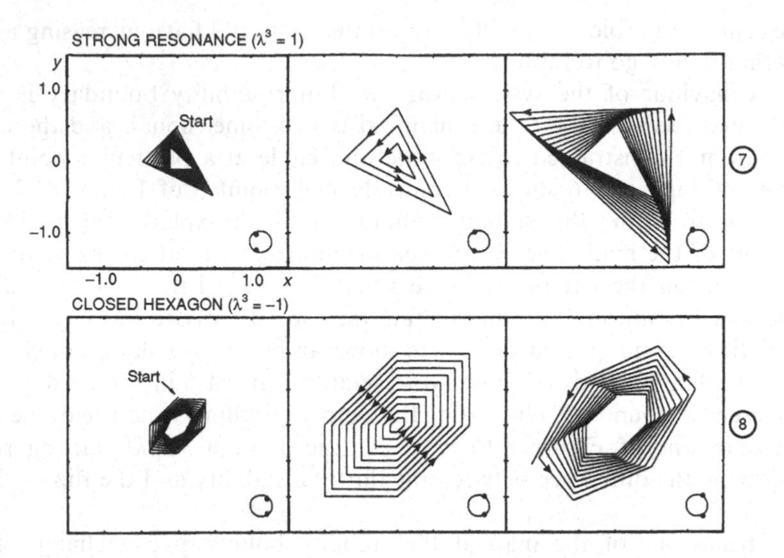

Figure 8.17 Two closed orbits in a linear flutter

8.5 BIFURCATIONS OF TWO-DIMENSIONAL MAPS

We have seen in the previous section the various ways in which a linear two-dimensional map can lose its stability. The next step is to consider the bifurcational response of parameterized maps in the presence of appropriate nonlinearities.

As we have clearly observed, the divergence and flip bifurcations are in essence one-dimensional phenomena, and can therefore be adequately studied by restricting attention to the one-dimensional centre manifold where the analyses of Section 8.3 can be invoked. At a flutter bifurcation, on the other hand, the centre manifold is two-dimensional, and further discussion is required.

Consider then the situation in which two complex conjugate eigenvalues cross the unit circle. The situation is now clearly analogous to the Hopf bifurcation for flows of a continuous dynamical system, and some form of rotating orbital behaviour can be expected; there are however a number of special cases not found in flows.

These relate to the rotation angle $2\pi\theta$ of the linearized map at the bifurcation point. If θ (mod 1) is irrational, the close analogy with the Hopf bifurcation of flows is preserved. If, however, θ is rational, certain resonant terms complicate the analysis.

If $\theta = 0$ we have the previously studied divergence with $\lambda = +1$ at a saddle node, while if $\theta = \frac{1}{2}$ we have the other one-dimensional phenomenon with real eigenvalues, the flip with $\lambda = -1$.

We next encounter two special cases, termed *strong resonances*, when $\theta = \pm\frac{1}{3}$ or $\pm\frac{1}{4}$ so that the eigenvalues are the third or fourth root of unity respectively. Here bifurcations to periodic orbits of $n = 3$ and $n = 4$ respectively are possible.

These resonance phenomena have been studied mainly for periodic orbits of a continuous flow using a Poincaré section; here we can have T-periodic orbits bifurcating into $3T$- or $4T$-periodic orbits. The main results were obtained by Iooss and Joseph (1977) and Chenciner and Iooss (1979); they are summarized for a rather restricted case in the book of Iooss and Joseph (1980). The bifurcating structures have also been analysed by Arnold (1977) and Takens (1974).

Away from these two special, and non-typical, strong resonances, and in the presence of suitable nonlinearities, a linear instability associated with complex eigenvalues will generate a bifurcating solution lying on a simple invariant closed curve, akin to the limit cycle of a Hopf flow bifurcation. In the familiar way, this bifurcating solution can be either stable and supercritical or unstable and subcritical. If our two-dimensional map be considered a Poincaré section of a three-dimensional flow, this bifurcation corresponds to the generation of an invariant torus from the instability of a periodic motion. In view of Neimark's pioneering work of 1964 on this phenomenon, we shall follow Abraham by calling this 'secondary Hopf bifurcation' a Neimark bifurcation.

A nice example of the Neimark bifurcation arising from problems encountered in biology and ecology is to be found in the response of the 'delayed logistic map'

$$x_{i+1} = \mu x_i(1 - x_{i-1})$$

As with differential equations, this single second-order equation can be rewritten as a pair of first-order equations by setting $x_{i+1} = y_i$ so that we have

$$\begin{aligned} x_{i+1} &= y_i \\ y_{i+1} &= \mu y_i(1 - x_i) \end{aligned}$$

This two-dimensional map, analysed in Guckenheimer and Holmes (1983), has a non-trivial path of fixed points given by

$$x = y = (\mu - 1)/\mu$$

which loses its stability at a supercritical Neimark bifurcation at $\mu = 2$ where the eigenvalues are the sixth roots of unity.

Some illustrated iterations for this are shown in Figure 8.18 for the values of μ and the bracketed starting values of x and y indicated. In the first diagram for $\mu = 1.2$ we see a node-like convergence to the primary fixed point, while in the second diagram for $\mu = 1.7$ the convergence has a spiralling character. At $\mu = 1.9$ the rate of convergence is noticeably slower, and in the fourth diagram at $\mu = 2.1$ we see the system moving outwards from near the unstable fixed point towards the attracting invariant closed curve. In the fifth diagram a second start at $\mu = 2.1$ shows the approach to the stable supercritical attracting curve from the outside. The last diagram shows the system spiralling outwards towards a larger invariant curve at a higher value of the control parameter.

The Neimark bifurcation from a cycle is included in our earlier summary diagram of Figure 7.3.

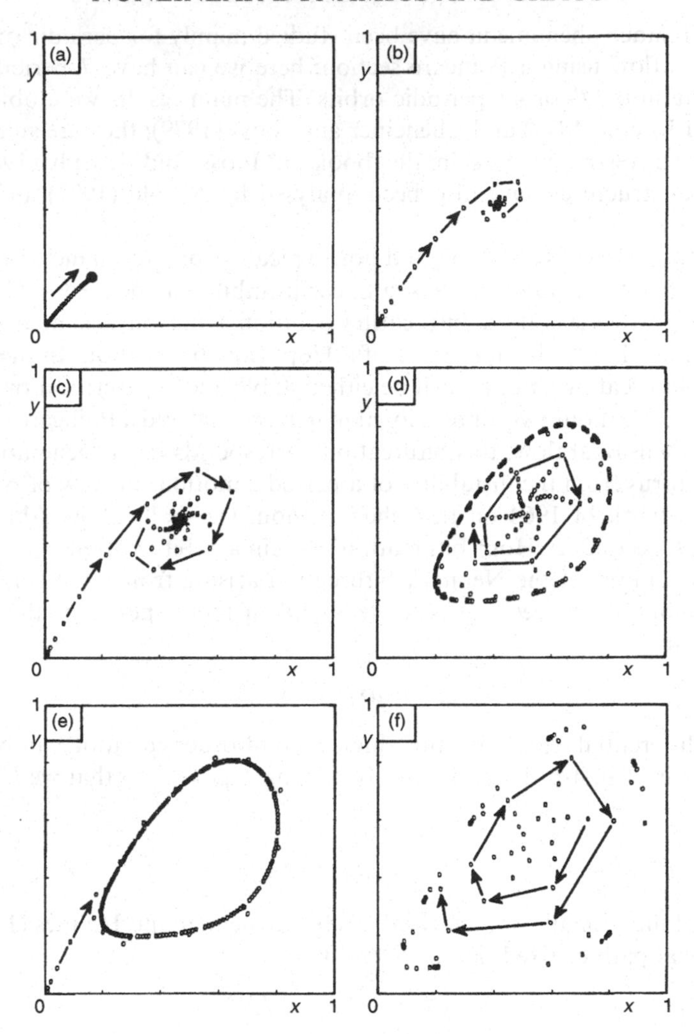

Figure 8.18 Trajectories near the Neimark bifurcation for the delayed logistic map: (a) $\mu = 1.2$, (0.01, 0.01); (b) $\mu = 1.7$, (0.01, 0.01); (c) $\mu = 1.9$, (0.01, 0.01); (d) $\mu = 2.1$, (0.5, 0.5); (e) $\mu = 2.1$, (0.01, 0.01); (f) $\mu = 2.2$, (0.5, 0.5)

8.6 BASIN CHANGES AT LOCAL BIFURCATIONS OF LIMIT CYCLES

So far in this chapter we have concentrated on how limit cycle bifurcations change the attractors in the phase portrait. Stable limit cycles attract nearby trajectories, the transients. Thus a full discussion of basins, filled with transients, is essential to complete the phase portraits of limit cycle bifurcations.

The cyclic fold portrayed in Figure 8.3 is closely analogous to the folded equilibrium path of Chapter 7. If we presented the cyclic fold as a bifurcation of a two-dimensional Poincaré map, the appearance would be entirely similar to

Figure 7.4, except that a trajectory of the two-dimensional map would jump on a sequence of discrete points along any curve in Figure 7.4. (For a picture of a nodal cycle about to collide with a saddle cycle sitting on a fractal basin boundary, see Figure 16.38; see also Figure 16.39.) Thus the inset (stable manifold) of the unstable saddle limit cycle would be two curves, each filled with an infinity of trajectories which settle onto the saddle cycle. No single trajectory asymptotic to the saddle separates nearby trajectories; only an ensemble of trajectories all asymptotic to the saddle can fill out a curve, which separates the flow into two basins.

As the control parameter μ or Λ in Figure 7.4(b), approaches its critical value μ^C in the cyclic fold, the stable limit cycle (attractor) and it basin boundary steadily come closer together. Just as in the equilibrium fold, the attractor approaches the boundary at the saddle. The structurally unstable phase portrait at the cyclic fold, where $\mu = \mu^C$, contains a residual basin of infinite area for a totally dissipative system. This residual basin is again the unusual two-dimensional inset of the critical unstable limit cycle; in the two-dimensional Poincaré surface of section, this structurally unstable basin portrait is analogous to that of the equilibrium fold in an oscillator.

For $\mu > \mu^C$ there is no nearby limit cycle, stable or unstable; all trajectories move off towards another remote attractor. The inevitable consequence of slowly increasing the control parameter μ through μ^C is a finite dynamic jump from the stable limit cycle which existed at subcritical values of μ.

So the attractor-basin phase portrait diagrams in Figure 7.4 and Figure 7.11 apply equally to unforced and to periodically forced oscillators experiencing loss of stability through a fold.

The basin description of the supercritical flip bifurcation is, like that of the supercritical Hopf, straightforward. As shown in Figure 8.6, there is a unique attractor throughout the sweep of the control parameter μ; the whole of the nearby region of phase space lies within a single basin of attraction for all values of μ. The supercritical flip bifurcation has no significant involvement with the basin structure, so the attractor-basin phase portrait diagram in Figure 7.11 is entirely appropriate to describe the supercritical flip of a stable limit cycle, as well as the supercritical Hopf bifurcation of a stable equilibrium.

The subcritical flip bifurcation, on the other hand, is associated with a radical change in basin structure, as seen in Figure 8.8. The unstable limit cycle at $\mu < \mu^C (p < 0$ in Figure 8.8) consists of a point on each branch of the parabolic path; in one-dimensional phase space, this $n = 2$ unstable orbit separates nearby trajectories and forms the basin boundary of the stable fixed point at the origin. So, as μ approaches μ^C, we observe the basin of attraction of the fixed point shrinking to zero around the point.

In a two-dimensional Poincaré section, the basin boundary would appear as a pair of infinite curves which together comprise the inset or stable manifold of the $n = 2$ unstable orbit. Each of these curves is an ensemble of trajectories asymptotic to the $n = 2$ unstable orbit. The points in a Poincaré section where the attractor meets the basin boundary as μ approaches μ^C are the $n = 2$ unstable orbit.

For $\mu > \mu^C$ there is no attractor in the locality, and all trajectories leave towards another, remote attractor. This dramatic change of basin structure generates an inevitable finite dynamic jump from the fixed point as μ is increased through μ^C. The attractor-basin phase portrait diagram in Figure 7.11 is entirely appropriate to describe the subcritical flip, as well as the subcritical Hopf.

In a similar way, the appropriate diagrams in Figure 7.11 aptly summarize the attractor and basin structures of the super- and subcritical symmetric or pitchfork bifurcations, as well as the Neimark bifurcations. In each case the supercritical version changes the attractor, while the basin boundary stays remote and uninvolved.

In the subcritical pitchfork, an attractor disappears as the control parameter μ is increased through μ^C; the basin boundary touches the attractor at the critical value μ^C. In one-dimensional phase space, each of two symmetric $n = 1$ unstable orbits separates nearby trajectories, and together they form the basin boundary. In a two-dimensional Poincaré section, the basin boundary is a pair of infinite curves, each of which is the inset or stable manifold of one of the two symmetric $n = 1$ unstable orbits. Each inset is an ensemble of trajectories asymptotic to an unstable orbit.

In the subcritical Neimark bifurcation, the basin boundary in a three-dimensional phase space is a single trajectory which forms an unstable invariant torus; this torus intersects the two-dimensional Poincaré section in an unstable closed invariant curve. As μ approaches the critical value μ^C, the entire basin boundary meets the limit cycle attractor, as seen in the two-dimensional Poincaré section and in the full three-dimensional phase space.

In all subcritical bifurcations, attractor-basin annihilation results in a finite dynamic jump to another, remote attractor.

9

Chaotic Behaviour of One- and Two-Dimensional Maps

9.1 GENERAL OUTLINE

We have seen in the previous chapter that algebraically defined mappings can exemplify important bifurcation behaviour; in addition they may possess chaotic attractors for certain ranges of control parameters. In fact, mappings are quite useful tools for studying chaotic behaviour; two-dimensional mappings are frequently invoked as models for the qualitative behaviour of Poincaré maps of forced oscillators. Two such examples are illustrated in Figure 9.1, the first related to nonlinear optics, and the second to a nonlinear mechanical oscillator.

Another important example of an algebraically defined two-dimensional mapping is given by the equations

$$\begin{aligned} x_{n+1} &= 1 - ax_n^2 + y_n \\ y_{n+1} &= bx_n \end{aligned} \tag{9.1}$$

This map, now called the *Hénon map*, was proposed by Hénon and Pomeau (1976) as a model of a Poincaré mapping of the Lorenz system of differential equations. It may be noted that the analogy is most apparent for very large Rayleigh number in the Lorenz model, a regime that will not be considered in this book. The Hénon map is also closely related to the Poincaré map of the Rössler band attractor and to the Smale horseshoe, both of which will be considered in detail in a subsequent chapter.

The Jacobian of the Hénon map is clearly just equal to the constant $-b$, so all values $|b| < 1$ correspond to volume-contracting dissipative systems. When $b = 0$ all points in the plane are mapped immediately onto the x axis; thus a special case of the two-dimensional map is the one-dimensional map

$$x_{n+1} = 1 - ax_n^2 \tag{9.2}$$

or equivalently

$$x_{n+1} = C - x_n^2 \tag{9.3}$$

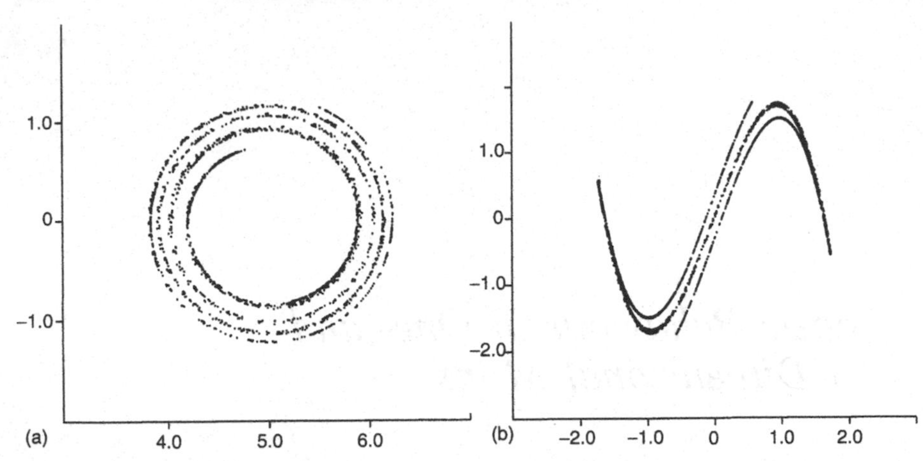

Figure 9.1 Chaotic attractors of two simple mappings: (a) a trigonometric map, $z_{n+1} = a + kz_n \exp(\mathrm{i}|z_n^2|)$, $a = 5$, $k = 0.2$; (b) a cubic map, $x_{n+1} = y_n$, $y_{n+1} = ay_n - y_n^3 - bx_n$, $a = 2.77$, $b = 0.2$. Part (a) reproduced, with permission, from Carr and Eilbeck (1984) and part (b) reproduced, with permission, from Holmes (1979)

after a change of scale. This one-dimensional map, known as the *quadratic map*, has strikingly rich behaviour, including chaotic behaviour. Because the map itself is so simply defined, it has been the object of much study. Another variant of the same map is the *logistic map*

$$x_{n+1} = \mu x_n(1 - x_n) \tag{9.4}$$

which arises directly as a model in population biology. The quadratic map and the logistic map are two examples of a much wider class of topologically equivalent *unimodal* one-dimensional maps, defined on a finite interval and having a single smooth maximum. As we shall see, all maps in this class can be expected to exhibit the same bifurcation patterns as a parameter such as μ or C is varied.

It should be noted that the two-dimensional Hénon map has one property lacking in the one-dimensional maps, namely *invertibility*. Equations (9.1) can be solved for a unique x_n, y_n for any values of x_{n+1}, y_{n+1}, so long as $b \neq 0$. However, when $b = 0$ the plane is mapped to a line, so that many points have the same image and no image has a unique inverse, or pre-image. The non-invertibility carries over to the one-dimensional maps (9.3) or (9.4) where a point may have two pre-images. This means that although solutions of ordinary differential equations and of many two-dimensional mappings can be uniquely followed backwards in time, solutions of the logistic or quadratic map cannot.

Nevertheless, it would seem that, in the Hénon map with small values of b, any attractors will lie at small values of the y coordinate, and final behaviour will be predominantly governed by the simpler equation (9.3). We may hope that, for small b, most attractors and bifurcations will follow the patterns of one-dimensional maps, which although rather complex are still simpler than the general two-dimensional case; see for example Tresser *et al.* (1980b) and Holmes and Whitley (1984). Thus it may be quite useful to summarize the behaviour of higher-dimensional maps or differential equations by one-dimensional maps. We shall see how a one-dimensional map may be extracted from a differential model in studying the Lorenz system; for the moment we concentrate on one-dimensional maps in their own right.

Other one-dimensional maps are also of considerable interest, notably the *circle map* defined by

$$\theta_{n+1} = \theta_n + \frac{K}{2\pi} \sin 2\pi\theta_n + \Omega \tag{9.5}$$

with two parameters K and Ω. This map is periodic in the angular variable θ, and may thus be considered as a mapping of the circumference of a circle onto itself. Such mappings are of great interest as models for nonlinear interaction of oscillations at two different frequencies, represented here by 2π and Ω. Thus equation (9.5) may stand for a simplification of the behaviour of the forced pendulum equation (6.13) or of the forced Van der Pol systems (6.6) or (6.7); parameters K and Ω correspond to the amplitude and frequency of the forcing oscillation. The specific map (9.5), and more general maps with the sine function replaced by a smooth periodic function, have been studied since Poincaré; more recent contributions include Arnold (1965), Hermann (1977), Aronson *et al.* (1982), Rand *et al.* (1982), Bak *et al.* (1984) and MacKay and Tresser (1984).

The circle map (9.5) can also be embedded in a two-dimensional map

$$\begin{aligned} r_{n+1} &= (1 - A)r_n + \frac{K}{2\pi} \sin 2\pi\theta_n + A\Omega \\ \theta_{n+1} &= r_{n+1} + \theta_n \end{aligned} \tag{9.6}$$

For $A = 1$ this reduces to the circle map; the limit $A = 0$ is equally interesting, since it represents a discrete mapping form of a two-variable area-preserving Hamiltonian system. The parameter A here represents a sort of dissipation. The Hamiltonian map has also been extensively studied, and is known as the Chirikov or *standard map*. Again, the two-dimensional map (9.6) is invertible except in the one-dimensional case $A = 1$.

These and other explicitly defined mappings have already proved their value as model dynamical systems, because they exhibit much of the behaviour of Poincaré maps of differential equations. The relative simplicity of explicit maps makes them amenable to detailed mathematical analysis, and also permits extensive numerical studies even with modest computing facilities. Many iterations of maps like the Hénon map can be computed in a fraction of the time required to compute a single Poincaré mapping point of a differential equation. This speed can be exploited by students as well as researchers, particularly where interactive computing facilities can be adapted to change starting points or parameter values and recompute. Simple mappings and the computer combine to create a valuable experimental laboratory for nonlinear dynamics. The book of Abraham, Gardini and Mira (1997) presents analyses of a great variety of explicit two-dimensional maps. The study of Holmes and Whitley (1984) concentrating on the Hénon map is also highly recommended.

9.2 THEORY FOR ONE-DIMENSIONAL MAPS

Before considering the general theory of one-dimensional maps, let us briefly consider the behaviour of a specific map, the logistic map (9.4). This map arises in population biology, as a model of fluctuations in the population of a single species with non-overlapping generations. Here we assume that the population at generation $(n+1)$ is uniquely determined by the population of the preceding generation according to some law

$$x_{n+1} = F(x_n) \tag{9.7}$$

The simplest such law is the Malthusian assumption of constant reproduction rate $F(x) = \mu x$. The exponential growth of population when $\mu > 1$ can be conveniently seen by the graphic iteration depicted in Figure 9.2(a), making use of the *first bisectrix* drawn at 45°.

A more interesting model with complicated behaviour is obtained by supposing that reproduction rate declines as population becomes large. A similar pollution effect occurs in a macroeconomic model studied by Stutzer (1980). The population model in simplest form is the logistic map (9.4), with population size renormalized to lie in the unit interval. With $F(x) = \mu x(1-x)$ the behaviour can be simple or complicated, depending on the value of the control μ. For $\mu \leq 3$, all initial population sizes evolve towards a unique stable equilibrium, which lies at the intersection of $F(x)$ with the first bisectrix. This is illustrated in Figure 9.2(b). At $\mu = 3$, when the slope of $F(x)$ at the equilibrium equals -1, the equilibrium bifurcates by period doubling; the resulting period 2 attracting limit cycle is illustrated in Figure 9.2(c). The size of the limit cycle increases continuously with μ beyond $\mu = 3$. Additional period doublings occur as μ increases further, and for typical values of μ near 4, numerical computation gives apparently non-periodic orbits as in Figure 9.2(d).

A reasonably complete description of the behaviour of the logistic map (9.4) or indeed of typical one-dimensional maps (9.7) with a single smooth maximum has been formulated; the standard reference is Collet and Eckmann (1980b), while Holmes and Whitley (1984) contains a helpful summary. Suppose that the continuous function F maps the interval $[-c, c]$ into itself, that F has only one critical point (a maximum) at $x = 0$, and that it is monotonically decreasing for $x < 0$. Additionally, suppose that the derivative F' is such that $|F'|^{1/2}$ is a convex function on $x < 0$ and $x > 0$. The condition on the derivative of F is equivalent to the requirement that the Schwarzian derivative

$$S(F) = \frac{F'''(x)}{F'(x)} - \frac{3}{2}\left(\frac{F''(x)}{F'(x)}\right)^2 \tag{9.8}$$

is negative on $x < 0$ and $x > 0$.

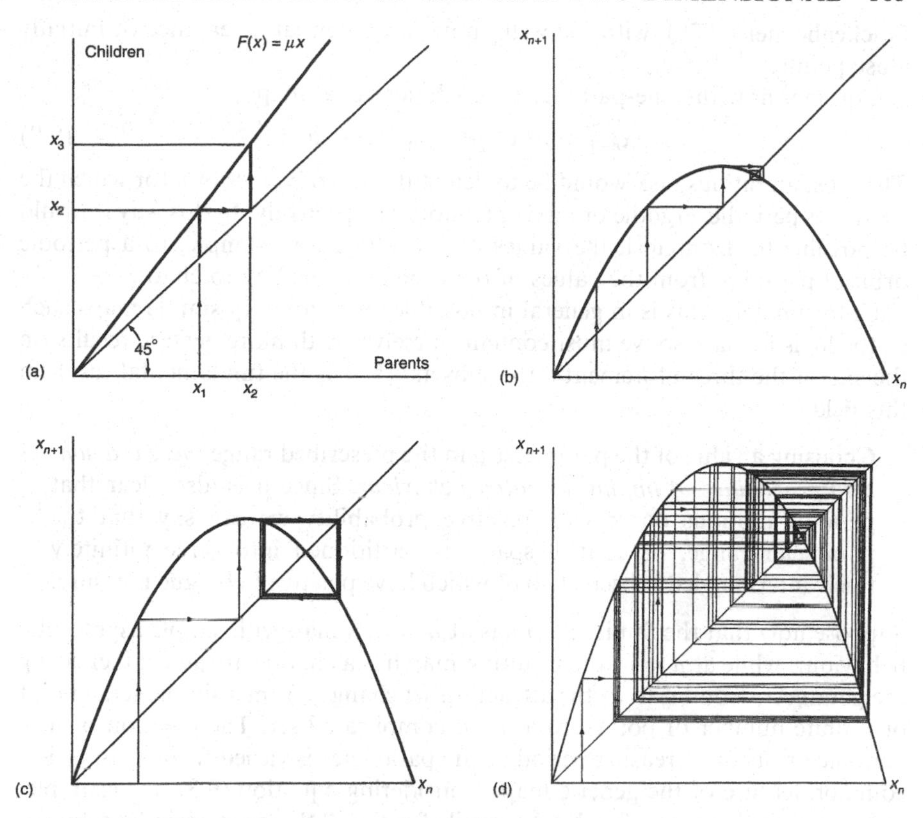

Figure 9.2 Behaviour of iterated maps. Part (a) shows the geometric construction of iterates. The other parts show logistic maps for different values of μ: (b) $\mu = 2.5$, a transient settles to an attracting equilibrium; (c) $\mu = 3.1$, an attracting limit cycle; (d) $\mu = 3.8$, aperiodic behaviour

This condition might seem to be quite unnatural, but it has a surprising role to play in the general theory. In fact, the attempts to present such a general theory of one-dimensional maps go back to Julia (1918) and Fatou (1919), while the first to discover the importance of the Schwarzian derivative was Singer (1978). He demonstrated, under the assumption of negative Schwarzian derivative, that the map can have *at most* one *stable* periodic orbit, and that *almost* all the points of the range $[-c, c]$ have the same asymptotic behaviour. Here and henceforth 'almost' means that a set of zero Lebesgue measure (observable with zero probability) is excluded; the *non-wandering* set can be a periodic orbit, namely a finite number of points, or a more complicated set, as for example a Cantor set. If the attracting set is not a periodic orbit, the motion of the trajectory on it can be either ergodic or mixing, in the following sense. In the *ergodic* case, the attracting set is an infinite one, and there is no sensitive dependence on initial conditions so that the orbits of two close initial points remain close. In the *mixing* case the attracting set is a *chaotic attractor*, and there is sensitive dependence on initial conditions, according to the definition of

Guckenheimer (1979), with the well-known exponential divergence of initially close points.

Consider now the one-parameter one-dimensional map

$$x_{n+1} = F(x_n, \mu) \quad \mu_0 < \mu < \mu_1 \tag{9.9}$$

The most ambitious task would be to detect the sets of values of μ for which the motion is periodic, ergodic or mixing (chaotic) respectively. In this way it would be possible to distinguish the values of μ leading, for example, to a periodic orbit of period n, from the values of the parameter leading to chaos.

Unfortunately, this is in general impossible, even for very simple maps such as the logistic map, so we must content ourselves with more generic results on the *size* of the three aforementioned subsets. Here is the fundamental result in this field:

> Choosing a value of the parameter μ in the prescribed range *there is a non-zero probability of mixing (chaotic) behaviour*. Since it is also clear that periodic motions occur with positive probability we can say that the parameter range, or control space, is partitioned into three infinitely finely intermingled subsets, two of which have positive Lebesgue measure.

Suppose now that the family (9.9) has at $\mu = \mu_0$ a map with simple asymptotic behaviour, while at $\mu = \mu_1$ the resulting map has a chaotic attractor: increasing the parameter from μ_0 to μ_1 the attracting set changes from a simple set, a point or a finite number of points, to a more complicated set. The creation of new periodic orbits of increasing period as the parameter is varied from μ_0 to μ_1 is a common feature of the generic map. Considering equation (9.3), for example, at $C = 0$ there is a single fixed point, while for $C = 2$ there are periodic orbits of arbitrarily long period. In fact, at $C = 2$ the map $F^n(x)$ has exactly 2^n fixed points for every $n > 0$. In Figure 9.3 the graphs of several iterates $F^n(x)$ are shown and the fixed points are given, of course, by the intersection of the graph of the function with the 45° line.

The existence of more than one fixed point for one μ value by no means contradicts the aforementioned theorem on the existence of no more than one *stable* periodic orbit. In fact, it can be shown that the quadratic map (9.3) at $C = 2$ has no stable periodic orbit at all.

The order in which these periodic orbits appear and the manner in which they emerge can be completely described by using *symbolic dynamics* and *kneading theory*, as described for example in Collet and Eckmann (1980b) and Holmes and Whitley (1984); see Hao and Zheng (1998). An important result is the Sarkovskii theorem, described as follows.

Consider the following order

$$3 \to 5 \to 7 \to 9 \to \ldots \to 2 \times 3 \to 2 \times 5 \to 2 \times 7 \to 2 \times 9 \to \ldots$$
$$\to 2^n \times 3 \to 2^n \times 5 \to 2^n \times 7 \to 2^n \times 9 \to \ldots \to 2^n \to \ldots \to 16 \to 8 \to 4 \to 2 \to 1 \tag{9.10}$$

If F has a periodic orbit of period p, and $p \to q$ in this order, then F has a periodic orbit of period q.

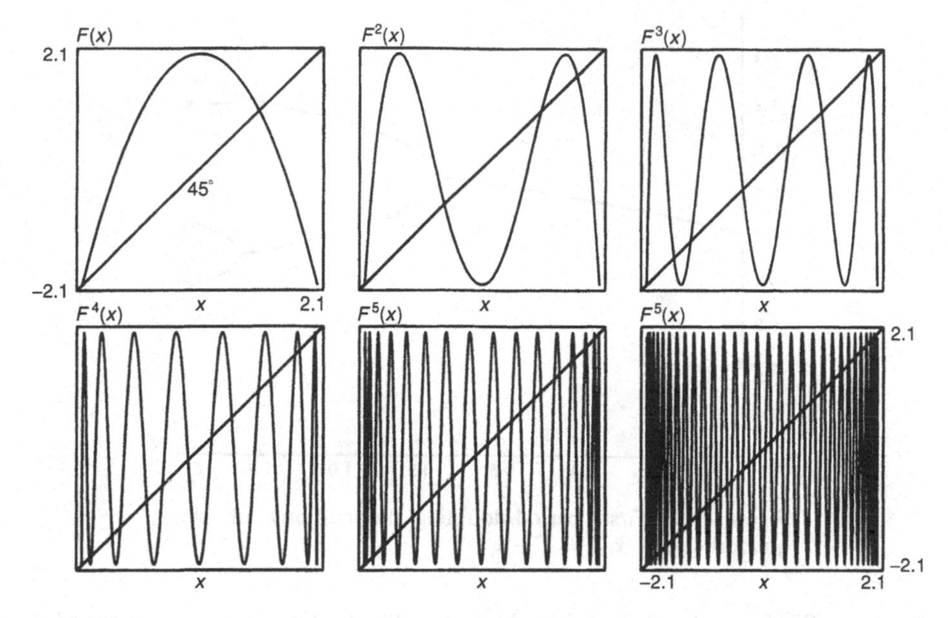

Figure 9.3 Graphs of the quadratic map and its first five iterates: $x_{n+1} = F(x_n)$; $F(x) = C - x^2$, $C = 2$; $F^2(x) \equiv F[F(x)]$, etc.

Hence, if a family of maps has a 3-periodic orbit, it also has orbits of arbitrarily long period, according to the famous statement 'period three implies chaos' of Li and Yorke (1975). This should be taken to mean that period 3 implies chaotic transients (in the sense of a horseshoe; see Chapter 12). The attractor may or may not be chaotic.

The Sarkovskii theorem is quite general, because it applies to all continuous maps of $\mathbb{R}$, not just to maps with a single critical point. On the other hand, it predicts only the order of first appearance of the periodic orbits, so it does not completely describe the location of *all* periodic orbits, because a higher periodicity can have several distinct orbits. For example, it is possible to show that in the quadratic map (9.3) there can be six orbits of period 5.

The existence of periodic orbits of arbitrarily long period does not necessarily imply chaotic attractors, which exhibit sensitive dependence on starting conditions; hence the routes to chaos remain to be explored.

9.3 BIFURCATIONS TO CHAOS

There are essentially two ways in which chaotic motion can arise in one-dimensional maps, each one associated with a particular local bifurcation.

One well-known route to chaos occurs by repeated flip bifurcations, as at the end of the Sarkovskii sequence. A fixed point becomes a periodic orbit of period 2 at $\mu = \mu_1$, then this 2-periodic orbit flips to a 4-periodic orbit at $\mu = \mu_2$, and so on, as is shown in Figure 9.4. The sequence $\{\mu_1, \mu_2, \ldots\}$ has a finite accumulation point μ_∞ involving an infinity of periodic orbits; remember

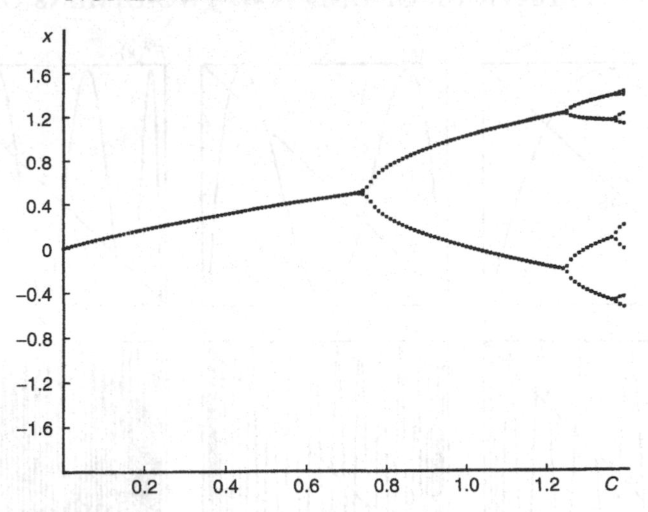

Figure 9.4 First period-doubling bifurcations of the quadratic map: $x_{i+1} = C - x_i^2$

that each flip leaves behind an unstable orbit. The motion at $\mu = \mu_\infty$ is ergodic (not mixing), and the non-wandering set contains a Cantor set, as shown by Misiurewicz (1981). Mixing occurs just past μ_∞.

Universal features of this scenario were discovered by Feigenbaum (1978), and later proved by Collet *et al.* (1980) and by Lanford (1982b).

The ratios

$$\delta_i = \frac{\mu_i - \mu_{i+1}}{\mu_{i+1} - \mu_{i+2}}$$

converge to the universal number $\delta_\infty = 4.66920\ldots$ as i tends to ∞. Universal means that δ_∞ is independent of the particular map, and careful computations have shown its validity for a number of specific examples. A period-doubling sequence also occurs for example in the impact oscillator, where the Feigenbaum number is approached very quickly; see Chapter 15. Convergent Feigenbaum cascades of flip bifurcations have also been observed in a number of experimental studies, e.g. Shaw (1984) and Roux *et al.* (1983). Other universal aspects of the subharmonic cascade include the power spectrum near μ_∞ (Feigenbaum 1979).

Successive flip bifurcations in a period-doubling sequence can be detected by inspecting the graphs of iterates $F^m(x)$. Just as the first period doubling is marked by $F(x)$ crossing the first bisectrix with slope -1, so the second flip occurs when $F^2(x)$ crosses its bisectrix with slope -1, and so on.

It is important to note that the beginning of a Feigenbaum cascade does not always mean that the cascade is completed; there may be only a finite number of period doublings, followed for example by undoublings or by other bifurcations. An example is illustrated in Figure 9.5 for a one-dimensional map

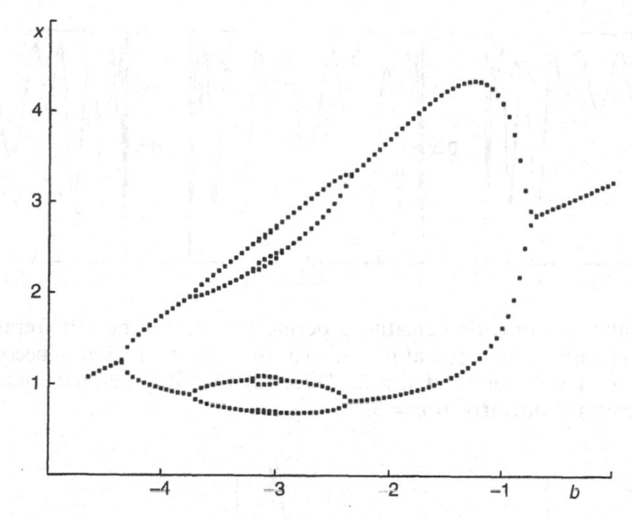

Figure 9.5 Feigenbaum trees: an example of an incomplete period-doubling cascade: $x_{i+1} = b + ax_i/(1 + x_i^2)$, $a = 11.5$

$$x_{n+1} = b + a\frac{x_n}{1 + x_n^2} \tag{9.11}$$

studied by Bier and Bountis (1984); here the parameter a is fixed and b is varied. For other values of a the subharmonic cascade completes and chaotic behaviour is observed. In other systems an important reason for incomplete period-doubling cascades is that the dynamics cannot be summarized by any one-dimensional map but is essentially multidimensional; in such cases other bifurcations to chaos are more likely to be observed; we will return to this point in Chapter 13.

The second route to chaos in maps of the interval involves the occurrence of a single saddle node bifurcation, and is also conveniently analysed using the graph of the iterated map $F^m(x)$. This bifurcation is one of three types of transition to chaos by *intermittency* first studied by Pomeau and Manneville (1980). Suppose for example we consider the range of μ near 3.75 for the logistic map, where a stable period 5 orbit exists. In Figure 9.6 we see graphs of $F^5(x)$ for three values of μ. For $\mu = 3.7$ the graph cuts the bisectrix once (apart from $x = 0$) at the unstable fixed point of F. At a slightly larger value $\mu_b \simeq 3.73775$, the graph of $F^5(x)$ becomes tangent to its bisectrix at five locations; for $\mu > \mu_b$ two fixed points of F^5 appear, one stable, the other unstable. This is a saddle-node bifurcation. When μ is slightly less than μ_b, the impending tangency results in trajectories exemplified by Figure 9.7, in which the staircase contains many steps for μ close to μ_b. The resulting observed behaviour is long intervals of nearly period 5 cycles interspersed with bursts of aperiodic behaviour, hence the name intermittency. As the control parameter μ *decreases* away from the saddle node, the nearly periodic intervals become shorter on the average, and aperiodic behaviour predominates. Pomeau and Manneville showed that the

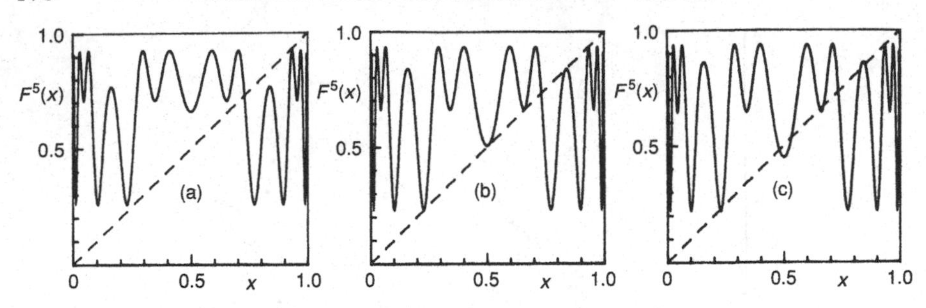

Figure 9.6 Tangent bifurcation creating a period 5 orbit: (a) the fifth iterated map has one intersection with the bisectrix at $\mu = 3.7$; (b) the fifth iterated map becomes tangent to the bisectrix in five points at $\mu = 3.73775$; (c) the fifth iterated map has eleven intersections with the bisectrix at $\mu = 3.75$

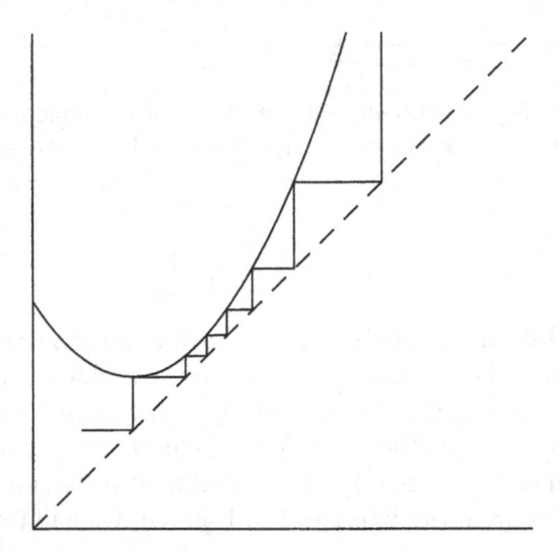

Figure 9.7 Intermittency mechanism for one-dimensional maps near tangent bifurcation ($\mu = 3.73773$): the system is forced to spend many iterations in passing through the narrow channel

normal form of the saddle-node bifurcation implies a certain scaling law for the average interval of nearly periodic oscillation in terms of $(\mu - \mu_b)$.

Behaviour described by the Pomeau–Manneville theory has been observed in experiments. For example, Perez and Jeffries (1982) studied the behaviour of a driven nonlinear semiconductor oscillator near a saddle node; the resulting intermittency is nicely illustrated by means of oscilloscope photographs.

9.4 BIFURCATION DIAGRAM OF ONE-DIMENSIONAL MAPS

The overall patterns of bifurcation in typical unimodal maps of the interval are topologically equivalent as shown by the kneading theory. It is therefore helpful to compute a diagram for one such map that can serve as a model for all. This reference diagram is given in Figure 9.8 computed from the logistic map; this diagram is analogous to Figure 9.4 extended to $C = 2$; the parameter μ of equation (9.4) varies from $\mu = 3$ at the left to $\mu = 4$ at the right of Figure 9.8.

The principal features we wish to understand in this diagram are the vertical windows that correspond to ranges of the parameter μ in which the attractor is a stable periodic motion. The most notable of these, near the right of the diagram, contains a period 3 limit cycle. To understand the windows is to know the order in which various periodicities occur as the parameter changes, and for each window to identify the bifurcations that begin and end the window, together with the fine structure in a typical window.

The order of occurrence of stable periodic orbits is predicted in part by the Sarkovskii sequence; for example, a period 5 window can just be seen to the left of the period 3 window in Figure 9.8. However, as mentioned earlier, there can be several occurrences of the same periodicity, but only the first occurrence appears in the Sarkovskii sequence. The full description of periodic windows was given by Metropolis *et al.* (1973) using combinatorial arguments; they also verified their predictions for several explicit maps and named their predictions the *U-sequence*, to suggest universality. This is something of an overstatement, since maps with more than one maximum or with a contorted dependence on the parameter will exhibit modifications of this sequence; see for example Brorson *et al.* (1983). One might instead think of the term U-sequence standing for unimodal map behaviour. The U-sequence has been observed in the behaviour of differential equations, for example in a chemical oscillator model by Tomita (1982). This is strong evidence that the dynamics of the Poincaré map could be approximately reduced to a unimodal map.

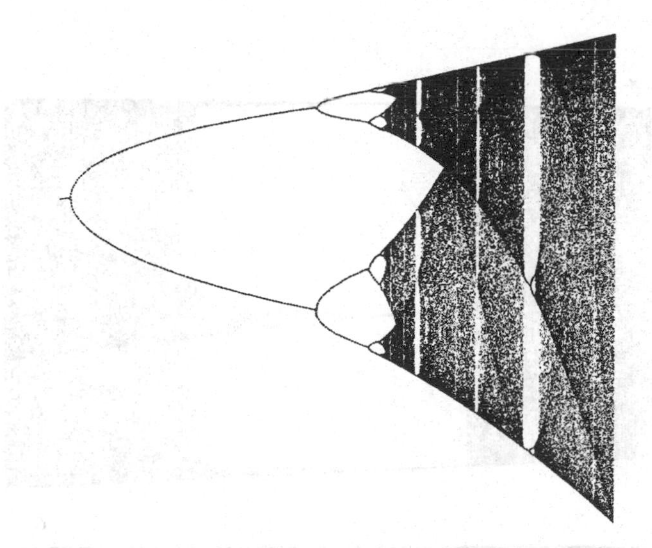

Figure 9.8 Bifurcation diagram of the logistic map: for an equivalent quantitative picture, see Figure G.7 in the Glossary, page 390. Courtesy of J. P. Crutchfield

Let us now consider the behaviour in a typical window of periodic behaviour, following Grebogi *et al.* (1983b). In Figure 9.9 we see part of the bifurcation diagram of the quadratic map greatly enlarged along the parameter axis. The period 3 window is seen in detail, with the abrupt appearance of the stable period 3 orbit near $C = 1.75$. This is the saddle-node or tangent bifurcation, which also creates a period 3 unstable orbit. From the normal form of the saddle node, we expect that the paths of the unstable and stable period 3 orbits follow a parabola opening to the right. Just to the left of the window, there is some evidence that the density of points concentrates near the impending period 3 orbit, according to the intermittency theory of Pomeau and Manneville.

Within the window we see what appear to be three miniature copies of the entire diagram of Figure 9.8; indeed the arrow identifies a period 3 superimposed on the base period 3, or period 9 window. Nearby, we see chaotic behaviour confined to three narrow bands; the bands are always visited in the same sequence as the period 3 orbit, but the location within bands is erratic; we may follow Lorenz (1980) and call this *noisy periodicity*.

Finally, at the end of the period 3 window, just past $C = 1.79$, we see a bifurcation from a chaotic attractor in three narrow bands to a chaotic attractor filling the entire interval. (Remember that these diagrams show only a finite sample of points whenever a non-periodic attractor is present.) This bifurcation is called an *interior crisis* by Grebogi *et al.*, who showed that it occurs when the unstable period 3 orbit from the saddle node just touches the chaotic three-band attractor. This occurs at a precise value of C that marks a discontinuous jump in size of the chaotic attractor. There is also evidence in the diagram that the density of attractor points in the large attractor near the crisis

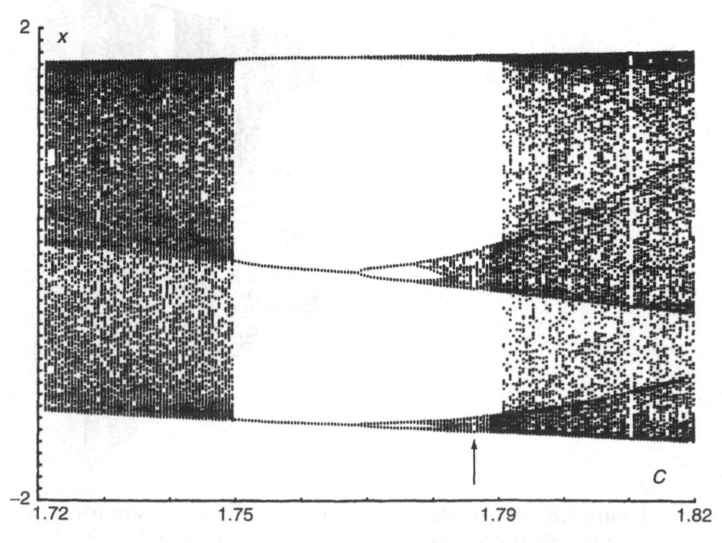

Figure 9.9 Enlargement of the period 3 window of the quadratic map

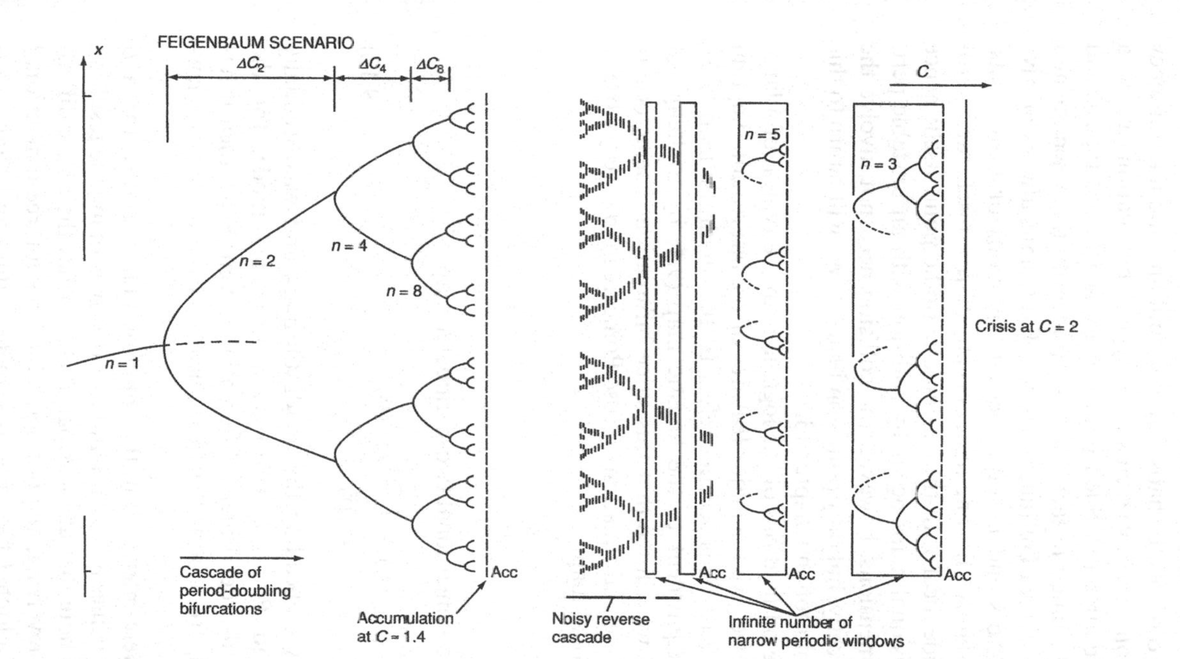

Figure 9.10 Schematic response of the quadratic map, $x_{n+1} = C - x_n^2$. Figure G.7 of the Glossary (page 390) shows the corresponding quantitative picture. In the Feigenbaum scenario, $\Delta C_k/\Delta C_{2k}$ tends to the universal number 4.66920... as k tends to infinity at the accumulation point ($C \approx 1.401$). After this point, we have an aperiodic (i.e. not periodic) regime desiccated by an infinite number of periodic windows. In this regime, a sub-set of positive non-zero Lebesgue measure of parameters has a *chaotic attractor* with mixing, sensitive dependence on initial conditions, and a broadband spectrum: this includes the chaos at $C = 2$. So choosing C at random (in the interval 0 to 2, say) there is a positive, non-zero probability of encountering a chaotic attractor. Each window opens with a saddle-node fold: under decreasing C this is the *intermittency explosion (map)* of Pomeau and Manneville (page 363). Each contains a cascade to chaos, and closes with a *chaotic-saddle explosion*, the (chaotic) interior crisis of Grebogi *et al.* (page 364). At $C = 2$ we have a *regular-saddle catastrophe*, the (regular) boundary crisis of Grebogi *et al.* (page 368), where the chaotic attractor touches the directly-unstable $n = 1$ solution (Figure G.7): beyond this, all iterations go to infinity. In the noisy reverse cascade, we see the *band merging* discussed by Lorenz (page 362)

concentrates in the three bands, and gradually spreads out, indicating another form of intermittency. However, we have no reason to expect universal statistical behaviour in this case, which is not one of the types of intermittency examined by Pomeau and Manneville.

Thus in addition to two routes to chaos—period doubling and intermittency by saddle-node bifurcation—we have seen a third type of bifurcation causing a jump in size of a chaotic attractor. Such jumps or *explosions* are typical and common in nonlinear dynamics; in fact, the first to identify this phenomenon was Ueda (1980b) in the forced Duffing oscillator. Explosions always involve collisions between attractors and unstable periodic motions or their insets, which are basin boundaries. In one-dimensional maps, insets are zero-dimensional and hence identical with the unstable periodic point itself; hence explosions in one-dimensional maps occur by collision with an unstable periodic point. In higher dimensions, however, the collision need not involve the unstable period point directly, but only points on its inset. We will return to this more general type of explosion in Chapter 13.

As a reminder that the bifurcations of the logistic map are typical of a large class of unimodal maps, we give the major features of the bifurcation diagram schematically in Figure 9.10. This schema refers to the quadratic map (9.3), which we know to be equivalent to the logistic map (9.4) by a change of variables. In Chapter 12 we will see that this same bifurcation scheme occurs in a simple system of differential equations whose Poincaré map is very close to a one-dimensional unimodal map.

9.5 HÉNON MAP

We consider now the two-dimensional two-parameter map

$$\begin{aligned} x_{i+1} &= 1 + y_i - ax_i^2 \\ y_{i+1} &= bx_i \end{aligned} \tag{9.12}$$

This can be obtained by embedding the quadratic map, and was originally proposed by Hénon and Pomeau (1976) as a simplified model for the Poincaré map of the third-order Lorenz differential system. Since then it has been extensively studied, from both the theoretical and the numerical viewpoints.

Hénon and Pomeau were interested in the study of a chaotic attractor, and they fixed the parameter values at $a = 1.4$ and $b = 0.3$ to examine the resulting map. It is worthwhile to note that at these parameter values the above diffeomorphism is orientation-reversing, and not derivable from a three-dimensional flow. Thompson (1982) adopted the same parameter values and studied numerically the sensitive dependence on initial conditions. The results of this study are nicely shown in Figure 9.11, from which it seems that the existence of a chaotic attractor could be assured. On the other hand, Newhouse (1979) and C. Robinson (1983) showed that near $a = 1.392$ and $b = 0.3$ there should be an infinity of periodic orbits with very small basins of attraction. Hence the

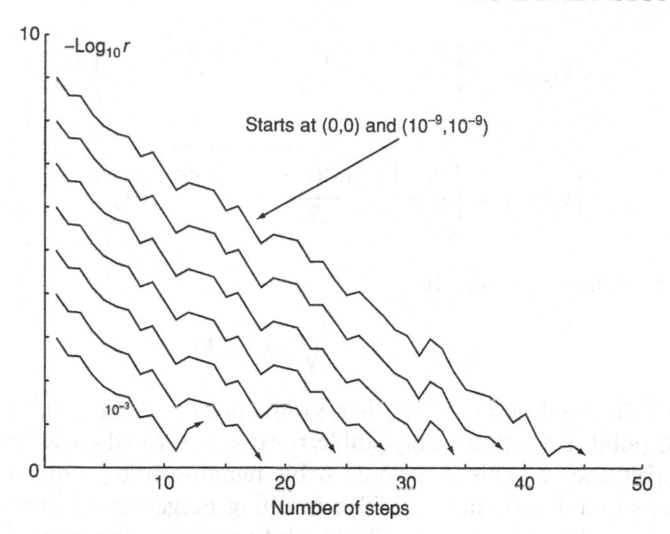

Figure 9.11 Hénon chaotic attractor: a log plot of the separation against the number of steps for a series of close starts near the origin (from which trajectories rapidly settle on the attractor). Compare with Figure 16.8

exponential divergence of initially close starting points could be due to their belonging to two different basins of attraction. The difficulty of distinguishing between a true strange attractor and an infinity of periodic orbits with very small basins of attraction illustrates the complexity of two-dimensional map behaviour.

A piecewise linear version of the Hénon map was proposed by Lozi (1978), and Misiurewicz (1980) was able to demonstrate that the Lozi map has a true chaotic attractor. As pointed out by Holmes and Whitley (1984), the existence of a chaotic attractor for the Lozi map does not imply the existence of a chaotic attractor for the original Hénon map. Furthermore the Hénon map appears to be more typical of Poincaré maps of differential equations, so important questions remain open.

Curry (1979) was the first to observe that there exists a range of parameter values for which two chaotic attractors coexist, illuminating a major difference between one- and two-dimensional maps. Simó (1979) fixed the parameter value of $b = 0.3$ and extensively studied the attracting sets for various values of the remaining parameter a, while Grebogi *et al.* (1983b) studied the case of an almost one-dimensional map: they fixed b at $-$ 0.025 and were able to detect discrepancies between the one-dimensional and the almost one-dimensional map.

Recently Holmes and Whitley (1984) constructed a partial bifurcation set in the parameter plane (a, b) and contrasted the very complex behaviour of two-dimensional maps with one-dimensional maps.

In the following we fix b at 0.3 and allow a to vary. The fixed points of the map are given by

$$(x_1, y_1) = \left(\frac{b - 1 + \sqrt{[(b-1)^2 + 4a]}}{2a}, bx_1 \right)$$
$$(x_2, y_2) = \left(\frac{b - 1 - \sqrt{[(b-1)^2 + 4a]}}{2a}, bx_2 \right) \tag{9.13}$$

and the eigenvalues are given by

$$\lambda_{1,2} = -ax \pm \sqrt{(ax^2 + b)}$$

For $b = 0.3$ the fixed point (x_2, y_2) is a saddle point for every value of a. The other fixed point is asymptotically stable for $a < 0.3675$ where a flip into a 2-periodic orbit occurs. This 2-periodic orbit remains stable until $a = 0.9125$, where it flips to a 4-periodic orbit. The resulting Feigenbaum scenario can be seen in Figure 9.12 and in Table 9.1. The bifurcation diagram of Figure 9.12 indicates some of the bifurcations, showing for example multiple attractors in the range $1.06237 < a < 1.08$. Some of the corresponding attractors are shown in Figure 9.13. We see some similarity between Figure 9.12 and Figure 9.8 but the small jump to the $n = 3$ stable orbit indicates additional multiple attractors and hysteresis in the Hénon map, which would require more detailed study. Such bifurcations involve truly two-dimensional structure of invariant manifolds as studied by Simó; simply locating the unstable periodic orbits, without knowing their insets and outsets, may be insufficient to understand the bifurcation behaviour.

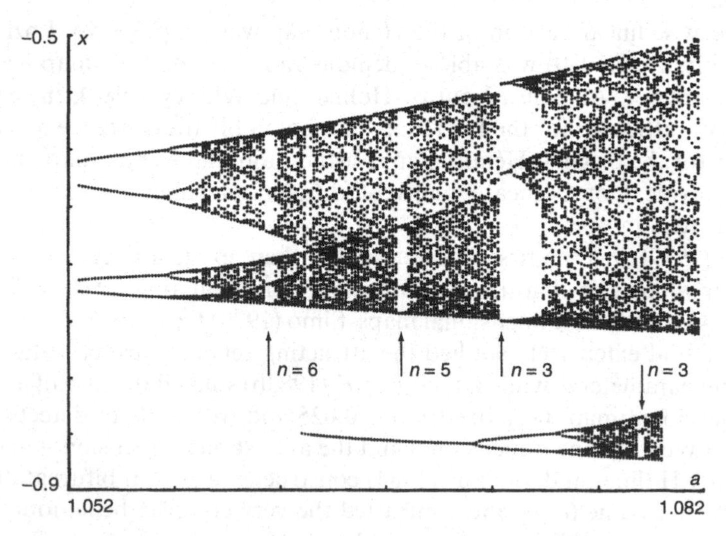

Figure 9.12 Projected bifurcation diagram of the Hénon map ($x_{i+1} = 1 - ax_i^2 + y_i$, $y_{i+1} = bx_i$) in the parameter range $b = 0.3$, $1.052 \leq a \leq 1.082$

Table 9.1 Convergence of a Feigenbaum cascade

n	Period 2^n	a_n	δ_{n-1}
0	1	−0.1125	
1	2	0.3675	
2	4	0.9125	4.844
3	8	1.026	4.3269
4	16	1.051	4.696
5	32	1.056 536	4.636
6	64	1.057 730 83	4.7748
7	128	1.057 980 893 1	4.6696
8	256	1.058 034 452 15	4.6691
9	512	1.058 045 923 04	4.6691
10	1024	1.058 048 379 80	4.6694
11	2048	1.058 048 905 931	

After Derrida *et al.* (1979).

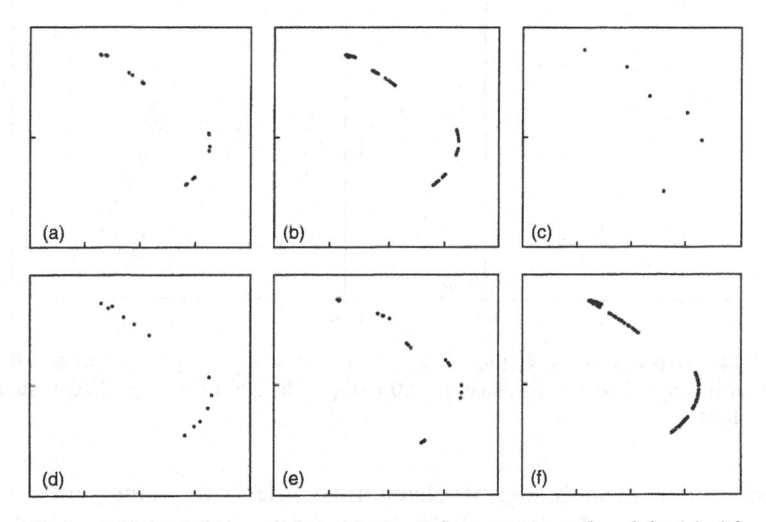

Figure 9.13 Phase portraits of the Hénon map at $b = 0.3$: (a) 16-periodic orbit at $a = 1.054$; (b) chaotic attractor at $a = 1.07$; (c) coexisting 6-periodic orbit at $a = 1.07$; (d) 12-periodic orbit at $a = 1.0725$; (e, f) coexisting chaotic attractors at $a = 1.079$

The classical Hénon attractor with $a = 1.4$ is shown in Figure 9.14, with a magnification sequence, reprinted from Thompson (1982). These show the fractal layers common to chaotic attractors.

The Hénon map is a convenient *archetype* for chaotic attractor dynamics in two-dimensional maps, related to the simple stretching and folding of the one-dimensional logistic or quadratic map. Simulations of the Hénon map provide ready and abundant evidence of chaotic attractors. However, this numerical evidence does not constitute a rigorous mathematical proof that chaotic attractors exist: digital simulations always use finite-precision arithmetic, so even

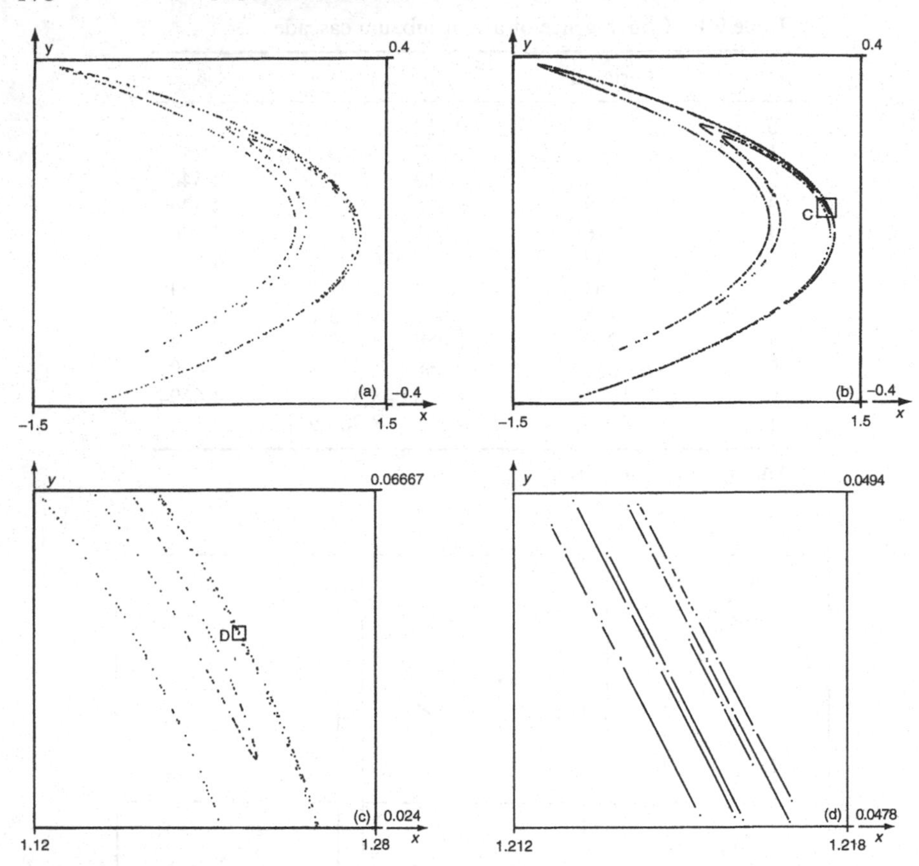

Figure 9.14 Hénon chaotic attractor ($x_{i+1} = y_i + 1 - ax_i^2$, $y_{i+1} = bx_i$) at $b = 0.3$ and $a = 1.4$ with magnification sequence: (a) 600 steps, (b) 2000 steps, (c) 10 000 steps, (d) 1 00 000 steps

a trajectory that strongly suggests chaos must, after a very long period, repeat itself. Benedicks and Carleson (1991) bolstered the experience of simulations with a mathematical proof that a truly non-periodic orbit exists for very small values of b. This implies the Jacobian is small, and the map is nearly one-dimensional; in forced oscillators, that would correspond to systems with strong dissipation. But for Hénon's value $b = 0.3$, or for negative b values corresponding to moderate or light dissipation, there is as yet no rigorous proof of a chaotic attractor.

As an archetypal two-dimensional map, the Hénon map is potentially useful in understanding Poincaré maps of forced oscillators. Note that behaviour similar to forced oscillators can only be expected of the Hénon map when the parameter b is negative. For positive values of b, the Hénon map is an example of an *orientation-reversing* map: an outline of a left hand is mapped to an outline of a right hand. Poincaré maps of forced oscillators are always *orientation-preserving*.

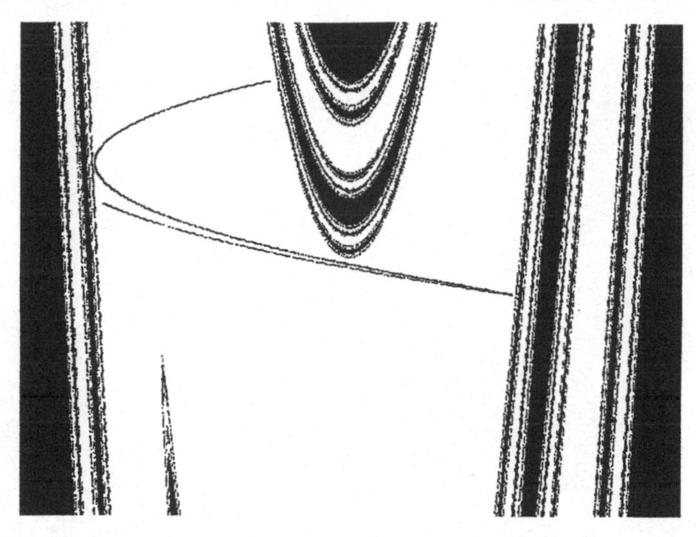

Figure 9.15 Phase portrait of a chaotic attractor and its basin, in the Hénon map with $a = 1.83$ and $b = -0.1$. For this value of b, the layers of the chaotic attractor are compressed, but there is clear evidence of a fractal basin boundary

In addition to chaotic attractors, the Hénon map provides abundant evidence of fractal basin boundaries. Figure 9.15 shows both a chaotic attractor and a fractal basin boundary in the Hénon map, for $a = 1.83$ and $b = -0.1$. Because the magnitude of b is small, the fractal layers of the attractor are strongly compressed. The dark region consists of initial conditions which diverge to infinity, while the white region is the basin of the chaotic attractor. In this figure the phase space is shown flipped, with the y coordinate increasing downward, in order to emphasize the similarity with Figure 6.19. The remarkable qualitative similarity of the two figures illustrates the usefulness of studying the Hénon map. For more on fractal basin boundaries in the Hénon map, see for example Grebogi *et al.* (1987) and Stewart, Ueda *et al.* (1995).

Part III

Flows, Outstructures and Chaos

10

The Geometry of Recurrence

In this chapter we review the basic sets—attractors, saddles, and repellors—from the mathematical viewpoint of hyperbolic structure.

10.1 FINITE-DIMENSIONAL DYNAMICAL SYSTEMS

The first part of this book concentrated on what may be the most important type of simple dynamical system, the nonlinear oscillator represented by a second-order ordinary differential equation. In addition to the free and forced oscillators already discussed, coupled systems of oscillators may also be studied as dynamical systems. In fact, the simple forced oscillator is actually a system of two oscillators coupled by a simple master–slave relation: the forcing term may be regarded as the output of a master oscillator that is undisturbed by the nonlinear oscillator it drives. Obvious extensions of the sinusoidally forced oscillator include giving the master oscillator a chaotic output, introducing a two-way coupling between the oscillators, and considering more than two coupled oscillators. This leads to the notion of *complex dynamical systems* built of units that are themselves simpler dynamical systems; see for example Abraham (1986), Rosen (1970) and Mees (1981). Recent research includes studies of coupled map lattices by Keeler and Farmer (1986), and Abraham *et al.* (1991). See Kaneko and Tsuda (2000) for application in biology, and Puu (2000) for examples in economics.

The second part of this book introduced the study of iterated mappings as discrete dynamical systems. Mappings come from continuous time systems by Poincaré section; iterated maps of an interval arise by another reduction pioneered by Lorenz, to be studied in the next chapter. Mappings may also occur directly as models, for example in population dynamics.

The full scope of dynamical systems theory might be described as deterministic initial value problems. This includes evolutionary laws embodied in iterated mappings, and initial value problems for both ordinary and partial differential equations, whether expressed as single differential equations or as systems of equations. The aim of qualitative dynamics in the spirit of Poincaré is to study in phase space the structures that govern recurrent behaviour in these initial value problems.

The study of partial differential equations also uses ideas of qualitative, geometric dynamics, as for example in Holmes *et al.* (1998), Mori and Kuramoto (1998), Scott (1999) and Vallos (1986). The phase space of a partial differential equation must have an infinite number of dimensions, which can certainly defeat geometric intuition. The successes of qualitative geometric ideas in partial differential equations often occur in applications where dissipation plays an important role. The effect of dissipation is to contract volumes in phase space, and it turns out that this effect is a powerful one, often reducing the final motions of a dissipative system to a subset of phase space that is finite-dimensional. The existence of laminar flow in viscous fluids is one confirmation of this fact. Recent experimental studies of transition from laminar to turbulent flow, reviewed by Gorman *et al.* (1980), study in detail how successive destabilizations of laminar flow push the final motions of a fluid into subsets of phase space of gradually increasing dimension. A variety of experimental evidence now indicates the importance of low-dimensional behaviour in weak fluid turbulence (e.g. Brandstäter *et al.* 1983), as well as in other continuum mechanics phenomena (e.g. Held *et al.* 1984).

Rather than tackle the full complexity of infinite-dimensional phase space, we assume from the outset that an initial value problem has been set in a phase space with a finite number of dimensions. In this chapter we describe with reasonable generality the types of recurrent behaviour that can occur, and the geometric phase-space identification of their stability.

First-order systems

A finite-dimensional *dynamical system* must first be a system whose state at any instant can be completely characterized by a set of scalar observables $x_1, x_2, \ldots, x_n$. This set is of course fixed in advance, and must always characterize the system throughout its evolution. One evolutionary history of the system is then given by time series $x_1(t), x_2(t), \ldots, x_n(t)$; these functions of time trace out a trajectory in n-dimensional phase space.

A dynamical system is *deterministic* if its evolution is always completely determined by its current state and past history. The laws governing this evolution may be expressed as equations giving the rates of change of the observables $x_1, x_2, \ldots, x_n$ in terms of the observables themselves. We will consider systems whose evolutionary laws can be written as a system of ordinary differential equations

$$\begin{aligned} \mathrm{d}x_1/\mathrm{d}t = \dot{x}_1 &= f_1(x_1, x_2, \ldots, x_n, t) \\ \dot{x}_2 &= f_2(x_1, x_2, \ldots, x_n, t) \\ &\vdots \\ \dot{x}_n &= f_n(x_1, x_2, \ldots, x_n, t) \end{aligned} \tag{10.1}$$

Although noise and uncertainty can be considered as additional factors, the basic system is defined without appeal to stochastic input of any kind.

Recall that this system of first-order equations is sufficiently general to include the second-order oscillators and other higher-order equations as special cases. In the example of a second-order equation

$$\ddot{x} + \dot{F}(x) + G(x) = f(t) \tag{10.2}$$

where $\dot{F}(x) = F'(x)\dot{x}$ by the chain rule, transformation to a first-order system may be effected by introducing a new variable $y = \dot{x}$. The resulting first-order system is

$$\begin{aligned} \dot{y} &= -F'(x)y - G(x) + f(t) \\ \dot{x} &= y \end{aligned} \tag{10.3}$$

which is clearly of the class described by equations (10.1). Such transformations are by no means unique; for the case where equation (10.2) represents a Van der Pol oscillator, one may choose y so that

$$\begin{aligned} \dot{x} &= y - F(x) \\ \dot{y} &= -G(x) + f(t) \end{aligned} \tag{10.4}$$

and in fact this is the reason for writing the friction term in equation (10.2) above as $\dot{F}(x)$.

We have already restricted our attention to dynamics described by a finite number of observables. There are systems of ordinary differential equations whose right-hand sides depend on delayed-time values of the observables, and integro-differential equations depending on integrals over past history, but these have infinite-dimensional phase space despite the finite number of observables. Such systems are beyond the scope of this book; see Farmer (1982) for an example.

The vector field

If the n scalar observables x_j completely characterize a system at each instant and determine its evolution, then the geometry of phase space contains important information about behaviour. The first step in obtaining this information is to identify the laws of evolution contained in the differential equations with a *vector field* in phase space. This means attaching to each point $\{x_j\}$ in phase space a vector with tail at $\{x_j\}$; the head of the vector is displaced from the tail by the values of $\{f_i\}$ evaluated at $\{x_j\}$ and time t. Geometrically, the laws of evolution indicate the direction and speed of evolution of the system away from each point in phase space. A trajectory in phase space is formed by piecing together infinitesimal steps in the direction indicated by the vector field. The result is a curve whose tangent at each point is always aligned with the vector field. Figure 10.1 shows an example of a vector field and a trajectory in a three-dimensional phase space. The solid dot is the tail of each vector with coordinates x_1, x_2, x_3, while the legs on each vector indicate the vector's vertical component f_3, together with a horizontal projection, which is the resultant of components f_1 and f_2.

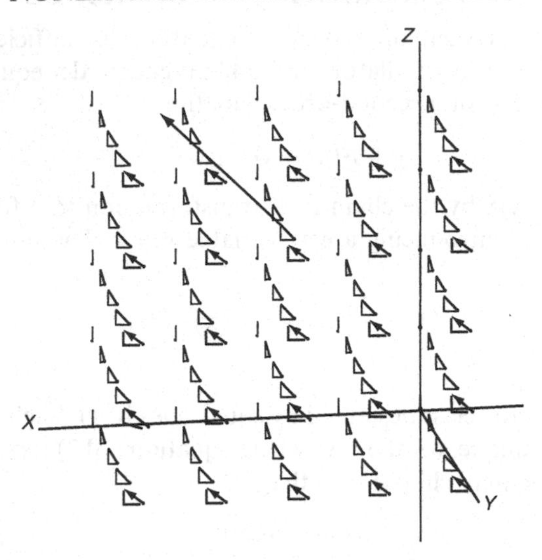

Figure 10.1 A three-dimensional vector field and part of a trajectory tangent to the vector field

The non-crossing trajectory property

The vector field defined by $\{f_i\}$ is said not to depend explicitly on time if each f_i is a function of the $\{x_j\}$ only; that is, the evolution vector at any point in phase space does not change in direction or magnitude with time. This means there are no time-varying forces acting on the dynamical system from outside; such ordinary differential equations are referred to as *autonomous systems*. The stationarity of the vector field has an important geometric consequence, the non-crossing trajectory property. No two trajectories $X(t) = \{x_j(t)\}$ and $\widetilde{X}(t) = \{\tilde{x}_j(t)\}$ of an autonomous system will cross through the same point in phase space in finite time. If they did, the point of crossing would be a single state of the system from which two different evolutions are possible. But there is only a single direction of evolution from a precise point in phase space. Combining this fact with the mathematical assumption that the evolution vector changes smoothly across the phase space (i.e. the head of the vector moves smoothly as the tail is moved), it can be proved (e.g. Rosen, 1970; Coddington and Levinson 1955) that only one trajectory passes through each point. Therefore no two trajectories may cross. The distinction between two trajectories crossing *through* a point in finite time, which is not allowed, and two trajectories asymptotically approaching an equilibrium point as $t \to \infty$, which is allowed, must be firmly kept in mind.

As one example of the importance of the non-crossing trajectory property, consider an autonomous dynamical system whose phase space is the Cartesian plane, and a trajectory that forms a closed loop in the plane as in Figure 4.3. This is the only type of self-intersecting trajectory (as distinct from asymptotically self-approaching trajectories of a saddle connection, e.g. in Figures 3.15

and 3.19). Any closed-loop trajectory of an autonomous system corresponds to a periodic motion. If the system starts from a state $X(t_0) = \{x_j(t_0)\}$ and returns to the same point after a time interval T, then evolution from $t_0 + T$ is the same as from t_0, and so $X(t+T) = X(t)$ for all $t \geq t_0$. If the phase space is the Cartesian plane, such a periodic loop organizes the phase space by separating it into two regions, an interior and an exterior. By the non-crossing property, any trajectory lying partly inside the loop must lie entirely in the loop for all time.

This is a key example of how geometric structures in phase space can organize behaviour. From a simple qualitative fact (the existence of a periodic motion), we can draw a conclusion about many other trajectories, and for each such trajectory the conclusion remains valid as t goes to infinity.

The non-crossing trajectories property leads to so many important conclusions, even in higher-dimensional phase spaces, that it becomes important to construct a phase space for *non-autonomous* systems where the non-crossing property holds. The usual analytical trick for doing this is to consider an augmented phase space for a non-autonomous system; this phase space shall have $(n+1)$ dimensions whose coordinates are the $\{x_j\}$ and time t. As a result, any point in the augmented phase space has a unique trajectory passing through it. By the device of introducing a formal state variable x_{n+1} that satisfies the differential equation $\dot{x}_{n+1} = 1$, the non-autonomous system of n equations (10.1) can be converted to $(n+1)$ equations that are autonomous:

$$\begin{aligned} \dot{x}_1 &= f_1(x_1, x_2, \ldots, x_n, x_{n+1}) \\ \dot{x}_2 &= f_2(x_1, x_2, \ldots, x_n, x_{n+1}) \\ &\vdots \\ \dot{x}_n &= f_n(x_1, x_2, \ldots, x_n, x_{n+1}) \\ \dot{x}_{n+1} &= 1 \end{aligned} \tag{10.5}$$

Geometric dynamics begins with the identification of a phase space in which the evolution of the dynamical system is completely characterized by a *stationary* vector field, and examines the *topological* structures formed by trajectories in that phase space. Because the vector field determines trajectories that fill the phase space and never cross, like the streamlines in a moving fluid, the vector field is sometimes said to define a *flow* in phase space. The major conclusions of geometric dynamics concern the *final-time* behaviour of the dynamical system after initial transients die away.

10.2 TYPES OF RECURRENT BEHAVIOUR

Final dynamic behaviours have interesting structure if they involve some form of recurrence. A keystone of qualitative dynamics is a precise description of what we mean by recurrent behaviour, and a systematic way of separating recurrent final behaviour from start-up transients. A mathematically general definition of recurrence would rely on the fundamental topological idea of a system of *neighbourhoods* of a point in phase space. Since this book deals only with finite-dimensional phase space, it is no loss of generality to side-step neighbourhoods and define recurrence directly in terms of a *distance* defined in the phase space. Two common examples of a distance function are

$$d(X,\tilde{X}) = \sum_j |x_j - \tilde{x}_j|$$

and the Euclidean metric function

$$d(X,\tilde{X}) = \left(\sum_j (x_j - \tilde{x}_j)^2\right)^{1/2}$$

Whatever distance function we choose, it makes no difference in the qualitative geometric conclusions to be drawn. One must only take care that the distance function involves all coordinates (so that the distance from X to $\tilde{X}$ is zero only if $X = \tilde{X}$); that the distance $d(X,\tilde{X})$ from X to $\tilde{X}$ is the same as $d(\tilde{X},X)$ from $\tilde{X}$ to X; and that the triangle inequality always holds:

$$d(X,Z) \leq d(X,Y) + d(Y,Z)$$

In terms of this notion of distance in phase space, a particular state of a dynamical system is *recurrent* if, by waiting long enough, we are assured of seeing the system return arbitrarily close to that state. If the system returns *arbitrarily* close to a recurrent state once, then we can wait again and find inductively that it will be found arbitrarily close to the recurrent state again and again. The system need not, however, settle down near the recurrent state. It may merely make a close but brief encounter and then spend long times away before returning close again. It may be helpful to think of the example of recurrence in the Sun–Earth–Moon system, where a recurrent state might be total eclipse of the Sun; in fact, celestial mechanics was Poincaré's chief motivation for developing methods of analysis based on the geometry of phase space.

This notion of recurrent state can be made precise mathematically. It serves to identify the *observable* recurrence in dynamical systems. However, it is not adequate to describe all phenomena associated with recurrence; for this one needs the mathematical concept of a non-wandering state. Whereas the recurrent state comes back arbitrarily close to itself, a *non-wandering state* has arbitrarily close states that return arbitrarily close. This is more general than a recurrent state; any recurrent state is a non-wandering state, but not vice versa. This distinction, introduced formally by Birkhoff, is subtle but not vacuous. The most important example of a non-wandering state that is not recurrent is furnished by the homoclinic trajectories, so named by Poincaré. A clear understanding of the non-wandering nature of points on a homoclinic trajectory leads inevitably to the heart of chaotic dynamics, which we hope at least to make accessible to the reader in the next few chapters.

Basic or decomposable?

One would like to extend the idea of a recurrent state to a concept of recurrent behaviour: a dynamical system would be said to engage in *recurrent behaviour* if every state (after a certain transient, perhaps) is a recurrent state. One would then like to know what are the basic types of recurrent behaviour. Unfortunately, attempts to define precisely which chaotic motions should qualify as a 'basic type of recurrent behaviour' have so far been only partly successful, and seem to lead to very difficult mathematical questions. As a provisional definition, let us say that a *basic recurrent behaviour type* is an ensemble of recurrent states linked together by a single trajectory of the dynamical system. In a phase space with competing attractors, for example, each attractor should be a basic type that is a transitive set. By *transitive* we mean that one trajectory can come arbitrarily close to every point in the attractor; such a trajectory is *dense*.

For example, the Poincaré sections of chaotic attractors in Chapter 6 were constructed by plotting the return points of a *single* steady-state trajectory. Similar pictures could have been obtained by scattering a cloud of starting points over the (x, y) plane at $t = 0$ and integrating all trajectories forwards to steady state. However, it is entirely possible that an attracting set computed from a cloud consists of several pieces that appear to touch but in fact are separated by tiny gaps; a single trajectory continued in time would wander over one piece but never reach other pieces. Such an attracting set is *decomposable* into smaller, transitive subsets. Computing only one trajectory for a long time has the advantage of depicting a transitive attractor, provided the numerical error is small enough. But usually we cannot prove that chaotic behaviour is transitive; we can only say that numerical evidence strongly suggests that it is.

In mathematical papers it is not uncommon to reserve the word *attractor* for the transitive set in phase space ultimately filled by a single steady-state trajectory determined without error or uncertainty; final motions subject to noise or approximation, or found by the cloud method, are said to define an *attracting set*, and could be decomposable. Because zero error cannot be achieved in practice, we have not maintained this fine distinction of terms, but we do use single trajectories rather than clouds in computing attractor portraits.

Hierarchy of recurrence

The simplest recurrent behaviour is equilibrium. If a dynamical system stays at some fixed state X^* for all time, recurrence is satisfied because the system 'returns' to X^* by always being there. Depending on how state variables are chosen, equilibrium can include the state of rest as well as equilibrium (steady) motion, such as a motor turning at constant speed. In order for X^* to be an *equilibrium point*, the vector field that describes the evolution must attach the null vector to the point X^*, i.e. $f_i(x_1^*, x_2^*, \ldots, x_n^*) = 0$ for each i. Then X^* is a *fixed point* of the vector field, also called a singularity.

There are two basic types of approach to equilibrium, direct or oscillatory. Typical time series for a scalar observable, say $X_1(t)$, as $X \to X^*$ might look like Figure 10.2. These are stable equilibria; unstable equilibria are also recurrent states because they can theoretically last forever if nothing disturbs the system. If $f(X^*) = 0$, then mathematically the solution of the initial value problem $X(0) = X^*, \dot{X} = f(X)$ is given by $X(t) = X^*$ for all t; the stability of the system at X^* is a separate issue determined by whether the system can compensate for small disturbances. Recognizing unstable recurrence is essential to building a complete qualitative portrait of a dynamical system. Even when the unstable recurrent state is unlikely to be observed, it is a key element in the geometry of phase space, since its presence and structure influence states that are observed.

The next basic type of recurrence is *periodic* motion. In this case the system returns arbitrarily close by returning exactly, always with period T. In an observable time series, periodic motion might look like Figure 10.3. In phase space, a periodic motion traces out a closed-loop trajectory, either in the plane or in a higher-dimensional phase space. If trajectories starting near the loop always approach it ever more closely, it is a stable limit cycle. Even if the reverse is true (so that small disturbances cause the system to wander from

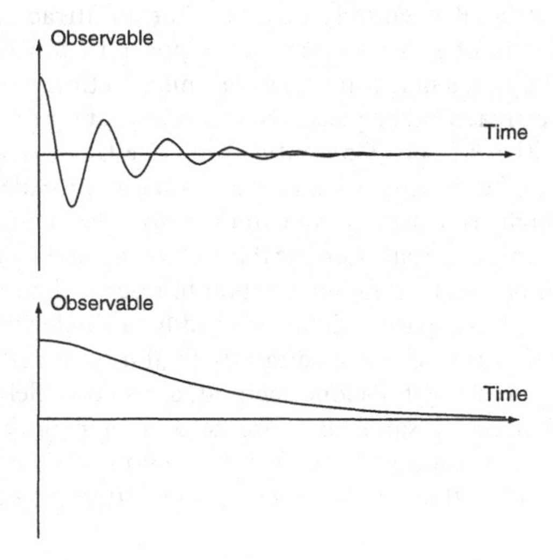

Figure 10.2 Two types of approach to equilibrium, as seen in typical time series

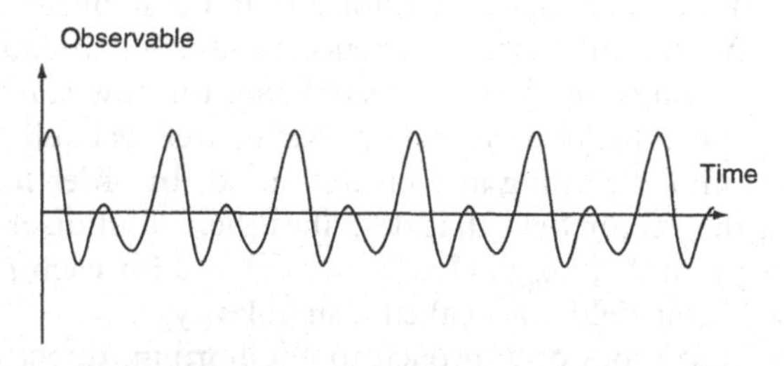

Figure 10.3 Time series of a periodic motion

the loop and not return), an unstable limit cycle is still periodic recurrent behaviour.

Another basic type of recurrence is quasi-periodic motion. In this case, a periodic motion is modulated in some way by a second motion, itself periodic but with a different period. An example would be a steady audible tone broadcast over a radiofrequency carrier wave. If the carrier frequency is an exact multiple of the audio frequency, the net broadcast signal is periodic; but if the ratio of frequencies is not a ratio of integers, the signal is *quasi-periodic*. This is recurrence that is not precisely repetitive; we need the general concept of recurrence to identify a quasi-periodic motion as post-transient behaviour. Since a combination of frequencies can produce time series whose regularity is not immediately obvious, the power spectrum is helpful in identifying quasi-periodic motion. The spectrum should consist of sharp peaks at each of the fundamental frequencies, together with their overtones and beat frequencies; all observed peaks should be explainable as overtones or beat frequencies. An example of a quasi-periodic spectrum is shown in Figure 10.4, taken from experiments on Taylor–Couette flow of a fluid by Fenstermacher *et al.* (1979). Notice that here the peaks are not mathematically sharp but instrumentally sharp, and there is a low level of background noise in the spectrum as is inevitable in experimental measurement. However, all prominent features of the spectrum are explained as combinations of two fundamental frequencies labelled ω_1 and ω_3; some of these combinations are identified in the figure.

The number of fundamental frequencies needed to explain observed peaks in the power spectrum might be called the degree of quasi-periodicity; in principle it could be greater than 2. Geometrically, degree 2 quasi-periodic final motion fills the surface of a torus in the appropriate phase space (imagining Figure 6.5 with incommensurate frequencies, for example); higher degrees fill hypertori.

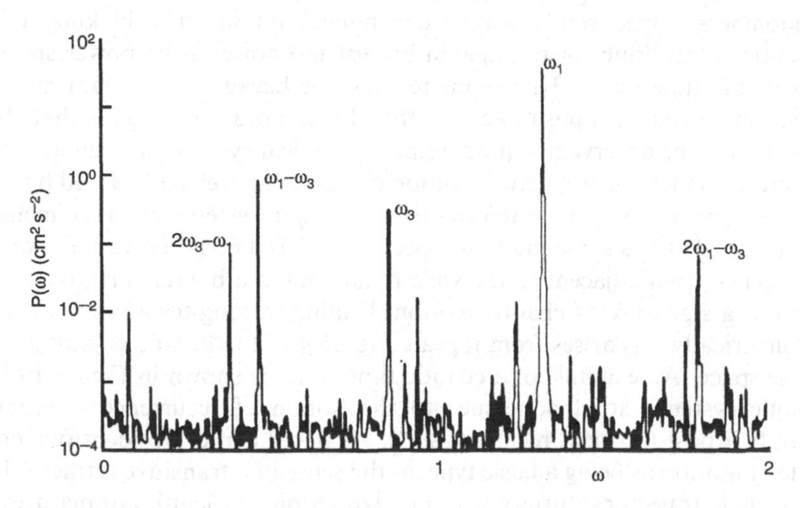

Figure 10.4 Experimental power spectrum of quasi-periodic motion. Reproduced, with permission of Cambridge University Press, from Fenstermacher *et al.* (1979)

A remarkable theorem in qualitative dynamics, proved by Ruelle, Takens, and Newhouse, suggests that quasi-periodicity of degree greater than 2 may be difficult to observe in nature. In Newhouse *et al.* (1978) it is proved that any dynamical system with a quasi-periodic 3-torus attractor can be perturbed by an arbitrarily small change in the dynamics to a system with a chaotic attractor. This supplements the theorem of Ruelle and Takens (1971), which established that systems with 4-torus attractors are even more easily perturbed into chaos. In both proofs, it is rigorously established that all nearby trajectories on a perturbed chaotic attractor diverge exponentially. This is at odds with a long-standing conjecture of Landau, who proposed that turbulent fluid flow is a state reached by successive transitions that increase the degree of quasi-periodicity step by step to infinity. That is, as the dynamical system is changed to produce additional instability, the result at each stage would be more sharp peaks in the spectrum due to a new fundamental frequency unrelated to the previously observed peaks. Recent experimental studies also contradict this conjecture, and hence tend to confirm the importance of the Ruelle–Takens–Newhouse theorem; after degree 2 quasi-periodicity, further destabilization often leads to another type of recurrent behaviour that is not degree 3 quasi-periodicity. However, it is by no means impossible to observe quasi-periodicity with three incommensurate frequencies; see for example Gorman *et al.* (1980) and Tavakol and Tworkowski (1984). As we shall see in Chapter 13, quasi-periodic behaviour is more typical of systems with weakly coupled modes of oscillation, while chaotic behaviour is characteristic of strongly coupled modes.

The latest addition to the list of recurrent behaviour types has been called chaotic motion or motion on a strange attractor. Although one way to define chaos is in the negative—recurrent behaviour that is not an equilibrium, a cycle or even a quasi-periodic motion—there is more to be said about chaos. Chaotic motion has some aspect that is provably as random as a coin toss. The randomness arises from sensitive dependence on imperfectly known initial conditions, resulting for example in broadband noise in the power spectra of observable time series. This seems remarkable because the dynamical system needs no stochastic input to achieve this. Even more surprising is that chaotic motions can be observed in quite simple dynamical systems, as seen in previous chapters. In fact, the negative definition of chaos can well be replaced by a more positive one: chaos is recurrent motion in *simple* systems or low-dimensional behaviour that has some random aspect as well as a certain order. Exponential divergence from adjacent starts while remaining in a bounded region of phase space is a signature of chaotic motion, leading to long-term unpredictability. Geometrically this arises from repeated *folding* of the bundle of trajectories in phase space. An example of a chaotic time series is shown in Figure 10.5. For chaotic systems, strict adherence to the notion of recurrence is needed to identify a post-transient recurrent state. As noted earlier, to identify a chaotic state of motion as being a basic type, in the sense of a transitive attractor linked by a single trajectory, turns out to involve serious difficulties of mathematical analysis. The chaotic motions of experimental studies, or numerical experiments based on digital or analogue computation, do not always suggest

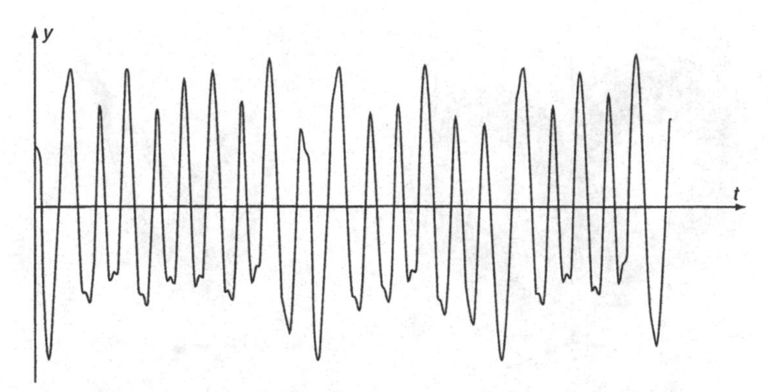

Figure 10.5 Time series of a chaotic motion, computed from the velocity-forced Van der Pol oscillator equations (6.7)

mathematical models for which properties like transitivity, or geometric divergence of all nearby trajectories, can be proved. Fortunately, applied dynamics presents a number of simple dynamical systems for study, many of which have chaotic regimes. The structure of invariant manifolds is a primary tool for understanding chaotic behaviour.

One reason for asking whether a chaotic motion is composed entirely of linked recurrent states is to help formulate a notion of dynamic stability for chaos. We know when an equilibrium is stable, and when a cyclic motion is stable, but how are we to know if a chaotic motion is stable, i.e. self-restoring in the aftermath of small disturbances? The question itself at first sounds peculiar. For evidence that a chaotic motion might in some sense be stable, it helps to look at a chaotic motion in the full phase space. Figure 10.6 shows trajectories of the Lorenz system

$$\begin{aligned} \dot{x} &= -\sigma x + \sigma y \\ \dot{y} &= Rx - y - xz \\ \dot{z} &= -bz + xy \end{aligned} \tag{10.6}$$

where R, σ and b are fixed parameters. Each view shows part of a final, post-transient trajectory in x, y, z state space seen from a different angle. Trajectories in the top row are for $R = 26$, while the middle and bottom rows are for $R = 28$ and $R = 30$, respectively. The differences are mainly due to the sample effect of finite trajectories; otherwise the similarity is so apparent that we suspect this recurrent behaviour might be a basic type, and that final motions lie in a subset of phase space, the attractor, which is in some sense stable.

Figure 10.6 represents just one of a number of known chaotic attractor types in three-dimensional phase space; we have already seen different attractor forms in the driven Van der Pol and Duffing oscillators, and a certain number of other different types have been discovered; see for example Arnéodo *et al.* (1981b), Marzec and Spiegel (1980), Leipnik and Newton (1981) and Rössler (1981). Although a few taxonomic principles can be suggested, they are far

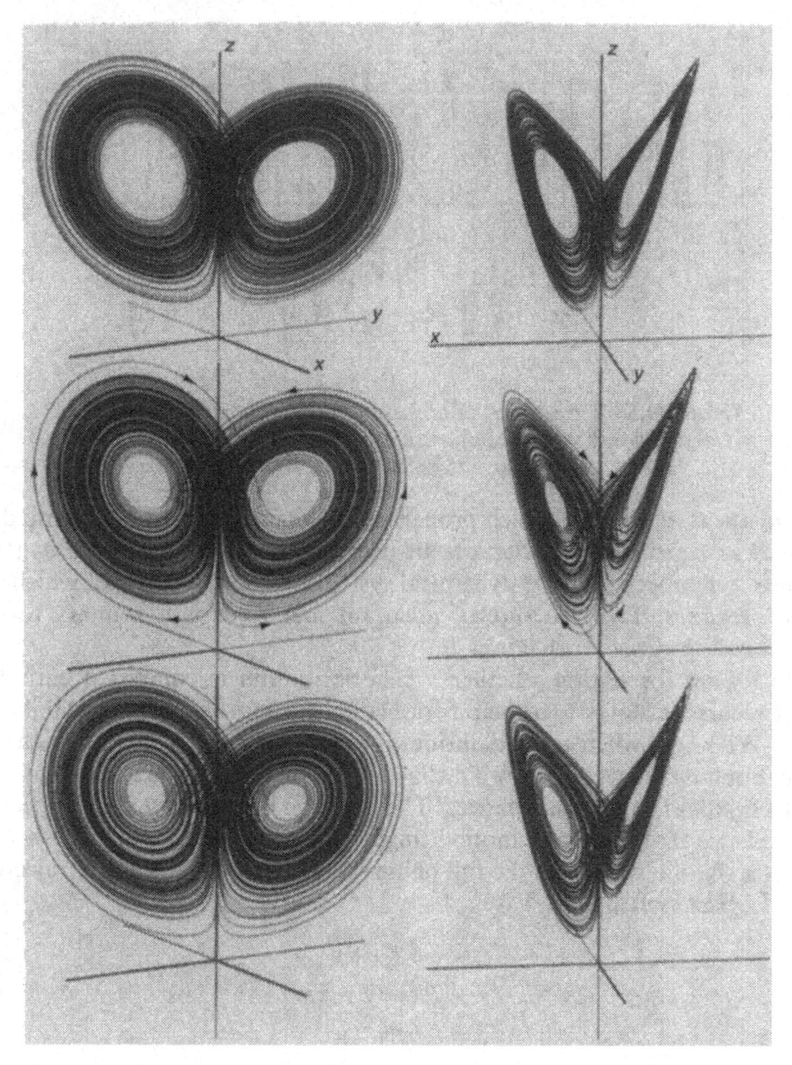

Figure 10.6 Trajectories of the Lorenz equations (10.6) in three-dimensional phase space: (top) $R = 26$, (middle) $R = 28$, (bottom) $R = 30$

from giving a complete classification of even the known chaotic attractors in three dimensions. Furthermore, Rössler (1983) has observed that four-dimensional phase space can in principle have attractor forms not realizable in three dimensions. It should therefore come as no surprise that the stability of chaotic behaviour is imperfectly understood. In fact, we cannot even appreciate what a general theory must involve without a complete picture of the phase-space structure of stable and unstable equilibria and cycles. In particular, unstable equilibria or cycles can always be found in the full phase portrait of a chaotic dynamical system, and certain special trajectory structures—invariant manifolds—associated with these unstable fixed points or cycles govern the geometric organization of phase space. This much is known, from the work of Poincaré, Birkhoff, Smale, Abraham and others, and this deeper geometric view is the subject of the next three chapters.

10.3 HYPERBOLIC STABILITY TYPES FOR EQUILIBRIA

Although an equilibrium is identified by the condition that the vector field vanish at a point, a full phase-space picture of the different types of equilibria emerges only by considering the structure of trajectories nearby. Figure 10.7 shows a number of trajectories near stable equilibria of dynamical systems in the plane; the two types are a direct and oscillatory approach as in Figure 10.2. The trajectories are paths in the phase space traced out by integrating a system of two first-order differential equations; the outer end of each curve represents an initial state that evolves along the curve with time, converging as $t \to \infty$ to the stable equilibrium.

These portraits represent the two typical kinds of stable equilibria, the node and the spiral, in a phase space of two dimensions, both for the nonlinear oscillators and for the general class of two-dimensional autonomous systems of first-order equations.

It should be clear that the observable versus time views of Figure 10.2 are simply reduced views of Figure 10.7; the observable of Figure 10.2 could be x or y or any well-defined functional combination of the two.

The basic types of equilibrium can be identified by a local linearized version of the dynamics near the point of equilibrium. That is, the nonlinear equations are replaced by approximate linear equations; this approximation is only correct in a small region of phase space surrounding the equilibrium point. For example, the equations

$$\dot{x} = -0.1x + 1.1y + O(x^2 + y^2)$$
$$\dot{y} = 1.1x - 0.1y + O(x^2 + y^2)$$

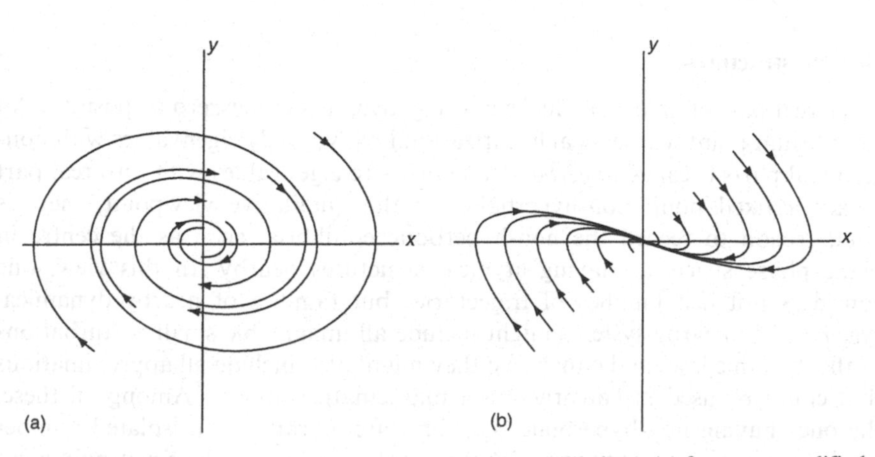

Figure 10.7 Phase portraits of two types of stable equilibrium: (a) focus, exemplified by $\dot{x} = y$, $\dot{y} = 0.2y - 0.6x$; (b) node, exemplified by $\dot{x} = y$, $\dot{y} = -y - 0.24x$

can be viewed in a small region of the fixed point at the origin as equivalent in structure to their linear part only. Linear algebra gives eigenvectors $u = x + y, v = x - y$, and the linearly equivalent system

$$\dot{u} = u + O(u^2 + v^2)$$
$$\dot{v} = -1.2v + O(u^2 + v^2)$$

From this we deduce that trajectories near the origin are drawn towards the eigenvector u, and along it away from the fixed point. That is, the origin is in this case a saddle.

The general procedure for local linear analysis of equilibria is to locate equilibrium points of an nth-order system by finding the roots of the simultaneous equations

$$f_1(x_1, x_2, \ldots, x_n) = 0$$
$$f_2(x_1, x_2, \ldots, x_n) = 0$$
$$\vdots$$
$$f_n(x_1, x_2, \ldots, x_n) = 0$$

Then find the linear part of these functions near each equilibrium point, for example the terms up to first order in a Taylor series about the point. This gives the local linear approximation of the dynamics near the fixed point.

This local linearization works for either a stable or an unstable fixed point. The stability type of the equilibrium can be determined by finding the eigenvalues of the matrix representing the linearized dynamics. Assuming no real part is zero, let the *index* of an equilibrium point be the number of eigenvalues with positive real part. If this index equals zero, then in the linear approximation the equilibrium is stable. Increasing values of the index correspond to increasing degrees of instability near the equilibrium point; we might therefore refer to this integer as the *instability index*.

Generic structures

If the real part of an eigenvalue is not negative, it is either zero or positive. An equilibrium point whose local linearization involves only eigenvalues with non-zero real parts is called *hyperbolic*; one or more eigenvalues with zero real part makes a fixed point non-hyperbolic. In the qualitative viewpoint, there is good reason to regard the non-hyperbolic equilibria, such as the centre in plane phase space, as having atypical structures nearby. In this view, one considers not just families of trajectories but families of nearby dynamical systems. The nearby systems might include all imaginable small perturbations of the dynamic laws and equations; they might also include all approximations that could be used in constructing a mathematical model. Among all these, the ones having non-hyperbolic equilibria are as rare as an isolated number (such as zero) in a continuum. That is, only the hyperbolic fixed points are *generic*.

The generic view of eigenvalues at an equilibrium point also explains why in Figure 10.7 the trajectories are pinched to a bundle in the case of the node, i.e. real eigenvalues. This happens because the two eigenvalues are typically unequal. The local linear approximation gives

$$\dot{u} = \lambda_1 u$$
$$\dot{v} = \lambda_2 v$$

near the fixed point, with negative real eigenvectors λ_1, λ_2. If for example $\lambda_1 < \lambda_2 < 0$, then u is damped out more rapidly than v at $t \to \infty$, and trajectories approach the eigenvector v more rapidly than they approach the fixed point along the direction of v. The picture of Figure 10.7 thus obtains at a typical node with generic, unequal eigenvalues.

Once the instability index of an equilibrium point is determined by counting the signs of the real parts of the eigenvalues, additional information can be found in the imaginary parts of the eigenvalues. If the ordinary differential equations are written, as we have assumed, for real quantities, the local linearizations must have only real coefficients, and so complex eigenvalues occur only in conjugate pairs. Figure 10.8 shows a familiar sequence of different dynamics, each having an equilibrium with complex conjugate eigenvalues. The clockwise angular speed on each trajectory in each system is the same, because the imaginary parts of the eigenvalues are the same; but the trajectories wind down rapidly, slowly, or not at all depending on the real part of the eigenvalues. In either of the first two cases, small changes in the dynamics can change the

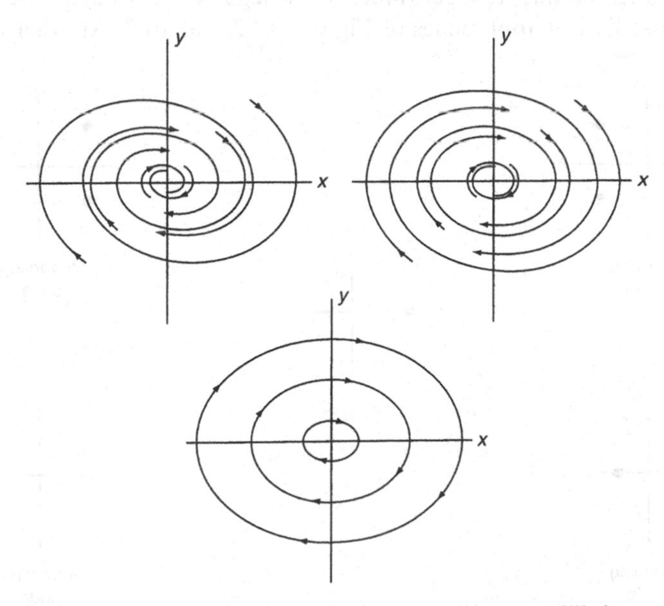

Figure 10.8 Varying dynamics near a focal equilibrium, changing to a non-generic centre: $\dot{x} = y$, $\dot{y} = -ay - 0.6x$ with $a = 0.2, 0.1, 0$

rate of decay but not the qualitative picture: trajectories of a perturbed system still spiral inwards. In the third case, however, any perturbation of the dynamics will break the closed loops and cause trajectories all to spiral inwards or all to spiral outwards from the equilibrium. The two inward-spiralling dynamics can also be qualitatively represented by any sufficiently good approximating system, while the third case will differ qualitatively from its approximations. Thus the dynamics near a hyperbolic fixed point are *structurally stable*, while the non-hyperbolic fixed point is not structurally stable. This assures us that the local linearization is a valid approximation for hyperbolic fixed points in any number of dimensions.

In sum, the generic fixed point is structurally stable. The fact that generic and structurally stable are equivalent for fixed points is not a definition, but an important theorem. Generic refers to the fact that hyperbolic fixed points (with eigenvalues having non-zero real parts) are the common rule, while zero real parts are the rare exception. Structural stability is determined from the effect of small perturbations on the nearby qualitative behaviour. Both of these notions can be made precise mathematically, and extended to more general recurrent behaviour. For periodic cycles it again turns out that generic and structurally stable are equivalent properties, but for quasi-periodic and chaotic behaviour this is no longer the case.

Considering only the (generic) hyperbolic fixed points, five types of equilibrium are possible for dynamics in a two-dimensional phase space. Figure 10.9 shows their eigenvalues located in the complex plane. Two types have instability index 0: one type with two negative real eigenvalues, the other type having a pair of complex conjugate eigenvalues with negative real part. These are the direct and oscillatory approaches of Figures 10.2 and 10.7. Another two types

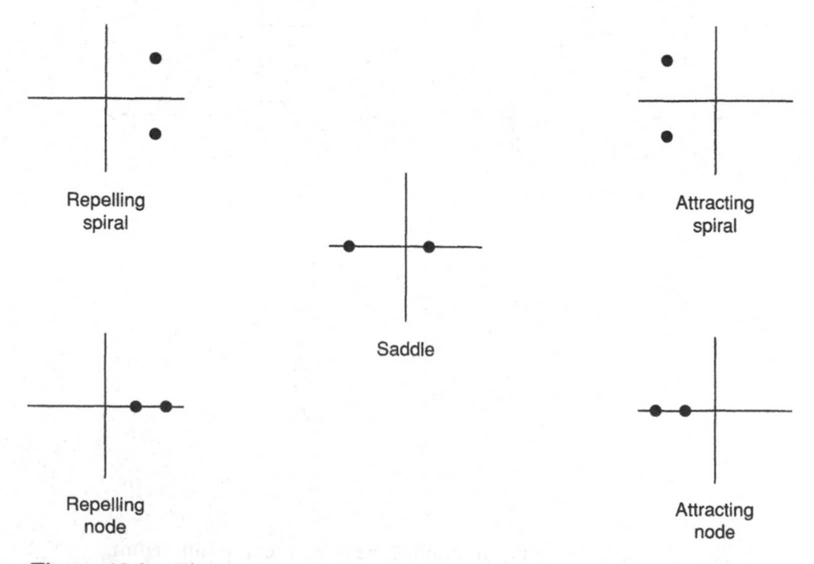

Figure 10.9 Eigenvalues in the complex plane for the generic types of planar equilibria

have instability index 2, and are just the first two types with signs reversed. These unstable equilibrium types are *repellors*, while the first two are *attractors*. The other type of equilibrium is the *saddle*, which in two-dimensional state space has index 1 (two real eigenvalues, one negative and one positive).

Saddle points and invariant manifolds

Figure 10.10 shows the phase portrait near a saddle fixed point in the plane. In addition to a number of trajectories nearby, four special trajectories asymptotic to the saddle point are indicated. Two are asymptotic as $t \to +\infty$, and ultimately lie in the direction of the incoming eigenvector having negative real eigenvalue; we call these asymptotic curves the *inset* of the saddle. The other two are asymptotic to the saddle as $t \to -\infty$, are tangent to the outgoing eigenvector with positive eigenvalue, and form the *outset* of the saddle. Note that the inset is conveniently found by starting at a point near the saddle on the incoming eigenvector, and integrating backwards in time. Integrating these trajectories over all time both towards and away from the saddle gives *invariant manifolds*.

It should be noted that referring to insets and outsets as *the invariant manifolds* is an abuse of terminology; in fact, any subset of phase space made up entirely of trajectories extending to $t \to -\infty$ and $t \to +\infty$ is invariant under the flow in phase space. However, the special invariant manifolds that do not wander to infinity but approach either a bounded α-limit set (as $t \to -\infty$) or a bounded ω-limit set (as $t \to +\infty$) are essential to concise description of the topological structure in phase space. Unless they are themselves repellors or attractors, these special invariant manifolds terminate at saddle-type

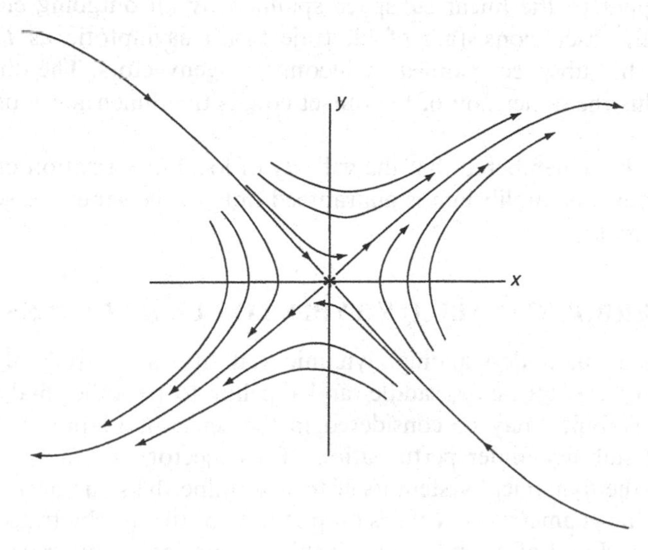

Figure 10.10 Phase portrait of a saddle equilibrium, exemplified by $\dot{x} = y$, $\dot{y} = 0.2y + x - x^3$

singularities. We may also follow Zeeman and refer to insets and outsets as the *outstructures*.

A saddle point is particularly important in any phase portrait because only special trajectories are asymptotic to it; other trajectories may come close but veer away. In contrast, *all* nearby trajectories are asymptotic in the case of an attractor (as $t \to +\infty$) or a repellor (as $t \to -\infty$). As Figure 10.10 illustrates, the general shape of all trajectories near a saddle point can be inferred if only the asymptotic trajectories, the inset and outset, are known. This organizing property of invariant manifolds depends on the non-crossing trajectory property, which holds only in the full phase space. The importance of saddles and their invariant manifolds will be studied extensively in Chapter 11 through the example of the Lorenz system.

Equilibria in higher-dimensional phase space can be similarly classified by their instability index. In three dimensions, equilibria may have the local structure of any two-dimensional equilibrium plus inward or outward dynamics along the direction of the third eigenvector. Thus, for example, there are three-dimensional saddle-type equilibria with spiral dynamics, either spiralling inwards but unstable in the direction of the third eigenvector, or spiralling outwards but drawn inwards along the third eigenvector. Saddles with three real eigenvalues are equally possible. The local geometry of equilibria in the three-dimensional phase space is shown in Figure 3.10. A fourth dimension adds the new possibility of crossed spirals, and so on. As the dimension increases, the possible saddle types proliferate.

In higher dimensions, saddle fixed points have invariant manifolds that may be curves, smooth surfaces or hypersurfaces. The outset manifold contains all trajectories asymptotic to the saddle as $t \to -\infty$, and its dimension is equal to the index, which is the number of outgoing eigenvectors. The outset manifold is tangent to the linear subspace spanned by all outgoing eigenvectors. Likewise the inset, consisting of all trajectories asymptotic as $t \to +\infty$, is tangent to the subspace spanned by incoming eigenvectors. The dimension of the inset plus the dimension of the outset equals the dimension n of the phase space.

It should be remembered that the validity of local linearization of nonlinear dynamics near an equilibrium is guaranteed only in the generic case of hyperbolic fixed points.

10.4 HYPERBOLIC STABILITY TYPES FOR LIMIT CYCLES

The next step in understanding dynamic stability is to apply the ideas of hyperbolicity and attractor, saddle, and repellor to periodic final behaviour. Here, too, stability may be considered in the sense of Liapunov, that is the asymptotic stability under perturbation of a trajectory, or more broadly, by perturbing the dynamical system itself to determine the structural stability. In either case the geometric view relies on portraits of the nearby trajectories, in a full neighbourhood of a limit cycle. Local information about trajectories near any one point along the limit cycle trajectory will not determine the stability of the full cycle. For example, the flow of air in the cylinder of an internal combustion engine might be stable during the exhaust part of the cycle but unstable on intake. To determine if the engine runs properly requires knowledge of the full cycle.

The stability of a closed-loop trajectory is determined by following nearby trajectories in phase space through a full period of the closed loop; if all nearby trajectories return near their starting points in state space but *closer* to the loop, then the loop is a stable limit cycle. Figure 10.11 shows one such nearby trajectory. Although the vector field may cause a temporary destabilization, the net change during one trip round the closed loop is towards the closed loop. If all nearby trajectories approach closer after every trip around, the closed loop is an attractor.

To determine precisely when a nearby trajectory has made one full trip, and hence whether it is approaching, it is convenient to use the Poincaré section, cutting across the loop at an arbitrary point. An example is the heavy arc in Figure 10.11. This surface need not be orthogonal to the closed-loop trajectory, but it is essential that the surface not be tangent to the curve nor to any nearby trajectory. We say the surface cuts the bundle of trajectories in a *transverse section.* The surface must have codimension 1, that is, its dimension must be one less than the dimension of the phase space.

Linearization via the Poincaré map

In addition to marking the point of full return for nearby trajectories, the Poincaré section serves to construct a linear approximation of the structure of trajectories near the closed loop. This linearization will be local in the sense that it describes the structure of trajectories in a small sleeve enclosing the loop; but

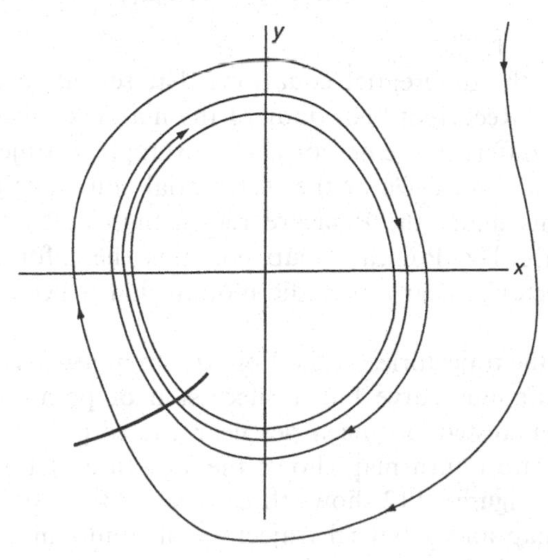

Figure 10.11 Trajectory approaching a limit cycle in plane phase space

it is not local in the sense of a short segment of the sleeve. In other words, it is not enough to linearize the vector field near any single point.

The simplest possible linearization of the structure of trajectories near a closed loop is obtained by constructing a new function, the return map on the surface of transverse section. The Poincaré map takes a point A in the Poincaré section near the periodic trajectory, and maps A to another point B in the section by following the unique dynamic trajectory through point A once around to the next time that trajectory crosses the surface. The next point of intersection is the image B under the Poincaré map of point A. Figure 8.1 shows an example of a trajectory near a closed loop that constructs the Poincaré return map of point A.

The linearized structure of trajectories near the loop can now be studied through the agency of the Poincaré map. First we need an $(n-1)$-dimensional coordinate system in the surface of section. It may be possible to choose the surface conveniently so that its coordinates are just the n coordinates of phase space with one coordinate fixed. Let us suppose this can be done, so that the Poincaré map is in principle a function

$$\begin{aligned}
\tilde{x}_1 &= p_1(x_1, x_2, \ldots, x_{n-1}) \\
\tilde{x}_2 &= p_2(x_1, x_2, \ldots, x_{n-1}) \\
&\vdots \\
\tilde{x}_{n-1} &= p_{n-1}(x_1, \ldots, x_{n-1})
\end{aligned}$$

taking a point X with coordinates $x_1, x_2, \ldots, x_{n-1}$ to $\tilde{X}$ with coordinates $\tilde{x}_1, \tilde{x}_2, \ldots, \tilde{x}_{n-1}$. The point where the periodic loop cuts the surface is a fixed point of this mapping, that is, a point X^* such that

$$x_j^* = p_j(x_1^*, x_2^*, \ldots, x_{n-1}^*)$$

for $j = 1, 2, \ldots, n-1$.

This map, like the differential equations that served to construct it, is a dynamical system. Recurrent behaviour of the map is equivalent to recurrent behaviour in the differential equations. For example, a trajectory that ultimately approaches a limit cycle of the differential equations corresponds to a sequence of points under the Poincaré return map that approach the fixed point of the map. The Poincaré map contains the information about the structure of trajectories near a periodic motion that is needed to characterize the recurrence.

Now consider the trajectories of the Poincaré map itself. Here a trajectory is no longer a continuous curve but a succession of points in the surface of section. We are interested in trajectories near a fixed point. The nearby linear part of the Poincaré return map shows the structure of trajectories that are sufficiently close. Figure 10.12 shows three types of fixed point that can occur for a Poincaré map and a typical trajectory of points in a two-dimensional surface of section. Each type of trajectory corresponds to a different type of periodic motion. In each case the origin of the coordinate axes marks the

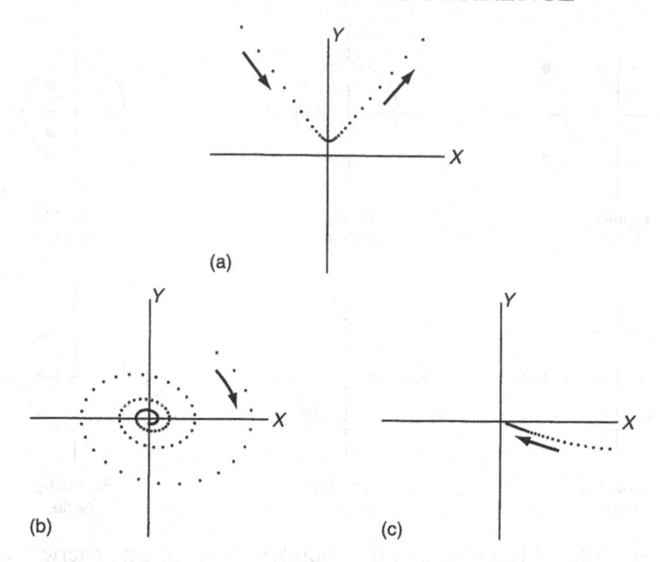

Figure 10.12 Trajectory of a Poincaré map near three types of generic fixed point: (a) saddle, (b) focal attractor and (c) nodal attractor

location of the fixed point of the map, where the periodic cycle crosses the surface of section. The three types of structure are (asymptotically) straight, saddle, and spiral. In the asymptotically *straight* or *nodal* case, the nearest points of the sequence lie on a straight ray through the fixed point. The sequence of points progresses towards the fixed point if the cycle is stable. The same straight trajectory can also be seen near a repelling cycle, in which case the sequence of points moves away. The *spiral* trajectory may also either approach an attracting stable fixed point or move away from a repelling unstable one, but in either case return points wind around the fixed point at an asymptotically constant rate. The third fixed point in Figure 10.12 corresponds to an unstable cycle of *saddle* type; a typical trajectory at first progresses towards the fixed point and slows down but then veers away.

These structures can be identified from the linear part of the Poincaré map near the fixed point, characterized by eigenvalues and eigenvectors. An attracting limit cycle is necessarily one whose eigenvalues lie inside the unit circle in the complex plane; a repelling cycle has all eigenvalues outside the unit circle. Saddle cycles have some (but not all) eigenvalues with magnitude less than 1. An eigenvalue with magnitude exactly equal to 1 corresponds to a non-hyperbolic cycle; this case is rare (non-generic) and its structures are changed by the smallest perturbation of the dynamics (structurally unstable). As with equilibria, the *hyperbolic limit cycles* are structurally stable in the sense that a slightly perturbed dynamical system will have limit cycles with the same stability type as the unperturbed system. Hyperbolic cases for two-dimensional Poincaré section are summarized by Argand diagrams in Figure 10.13. Three-dimensional phase-space views of attracting types were shown in Figure 4.2.

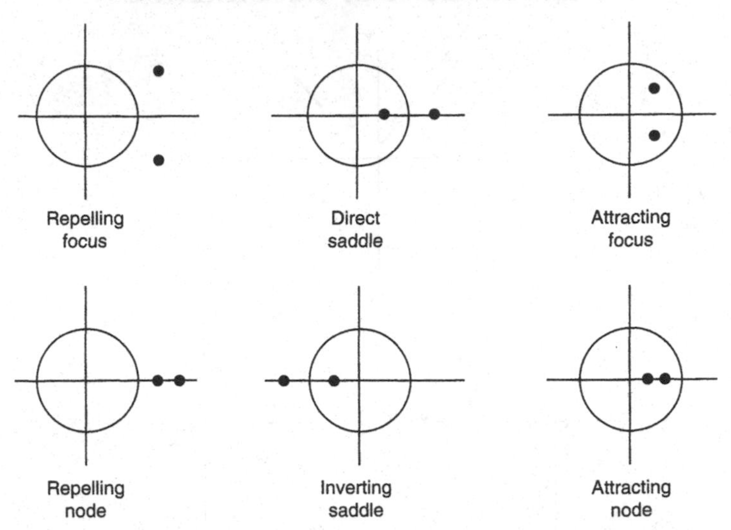

Figure 10.13 Eigenvalues in the complex plane for six generic fixed points of a planar Poincaré mapping, corresponding to generic limit cycles in three-dimensional phase space

As with equilibria, an instability index may also be defined for cycles, this time by counting the eigenvalues outside the unit circle. Figure 10.13 indicates the cases with positive real part. For maps, an eigenvalue with negative real part indicates that, near the fixed point, successive return points fall alternately on opposite sides of the fixed point. (This is a variation not found in equilibria.) An example not in Figure 10.13 is the nodal attracting cycle with two negative real eigenvalues inside the unit circle (cf. Figure 4.2). Another important case of negative multipliers is the inverting saddle, which is shown in Figure 10.13; the inverting saddle is created for example by a supercritical flip bifurcation, and is present in folded band attractors.

Saddle cycles and phase portraits

The limit cycle of saddle type has special trajectories asymptotic to it as $t \to +\infty$, forming an inset, and trajectories asymptotic to it as $t \to -\infty$, which form the outset of the saddle cycle. Viewed in the Poincaré section, the inset trajectories approach the incoming eigenvector, while the outset departs along the outgoing eigenvector. Figure 10.14 shows an example. Notice that many sequences of points are shown in the inset and in the outset; there are in fact infinitely many asymptotic trajectories, all ultimately tangent to the eigenvectors, which fill in the space between points of the trajectories shown. Thus the inset and outset fill smooth curves in the Poincaré section, and smooth two-dimensional surfaces in three-dimensional space.

Figure 10.14 also shows a pattern of straight lines connecting a ring of points to their images under the Poincaré map. Note that traversing the ring

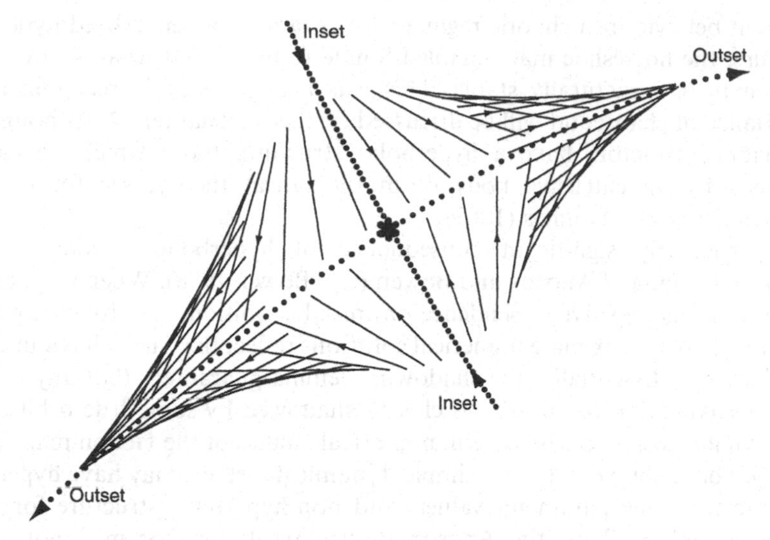

Figure 10.14 Poincaré section of a saddle limit cycle with inset, outset and some representative point mappings

clockwise, the direction of mapping lines rotates once anticlockwise. This topological property is useful in searching for saddles; the underlying theory can be found in Lefschetz (1957), Birkhoff (1913), Smale (1967) and Hsu (1980).

These invariant manifolds may seem rather abstract at this point, but we will see in the next chapters how crucial they are to understanding nonlinear dynamics and chaos. For example, although we did not identify them, saddle cycles were present in all the forced oscillator systems of Chapter 6. Following their outsets leads us to the attractors; in cases where competing attractors exist, the separatrix between basins is an inset. In general, the complete *attractor-basin* or *AB portrait*, in the terminology of Abraham (1979), in finite-dimensional phase space hinges on the invariant manifolds of saddle cycles and saddle equilibria. Furthermore, the structure of chaotic attractors is determined by intersections of insets and outsets, as we will see in the next two chapters. And finally, the bifurcations of chaotic attractors can only be completely understood in terms of insets and outsets. A striking example, the blue sky catastrophe for the Birkhoff–Shaw bagel, will be discussed in Chapter 13.

10.5 IMPLICATIONS OF HYPERBOLIC STRUCTURE

By concentrating on hyperbolic equilibria and cycles, we can characterize stability in a precise and systematic way. This characterization has both an analytic and an equivalent geometric interpretation. The notion of a nearby hyperbolic structure can also be defined for non-periodic recurrence. Although hyperbolic and generic are not synonymous for non-periodic recurrence, hyperbolicity can be helpful in understanding the structure of recurrent behaviour in chaotic regimes. For example, a generalized hyperbolic structure, the horseshoe map, enabled Smale to prove that what is now called chaos may be structurally stable. This was a major step in recognizing the importance of chaos, and will be discussed further in Chapter 12. Although not all chaotic attractors have a hyperbolic structure, those which do can be described by an extensive body of mathematical theory; see for example Guckenheimer and Holmes (1983).

A particularly significant consequence of hyperbolic structure is the shadowing lemma of Anosov and Bowen (e.g. Bowen 1978). When a dynamical system exhibits sensitive dependence on initial conditions, the following question arises: do approximate numerical solutions reflect the true behaviour of the actual system? Essentially, the shadowing lemma guarantees that any reasonable approximation to an orbit is closely shadowed by some true orbit of the same dynamical system. By careful numerical studies of the Hénon map, Lai *et al.* (1993) have shown that even simple dynamical systems may have hyperbolic structure for some parameter values, and non-hyperbolic structure for other parameter values. Thus the Anosov–Bowen result may or may not apply. Hammel *et al.* (1987) have given a computer-assisted proof that an approximate orbit of a non-hyperbolic system, the logistic map, can be shadowed by a true orbit for a long time. How long depends on the approximation: the proof stops when the orbit visits a neighborhood where the structure is non-hyperbolic. These results suggest that numerical orbits subject only to round-off error are close to true orbits for millions of iterations. But observed orbits subject to experimental noise and real-world disturbances might need to be interpreted with some caution.

Hyperbolic structure is an important consideration in applying other numerical methods to chaotic systems, such as reducing noise in a chaotic signal (Abarbanel 1996).

In higher-dimensional phase space, non-hyperbolic structure can be a more serious problem. When one of the Liapunov exponents fluctuates near zero, shadowing fails so dramatically that such systems have been called *unmodelable*; see Kostelich *et al.* (1997).

Although 'non-generic' and 'non-hyperbolic' are negative terms, it should not be inferred that non-generic structures are uninteresting. On the contrary, non-generic structures occur as a dynamical system goes through a bifurcation from one generic structure to a different one. This suggests a systematic approach to bifurcations begun in Chapter 7 and pursued in Chapter 13. But first, let us examine how the hyperbolic structures of points and cycles are used in constructing a full phase portrait of a dynamical system with a chaotic attractor, the Lorenz system.

11

The Lorenz System

Among the first to observe and describe chaotic solutions in a simple differential equation was E. N. Lorenz, a student of Birkhoff. In 1963 Lorenz published a paper describing the numerically observed behaviour of solutions of a system of three first-order ordinary differential equations with simple nonlinearities:

$$\begin{aligned} \dot{X} &= -\sigma(X - Y) \\ \dot{Y} &= RX - Y - XZ \\ \dot{Z} &= XY - bZ \end{aligned} \tag{11.1}$$

Lorenz noticed exponential divergence of nearby initial conditions, and interpreted this as evidence for the fundamental difficulty of predicting the evolution of a turbulent fluid flow from necessarily imprecise knowledge of initial conditions. The equations themselves have since been the object of much study, and a fairly complete picture of their phase-space structures has been assembled, at least in certain ranges of the parameters, σ, b and R.

11.1 A MODEL OF THERMAL CONVECTION

Lorenz was motivated by the problem of weather forecasting. Part of the turbulent motion of the atmosphere is caused by thermal convection, when air warmed near the Earth's surface rises. The resulting convection currents may spontaneously organize themselves in convection cells, either as long cylindrical rolls or in some cases forming a pattern that resembles a honeycomb when viewed from above. In a honeycomb cell, warm fluid rises in the central portion while cold fluid descends near the edges of the cell. In a cylindrical roll, fluid circulates around the horizontal axis of the cylinder. Honeycombs of nearly hexagonal *Bénard cells* can be observed in the laboratory in a shallow heated pan; for a photograph and more information, see Velarde and Normand (1980). The effect of convection cells on the Earth's surface can be seen in the formation of sand dunes in the desert, as illustrated in Abraham and Shaw (1985).

Lorenz obtained his equations from an idealized mathematical model of thermally driven fluid convection first considered by Rayleigh. This model might represent a vertical plane cross-section perpendicular to the axis of a cylindrical convection roll. A viscous, thermally conducting fluid in a two-dimensional rectangular flow region is heated uniformly along the bottom edge of the rectangle in such a way that the temperature difference ΔT between fluid at the top edge and fluid at the bottom edge is always constant (see Figure 11.1). This viscous flow and the conduction of heat through the fluid can be described by partial differential equations, which Lorenz used as a starting point for deriving a simpler, approximate model using Fourier series. The unknowns in the partial differential equations are the field of fluid velocity $\mathbf{v}$ in the box and the temperature field T. Because the problem is assumed to be two-dimensional, i.e. everything is uniform in the third dimension, the incompressibility condition $\nabla \cdot \mathbf{v} = 0$ implies that the vector velocity field $\mathbf{v}$ can be replaced by a scalar field ψ, the stream function, where $\mathbf{v} = (-\partial\psi/\partial y, \partial\psi/\partial x)$. Suppose that the instantaneous stream function ψ and temperature T can be represented inside the rectangle by Fourier series in the space coordinates. One can substitute the presumed Fourier series expansions—with time-dependent coefficients—into the partial differential equations. This had been done by Saltzman (1962), who found a system of ordinary differential equations, one for each of the time-dependent coefficients; these determine the temporal evolution of all coefficients.

Since the resulting set of coupled ordinary differential equations is infinite in number, this is not much of a bargain, unless there is reason to believe that the original problem can be reasonably well described by a limited number of low-frequency modes in the Fourier series expansion. In this case we might discard all but a finite number of the ordinary differential equations for the coefficients. Based on Saltzman's observation that numerical solutions of these ordinary

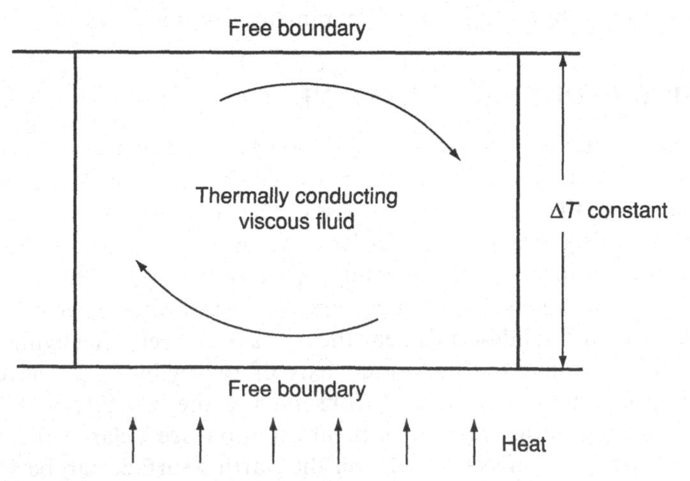

Figure 11.1 An idealized rectangular region of thermal convection modelled by the Lorenz equations. The bottom of the rectangle is heated in such a way that the temperature difference between top and bottom remains constant

differential equations sometimes settled into final behaviour wherein only a few modes continued to oscillate (the remaining modes having reached equilibrium values), Lorenz decided to focus on the dynamics of the first three modes of the Fourier series.

The first of these three modes is the first term in the expansion of the stream function ψ, representing convective motion. This first mode corresponds to convective circulation in a single large eddy that fills the rectangle, as suggested in Figure 11.1. When the coefficient $X(t)$ of this first mode is positive, the circulation is clockwise with speed proportional to X; negative X corresponds to anticlockwise circulation. This is the only mode of fluid motion in the Lorenz model, so we are assuming our fixed rectangle is always a cross-section of exactly one convective roll.

The second mode with coefficient $Y(t)$ is the first term in the Fourier series expansion of the temperature, and describes a temperature distribution with fluid warmer on one side of the rectangle. If X and Y have the same sign, the warmer fluid is on the side that is rising in the convective eddy, while the colder fluid is descending on the other side. These two modes have the same left-right symmetry as the rectangle itself, and as a result any particular values X, Y describe a state of flow and temperature that is a mirror image of $-X$, $-Y$; this fact can be verified directly in the Lorenz equations above, because changing the signs of X and Y just changes the signs of $\dot{X}$ and $\dot{Y}$.

The final mode with coefficient $Z(t)$ affects the vertical profile of temperature. When $Z = 0$, the temperature along every vertical line changes linearly from top to bottom. When Z is positive, the gradient of temperature is steeper near the top and bottom of the rectangle and more gentle (or even reversed) in between.

The ordinary differential equations for X, Y and Z were obtained by Lorenz (following Saltzman) to describe the dynamics of flow and temperature in the rectangle if the fluid recognized only these three modes of action. The constants in the differential equations are σ, the Prandtl number of the fluid; R, a dimensionless Rayleigh number representing a ratio of ΔT (the driving force) to damping caused by viscosity and thermal conduction; and $b = 4/(1 + a^2)$ where a is the aspect ratio of the rectangle. Lorenz chose values for each of these, $\sigma = 10$, $b = 8/3$ and $R = 28$, which resulted in solution behaviour since named chaotic. The value $\sigma = 10$ is too high for dry air, and would correspond to cold water as fluid; $b = 8/3$ yields a rectangle of proportions that minimize the value of ΔT needed for onset of convection. In addition to this single value of R, we will also consider in this chapter a family of dynamical systems obtained by varying the parameter R, the ratio of driving to damping.

11.2 FIRST CONVECTIVE INSTABILITY

The first test of the Lorenz equations is to see whether they represent the first instability of thermal convection: for low values of R (i.e. low thermal driving or high viscosity and thermal conductivity), the rest state $X = Y = Z = 0$ should be a stable equilibrium, but as R increases this equilibrium should lose stability as the influx of heat produces a convective roll. This happens when the buoyancy of warm fluid overcomes damping forces, and the warm fluid rises to the top to be cooled and descend again. An analysis of this instability was carried out by Lord Rayleigh, who showed that the onset of convection is predicted by a dimensionless function of fluid properties and thermal driving corresponding to $R = 1$ in equations (11.1).

From the differential equations for X, Y and Z, it is clear that the rest state (0, 0, 0) always gives zero time derivatives, and hence is an equilibrium state for any values of σ, b and R. The stability of this equilibrium is determined by linearizing the right-hand sides near (0, 0, 0) and finding the eigenvalues λ of

$$\begin{vmatrix} -\lambda-\sigma & \sigma & 0 \\ R & -\lambda-1 & 0 \\ 0 & 0 & -\lambda-b \end{vmatrix} = 0$$

One eigenvalue $\lambda = -b$ corresponds to the eigenvector (0, 0, 1), which means that any small Z mode disturbance from the rest state is always damped out since b is positive. The other two eigenvectors lie in the (X, Y) plane and have eigenvalues satisfying

$$\lambda^2 + (\sigma + 1)\lambda + (1 - R)\sigma = 0$$

If σ is fixed, a root may only change sign as R passes the value 1. For R near zero the eigenvalues are near $-\sigma$ and -1; in this case the origin has a *hyperbolic* structure, which is attracting along all three eigenvectors. There are no spiral dynamics because all eigenvalues are real. Some typical trajectories are shown in Figure 11.2.

Now consider the behaviour of the system (11.1) near the origin as R increases. As R approaches 1, the two most negative eigenvalues (near $-b$ and $-\sigma$) and their eigenvectors change only slightly, but the least negative

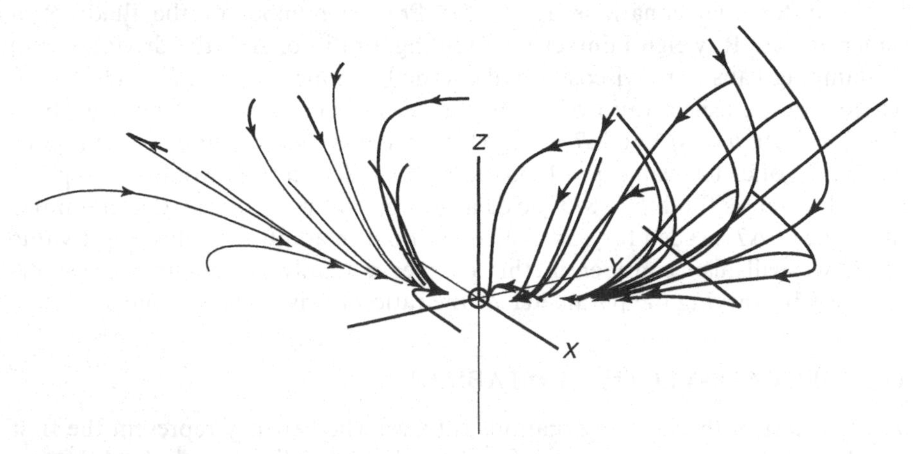

Figure 11.2 Trajectories in the three-dimensional phase space of the Lorenz equations (11.1). Here the parameter R is less than 1, and all trajectories approach stable equilibrium at the origin. Thicker curves are closer to the viewer

eigenvalue increases continuously until, at $R = 1$, it reaches zero and becomes positive for R greater than 1. The instability index of the fixed point at the origin changes from 0 to 1, and the equilibrium has become unstable, changing to a saddle point. A bifurcation has occurred, giving rise to qualitatively different final behaviour. From the right-hand side of equations (11.1) we find that in addition to the fixed point (0, 0, 0) there are now two additional fixed points whose coordinates satisfy $X = Y$, $Z = X^2/b$ and $X[(R-1) - X^2/b] = 0$. These equations have non-zero real solutions when R is greater than 1. The new pair of solutions corresponds to a pair of equilibrium points that emerge from (0, 0, 0) as R passes the value 1.

To determine the stability of the new fixed points, we find the linear approximation of the vector field near them; the linear part here turns out to have three negative real eigenvalues for R just greater than 1. This is not surprising because two of the eigenvalues and eigenvectors of the fixed point at the origin have changed little as R passes 1, so when the new fixed points are still close to the origin, continuity in phase space suggests that the vector field should point inwards along directions that nearly parallel the inward eigenvectors at the origin. The third eigenvector at the origin, whose eigenvalue changes sign in the bifurcation at $R = 1$, is in fact the direction along which the new fixed points emerge. This eigenvalue is tangent to the *centre manifold.* Along this direction the vector field switches, as R passes 1, from pointing towards the origin to pointing away, and hence towards the new fixed points. The situation is illustrated schematically in Figure 11.3, where for simplicity the two always-attracting directions are projected down to one. These are stages of the pitchfork bifurcation of one stable equilibrium producing a pair of new stable equilibria and leaving a saddle equilibrium behind.

Because the rest state is no longer stable, the Lorenz model of the rectangular fluid cell now has dynamics that eventually settle to one or the other of the two new fixed points, each corresponding to a steady convective circulation. The two new fixed points are symmetric, because the two final steady motions are left-right mirror images of each other. Which of the two final motions is observed depends only on the symmetry breaking implicit in a given set of initial conditions.

Since the three-mode Lorenz model passes this elementary test of describing (at least qualitatively) the onset of first instability in the fluid, we will follow Lorenz in studying the behaviour of the model for much larger values of R

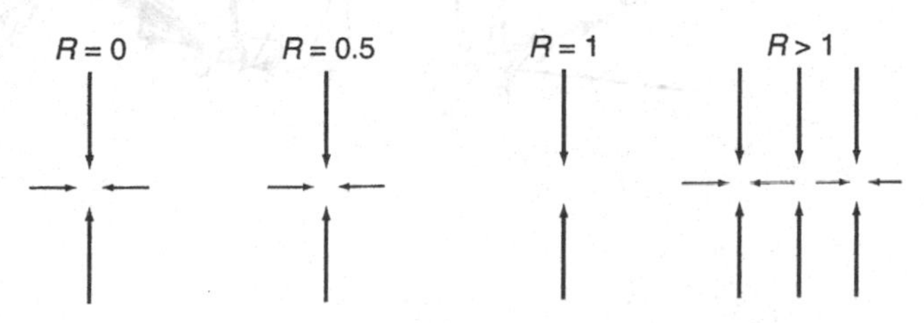

Figure 11.3 Schematic diagrams to represent the supercritical pitchfork bifurcation of equilibria at $R = 1$ in equations (11.1)

where more complicated, non-periodic behaviour appears. Before doing so, however, let us complete our portrait of the dynamics for R just greater than 1 by looking at the basins of the two attracting fixed points.

Any solution trajectory of the Lorenz equations for R just greater than 1 eventually settles to one or the other of the two attracting fixed points. The existence of only two stable fixed points suggests this is true, but to be sure, we need to try a convincingly large sample of initial conditions and verify that there are no other attractors, periodic or otherwise. Figure 11.4 shows a sample batch of trajectories, all settling to one of the two stable equilibrium points. The state space appears to be neatly divided into two regions, each consisting of states that are eventually drawn to one of the attracting fixed points. These regions are the *basins* of the attractors, also called domains of attraction or catchment areas. The two basins make up all of the phase space, except for an infinitely thin surface that is the common boundary of the two basins. This surface, the *separator*, must pass through the unstable saddle equilibrium point at the origin; states near the origin diverge in the direction of either attracting fixed point, as in Figure 11.4.

In the linear approximation of the dynamics about the saddle point, trajectories leading straight away from the origin lie on the eigenvector with positive eigenvalue. Conversely, trajectories starting on any linear combination of the other two eigenvectors (with negative eigenvalue) would, in the linear approximation, end up settling to the origin. Of course, a starting point even slightly off the plane of the two inward eigenvectors would, in the linear approximation, begin by moving towards the origin but, as the outward eigenvector begins to exert its influence, the trajectory would be swept away towards an attracting fixed point (cf. Figure 10.10). Thus in the linear approximation near the origin, the separator of the two basins is the plane of the two eigenvectors with negative eigenvalues.

Once this is recognized, the true separator (without linearization) can be easily constructed: take all linear combinations of the two *inward* eigenvectors

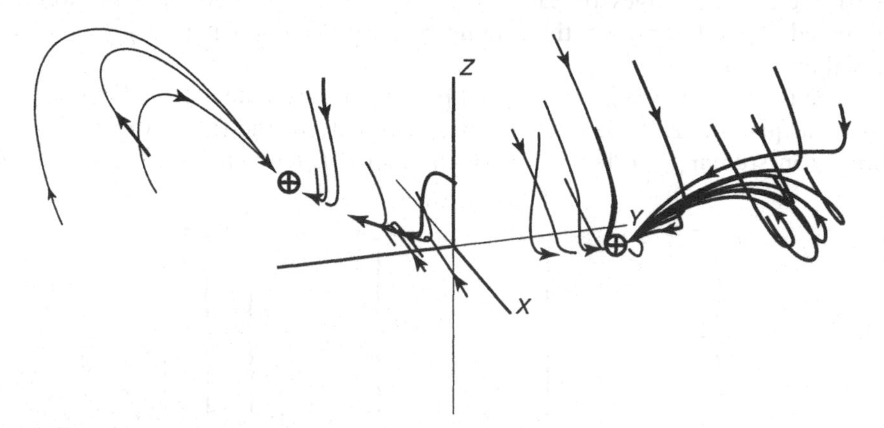

Figure 11.4 Trajectories of the Lorenz system attracted to the two new stable equilibria, for R just greater than 1

and, starting from an infinitesimal displacement in those directions, integrate *backwards* in time. The ensemble of integral curves thus determined fits together to form a smooth invariant surface, the *inset* or *stable manifold* of the fixed point. This is the set of initial conditions which, if integrated forwards in time, will not veer off to either attracting fixed point, but instead settle precisely onto the saddle equilibrium as $t \to +\infty$.

In a general dynamical system, a hyperbolic saddle point has an inset whose dimension equals n (the dimension of the phase space) minus the index of the fixed point. The local span of all the *outward* eigenvectors integrated forwards in time generates another manifold, the *outset* or *unstable manifold* of the fixed point. The dimension of the unstable manifold equals the index of the fixed point, whereas the codimension of the stable manifold equals the index. In the Lorenz system with R just greater than 1, the outset of the origin consists of the two one-dimensional paths asymptotically leaving the origin along the outgoing eigenvector, and finally arriving asymptotically at the two attracting equilibria.

These outstructures of the origin are the key to understanding much of the geometric behaviour of trajectories of the Lorenz system, even at much larger R values where the inset no longer plays the role of separating two basins. The traditional terms 'stable manifold' and 'unstable manifold' can, however, be confusing: trajectories near a stable manifold fall *away* from the stable manifold, while trajectories near an unstable manifold typically fall *towards* the unstable manifold. Thus for the sake of both clarity and brevity we prefer Zeeman's terminology, and refer to the stable manifold of the origin as its *inset*; the unstable manifold will be called the *outset*. The construction of an inset by extending the inward eigenvectors to a global structure is of great practical use

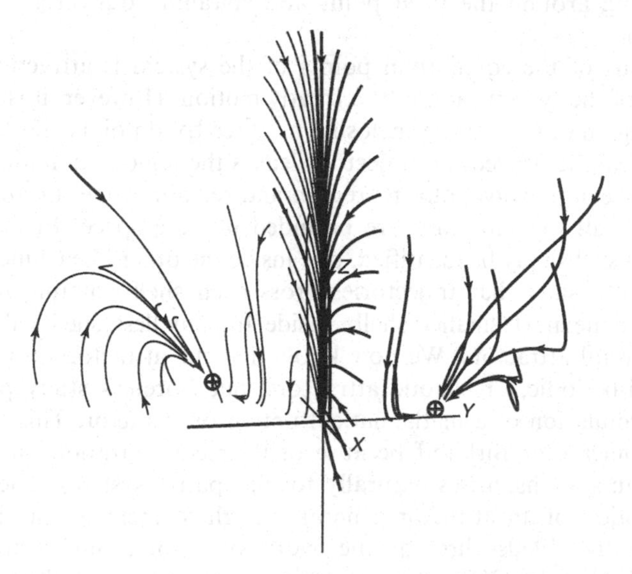

Figure 11.5 Trajectories moving towards the two attracting fixed points for $R > 1$, with the basin separator shown

when governing equations (or the complete vector field) for the dynamics are known, and can be integrated backwards in time. In general, an inset is identified as the collection of trajectories asymptotic to a saddle equilibrium (or a saddle limit cycle) as $t \to +\infty$. An outset is the trajectories asymptotic to a saddle as $t \to -\infty$. A basic theorem of differentiable dynamics asserts that insets and outsets are smooth manifolds. Figure 11.5 shows a sampling of curves lying in the inset of the origin, the separator, along with a sampling of solution curves moving through the two basins towards the attracting fixed points.

11.3 THE CHAOTIC ATTRACTOR OF LORENZ

Geometric outstructure analysis applies equally to the chaotic regime of the Lorenz system. Chaotic behaviour occurs at larger values of R, where increased thermal driving causes a greater degree of instability. At $R = 28$ (the value studied by Lorenz) the differential equations still have three equilibrium points. However, all three are now saddles. The origin has index 1, with straight dynamics as before. The other two fixed points, always a symmetric pair, now lie at a considerable distance from the origin; their Z coordinates are always $(R-1)$, while their X and Y coordinates are $\pm\sqrt{[b(R-1)]}$. The index of each of these symmetric fixed points is now 2; analysis of the linear part of the vector field near each fixed point shows one negative real eigenvector and a complex conjugate pair with positive real part. Trajectories near such a fixed point are first drawn along its one-dimensional inset towards the two-dimensional outset of the fixed point, which is a surface tangent to the eigenvectors with positive real part. Once near this outset, a trajectory moves along it, circulating around the fixed point and spiralling outwards, as shown in Figure 11.6.

Since none of the equilibrium points of the system is attracting, the final behaviour of the system cannot be a steady motion. However, it is not difficult to find a region of state space enclosing all three fixed points and large enough so that (as can be proved) no trajectory leaves the region: all initial conditions outside the region evolve into the region and remain inside for all subsequent time. Hence all final motions are bounded in state space. In fact, a nested sequence of shells may be identified, one inside the other like Chinese boxes, or Russian dolls, such that trajectories cross each shell pointing inwards and continue to the next smaller shell. Inside the smallest shell will be one or perhaps several attractors. We now know that the attractor(s) might be periodic, quasi-periodic, or chaotic attractors, but Lorenz's study preceded the general formulation of a mathematical notion of attractor. This concept, not used by Poincaré or Birkhoff because of their concentration on *conservative* Hamiltonian systems, arises naturally for dissipative systems. The mathematical conception of an attractor general enough to include chaotic attractors evolved in the 1960s through the work of Thom, Smale and Williams (Thom 1975; Smale 1967; Williams 1974), and is based on the intuitive idea of trajectories falling through nested Chinese boxes. In the case of the Lorenz

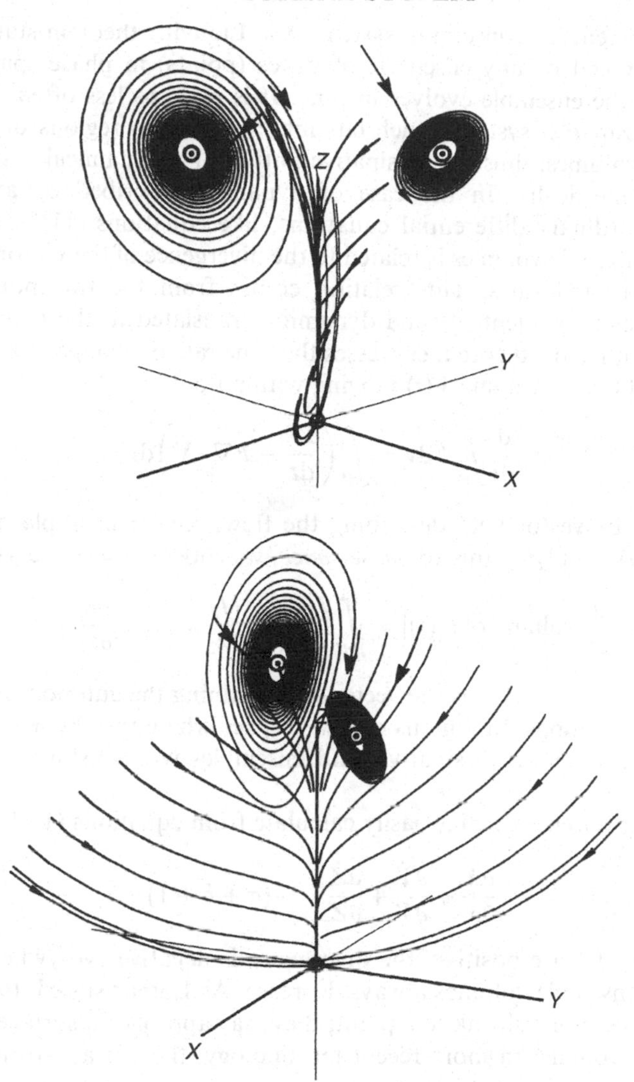

Figure 11.6 Trajectories of the Lorenz system (11.1) near the spiral saddle points for $R = 28$, shown with part of the inset of the nodal saddle point at the origin

system, there is a large gap between the smallest easily provable box and the much smaller numerically observed region of final motion; the existence of intervening nested boxes is conjectural.

Although the general notion of a non-periodic attractor had not been formulated by mathematicians in 1963, Lorenz was fully aware that clouds of initial conditions for equations (11.1) must shrink towards some final asymptotic structure. His primary evidence for this involved a simple computation, related to Liouville's notion of evolving ensembles in phase space. For

Hamiltonian, energy-conserving systems, the Liouville theorem states that the volume occupied by any ensemble of states (points) in phase space remains constant as the ensemble evolves in time. It is perhaps less often emphasized that, for dissipative systems, such ensembles evolve to regions of constantly decreasing volumes; thus the dissipative nature of a dynamical system can be verified geometrically. In the case of dynamics described by autonomous systems of ordinary differential equations, like equations (11.1), the rate of change of ensemble volumes is related to the divergence of the vector field given by the right-hand sides. This relation comes from the transport theorem, familiar to every student of fluid dynamics, translated to the realm of phase space. The transport theorem expresses the time rate of change of any function F integrated over a volume $V(t)$ moving with a flow as

$$\frac{\mathrm{d}}{\mathrm{d}t}\int_V F\mathrm{d}v = \int_V \left(\frac{\mathrm{d}F}{\mathrm{d}t} + F\nabla \cdot \mathbf{V}\right)\mathrm{d}v$$

where $\mathbf{V}$ is the vector field describing the flow; see for example page 131 of Serrin (1959). Applying this to *phase space*, we choose $F = 1$ everywhere, so

$$\frac{\mathrm{d}}{\mathrm{d}t}[\text{volume of } V(t)] = \int_V \left(\frac{\partial f_1}{\partial x_1} + \frac{\partial f_2}{\partial x_2} + \ldots + \frac{\partial f_n}{\partial x_n}\right)\mathrm{d}v$$

where $\{f_i(x_1, x_2, \ldots, x_n)\}$ is the vector field defining the autonomous ordinary differential equations. In regions of phase space where the above divergence is negative, the system is dissipative, and ensembles will constantly decrease in volume.

Following Lorenz, we may easily calculate from equations (11.1) that

$$\frac{\partial \dot{X}}{\partial X} + \frac{\partial \dot{Y}}{\partial Y} + \frac{\partial \dot{Z}}{\partial Z} = -(\sigma + b + 1)$$

Because σ and b are positive, the divergence is negative everywhere in phase space, and ensemble volumes always decrease. As Lorenz stated, this need not imply that ensembles shrink to a point; they may approach a surface, or any set having zero volume. In more recent terminology, there is an attracting set of zero volume.

Mathematical analysis thus establishes the existence of three equilibria, none of which is attracting, even though all final behaviour settles to a bounded region of state space with zero volume. Local linear analysis gives the structure of trajectories near the three saddle equilibria. To obtain a complete picture of the final behaviour, some global information is needed about how these local structures join together. This information can at present only be obtained by studying numerical solutions of the differential equations and examining the geometry of resulting trajectories. Here theoretical knowledge is not self-sufficient, but it is nevertheless essential in interpreting the numerical results.

Let us follow a trajectory spiralling out from one of the two symmetric fixed points. Since each corresponds to a convective motion with warm fluid rising, these two saddle points lie over the first and third quadrants of the (X, Y) plane; label the two symmetric fixed points C+ and C− respectively. As a trajectory spirals out from C+, it continues for many turns close to a two-dimensional surface, the outset of C+. Near C+ or C−, their outsets are inclined to each other so that, seen edge on, as in the bottom in Figure 11.6, they form a V. When the outward spiral from C+ reaches more than half the distance to C−, a change occurs; the trajectory now comes under the influence of C−. Travelling on the downward part of its last loop around C+, the trajectory drops into the region where the inclined two-dimensional spiral outsets come together at the bottom of the V. The trajectory now veers off towards C− rather than returning for another trip around C+. It rises up around C− and then begins an outward spiral along the outset of C−. When this spiral again reaches the halfway point, the trajectory switches back to spiral out from C+. This spiralling out and switching over continues forever. At these values of the parameters R, b and σ, no final attracting periodic motion is ever observed numerically. The result is shown in Figure 10.6.

The number of turns around each side before crossing over is not fixed, nor does it appear to settle into any pattern; in fact, it is both deterministic and unpredictable. When the trajectory crosses over is always determined by its relation to a well-defined surface, which separates a spiral returning around the same side from one crossing over. This surface, partly shown in Figure 11.6 by segments of curves lying on the surface, is the two-dimensional inset of the saddle point at the origin. A trajectory descending near the inset of the origin comes under the influence of the one-dimensional outset of the origin, which pulls it back around the same spiral if it falls on the near side of the two-dimensional inset, or pushes the trajectory across to the other spiral if it descends on the far side of the two-dimensional inset of the origin. We note that for R near 28 the two-dimensional inset of the origin is not a simple vertical wall as at R just greater than 1; the global geometric configuration is best appreciated via the illustrations in Abraham and Shaw (1985), and a computer-animated film by H. B. Stewart (1984). Although the two-dimensional inset of the origin guarantees strict determinism in each decision to cross over or not, nevertheless, if we note which side, C+ or C−, is visited on each successive turn, the sequence ultimately looks as random as a coin toss (Ford 1983).

To understand this seeming paradox, Lorenz reduced the behaviour on successive cycles to its simplest description, an iterated one-dimensional mapping. The function that defines the mapping can be determined by plotting the outward distance from a spiral node on one turn versus the distance on the previous turn.

This distance of course varies continuously through one turn and there are many equivalent ways of assigning a single distance to one turn; Lorenz chose to mark the highest point on each turn, that is, the maximum value of the Z coordinate. Figure 11.7 shows the result obtained by Lorenz.

To a very good approximation all points appear to lie on a single curve, that is, the peak value of Z on one turn appears to be uniquely determined by the maximum on the previous turn. On the left of Figure 11.7, the points

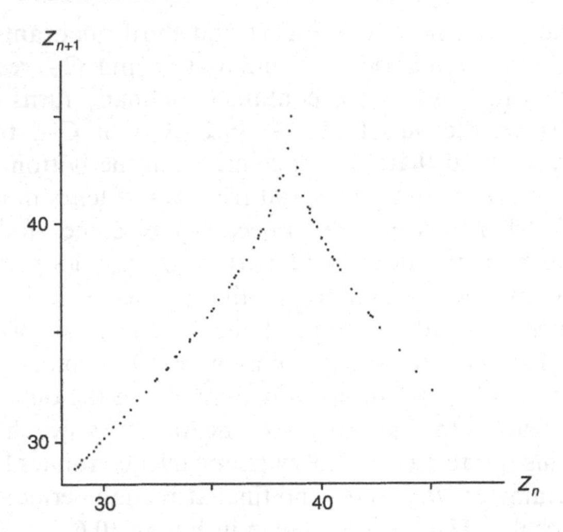

Figure 11.7 A one-dimensional map of successive maximum Z values of a final trajectory of equations (11.1). Reproduced, with permission of the American Meteorological Society, from (Lorenz 1963)

correspond to trajectories starting near C+ or C− and spiralling out, and the ordinate is always greater than the abscissa. In the middle, the graph is slightly misleading because there is discontinuous behaviour in the Z maximum of three-dimensional trajectories at the cusp of the graph. Just to the left of the high point, trajectories return on the same side of the two-dimensional inset, while just to the right they cross over and attain nearly the same maximum Z coordinate, but at opposite X, Y coordinates. This discontinuity of crossing over is not seen as a discontinuity in the graph of Figure 11.7, but a graph defined in a slightly different way would show discontinuity; see for example Guckenheimer and Holmes (1983).

In either case, points near the cusp represent trajectories coming very close to the two-dimensional inset of the origin, which then rise to very high maximum Z because they come very close to the one-dimensional outset of the origin. The one-dimensional outset loops around the spiral nodes at the greatest possible distance for a final motion. This great turn of the outset of the origin then crosses over and reinserts (into a two-dimensional spiral outset) as close as possible (for a final motion) to the spiral node, as indicated in Figure 11.8. Thus the points on the far right in Figure 11.7 have low ordinates.

From the Lorenz map in Figure 11.7 we can deduce why the sequence of crossings must ultimately be unpredictable. Two nearby final trajectories of the Lorenz system will be separated from each other by a distance that we may note as the difference of abscissae in Figure 11.7 at the beginning of one turn. At the end of that turn, their separation is given by the ordinates, and is now equal to the former separation times the slope of the Lorenz map function.

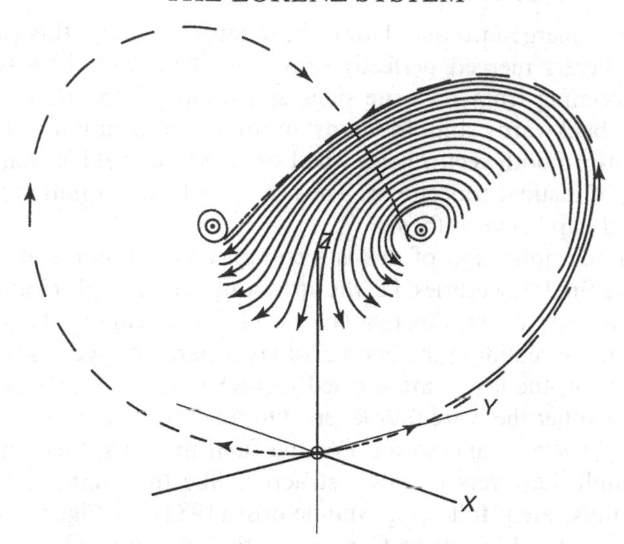

Figure 11.8 The outset of the origin (broken curves) reinserts near opposite spiral fixed points

This slope is always greater than 1, so nearby trajectories continually spread apart as they evolve. This spreading continues up to the point where trajectories fall on opposite sides of the inset of the origin, after which they spread as far apart as possible within the bounded attractor.

This spreading of nearby trajectories does not contradict the continually diminishing volume of a cloud of evolving states. Volumes in phase space are pressed into a spiral outset faster than they spread within the spiral. But the spreading in at least one direction has a profound effect when observing trajectories with less than infinite precision. Any uncertainty in the starting position of a trajectory leads to an error in the evolving position that increases exponentially as the true and approximate trajectories spread apart. The dynamics *depend sensitively* on *initial conditions*. The Lorenz map shows how the uncertainty grows geometrically, because it is multiplied on each full turn by the slope of the function in Figure 11.7. This means that predicting a later state of the system with specified accuracy will require that uncertainty of initial conditions decrease as an exponential function of the timespan of prediction. The implications of this for weather forecasting, assuming that real fluid turbulence has similar sensitivity to initial conditions, are discussed in Lorenz (1963, 1964).

The exponential separation of nearby trajectories is a basic property that distinguishes chaotic motion from periodic or quasi-periodic motion. If trajectories continually spread apart and yet stay within a bounded region of state space, there must at some point be a folding back together. In the Lorenz system this is achieved when the two halves of a cloud of trajectories are split apart by the saddle at the origin, come around over the spirals and layer together as they descend again. As the two layers approach the origin again, they appear to merge into one. However, strictly speaking, this cannot be the case. If the layers merged perfectly into one, there would be two different trajectories coming from opposite sides and joining into one trajectory. The uniqueness (backwards and forwards in time) of solutions of differential equations with smooth vector fields (the Lorenz vector field is analytic) forbids the merging of distinct trajectories. Thus the two layers coming together must retain a small gap between them.

The ultimate implication of this is a complicated structure for the set containing all the final trajectories, i.e. the attracting set. A single cloud splits apart and returns as two layers; this two-layered cloud is again split apart into two pieces, each piece retaining the two narrowly separated layers. As these pieces make a full turn, the layers are pressed further together, but when the pieces come back together there are four layers. On the next turn we find, in the same thickness, eight layers, and so on. For the final motions, then, cutting transversely through the layers reveals a structure like the *Cantor set* whose construction is illustrated, following Mandelbrot (1983), in Figure 11.9. At each stage in the construction of the Cantor set, the middle third is removed from each of the intervals present at the previous stage. If the original interval is [0, 1], the final set can be described as all numbers in [0, 1] whose expansion in base 3 notation contains 0s and 2s but no digit 1 in any position.

Although the proportions in this construction are not the same as in a section through the layers in the Lorenz system, we can use the Cantor set construction to verify that a set may have such an infinitely layered structure and still have zero volume. The total length of all intervals removed from Figure 11.9 is

$$\text{length removed} = \tfrac{1}{3} + 2\left(\tfrac{1}{3}\right)^2 + 4\left(\tfrac{1}{3}\right)^3 + 8\left(\tfrac{1}{3}\right)^4 + \ldots$$

whose sum is $\frac{1}{3}/(1-\frac{2}{3}) = 1$; the length remaining must therefore be zero. Thus a Cantorian cross-section may be consistent with the earlier conclusion that final motions in the Lorenz system occupy a set with zero volume. Since we now have a fairly clear picture of this set, we may give it a name: the Lorenz

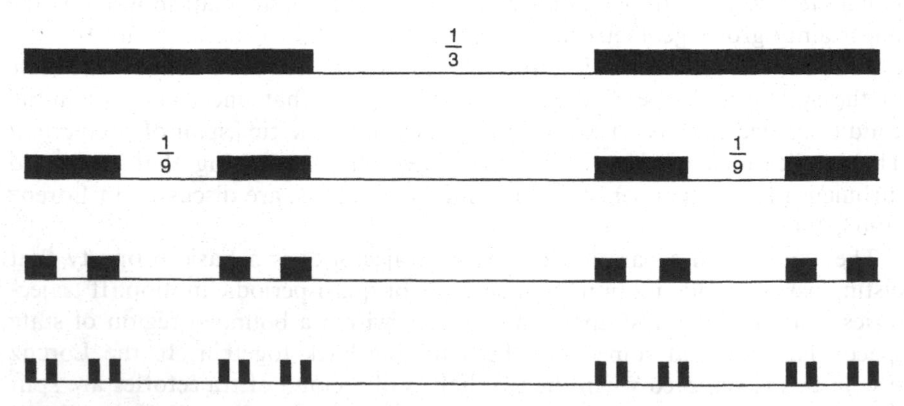

Figure 11.9 Diagram showing the first stages in the construction of the triadic Cantor set

attractor. This attracting set in the Lorenz system is an example of a chaotic attractor, often referred to in the literature as a strange attractor because of the complex, infinitely nested layer structure that seems to be a universal feature of chaotic attractors. Once it is appreciated that chaotic motions produce spreading trajectories, that spreading in a bounded region leads to folding, and that the folding is recurrent, the adjective 'strange' may seem unnecessary.

Although the detailed structure of the Lorenz attractor may seem complicated, the attractor is built from relatively simple geometric structures (Abraham and Shaw 1985). The essential ingredients are saddle equilibria with two-dimensional spiral outsets, each outset intersecting *transversely* with the two-dimensional straight inset of a third saddle point; and an organized *reinsertion* from each spiral outset to the other, guided by the one-dimensional outset of the origin. By composing the Lorenz attractor from these building blocks, we reinforce the impression that observed behaviour in the Lorenz system will be qualitatively unaffected by small perturbations, either in the initial conditions or the equations themselves. Of course we know that a particular trajectory over a specific timespan is quantitatively very sensitive to disturbances; but once having seen the underlying structure, we may hope that small disturbances merely induce the system to stray to a different path with the same or a nearby equivalent attracting structure.

The most exacting topological analysis of the Lorenz attractor (Williams 1979; Guckenheimer and Williams 1979; Guckenheimer and Holmes 1983) suggests that, contrary to this hope, there is qualitative structure in the Lorenz attractor, which can change due to small perturbations of the system, such as a small change in the value of R. Such detailed changes are evidenced, for example, by following the one-dimensional outset of the origin to its injection close to a spiral node and then counting the number of turns made before the next crossing to the other spiral node. For R around 28, the number of consecutive turns made by the one-dimensional outset is the greatest possible number of consecutive turns on one side, since no other final trajectory is reinserted closer to a spiral node. (This maximum number of consecutive turns is the only restriction on the claim that sequences of turns are ultimately as random as a coin toss.) If at $R = 28$ the one-dimensional outset makes 24 turns before crossing over, then for a somewhat larger value of R the outset makes only 23 turns before crossing over. By implication there is a unique R value in between such that after 23 turns the outset neither crosses over nor continues for a 24th turn, but finds itself landing precisely on the origin along its inset. In a strictly rigorous topological view these are qualitatively different structures.

The geometric model of the Lorenz attractor studied by Guckenheimer and Williams (1979)—see also Afraimovich *et al.* (1977)—provides much mathematical insight into the behaviour of solutions of the Lorenz equations (11.1). However, the geometric model is an idealization of what one observes in solutions of (11.1), and does not give a direct proof that the Lorenz system has a chaotic attractor. Recently, a rigorous proof that the Lorenz attractor exists has been announced by W. Tucker; see Stewart (2000) or Viana (2000) for more information.

11.4 GEOMETRY OF A TRANSITION TO CHAOS

Although the chaotic attractor of Lorenz is topologically speaking structurally unstable, the qualitative changes under perturbations near $R = 28$ have minimal impact on the observable dynamics, and pale when compared to the differences in behaviour from $R = 2$ to $R = 28$. It is natural to ask whether in the intervening range of R there are major qualitative changes at specific R values that account for the great differences in the observable dynamics. To answer this, we follow Kaplan and Yorke (1979b) and return to lower R values to consider the outstructures, that is, the insets and outsets of the equilibrium points. Their topology governs the major bifurcation events. A similar transition can also be observed by varying σ, as discovered by Afraimovich *et al.* (1977).

After the pitchfork bifurcation at $R = 1$, the new attracting fixed points have insets consisting of their three-dimensional basins while their outsets are empty. The only interesting outstructures are the inset and outset of the saddle point at the origin. For R just greater than 1, the outset is a slightly curved path leading from the straight saddle point to the straight attracting fixed points. The inset also has a slight curvature but for practical purposes can be regarded as a flat sheet. As R reaches about 1.346, the attracting nodes switch to spiral dynamics; each spiral plane is transverse to the inset of the origin and slightly tilted from the Z axis. Because the third (real) eigenvalue at C+, C− has large magnitude, trajectories in the basins are pulled quickly to the spiral plane, and then more slowly to C+ or C−. With further increase in R to 10 and beyond, this spiral behaviour becomes more pronounced. The imaginary parts of the complex conjugate eigenvalues increase in magnitude while the real part becomes less negative. As a result, the outset of the origin makes wider turns and settles more slowly to C+ or C−. This is shown in Figure 11.10, where the outset of the origin appears as a broken curve.

Increasing R also causes a growing deformation of the two-dimensional inset of the origin. If we imagine within this infinite inset a rectangular piece whose bottom edge sits on the (X, Y) plane, then near its upper corners this rectangle would bend in opposite directions. The vertical centreline of the infinite sheet remains coincident with the Z axis, as the equations confirm: $X = Y = 0$, $Z = Z_0 e^{-bt}$ is always a solution. On either side of the Z axis, the sheet pulls over as if each half would wrap around one of the attracting spiral equilibria. Figure 11.10 shows full curves that are samples of the trajectories making up the inset, thus indicating the locus of this two-dimensional smooth invariant manifold.

As the spiralling one-dimensional outset opens up and the two-dimensional inset bends down, a collision of the two outstructures appears inevitable. The collision occurs when R reaches about 13.926..., a value that can only be determined numerically. This collision causes remarkable changes in the

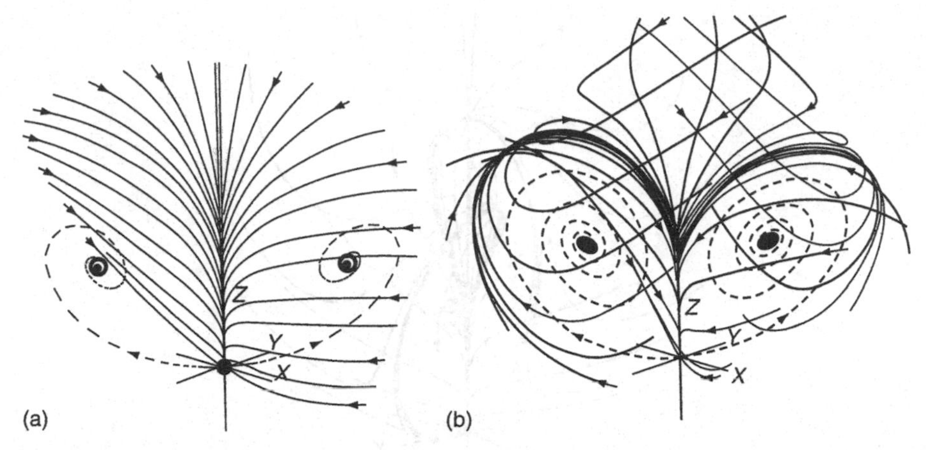

Figure 11.10 Outstructures of the saddle fixed point at the origin: (a) for $R = 5$, (b) for $R = 10$

outstructures and in the observable dynamics. The two-dimensional inset must touch the one-dimensional outset of the origin along its entire length at this one value of R; uniqueness of trajectories rules out an isolated point contact. (More generally, any two *invariant manifolds* of a dynamical system, i.e. manifolds made up of complete infinite-time trajectories, must intersect in an invariant manifold.) This means there is a trajectory starting from the origin along the outgoing eigenvector that loops around C+ or C− and returns to the saddle point in the plane of the inward eigenvectors; the outset of the origin lies in the basin separator and does not spiral down to C+ or C−. A trajectory that approaches the same equilibrium (necessarily a saddle) for $t \to -\infty$ and for $t \to +\infty$ is called a *homoclinic connection*.

The homoclinic connection in the Lorenz system (actually a double homoclinic connection) is shown in Figure 11.11. This trajectory is unstable in two senses. First, nearby trajectories are not drawn asymptotically towards the homoclinic trajectory, unless they lie precisely in the inset of the origin. The slightest perturbation off the inset causes trajectories eventually to move away from the homoclinic trajectory and settle to C+ or C−. Thus the homoclinic connection inherits part of the saddle structure of the fixed point at the origin.

The homoclinic connection in the Lorenz system also has a second, structural instability. Geometrically, this is because the dimension of the inset plus the dimension of the outset just equals 3, the dimension of the phase space. With zero-dimensional intersection disallowed (by the non-crossing trajectory property), only a relatively rare one-dimensional intersection can happen. (If a curve may cut *through* a surface in 3-space, an intersection is easily arranged, but if the curve may only touch by lying *entirely* in the surface, non-intersecting arrangements are overwhelmingly more likely than intersection.) As a result, any small change in R destroys the homoclinic connection, as does any perturbation of the vector field. A non-transverse intersection of outstructures cannot be structurally stable.

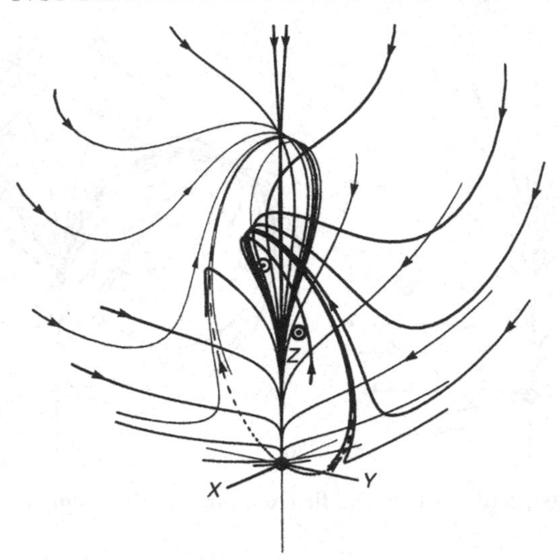

Figure 11.11 A homoclinic connection (broken curve) of the Lorenz system, occurring at $R = 13.92656\ldots$

Because of its structural instability, the homoclinic trajectory itself will not be observed in the Lorenz equations or in any physical system they describe. The evidence for it will be indirect; for some values of R the outset will return very close to the origin on one side of the inset, and for other values the *same* branch of the outset returns on the other side of the inset. A return *in* the inset occurs by implication for some value of R in between.

The structurally unstable homoclinic connection signals a *bifurcation* at $R = 13.926\ldots$. (We use bifurcation in the general sense of any *qualitative* change in the final behaviour of a dynamical system.) The one-dimensional outset of the origin lies outside the bent two-dimensional inset for R larger than the bifurcation value, and ends up crossing to the side opposite from the one it started on, settling to the opposite attracting spiral node (see Figure 11.12).

After the homoclinic connection, all trajectories still eventually settle to one of the two attracting fixed points C+ or C−, but many trajectories cross over and back repeatedly before settling down. Such trajectories look initially like the final motions at $R = 28$, with a seemingly patternless sequence of crossings. These chaotic initial transients may go on for long times, depending on the initial conditions; the average length of initial chaotic transients increases with R. This is called the preturbulent or prechaotic regime of the Lorenz system.

In this regime, initial conditions eventually settling to C+ may be finely intermingled with conditions that settle to C−. Although there are still two attracting fixed points that account for all observed final motion, with two corresponding basins, the two basins are now intertwined in a complicated three-dimensional way; thus the inset (which is still the basin separator) also

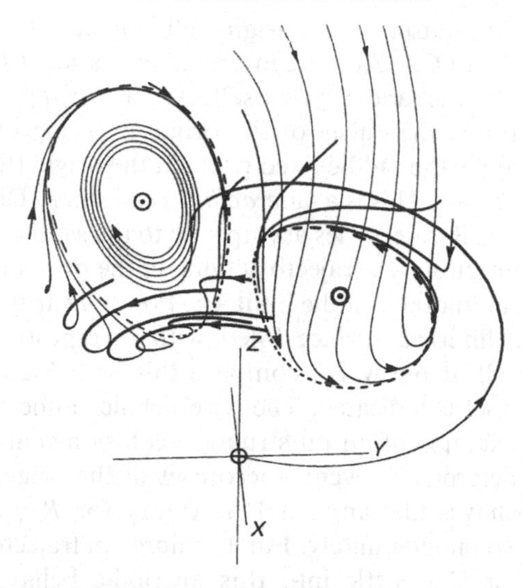

Figure 11.12 The outset of the origin for equations (11.1) with $R = 20$ and the pair of saddle cycles (broken curves) and their insets

has a complicated topology. This intertwining structure can be appreciated in the illustrations of Abraham and Shaw (1985) and in the film by H. B. Stewart (1984).

Figure 11.12 also shows two closed loops indicated by broken curves. Chaotic initial transients come close to, but just outside, these closed loops. When a chaotic transient crosses over and falls inside one of these loops, it then spirals down to the fixed point on that side without further crossings. These two closed loops are limit cycles of saddle type; they approach the homoclinic connection if R is decreased back to 13.926.... The two saddle cycles must therefore be regarded as by-products of the homoclinic bifurcation. The creation or destruction of a limit cycle in a homoclinic connection is an important type of bifurcation discussed in Chapter 13.

These saddle cycles play an essential role in the formation of the chaotic attractor. If we increase R towards 28, the saddle cycles contract in size around the spiral attracting points. As a result, the spiral terminations of the one-dimensional outset of the origin are squeezed by the shrinking saddle cycles. By algebraic analysis of the Lorenz equations, at $R = 24.74\ldots$ the real part of the complex eigenvalues at C+ and C− goes to zero; the saddle cycles collapse onto the fixed points. This is a Hopf bifurcation in its catastrophic (subcritical) form. The collision of the saddle cycle with the attracting focus leaves only a saddle point. By this value of R something must have happened to the squeezed outset.

In fact, something happens before the Hopf bifurcation, at around $R = 24.06$. For R less than this value, the outset of the origin passes close to the saddle cycle and lands inside, spiralling down to C+ or C−. For R greater than about 24.06, the outset of the origin falls outside the saddle cycle, and spirals away from C+ or C−. At some intermediate value of R, the outset of the origin must settle onto the saddle cycle itself. At this R value, which can only be determined numerically, the outset of the origin is asymptotic to two distinct limit sets of saddle type: the saddle fixed point at the origin (for $t \to -\infty$) and a saddle cycle (as $t \to +\infty$). This is a *heteroclinic connection*. Defining the *inset* of the saddle cycle to be all trajectories asymptotic to the saddle cycle for $t \to +\infty$, the heteroclinic connection is a trajectory lying in the outset of one saddle limit set and in the inset of another saddle limit set. For all nearby R, the inset of the saddle cycle is a cylindrical surface filled with asymptotic trajectories, with the saddle cycle itself at the waist; compare this with Figure 11.12 in which half of each cycle inset is indicated. The heteroclinic connection in the Lorenz system is another example of an outstructure collision that signals a bifurcation. Beyond this heteroclinic event, the outset of the origin spirals out from C+ or C− and behaves like any final trajectory for $R = 28$, crossing over, spiralling out, and so on indefinitely. Furthermore, all trajectories, except those starting near C+ or C−, settle into this aperiodic behaviour. That is, the chaotic transients become a chaotic attractor because of the heteroclinic bifurcation.

The phase space portrait just past $R = 24.06\ldots$ differs from the portrait at $R = 28$ because of the two small basins enclosed by the tube-shaped insets (separators) of the saddle cycles. Initial states inside these small basins settle to C+ or C−, the attracting fixed points inside. But all other initial conditions lead to chaotic final motions on an attractor similar to the one at $R = 28$. Although the basins of C+ and C− are small, and will vanish at the Hopf bifurcation, this is a conceptually important example of coexisting basins for steady final motion and chaotic final motion. According to this model, steadily circulating flow in the thermal convection cell corresponding to C+ or C− would be stable barring disturbances of a certain finite size, which could cause the flow to become chaotic. These competing attractors have recently been observed by Gorman *et al.* (1984) in an annular convection loop.

The changes in phase portrait from $R = 2$ to $R = 28$ take place at three R values, which can be regarded as a single sequence of events. The first homoclinic event produces chaotic transients via complicated basin structure, and creates the saddle cycles. The heteroclinic event converts chaotic transients to chaotic final motions, and is not possible without the homoclinic event that produced the saddle cycles. The final Hopf bifurcation, while not strictly needed to produce a chaotic attractor, is certainly sufficient grounds to expect one in this case: by the time the saddle cycle shrinks to a point, the heteroclinic event must have occurred. The Lorenz attractor is created by a homoclinic connection throwing off saddle cycles that ultimately collapse on the outset of the origin to generate reinsertion.

These bifurcations are summarized in Figure 11.13, which describes the final behaviour types for various values of the control parameter R. There are additional bifurcations for very large R, but we do not discuss them here; see for example Sparrow (1982).

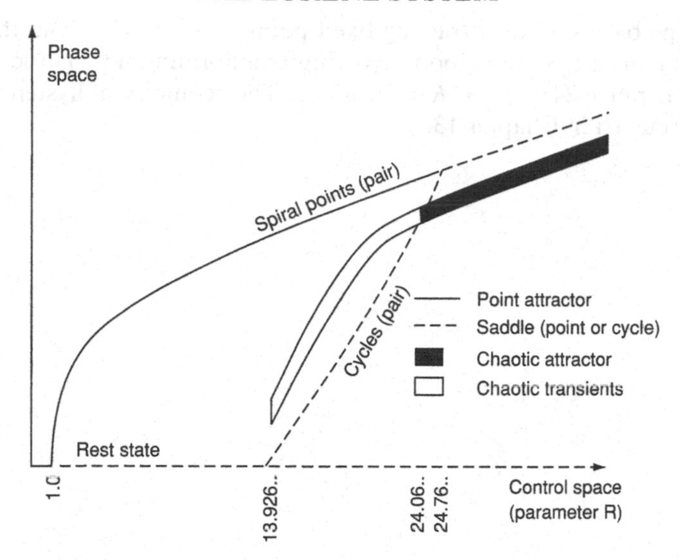

Figure 11.13 A partial control-phase portrait of qualitative behaviour of the Lorenz system

The three-dimensional phase-space geometry of the Lorenz model in the prechaotic regime is without doubt difficult to visualize. In addition to the figures presented here, the illustrations in Abraham and Shaw (1985) and the movie by H. B. Stewart (1984) will be helpful. Our purpose in describing this geometry has not been to discourage the reader, but rather to make some important points that are well illustrated by the Lorenz model.

First, the transition to a chaotic attractor can occur *abruptly*, at a single threshold value of a control parameter. In the Lorenz model this occurs by a heteroclinic connection.

Secondly, the transition to a chaotic attractor can occur at a parameter threshold where no local bifurcation (of fixed points or limit cycles) takes place. The heteroclinic connection is an example of a global phenomenon—an *outstructure collision*—causing a chaotic attractor to appear.

Thirdly, outstructure collisions do not always cause attractors to appear or vanish from a phase portrait. Outstructure collisions do signal some qualitative change in dynamics. The homoclinic connection in the Lorenz model marks a partition in the space of control parameters (such as R or σ) between simple and prechaotic dynamic regimes. Other possible consequences of outstructure collisions include a sudden jump in the size of a chaotic attractor, as in Simó (1979) and Ueda (1980b). Arnéodo *et al.* (1982) and Gaspard and Nicolis (1983) should also be consulted in this regard.

Finally, an abrupt transition to chaos can involve *hysteresis*. The geometry of phase space identifies *hysteretic transition* as any bifurcation in which an attractor appears or vanishes at a positive distance from other attractors. In the Lorenz system for certain values of R, the saddle cycles and their insets separate the basins of the attracting fixed points C+ and C− from the chaotic attractor. Thus a hysteresis loop involving equilibrium and chaotic dynamics exists in the range $24.06\ldots < R < 24.74\ldots$. The geometry of hysteresis will be further discussed in Chapter 13.

12

Rössler's Band

The folding structure underlying the Lorenz attractor, although relatively simple, is still somewhat special in that a symmetry is essential to the global structure. Small asymmetric perturbations do not change the essence of the attracting structure. Since symmetry is always a special attribute, it is natural to ask how a simple chaotic mixing occurs in nonlinear equations with no symmetry. An elegantly simple answer to this question was provided by Rössler.

12.1 THE SIMPLY FOLDED BAND IN AN AUTONOMOUS SYSTEM

Rössler actually was able to extract simpler, asymmetric attracting structures from the Lorenz attractor in two entirely different ways. One method resulted from studying the Lorenz system itself for values of R much larger than 28 (Rössler 1977); the other method was to recreate the folding effect of the Lorenz system, first abstractly as an exercise in the geometry of phase space, then by constructing the simplest nonlinear vector field capable of generating this new type of folding action (Rössler 1976a). Both of these procedures led Rössler to the same topological structure, which we call the *simply folded band.* We shall examine the synthetic method; this is equivalent in Rössler's words to constructing 'a model of a model'.

Rössler originally proposed the following equations, which yield the folded band:

$$\begin{aligned} \dot{x} &= -y - z \\ \dot{y} &= x + ay \\ \dot{z} &= b + z(x - c) \end{aligned} \tag{12.1}$$

This simple autonomous system has only a single nonlinear term, the product of x and z in the third equation. Each term in these equations serves its function in generating the desired global structure of trajectories. Considering the first two equations, let us for the moment suppose that z is negligibly small. Then the subsystem

$$\dot{x} = -y$$
$$\dot{y} = x + ay$$

can be transformed to the second-order linear oscillator

$$\ddot{x} - a\dot{x} + x = 0$$

With positive a, this oscillator has negative damping, and the origin is an unstable focus for $0 < a < 2$. Thus in the full system of three first-order equations, trajectories near the (x, y) plane spiral outwards from the origin. This produces spreading of adjacent trajectories, which is the first ingredient in the mixing action of chaos.

This spreading is achieved with only linear terms. But if the full system of three equations were linear, the spreading would merely continue as all trajectories diverge to infinite distance from the origin. To confine the spreading action within a bounded attractor, the nonlinear term is required. The constant c in the third equation acts as a threshold for switching on the nonlinear folding action. Considering the third equation alone, whenever the value of x is less than the constant c, the coefficient of z is negative, and the z subsystem is stable, tending to restore z to a value near $-b/(x - c)$. However, if x should exceed c, then z will appear in the third equation multiplied by a positive factor, and the previously self-restoring z subsystem diverges. Choosing $b > 0$ ensures that this divergence will be towards positive z.

The effect of this is shown in Figure 12.1. A trajectory spirals outwards while appearing to remain in a plane near to and parallel to the (x, y) plane. When x becomes large enough, the z subsystem switches on and the trajectory leaps upwards. Once z becomes large, the z term in the first equation comes into play, and $\dot{x}$ becomes large and negative, throwing the trajectory back towards smaller x. Eventually x decreases below c, the z variable becomes self-restoring, and the trajectory lands near the (x, y) plane again. Through the feedback of z to the $\dot{x}$ equation, trajectories are folded back and reinserted closer to the origin, where they begin an outward spiral once more.

The chaotic attractor of Figure 12.1 was named *spiral chaos* by Rössler (1967a). Since the same attracting structure can occur in other dynamical systems having no fixed point with spiral outset, we refer to this attractor type as a folded band.

For some values of the parameters a, b and c, it can happen that only a periodic attractor is observed. In this case, one observed trajectory spirals out and is thrown back repeatedly, and after a certain number of turns lands precisely upon itself. All other trajectories approach this unique periodic attractor asymptotically. Although the vector field was constructed to spread trajectories apart in one region of phase space, this does not guarantee chaotic final motion; the spreading effect may be cancelled during the folding. However, for substantial ranges of the parameter values, no such periodic attractor can be detected in numerical integrations, and in these ranges all trajectories eventually resemble Figure 12.1. Any two trajectories starting at nearby points

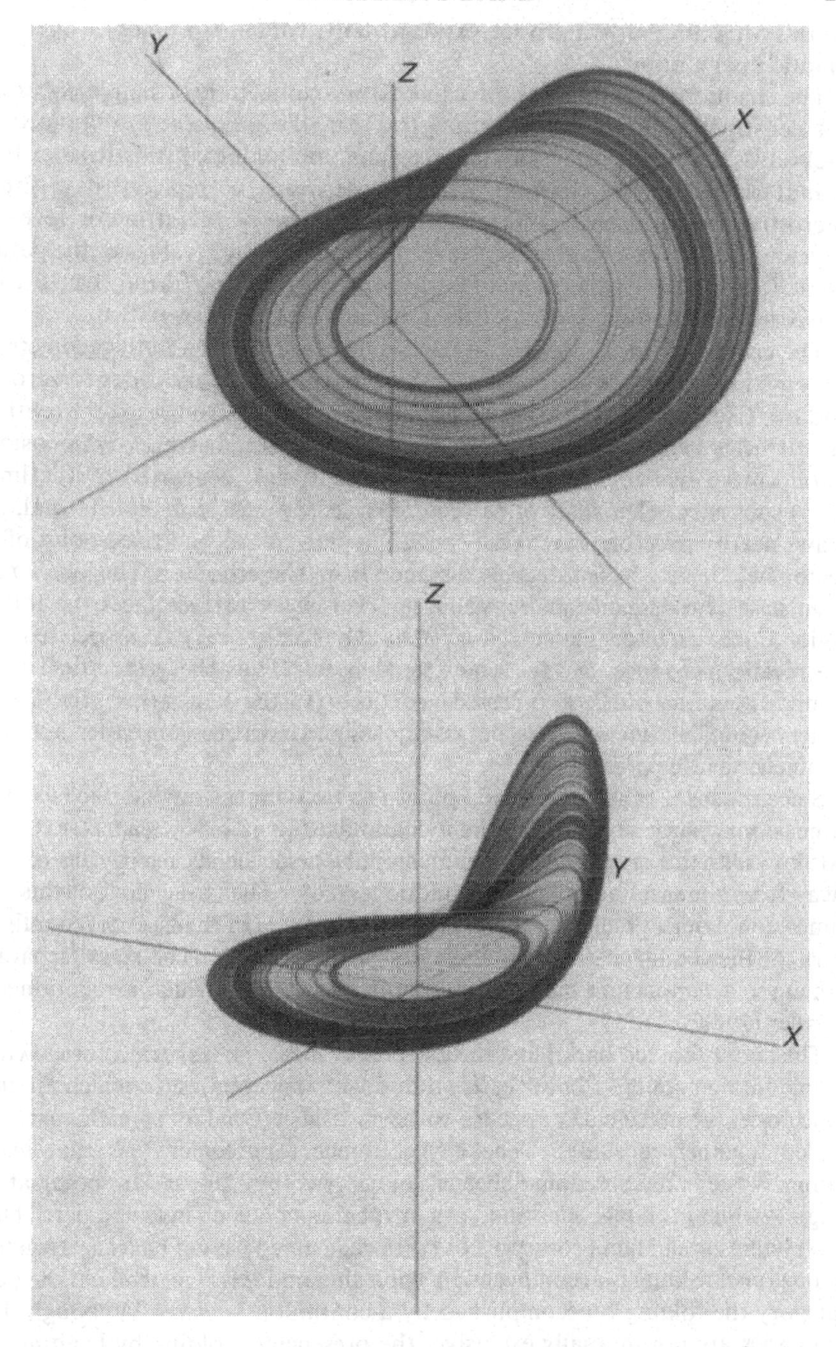

Figure 12.1 A post-transient trajectory of Rössler's equations (12.1) for the simply folded band attractor. Parameters are $a = 0.398$, $b = 2$, $c = 4$

on this attracting set will diverge exponentially, which is primary evidence of chaotic final motion.

The chaotic attracting structure can be approximated by a simple paper cut-out construction. As in the Lorenz attractor, the uniqueness of trajectories backwards and forwards in time implies that the folding of this Rössler band cannot be perfect; the thickness of the paper must be replaced by a fractal microstructure, and passing through the thickness of the attractor reveals a structure much like a Cantor set. Nevertheless, for typical values of the parameters a, b and c the folding is very nearly complete, and the fractal structure can only be seen if a small portion of the attractor is greatly magnified.

The genesis of this simple folding action can also be seen in the geometry of the outstructures of equations (12.1). Setting the right-hand sides to zero, we find two fixed points of the system, both of saddle type. One lies at the centre of the attractor, near the origin. The *central fixed point* has a two-dimensional spiral outset, which is essentially the outward spiral generated by the linear (x, y) subsystem. The inset of the central fixed point is one-dimensional, and draws nearby trajectories into the spiralling plane. A second fixed point of the vector field lies at a considerable distance from the attractor. This *outer fixed point* has a two-dimensional spiral inset. This inset surface faces the folding region of the attractor, in such a way that the folding may be associated with the rotation produced by the outer fixed point. Thus the vector field is the composite action of the two crossed vortices. (In the same terms, the Lorenz system would be viewed as a pair of misaligned vortices on either side of a symmetric saddle point.)

Since the inset of the outer fixed point is a two-dimensional surface in three-dimensional phase space, it is a likely candidate for a basin *separator*. On the side towards the origin, the one-dimensional outset leads nearby trajectories away from the inset and onto the chaotic attractor. The other half of this one-dimensional outset pulls trajectories on the other side of the inset off to infinity. Parts of these outstructures are indicated in Figure 12.2. The spiral separator can cause an important bifurcation in equations (12.1), which is mentioned in Rössler (1976b).

The simply folded band, like the Lorenz attractor, mixes trajectories with a simple folding action. In both cases, a bundle of trajectories (to which all other trajectories are attracted) appears to form a sheet that is spread apart and folded together repeatedly. The key difference topologically is that in the Lorenz system the spreading sheet is cut in two (by the saddle point at the origin), while in the Rössler band system there is no such cutting apart. These two systems are in fact prototypes of two broad categories of chaotic attractors. In one type, folding is accomplished by splitting and *layering*, while in the other category the folding is accomplished by a continuous *bending*. Although these categories are not mutually exclusive, the presence of folding by bending in a chaotic attracting structure may indicate greater complexity than is found in an attractor purely of the layering type. The extra complexity of bending is not so much a matter of overall appearance of the attractor shape as of its internal composition.

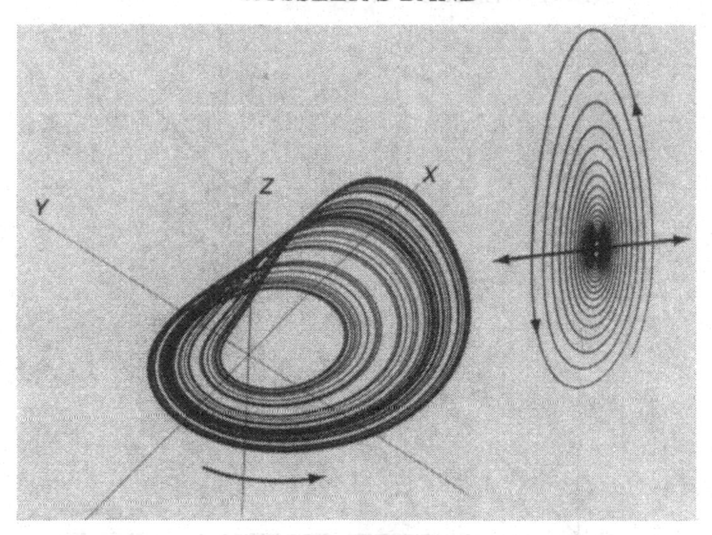

Figure 12.2 A trajectory on the attractor of equations (12.1), together with part of the spiral inset and straight outset of the outer fixed point

12.2 RETURN MAP AND BIFURCATIONS

To understand the nature of the simply folded band attractor, it helps to construct again a Poincaré return map. As in the Lorenz system, there is no fixed period of time on which to base the strobing; the return surface must be specified geometrically after careful inspection of the attractor itself. In the case of equations (12.1), we can see from Figure 12.1 that any half-plane whose edge is the vertical z axis appears to cut the bundle of trajectories transversely. Selecting one such vertical half-plane defines a Poincaré section of the attractor, a point set which under the first return map is carried around exactly onto itself. For some values of the parameters, all points in this Poincaré section take very nearly the same time to return to the half-plane (Farmer *et al.* 1980); but in other regions of control space there is no such coherence, and different first-return trajectories of the same system may require widely varying times to complete.

The half-plane $y = 0$, $x < 0$ cuts the attractor of Figure 12.1 in a Poincaré section, which would appear on this scale to be a line segment. Because the folding action is visibly complete at this location, none of the fine fractal structure is apparent. For all practical purposes, the point where a trajectory on the attractor crosses this half-plane can be identified by giving the distance from the centre, i.e. the x coordinate alone. This simple insight allows us to summarize the dynamics of the attractor quite accurately in terms of a one-dimensional mapping. One way to obtain this mapping is to find a trajectory on the attractor, and record x_{n+1}, the x coordinate of the $(n + 1)$th crossing of the Poincaré section, as a function of x_n, the x coordinate at the nth crossing. The result is shown in Figure 12.3, where x_{n+1} versus x_n is plotted for a number of

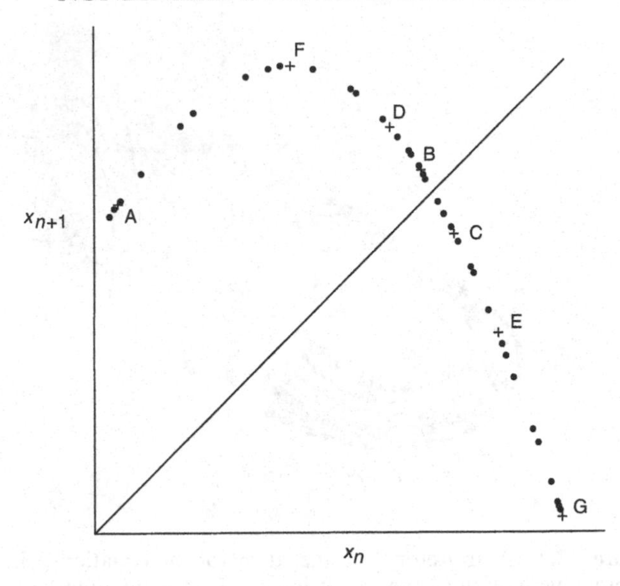

Figure 12.3 One-dimensional mapping points taken from a trajectory of the differential equations (12.1). Crosses A to G identify a sequence of seven successive points

crossings of the half-plane $y = 0$, $x < 0$. All points appear to lie on a simple smooth curve with a single maximum. A sequence of seven successive points are indicated in the figure and labelled A to G; the reader may check that each point can be obtained from the previous one via the usual construction with the diagonal.

Here is an example of the usefulness of understanding the behaviour of iterated one-dimensional maps. If the dynamics of the attractor in Figure 12.1 are well represented by the mapping suggested by Figure 12.3, then knowledge of iterated mappings of an interval should carry back to the differential equations (12.1). For example, we know the bifurcation diagram of the logistic map, given earlier in Figure 9.8, describes qualitatively the bifurcations of all typical one-dimensional maps having a single smooth maximum. This means that, if we vary a parameter in the system of equations (12.1), we expect to find, as in the logistic map, windows of periodic behaviour interspersed with parameter ranges where the final motion is aperiodic. Furthermore, parameter values giving periodic final motions with period 2, 5, 3, etc., should follow the same patterns established by the theory of one-dimensional mappings and confirmed by the logistic map diagram. Figure 12.4 shows final trajectories of equations (12.1) for a number of different values of the parameter a, always with $b = 2$, $c = 4$. The main periodic windows occur in the U-sequence observed for the logistic map. In fact, if the parameter values noted in Figure 12.4 are plotted against parameter values of the corresponding behaviour in the logistic map (taken for example from Figure 9.8 using a ruler), a nearly straight line will be obtained.

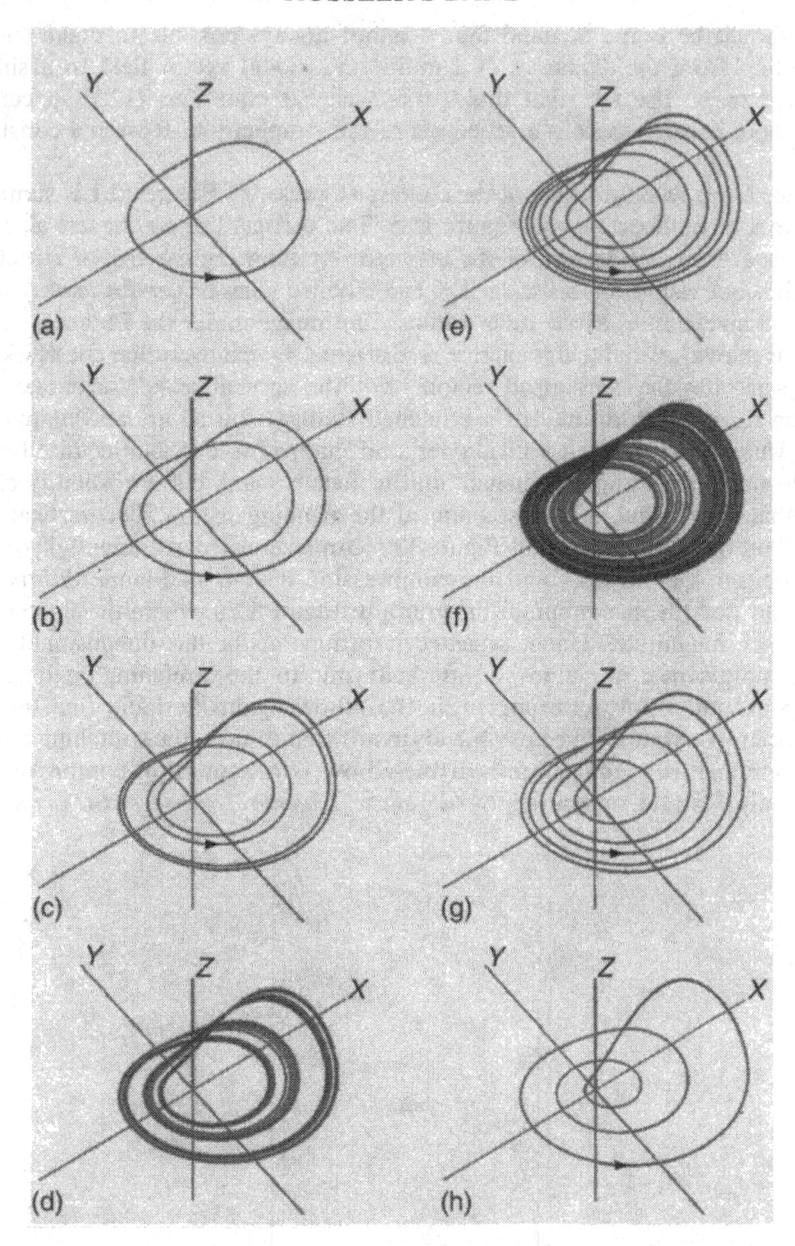

Figure 12.4 Final trajectories of equations (12.1) for different values of the parameter a: (a) limit cycle, $a = 0.3$; (b) period 2 limit cycle, $a = 0.35$; (c) period 4, $a = 0.375$; (d) four-band chaotic attractor, $a = 0.386$; (e) period 6, $a = 0.3909$; (f) single-band chaos, $a = 0.398$; (g) period 5, $a = 0.4$; (h) period 3, $a = 0.411$. In all cases $b = 2$ and $c = 4$

It should be borne in mind that it is not always possible to make such a reduction from the dynamics of a multidimensional vector field to a simple iterated map. The fact that this works well for equations (12.1) in certain regions of control space is a reflection of the simplicity of Rössler's construction.

The simple folding action of the chaotic attractor of Figure 12.1 is summarized in a more direct way in Figure 12.5. The vertical line on the left idealizes our usual Poincaré section of the attractor by omitting the fractal structure, which is not visible on this scale. The end labelled a lies nearer the centre, while the end labelled e is at the outer radius. The image under the Poincaré return map is shown at right; this picture is distorted by exaggerating the thickness that separates the overlapped regions, but the vertical scale is accurate. The region from a to b on the left is stretched slightly, shifted up, and mapped to a′b′, while de is stretched, folded over, and mapped to d′e′. Notice that bc and cd do not appear to be stretched, and in fact b′c′ and c′d′ are slightly closer together; this is due to compression in the bending region. The vertical positions on the right diagram of Figure 12.5 can be constructed directly from the mapping of Figure 12.3, and the compression near c′ in Figure 12.5 is seen to occur near the maximum of the graph in Figure 12.3, where the slope is less than 1 in magnitude. Under repeated iterations of the one-dimensional mapping, points near the elbow c′ are sent out to the stretching regions and jump around before returning near the elbow again. Periodic final motion will occur if a typical trajectory spends insufficient time in the stretching regions to overcome the compression near the elbow. The amount of compression or stretching can be measured by defining *Liapunov characteristic exponents*

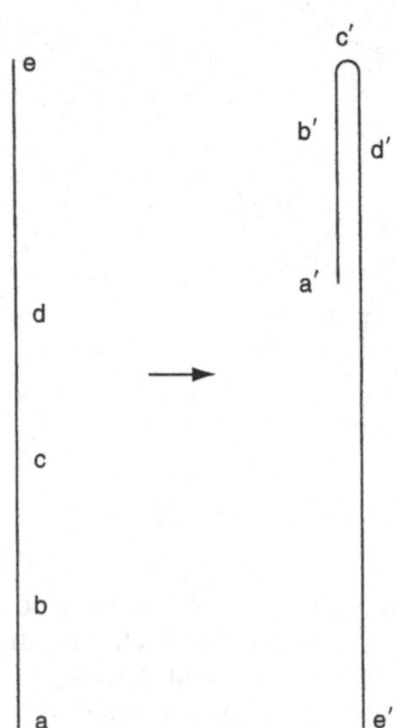

Figure 12.5 Diagram of the return map of the attractor in Figure 12.1, showing the folding of trajectories on the attractor

for recurrent trajectories on the chaotic attractor, analogous to the characteristic exponents for fixed points (or multipliers for cycles). For information about Liapunov exponents, see Oseledec (1968), Farmer (1982), Frederickson *et al.* (1983), Grassberger and Procaccia (1983) and Eckmann and Ruelle (1985).

From available evidence, it appears that compression in the bending region may be typical of the folding action in many chaotic attractors. Such compression need not occur in attractors of the layering type. (Remember that we are discussing compression along the attractor section; there is always contraction in the transverse direction, which forms the fractal layers.) In the Lorenz system, for example, the reduced map for $R = 28$ shown in Figure 11.7 always has slope greater than 1 in magnitude, because the change from positive to negative slope takes place at a discontinuity caused by the saddle separator through the origin. Therefore the distance between adjacent trajectories expands with each turn around either side of the Lorenz attractor.

In the Rössler folded band, however, trajectories only separate on average over a number of returns to the Poincaré section plane. Indeed, this spreading on the average is only an inference from numerical experiments, and is difficult to affirm in theory. Because of the compression region, it is no simple matter to prove that trajectories spread apart exponentially. They will only do so if typical trajectories spend enough time away from the compression region. The behaviour of the logistic map suggests that a very small change in a control parameter might suffice to change the final behaviour from periodic to aperiodic.

More generally, the existence of a compression region could mean that an attractor that appears to mix in a thoroughly aperiodic way might actually contain inside itself *stable* period motions. These cycles might have long periods and extremely narrow basins, and hence be easily overlooked in experiments or be masked by noise. Still, such pockets, or *sinks*, in an attracting set would make the recurrence non-transitive, that is, not linked together by a single dense recurrent orbit. This hindrance to a straightforward theory of basic recurrent behaviour types has been analysed by Newhouse, and the pockets are often called Newhouse sinks; see for example Newhouse (1980b). In some cases the size of the basin should decrease exponentially with the subharmonic number of the periodic orbit inside (Greenspan and Holmes 1982). This may explain why Newhouse sinks have not yet been observed in simulations; or it may be that the sinks only exist for very small ranges of parameters. There is reason to believe that they may occur in the Hénon map (C. Robinson 1983), and in forced oscillators, as discussed in Section 2.2 of Guckenheimer and Holmes (1983). Unfortunately, their Figure 2.2.4 is mislabelled; careful numerical studies show that the points labelled 'sinks' in that figure are actually unstable periodic orbits, not stable ones.

The fact that bending may lead to compression and Newhouse sinks causes a vexing problem in the theory of chaotic attractors. We earlier defined a chaotic attractor as a linked set of recurrent states. But if an 'attractor' includes a sink, the periodic attracting orbit inside the sink would be surrounded by a small basin filled with transients only. Then there could be no final-time recurrent trajectory linking the periodic orbit in the sink to the nearby chaotic final behaviour on the edge of the sink. Even if the sink is so small as to appear virtually within the chaotic attracting set (as is quite possible in theory), strictly speaking one would have to deal with two separate attractors. It has even been suggested (for example in connection with the Hénon attractor) that some chaotic attractors are in fact nothing but an ensemble of many long subharmonics, whose basins are so small and close together that the slightest amount of noise moves experimentally observed trajectories from one sink to another, giving the appearance of chaos. This suggestion is not so outlandish as it might seem; as we shall see in the next section, chaotic attractors can be expected to contain infinitely many subharmonics. Fortunately, in at least a few cases like the Lorenz attractor, we believe that there is no compression that could lead to sinks, only the normal transverse contraction to produce fractal layers. This means that all subharmonics that exist within the Lorenz attractor are unstable in theory as well as in practice.

The intermingling of periodic and chaotic attractors in a phase portrait via Newhouse sinks is postulated for a fixed set of controls in a dynamical system. Another type of intermingling is the possibility of windows of periodic and aperiodic final behaviour as a control parameter is varied; this intermingling in control space does occur without doubt, for example in the logistic map, and can be appreciated in the bifurcation diagram of Figure 9.8. Until more evidence and understanding of the phase-space intermingling with fixed controls is gathered, it must be remembered that what appears to be a single chaotic attractor could in reality be decomposable into smaller, theoretically separate pieces. Such complex internal structure is at least possible when chaotic mixing is achieved by continuous bending.

These possible complications mean that we abuse mathematical terms when we say, for example, that the chaotic attractor of the softening spring in Figure 6.10 is topologically equivalent to the attractor shown in Figure 12.1. In each case there are questions of topological structure that are not completely answered; but to the extent that each attractor has dynamics described by a one-dimensional map, the fine-structure questions for both should be similar.

12.3 SMALE'S HORSESHOE MAP

The folding action of Rössler's band, expressed in the diagram of the Poincaré mapping in Figure 12.5, represents the simplest known way in which differential equations can achieve chaotic mixing. Evidence to date indicates that simplicity is favoured not only by theoreticians but by nature as well: a number of instances of chaotic attractors topologically equivalent to Figure 12.1 have been identified in experiments. Examples include the Belousov–Zhabotinsky reaction reported in Roux *et al.* (1983), a study of chaos in a dripping faucet (Shaw 1984), and numerical experiments with an anharmonic oscillator in Huberman and Crutchfield (1979); equations (12.1) are only one way of observing the folded band.

The folded band is also important as a prototype of the way stretching and folding produce mixing in other chaotic attractors, such as the Birkhoff–Shaw bagel. This was appreciated by Smale, who studied a two-dimensional iterated mapping he called the horseshoe map, with a view to understanding behaviour of the forced Van der Pol equation. Although Smale's work pre-dates the discovery of the bagel attractor in a differential equation by Shaw and of the folded band by Rössler, the horseshoe map may be regarded in hindsight as a further idealization of the Poincaré map of Figure 12.5. This idealization allowed Smale to deduce rigorously some fundamental properties of chaos. The price paid for this is that the horseshoe map does not generate a chaotic attractor, but only prechaotic transients, as in the Lorenz system for $13.926\ldots < R < 24.06\ldots$

Smale's horseshoe map transforms a point set in the plane. To specify the transformation, we begin with a square as in Figure 12.6 on the left. First the square ABCD is distorted by uniform stretching in one direction to μ times its original height, and uniform contraction in the perpendicular direction to a fraction λ of its original width. The intermediate result is shown in the middle of Figure 12.6. Next this rectangle is bent into a U-shape; the final image A′B′C′D′ overlaps the original square ABCD in the shaded region on the right of Figure 12.6. Working backwards, the shaded regions indicate where the overlap region originates, i.e. the pre-image of the overlap region. In mathematical notation, let S be the square ABCD and f be the transformation; then the shaded regions on the right make up $S \cap f(S)$, while the shaded regions on the left make up $f^{-1}(S) \cap S$. Exactly how the mapping f works in the unshaded regions is not of primary importance, but in either half of $f^{-1}(S) \cap S$ the transformation f in coordinates shall be a linear transformation with eigenvalues $\mu > 2$ and $\lambda < \frac{1}{2}$.

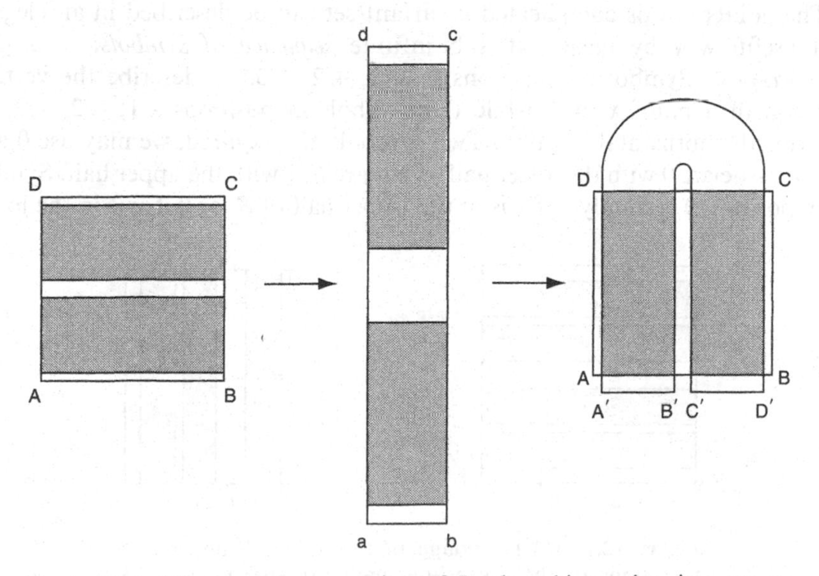

Figure 12.6 The horseshoe mapping of Smale, taking points in a square onto a U-shaped region overlapping the square

To analyse the recurrent behaviour of this mapping, we need to identify its *invariant set*, that is, a set $I \subset S$ such that $f(I)$ is identical with I itself. This can be done recursively. Starting with S we have $f(S) \cap S$ and by recursion can find $f(f(S)) \cap S$, denoted $f^2(S) \cap S$, which consists of four vertical strips as on the right in Figure 12.7. Evidently, by continuing this process we obtain $f^3(S) \cap S$ with eight vertical strips, and so on, so that

$$I^+ = \bigcap_{n=0}^{\infty} f^n(S)$$

consists of infinitely thin vertical strips over a Cantor set. Likewise,

$$I^- = \bigcap_{n=0}^{\infty} f^{-n}(S)$$

is composed of horizontal strips that intersect any vertical line in a Cantor set. Then inside S the invariant set is

$$I = I^- \cap I^+ = \bigcap_{n=-\infty}^{+\infty} f^n(S)$$

because any point in both I^+ and I^- is carried by f into another point $f(x)$ that is also in I. Thus the set I describes those points in S which, under infinitely repeated applications of the mapping f, may exhibit recurrent or non-wandering behaviour. Points in S that are not in I must eventually fall outside S under forwards or backwards iteration of f. An approximate picture of I is shown in Figure 12.8. From the invariance of I and the linearity of f on the halves of I comes a scaling property: any one of the four quarters of I can be stretched by μ vertically and λ^{-1} horizontally to give the same picture as I itself.

The points in this complicated invariant set can be described in an elegant and useful way by means of a bi-infinite *sequence of symbols*: $\cdots a_{-2}a_{-1}$ $.a_0a_{+1}a_{+2}\cdots$. Symbols in positions $0, +1, +2, +3, \ldots$ describe the vertical location of a point x in I, while the symbols in positions $-1, -2, -3, \ldots$ describe the horizontal location. Two symbols are required; we may use 0 and 1, and associate 0 with the lower half of square S, 1 with the upper half. Symbol a_0 in position 0 is simply 1 if x is in the upper half of S, or 0 if x is in the lower

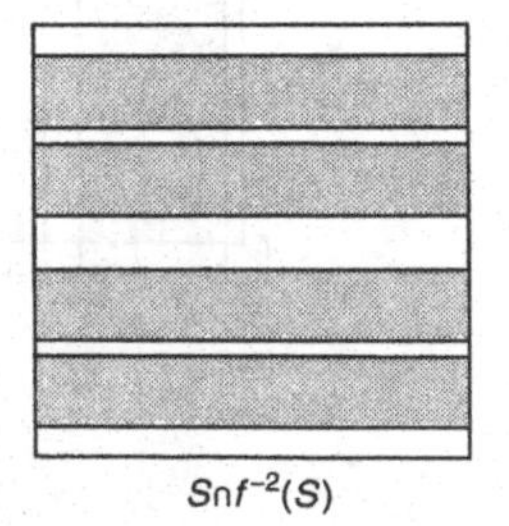

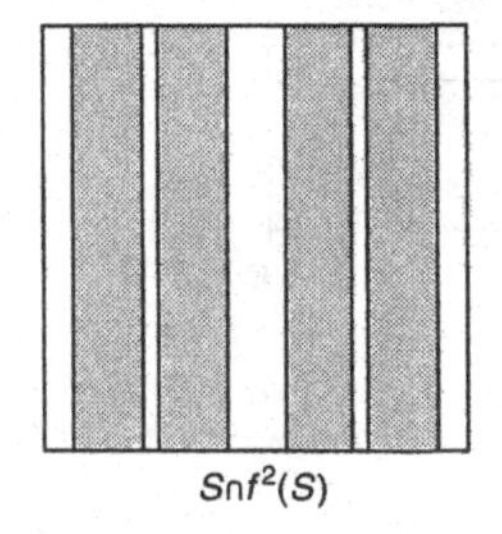

Figure 12.7 (a) Pre-image of the square S under two inversions of the horseshoe map. (b) The image of S after two iterations

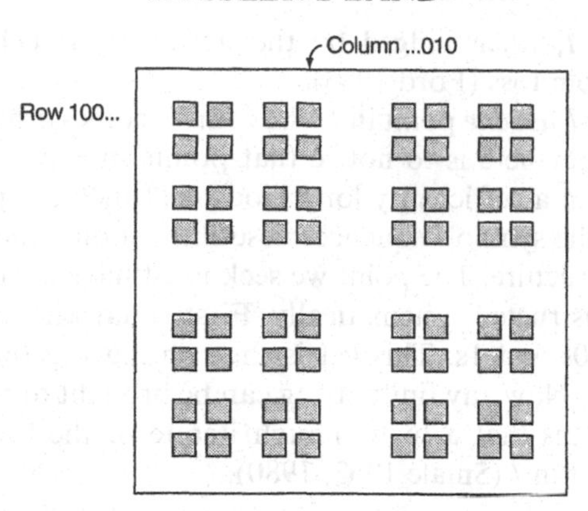

Figure 12.8 An approximate picture of the invariant set of the horseshoe mapping

half. Symbol a_{+1} in position +1 is determined from $f(x)$: 1 if $f(x)$ is in the upper half, 0 if in the lower half. Inductively, symbol a_{+i} is 1 if $f^i(x)$ falls in the upper half, 0 if in the lower half of the square. Thus, for example, all points in the top row visible in figure 12.8 have symbol sequences $\cdots a_{-2}a_{-1}.100\cdots$.

In the reverse direction, a_{-i} is 1 if $f^{-i}(x)$ falls in the upper half of the square, 0 if $f^{-i}(x)$ falls in the lower half. This specifies the *horizontal* position of the point x. For example, consider any point in the fourth column from the left visible in Figure 12.8. Applying the inverse map once puts these points in the lower half of S; applying the inverse a second time lands in the upper half; a third inversion lands in the lower half. Thus all points in this column have symbol sequences $\cdots 010.a_0a_{+1}a_{+2}\cdots$.

Constructed in this way, there is obviously a bi-infinite sequence for every point in I. Conversely, for every bi-infinite sequence of 0s and 1s there is exactly one point in I. Thus the bi-infinite sequences describe I precisely. Furthermore, the action of the horseshoe map is easily described in terms of the bi-infinite sequence: given x in I, form its symbol sequence, then *shift* each symbol one place to the left; the result is the symbol of $f(x)$.

The construction of symbol sequences enabled Smale to demonstrate some remarkable properties of the recurrent behaviour in I.

(i) There is a *countably infinite set* of points in I with *periodic* orbits. A point x in I has a periodic orbit if some iterate of the horseshoe map returns x precisely to itself, $f^n(x) = x$. Clearly, x is periodic if its symbol sequence is periodic: $a_i = a_{i+n}$ for all i. For each n, there are 2^n different bi-infinite sequences that are periodic with period n, or perhaps a divisor of n. Each such sequence identifies a periodic point in I. (All these periodic orbits are unstable, since the distance between any two nearby points increases by a factor $\mu > 2$ for each application of the horseshoe map, until the distance becomes comparable to the size of S.)

(ii) There is an *uncountably infinite set* of points in I that are *not periodic*. These are the points whose symbol sequences are not periodic. The orbit of the typical point in I, when judged by the upper/lower dichotomy, behaves as randomly as a coin toss (Ford 1983).

(iii) There is *at least one* point in I whose orbit comes *arbitrarily close* to every point in I. The key here is to notice that points in I are close if their symbol sequences agree in a sufficiently long block that includes position 0. In mathematical terms, the symbol sequences describe not only the point set I but also its topological structure. The point we seek must have a complicated sequence, but it can be constructed systematically. First enumerate in a sequential list all *finite* strings of 0s and 1s. This list is then strung together in one bi-infinite symbol sequence. Now any finite string can be brought to position 0 by enough shifting; this means that a high enough iterate of the horseshoe map comes close to any point in I (Smale 1963, 1980).

As a particularly important consequence of the third property, the invariant set cannot be split into two or more separate pieces, because at least one point eventually comes near every point of I. Thus the invariant set I of the horseshoe map fits our notion of a linked set of recurrent states, or a basic type of chaotic motion, because the orbit constructed in (iii) is dense.

Finally, Smale was able to show that all of these properties hold if the horseshoe map is distorted by a perturbation of small size but arbitrary shape. That is, the recurrent behaviour of the horseshoe map is *structurally stable*.

As mentioned above, the invariant set I of the Smale horseshoe mapping is not an attractor. This is because, among all possible starting points in the square S, those finally recurrent in S are a set of probability measure 0. With probability 1, a point chosen at random in the square will remain in the square only for an initial transient. This is similar to the prechaotic regime of the Lorenz equation, between $R = 13.926\ldots$ and $R = 24.06\ldots$, where symbol 0 would correspond to a turn around C−, while symbol 1 stands for a turn around C+. In the prechaotic Lorenz system, any finite sequence of turns can be observed; aperiodic sequences correspond to trajectories observed with zero probability. The Smale horseshoe theorem describes the recurrence of either chaotic or prechaotic motion.

The term 'horseshoe' is often used in the literature in a broader sense than just the mapping f defined above. Any mapping that returns a set into itself is an *endomorphism*. If an endomorphism sends all points to two bins, or a finite number of bins, then recurrence in the endomorphism (e.g. Poincaré map) is described by bi-infinite symbol sequences; one symbol is used for each bin. If it can be checked, as for the function f above, that the invariant set I contains a point for each symbol sequence, then the invariant set has the recurrence properties of Smale's horseshoe: infinitely many periodic orbits, uncountably many aperiodic orbits. Such mappings are also called horseshoes, or occasionally shoes. Horseshoes in this sense can be found in all known chaotic attractors of dissipative dynamical systems, and can be taken as the definition of *folding*.

For example, the Poincaré map shown in Figure 12.5 (representing the return map of Rössler's band) does not itself contain a horseshoe: a candidate square

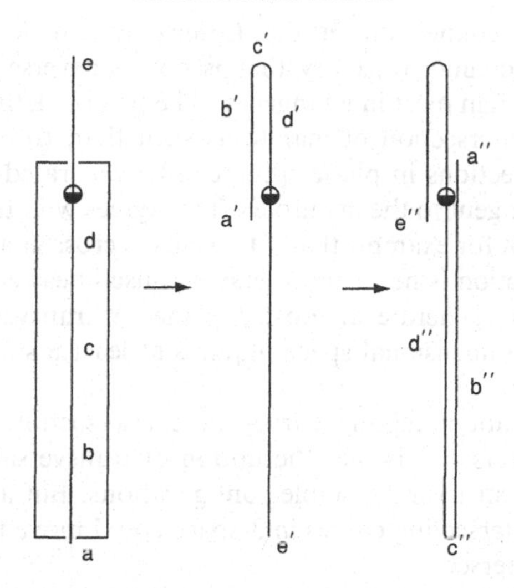

Figure 12.9 Two iterations of the return map shown in Figure 12.5. Points that start inside the rectangle return after two iterations to cover the full height of the rectangle twice over, indicating the presence of horseshoe-like dynamics

S could extend only from c' down to the level of a′, and such a square is not mapped to itself. However, the second return map, i.e. the return map iterated twice, does contain a horseshoe, in the more general sense of the term, as illustrated in Figure 12.9. Here the image of the band after one return shows one fold, while after two returns the first folding is compressed and a second folding occurs. Also shown in Figure 12.9 is the unstable (saddle) fixed point of the return map, represented by a half-filled circle. This unstable fixed point lies on the first bisectrix, i.e. the diagonal, in Figure 12.3. Any rectangle whose upper edge lies between the fixed point and image point a″ in Figure 12.9 (and whose bottom edge is above c″) encloses a mapped region analogous to the square S of the horseshoe mapping in Figure 12.6. The folding in Figure 12.9 may be compared with Figure 6.10.

12.4 TRANSVERSE HOMOCLINIC TRAJECTORIES

The Smale horseshoe construction is a paradigm for the action of chaotic attractors. Furthermore, the characterization of the horseshoe dynamics by a symbol sequence gives a precise description of the mixing behaviour. For these reasons, it is helpful to identify horseshoe-like folding dynamics when they are present. The existence of horseshoe-like dynamics is closely related, both in theory and in practice, to a specific geometric configuration of outstructures, the transverse homoclinic trajectory.

The homoclinic connection in the Lorenz system $R = 13.926\ldots$ is an example of a homoclinic trajectory that is not transverse, because the inset and outset of the origin meet in a tangency. The precise definition of *transverse intersection* is an intersection of manifolds such that, from any point in the intersection, all directions in phase space can be generated by linear combinations of vectors tangent to the manifolds. This agrees with the everyday notion of transverse, except for example that if two curves cross in a three-dimensional space, their intersection is never transverse because linear combinations of the two tangent vectors generate at most a plane. A transverse intersection of manifolds in three-dimensional space requires at least a surface intersecting a curve (Figure 12.10).

The reason for not accepting curve–curve intersection in 3-space in the definition of transversality is that the notion of transversality should correspond to persistent, structurally stable configurations. But a typical small perturbation of two intersecting curves in 3-space could move them apart so they would no longer intersect.

A surface may intersect a curve transversely in three dimensions. If the surfaces are outstructures, the intersection can only be a saddle point, as in Figure 3.10. Transverse homoclinic intersection arises only as suggested in the lower right of Figure 12.10.

Outstructure intersection involving two surfaces can happen for example between the inset and outset of a periodic limit cycle. Recall from Chapter 10 that a limit *cycle* of saddle type is equivalent to a saddle *point* of the Poincaré map; this saddle point has incoming and outgoing eigenvectors that form a cross-shaped pattern in the Poincaré section plane. An example is shown in Figure 12.11; a saddle point near the origin is shown in a Poincaré section of the vibrating buckled beam equation (6.12) studied by Moon and Holmes (1979),

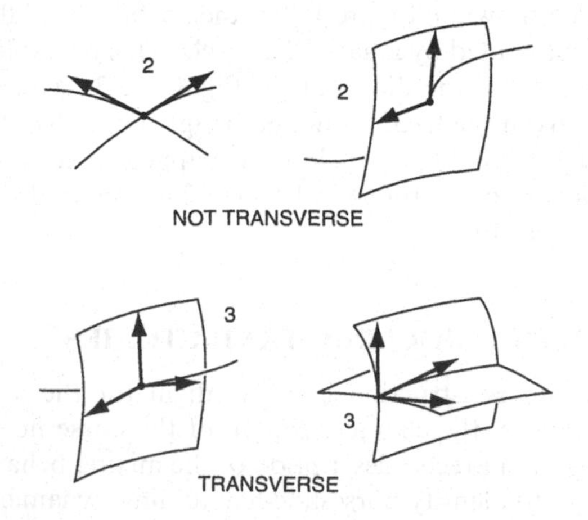

Figure 12.10 Intersections of curves and surfaces in three dimensions: heavy arrows indicate tangents, and their span must have dimension 3 for transversality

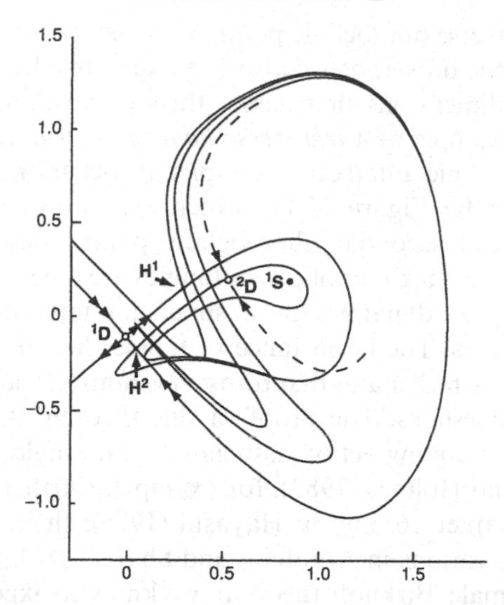

Figure 12.11 Poincaré section of the twin-well Duffing equation model for the vibrated buckled beam, showing the saddle cycle ^{1}D which dynamically separates the two potential wells, and parts of its inset and outset, with transverse homoclinic points. In this phase portrait the attractors are periodic (the stable cycle ^{1}S and its twin in the other well); there are no chaotic attractors but there are chaotic transients

Holmes (1979), Holmes and Whitley (1983) and Arecchi and Califano (1984); see also Section 2.2 of Guckenheimer and Holmes (1983). The parameters for Figure 12.11 are damping = 0.25, forcing magnitude = 0.215, and forcing frequency = 1.02; the section is taken at angle $t = 0$.

The figure shows parts of the inset and outset of the hilltop saddle cycle ^{1}D produced by mapping the local eigenvectors. Starting near the saddle cycle on the incoming eigenvector, points mapped backwards in time, as in Figure 5.10, generate a global inset. Points starting on the outgoing eigenvector are mapped forwards in time repeatedly to generate the global outset. A continuum of starting points fills the outset and inset to the appearance of continuous curves in the Poincaré section; in the full three-dimensional phase space these become two-dimensional surfaces. Also shown in Poincaré section are another saddle cycle ^{2}D, and a stable attracting cycle ^{1}S.

The inset and outset curves of the forced buckled beam in Figure 12.11 show a number of places where the inset appears to cross the outset. The original crossing at the saddle cycle is here not counted as an intersection, because neither the inset nor the outset includes the saddle cycle itself, only trajectories asymptotic to the saddle cycle. The inset and outset do intersect at point H^1, and at its next return to the Poincaré section H^2; both are homoclinic points. There are in fact an infinite number of them, which is a typical phenomenon:

the image of a transverse homoclinic point is another point that must also be in both the inset and the outset, hence also is a transverse homoclinic point. The full curve in three dimensions that passes through all these transverse homoclinic points is an example of a *transverse homoclinic trajectory*.

Transverse homoclinic intersections typically occur in complex recurring structures suggested by Figure 12.11, which are known as *tangles*. Tangles contain an infinity of secondary homoclinic points, described in detail by Birkhoff. Although they are complicated, tangles are a persistent phenomenon, because perturbing the dynamics by a small amount does not remove the transverse intersections. The importance of tangles lies in the Smale–Birkhoff homoclinic theorem, which states that transverse homoclinic intersections imply the existence of horseshoes. The proof of this theorem is basically a precise specification of the mapping action associated with tangles, and can be found in Guckenheimer and Holmes (1983), for example. Further pictures of tangles can be found in chapter 16, and in Hayashi (1975); their theory is presented with numerous illustrations in Abraham and Shaw (1985). See Figure 16.24.

Because of the Smale–Birkhoff theorem, we know to expect complex behaviour like the horseshoe whenever the outstructures of a dynamical system have transverse homoclinic intersections. It should be noted that a variety of transverse outstructure intersections have the same effect. For example, let A and B be two limit cycles; the Smale–Birkhoff theorem also applies if the inset of A intersects transversely with the outset of B, and the inset of B in turn intersects transversely with the outset of A. In fact, any *cycle* of transverse inset–outset intersections causes horseshoes.

Finally, as a further indication of the importance of horseshoes and transverse homoclinic intersections, we mention a theorem of Katok (1980). This involves the concept of *topological entropy*, which is one mathematical formulation of the notion of exponential divergence of nearby initial conditions. Katok shows that a large class of Poincaré mappings with positive topological entropy must have transverse homoclinic points. In other words, exponential spreading of nearby starts is linked to the existence of transverse homoclinic trajectories and horseshoes.

But remember that a horseshoe may correspond to either a chaotic attractor or a prechaotic behaviour in which aperiodic final motions exist but are observed with probability 0, so that folding and mixing are only observed in transients. The condition that a folding action have an attractor is that observed trajectories that fall off the invariant set I in Smale's construction somehow be reinserted into the square S. The bifurcations from non-attracting to attracting horseshoes may be either local or global, as we shall see in the next chapter.

12.5 SPATIAL CHAOS AND LOCALIZED BUCKLING

In recent years, new insights and phenomena have emerged from an application of the theory of homoclinic orbits to the localization of structural post-buckling patterns. This is made possible by the use of a static–dynamic analogy, in which an independent spatial coordinate of a long elastic structure is identified as time

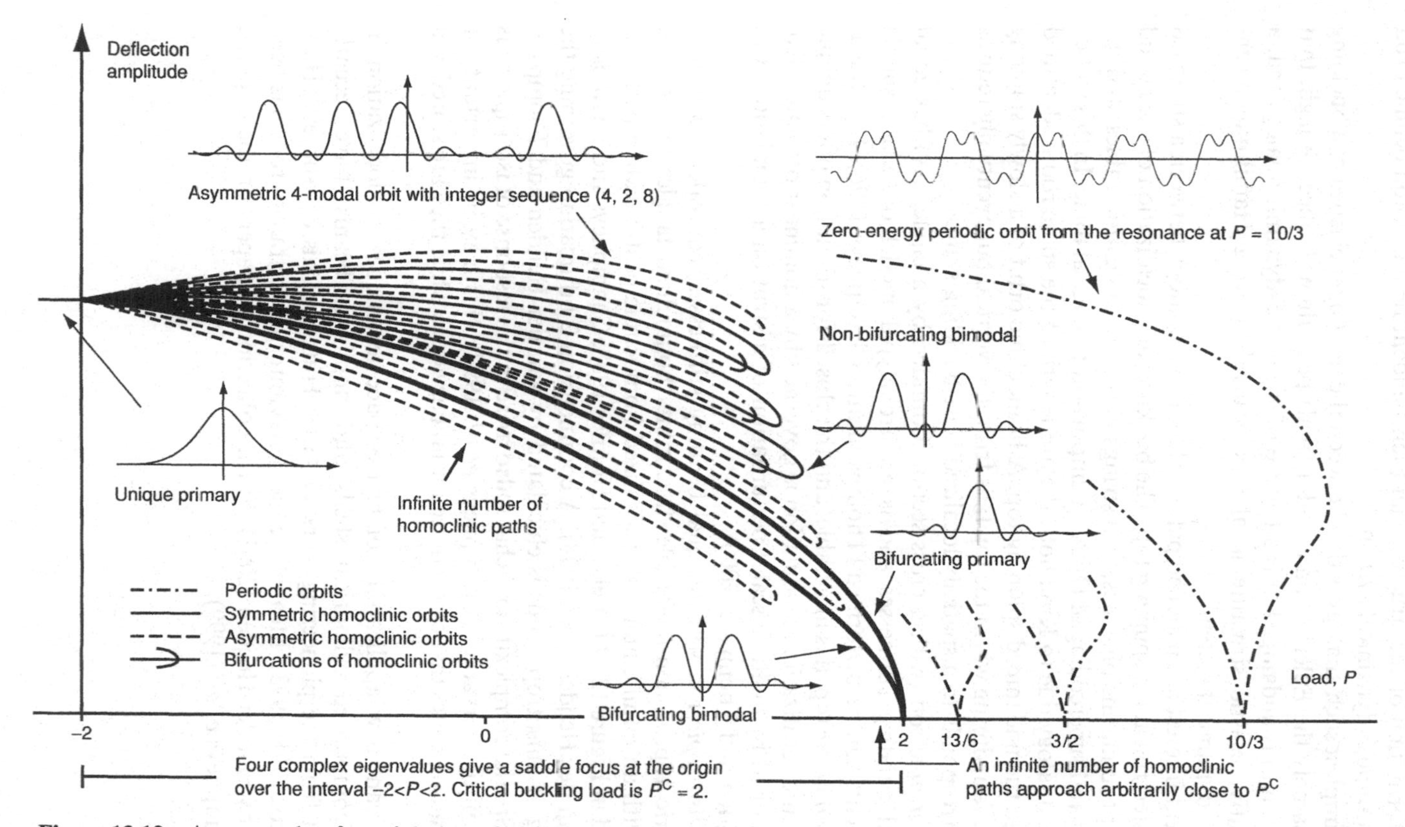

Figure 12.12 An example of spatial chaos in a beam on a nonlinear elastic foundation, showing an infinite number of localized homoclinic post-buckling paths (Buffoni *et al.* 1996). Take the equation $u_{xxxx} + Pu_{xx} + u - u^2 = 0$ and replace x by t to obtain an equivalent 4D dynamical system. Homoclinic orbits have the asymptotic property $u \to 0$ as $x \to \pm\infty$

in a real or imaginary equivalent dynamical system (Hunt *et al.* 1989). A localized deformation of the structure can then be identified as a homoclinic orbit of the nonlinear dynamical system.

The simplest such analogy is that between the nonlinear elastica of a buckling beam, namely the Euler column, and a (real) pendulum. Notice, though, that the pendulum is undamped: it is a feature of the static–dynamic analogy that a conservative elastic structure is always equivalent to a Hamiltonian (undamped) dynamical system.

Of great interest as an archetypal problem is the analogy between a stretched and twisted rod and a spinning top. The buckling and localization of such a rod is relevant to marine pipelines, drill strings and communication cables, and the writhing and supercoiling of DNA, an important biological topic. In this static–dynamic analogy due to Kirchhoff, the arc length of the static rod is identified as time in the undamped spinning top. A homoclinic orbit of a slowly spinning unstable top, during which the top falls from the vertical but eventually returns to it, corresponds to a spatially localized post-buckling state.

An *isotropic* rod (with a cross-section bounded by a circle, say) has equal principal bending stiffnesses and is an archetypal model on which simple experiments can be performed (Thompson and Champneys 1996). Equivalent to the *symmetric* top, it is integrable, and the closed-form solution offers unique insights into localization: modulation is governed by a nonlinear oscillator, like that obtained by multiple-scales perturbation of a strut on a foundation (van der Heijden and Thompson 2000).

An *anisotropic* rod with unequal bending stiffnesses, equivalent to a non-symmetric top, is non-integrable: it exhibits homoclinic tangles and spatial chaos with an infinite number of localized homoclinic post-buckling paths, as sketched in Figure 12.12 (Mielke and Holmes 1988; Champneys and Thompson 1996; van der Heijden *et al.* 1998). A codimension 2 bifurcation governing the buckling of anisotropic rods is elucidated by van der Heijden and Thompson (1998). Spatial complexity and chaos also arise in the response of isotropic rods with an initial curvature (Champneys *et al.* 1997); this may be important for cables and pipelines that can acquire initial curvature while they are stored in a coil.

Significant new applications of phase-space techniques, to localization in axially compressed cylindrical shells and nonlinear beams, have recently emerged from the pioneering work of Giles Hunt and his co-workers (Hunt and Neto, 1991, 1993). Meanwhile the behaviour of constrained rods has been studied by van der Heijden (2001). For a collection of papers on these topics, see Champneys *et al.* (1999).

13

Geometry of Bifurcations

Having examined several examples of nonlinear dynamics and their bifurcations, we now turn to a more systematic description of qualitative changes in the behaviour of typical nonlinear dynamical systems. Key concepts in classifying bifurcations are the distinction between *continuous* and *discontinuous*, or *catastrophic* bifurcations; and the difference between *local* and *global* bifurcations. Geometric analysis in phase space and control-phase space provides the essential means for making these notions precise. After reviewing the established theory of bifurcations of equilibrium and limit cycle attractors, we present a topological classification of the bifurcations of chaotic attractors, concentrating mainly on dynamical systems under a single control.

It should be noted that some dynamical systems (with one or more controls) exhibit sudden changes in behaviour that appear to be bifurcations, but (strictly speaking) are not bifurcations. These apparent bifurcations happen when the phase portrait depends continuously on the parameter, but changes very rapidly in an extremely narrow range of control values. Such events have been studied recently, and named *canards*; it appears that they will play a significant role in applications. For further information, the expository article by Diener (1984) should be consulted. Here we shall consider only bifurcations in the strict topological sense—a qualitative change occurring at a mathematically precise control threshold.

13.1 LOCAL BIFURCATIONS

Local bifurcations of a dynamical system with controls are the qualitative changes in the phase portrait that can be characterized near a single point in phase space. Equilibrium point bifurcations such as the fold and pitchfork are clearly included. By the device of Poincaré mapping the bifurcations of limit cycles, such as the cyclic fold and flip, can also be characterized near a point in the Poincaré section; so these also are considered as local bifurcations.

The important local bifurcations are all identified by an appropriate local simplification of the dynamics. This approximation accounts for the lowest-order part of the vector field near a point in phase space, as well as for the lowest-order part of its dependence on the controls. In other words, low-order approximation is carried out near a point in control-phase space.

For dynamical systems given explicitly by differential equations such as

$$\dot{x} = f(\mu, x) \tag{13.1}$$

(where x and the control μ are vectors), the local bifurcations can be studied by computation of normal forms that embody the local low-order approximations. The equilibrium points are first located by finding states $x_0(\mu)$ such that

$$f(\mu, x_0(\mu)) = 0$$

For example, if μ represents a single scalar control, these $x_0(\mu)$ are paths of equilibrium points in control-phase space. Next the linearization of the vector field near each $x_0(\mu)$ is examined; any value μ_0 where the linearization has eigenvalues with zero real part is non-hyperbolic, i.e. a candidate for a bifurcation point. Finally one must compute terms of the nonlinear function f that are of lowest order near $x_0(\mu_0)$ to obtain the local form of the bifurcation. This computation is systematic but often non-trivial. Examples beyond those of Chapters 7 and 8 can be found in Chapter 3 of Guckenheimer and Holmes (1983).

Similar computations for bifurcations of limit cycles are possible once an explicit mapping

$$X_{n+1} = F(\mu, X_n) \tag{13.2}$$

of the Poincaré section is known; that is, X_n is now a vector of coordinates in the surface of section. A dynamical system defined directly as a mapping is immediately suited to analysis of local bifurcations of limit cycles; as in the study of equilibrium points for (13.1), the steps are find fixed points, determine μ_0 where fixed points are non-hyperbolic, and compute lowest-order terms in the neighbourhood of bifurcation values. The Taylor series analysis of maps in Chapter 8 suggests how this local-form computation proceeds.

Analysis of local bifurcations of limit cycles for differential equations (13.1) is also possible, but one is hindered by the fact that it is usually difficult to find an expression F for a Poincaré mapping if (13.1) are the defining equations. Typically one must consider the mapping F to be defined only numerically. Despite this practical difficulty, local forms are important because the systematic nature of local-form computation makes it possible to enumerate and classify *all* local bifurcations, starting with the simplest and gradually increasing in complexity. The local bifurcations encountered in applications are those with the simplest local forms. Furthermore, these common local bifurcations all have simple distinctive geometries in control-phase space. Thus a local bifurcation in a real dynamical system can be identified on the basis of geometric evidence alone, by examining trajectories, even when equations for the dynamical system are not known. In principle, the only requirements for geometric analysis of bifurcations are a sufficient degree of control over the dynamical system, so that different initial conditions can be chosen at will, plus adequate visualization of dynamic trajectories in phase space. With practice, even these requirements can be minimized.

For these reasons, we shall review the classification of local bifurcations emphasizing qualitative geometry, while simply stating the results of normal-form computations.

The summary below follows Guckenheimer and Holmes (1983) and Abraham and Marsden (1978), both of which contain valuable illustrations; among other references for this material, we mention Andronov *et al.* (1966), which includes a number of examples in mechanics and circuit theory.

Fold catastrophe or saddle node

The *fold* bifurcation, in which a saddle-node pair of fixed points appear simultaneously as a single control passes a threshold, is in many ways the most fundamental bifurcation in nonlinear dynamics. We have already studied this bifurcation in Chapter 7, where its normal form was given as

$$\dot{x} = \mu - x^2 \tag{13.3}$$

For $\mu < 0$ all starts diverge to $x \to -\infty$ as $at \to \infty$, while for $\mu > 0$ a pair of equilibria exist, one attracting and the other unstable. The unstable equilibrium is a repellor in the one-dimensional centre manifold, or a saddle if the full multidimensional phase space is considered. Decreasing μ from positive values, these two equilibria approach and annihilate each other at $\mu = 0$. Increasing μ from negative values, the saddle-node pair suddenly materializes out of the blue. The portrait of this bifurcation in the (μ, x) control-phase space was indicated in Figures 7.3 and 7.4.

The fold bifurcation, also called the saddlenode, or limit point, or static creation, is the simplest form of local bifurcation in that it requires only a one-dimensional phase space (or centre manifold), only one control, and has only lowest-order terms. From consideration of a Taylor series, we known that addition of higher-order terms will not qualitatively change the local dynamics. More generally, a small perturbation of the dynamics will not alter the bifurcation qualitatively; in other words, the bifurcation is structurally stable. Thus the normal form (13.3) can be a valid local model of bifurcation behaviour in complicated dynamical systems.

An example of a fold occurring in elasticity theory is the snap-buckling of a support arch under a vertical load (Thompson 1982, p. 54ff.) If the vertical load is P, with P_c being the critical value at which the arch buckles, then μ in the normal form (13.3) will be proportional to $(P_c - P)$. From the theory of local bifurcations, one may expect that buckling occurs when an unstable equilibrium state of the arch approaches the unbuckled stable equilibrium state, and this is in fact the case.

It is interesting to note that, if saddle equilibria can be identified in the phase portrait of a dynamical system, the occurrence of a fold bifurcation can be predicted by examining phase portraits, without taking the control parameter beyond the bifurcation value. One might for example track the distance in phase space between a saddle and a node as a function of a control parameter. The mutual approach of a node and complementary saddle will typically lead to annihilation. Equation (13.3) indicates that the distance in phase space separating the saddle and the node will vary as $\mu^{1/2}$ near the bifurcation value $\mu = 0$. Because of this, predicting the critical value of μ by *linear* extrapolation always overestimates the margin of safety.

The fold bifurcation of a mapping describes the appearance of an attracting limit cycle together with a complementary limit cycle of saddle type. As we have seen in Chapter 8, its normal form is

$$X_{i+1} = X_i + \mu - X_i^2 \tag{13.4}$$

This bifurcation is qualitatively identical to the fold for equilibrium points if one makes the translation from map to differential equation, hence the names *cyclic fold* or *dynamic fold*. The fact that a stable limit cycle typically appears by a bifurcation involving a complementary saddle cycle was already known to Poincaré (1899).

The fold bifurcations of equilibria and of cycles are now recognized as the primary examples of discontinuous or catastrophic bifurcations in dynamics.

Discontinuous bifurcations

Thom's elementary catastrophe theory classifies the fold as the simplest of the catastrophes involving only equilibria. Although the elementary catastrophe theory deals only with the restricted class of gradient dynamical systems, the fold remains the simplest catastrophe even in a broader context.

A geometric interpretation of the notion of discontinuity in a fold catastrophe will be helpful. Referring for example to the saddle-node diagram in Figure 7.3, we may imagine scanning the various phase portraits for various values of the parameter μ. Viewing a vertical slice of this (μ, x) diagram sliding continuously to the left with decreasing μ, we see the attracting node and the saddle move towards each other, touch, and vanish. Every aspect of this animation changes continuously with μ, except at $\mu = 0$ where the attractor vanishes. The interrupted path of the attractor indicates a discontinuous bifurcation.

A contrasting case is the supercritical pitchfork bifurcation encountered in the Lorenz system, which is a *continuous bifurcation*. Although the attracting set bifurcates across the pitchfork, any arc in the (μ, x) control-phase space parametrized by μ and tracing a continuous path within the attractrix on one side of the bifurcation can be followed continuously across the bifurcation without leaving the attractrix. (By *attractrix* we mean the ensemble of attracting sets in the (μ, x) control-phase space.) We may take this as a definition of continuous bifurcation.

This definition also applies to bifurcations of periodic and chaotic attractors. A useful reformulation of this definition is based on the idea of a *control-phase function*, which maps each control value μ to the corresponding attractor-basin portrait. For one control, we have seen control-phase functions in the logistic map diagram of Figure 9.8, and schematically in the bifurcation diagram of the Lorenz system (Figure 11.13). It is also possible to consider control-phase functions with two or more controls; Figure 1.10 is an example, if we imagine a phase portrait attached to each point in the control plane. Other examples can be found in Guckenheimer and Holmes (1983, pp. 71–72) and in MacDonald and Plischke (1983). Using this notion of control-phase function, a *discontinuous* bifurcation occurs at a control value μ where the control-phase function is discontinuous (Zeeman 1982).

Schematic diagrams of continuous and discontinuous bifurcations at $\mu = \mu_0$ are shown in Figure 13.1. Heavy lines and shaded regions represent the attractrix. For any given μ, a horizontal slice of the attractrix yields the attractor(s). In the upper diagram, the locus in phase space of such a slice varies continuously with μ passing through μ_0; thus any attractrix path P can be extended continuously across $\mu = \mu_0$. In the lower diagrams, some or all attractrix paths Q cannot be continued, and the attractor bifurcation is discontinuous. See also Figure G.2 in the Glossary, page 373.

The concept of discontinuous bifurcation of non-equilibrium attractors was central to Thom's generalized catastrophe theory, and has also been appreciated in the former USSR, where the terms *safe* and *dangerous boundary* (in control space) are used; see for example Shilnikov (1976). We shall use the terms discontinuous bifurcation and *catastrophic* bifurcation synonymously for any bifurcation in which some attracting set paths are interrupted. A concise yet comprehensive theoretical framework for these ideas has been set forth by Abraham (1985).

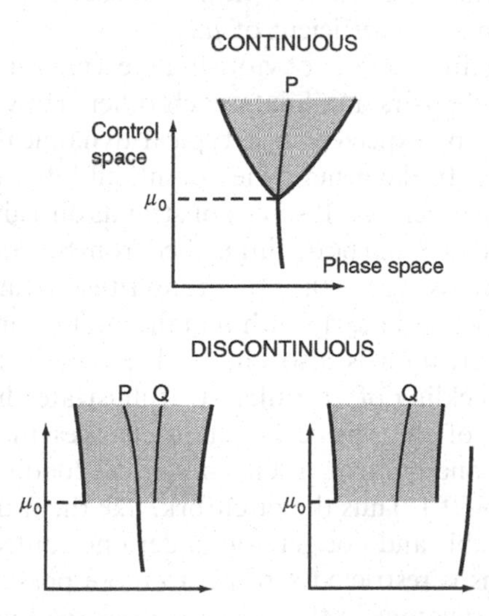

Figure 13.1 Schematic control-phase space diagrams of continuous and discontinuous bifurcations at $\mu = \mu_0$: attractrix paths P continue across the bifurcation threshold, while paths Q are interrupted

Transcritical and pitchfork bifurcations

In addition to the fold catastrophe, two further bifurcations of equilibria are frequently encountered that transpire in a one-dimensional centre manifold. As shown in Chapter 8, these bifurcations, the transcritical and the pitchfork, also arise naturally from the lower-order terms of a power series representation of a one-dimensional mapping; similarly one may consider a first-order differential equation and obtain (taking unit coefficients)

$$\dot{x} = \mu x - x^2 \tag{13.5}$$

as the simplest form of the *transcritical* bifurcation, and

$$\dot{x} = \mu x - x^3 \tag{13.6}$$

for the *pitchfork* bifurcation of equilibria. The differential and the discrete forms were summarized in Figure 7.3.

Both of these normal forms contain higher-order terms than the normal form of the fold catastrophe. In the case of the transcritical, the μx term is of higher order than the simpler μ dependence of (13.3). Adding a term $-\varepsilon$ to (13.5) will change the bifurcation from transcritical to fold (actually back-to-back folds separated by an interval $-2\sqrt{\varepsilon} < \mu < 2\sqrt{\varepsilon}$ with no attractor), no matter how small is ε. From the generic viewpoint, any typical perturbation of a dynamical system containing (13.5) would introduce a term $-\varepsilon$, so the transcritical bifurcation is not structurally stable and not intrinsically generic under one control. The transcritical bifurcation *is* structurally stable within a restricted universe of dynamical systems in which fold catastrophes are prohibited by replacing the lowest-order transversality condition (non-zero coefficient of μ) with a higher-order condition (non-zero coefficient of μx).

The transcritical bifurcation is obviously an exception to the claim that approaching saddle-node pairs annihilate each other. However, the transcritical bifurcation is not to be expected in a typical dynamical system unless some constraint is in effect. In the generic viewpoint, all bifurcations are fold catastrophes unless proven otherwise. If some constraints on a given dynamical system are unknown, the fold may still be distinguished from the transcritical bifurcation in principle without crossing $\mu = 0$; in the transcritical bifurcation the separation of saddle from node varies linearly with μ in the limit of small separation.

The pitchfork bifurcation is also changed qualitatively by a typical small perturbation. The buckling of an Euler strut illustrates how any defect in the supposed symmetry of the physical system changes the pitchfork to a fold catastrophe if only one control (such as vertical loading) is considered (see Thompson 1982, p. 35ff.). Thus the pitchfork, like the transcritical bifurcation, is not structurally stable and not generic under one control unless the universe of dynamical systems is restricted a priori to those possessing the appropriate symmetry. For imperfections in these bifurcations see Figure 7.8.

We note that the transversality conditions for the fold, transcritical, and pitchfork bifurcations may be interpreted geometrically with reference to a three-dimensional surface with coordinates $(\mu, x, f(\mu, x))$. The transversality condition for the fold refers to the slope of this surface at the bifurcation point along the direction of the μ axis. In the generic case of the fold, the surface crosses $f = 0$ in the most transverse way possible, while the non-generic transcritical and pitchfork bifurcations have surfaces tangent to the μ axis. Transversality conditions play an important role in more general settings, such as higher-dimensional control-phase spaces, which are more difficult to visualize; the mathematical theory can be found in Abraham and Robbin (1967).

The pitchfork bifurcation is part of a bifurcation scheme that *is* generic under *two* controls: the cusp catastrophe (Figure 7.8). This is nicely illustrated by considering both vertical loading and symmetry defect as independent controls in the buckling of an Euler support column, and is described in Thompson (1982).

The generic forms for discontinuous bifurcation of equilibria under one or more controls are described by Thom's elementary catastrophe theory for gradient dynamical systems. A complete theory of bifurcations in dynamics would attempt to classify the generic bifurcations, or at least the discontinuous ones, for limit cycles and chaotic attractors as well. At present this can only be done for dynamical systems under one control. For further discussion of elementary catastrophe theory, Poston and Stewart (1978), Thompson and Hunt (1984) and Zeeman (1977) contain readable presentations.

A notable feature of the pitchfork bifurcation is its occurrence in two distinct forms, depending on the sign of the cubic term in the normal form. Equation (13.6) describes the bifurcation of a stable, attracting equilibrium into a pair of attractors with a saddle as the by-product needed to separate the newly divided basin. As seen in Figure 7.3, the same geometric configuration occurs when the sign of x^3 is changed; however, attractors and saddles are interchanged. Following the attractor in this subcritical version, it collides with a symmetric pair of saddles and disappears, leaving only a single saddle point for $\mu > 0$. The path of the attractor is now interrupted, as in the definition of a discontinuous or catastrophic bifurcation. Thus the pitchfork has both a continuous or *subtle* form in which an attractor branches, and a discontinuous or *catastrophic* form in which symmetric saddles coalesce around and annihilate an attractor.

It may be asked whether the transcritical bifurcation (which has only one form) is subtle or catastrophic. Although the attractor appears to follow a continuous path, technically this path is interrupted at $\mu = 0$ where only a degenerate saddle point exists. According to a strict interpretation of the definition above, this means the bifurcation is catastrophic. However, the behaviour at a single mathematically precise value of a control is unlikely to be observed in any real dynamical system. From this point of view it is preferable to exclude removable discontinuities from consideration, so that the transcritical bifurcation becomes a continuous one.

Flip bifurcation

Among local bifurcations the flip has the distinction of being the simplest to have a mapping form but no corresponding differential equation form in

Figure 7.3. Of course the flip bifurcation does occur in differential equations, but only as a bifurcation of limit cycles; its centre manifold is a Möbius strip.

The diagram of the supercritical flip bifurcation is quite similar to the pitchfork, except that stable motions for $\mu > 0$ oscillate between the two parabolic branches, having twice the period of the attractor for $\mu < 0$. (We may still say that there are two attractors after the flip bifurcation if we agree to distinguish period 2 oscillations that are phase-shifted by one cycle.) Likewise the subcritical flip resembles the catastrophic form of the pitchfork, but with a period 2 saddle converting the period 1 attractor to a period 1 saddle. (As usual, the multidimensional saddle point appears as a repellor if only the centre manifold is considered.) The supercritical flip is a continuous or subtle bifurcation that doubles the period of an attracting limit cycle, leaving a period 1 saddle as by-product. The subcritical flip is a discontinuous or catastrophic bifurcation.

In contrast with the pitchfork bifurcation, the flip bifurcation *is* structurally stable under arbitrary small perturbations of the underlying dynamics, and thus occurs generically in dynamical systems with one control.

The subtle form of the flip bifurcation has already been seen in the logistic map and the Hénon map in Chapter 9, and the equations for Rössler's band in Chapter 12. The role of period-doubling cascades as a route to chaos has become widely recognized, but as we have seen, a single supercritical flip bifurcation does not always lead to an infinite cascade.

Although it would be possible to consider even higher-order normal forms of bifurcation with one control and one-dimensional centre manifold, such bifurcations are less important than those involving additional controls (e.g. elementary catastrophes) or more dimensions of phase space. We shall now turn in this latter direction, in order to understand bifurcations in which point attractors become cycles, and the transitions to chaos.

Hopf bifurcation

So far the local bifurcations in Figure 7.3 have governed the mutation of equilibria or of limit cycles. There is also a local bifurcation that transforms an equilibrium into a limit cycle, namely the Hopf bifurcation.

The continuous bifurcation of a point attractor to a stable limit cycle is the supercritical Hopf bifurcation. A two-dimensional centre manifold is a requisite for this bifurcation, and the normal form involves a pair of first-order equations. The equilibrium point loses stability when a complex conjugate pair of Liapunov characteristic exponents cross the imaginary axis in the complex plane, as in Figure 7.2(b). The transversality condition states that, as the control parameter varies uniformly, the exponents should cross the axis with non-zero speed.

The supercritical Hopf bifurcation gives the generic description of the *gradual* onset of a single mode of periodic oscillation. From the normal form one can deduce that the amplitude of oscillation initially grows like the square root of a typical control parameter. As an inevitable by-product of the appearance of the new stable limit cycle, an unstable equilibrium point is left in place of the formerly stable equilibrium, as indicated in the diagram of Figure 7.3. In the two-dimensional centre manifold, the unstable equilibrium appears as a spiral repellor.

Like the flip, the Hopf bifurcation also has a complementary subcritical form, also illustrated in Figure 7.3, in which an unstable equilibrium is converted to an unstable limit cycle, leaving a stable equilibrium of spiral type. Or, varying the control parameter in the opposite sense, an unstable limit cycle shrinks down around a stable spiral equilibrium and, upon contact, makes the equilibrium point unstable. In terms of attractrix paths, this is a discontinuous bifurcation. An example of this catastrophic form of the Hopf bifurcation was encountered in the Lorenz system at $R = 24.76\ldots$. An extensive study of the Hopf bifurcation can be found in Marsden and McCracken (1976).

The continuous growth of a second mode of oscillation from a limit cycle is described by the Neimark bifurcation for mappings with two-dimensional centre manifolds. Topologically, an attracting torus grows out of a stable limit cycle, leaving behind a spirally unstable limit cycle. Neimark bifurcation has a continuous and a complementary catastrophic form, but the complete description is complicated by the possible effects of resonances, as mentioned in Chapter 8.

The continuous growth of oscillatory modes in a dynamical system with one control is of course a commonly observed phenomenon. Abraham and Marsden (1978) thus refer to a Hopf bifurcation followed by a Neimark bifurcation as the *main sequence*.

Finite jumps from dangerous boundaries

The distinction between subtle and catastrophic bifurcations has very important implications from a practical operational point of view, and these are perhaps worth emphasizing here. For the local bifurcations under discussion (excluding therefore for the moment the explosions of Section 13.2), we can say that a *catastrophic* bifurcation occurs at a *dangerous boundary*, while a *subtle* bifurcation occurs at a *safe boundary*. Thus a real system *evolving* slowly along an equilibrium path, or along a trace of limit cycles, due to the *slow variation* of a control would experience a finite, rapid, dynamic jump to a remote attractor at a dangerous boundary but not at a safe boundary. In this context it is worth stressing that the type and form of the local bifurcation tells us *nothing* about the new remote final state after a catastrophic jump: a static fold could for example trigger a jump to a remote equilibrium, a remote limit cycle, or even a remote chaotic attractor. It could also be *indeterminate* (see Appendix page 359). This is in complete contrast to the subtle bifurcation, where the type of the bifurcation tells us precisely the qualitative form of the attractor to which an evolving system would make a gradual smooth transition.

13.2 GLOBAL BIFURCATIONS IN THE PHASE PLANE

In Figure 3.19 we saw how a qualitative change in a dynamical system, namely the introduction of damping, can be associated with a change in the topological configuration of the invariant manifolds of a saddle point. The phase portrait of the unforced oscillator with twin-well potential equation (3.40) has doubly asymptotic curves, or saddle connections, which are broken by the slightest amount of dissipation.

Likewise in Chapter 11 we found that profound changes in qualitative behaviour caused by varying the Rayleigh number in the Lorenz equations (11.1) are associated with no local bifurcation at all, but solely with the global topological configuration of the insets and outsets of saddle fixed points and/or saddle limit cycles. Bifurcations occur as a typical control parameter passes a threshold value, the threshold itself being identified for example by an outset curve lying within an inset surface.

A variety of types of global bifurcation exist, depending on the particular topological configuration of invariant manifolds involved. In general, any topological change in the configuration of invariant manifolds can be expected to cause some qualitative change in behaviour. In some cases, like the homoclinic connection in the Lorenz system at $R = 13.926\ldots$, the qualitative change dramatically affects the basin structure of the phase portrait without changing the attractors. Other global bifurcations create or destroy attractors, for example the Lorenz attractor that appears at $R = 24.06\ldots$ due to a heteroclinic connection. Which effect a given global bifurcation will have depends on the specific topological arrangement involved.

In this chapter we shall pass over the basin bifurcations to concentrate on prototypical examples of global bifurcations that create, destroy, or qualitatively change attractors. There are actually two ways in which this can happen. One is the global bifurcation proper, that is the qualitative change of invariant manifold topology. The second is a hybrid type, in which the bifurcation event is a local bifurcation of the catastrophic variety, whose full repercussions are determined by global structure of invariant manifolds.

These two types, the global and local/global, are most clearly seen in two fundamental examples drawn from the theory of dynamical systems with plane phase space. These two examples are well known in the literature; see for example Chapter VI of Andronov *et al.* (1966) and Chapter 6 of Guckenheimer and Holmes (1983). However, the physical and topological interpretation of the two fundamental examples—and particularly their implications for classifying the routes to chaos—are not yet widely appreciated. The implications of these two examples relate not only to the global versus local/global dichotomy, but to the topological dichotomy between intermittency and hysteresis. In the two examples of this section, these two dichotomies appear to be the same; not until we consider bifurcation of chaotic attractors will it become clear that these dichotomies are different and only partly overlapping.

Intermittency and mode locking

From the point of view of applications, the most common example of the local/global bifurcation occurs not in a differential system in the phase plane but in a two-dimensional Poincaré mapping. The flow and the mapping form are topologically equivalent, and we shall therefore concentrate on the example of a forced Van der Pol system

$$\begin{aligned} \dot{y} &= (1 - x^2)y - x + A\sin(1.1t) \\ \dot{x} &= y \end{aligned} \tag{13.7}$$

and study its Poincaré mapping in the plane section at $t = 0$. This system is equivalent to the acceleration-forced second-order equation (6.6); a similar bifurcation is observed in the system (6.7) with velocity forcing.

Figure 13.2 shows phase portraits of equation (13.7) at two different values of the forcing amplitude A; between these values, at $A \simeq 0.611$, a bifurcation occurs. The first phase portrait at $A = 0.61$ shows 500 returns of a trajectory on the attractor. Return points progress anticlockwise, advancing rapidly near the top of the attractor and more slowly in the region where points are visibly dense. As they wind repeatedly around, return points seem never to coincide precisely with any previous return point, even in accurate, high-order digital simulation. Return points appear to fill out a closed loop in the Poincaré section plane, a *drift ring*. There could in fact be frequency locking at some extremely slow subharmonic, but the simulations show incommensurate frequencies, i.e. quasi-periodic motion. The corresponding topological attractor in three-dimensional $(x, \dot{x}, \phi)$ space such as Figure 6.5 would be a torus.

In the second phase portrait, with $A = 0.62$, the return points shown are transients, with a single attracting fixed point of the return map marked by a

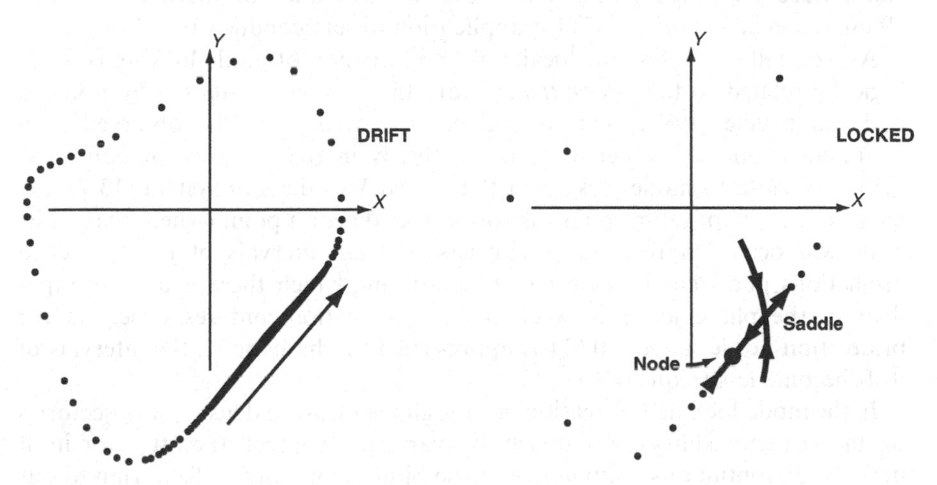

Figure 13.2 Phase portraits of the acceleration-forced Van der Pol equations (13.7) before and after mode locking. This is the *map explosion* of the Appendix

large dot in the figure. Experimenting with different initial conditions shows that all trajectories approach the attracting fixed point, which corresponds to mode-locked oscillation at the forcing frequency. All starting points progress quickly under return mapping towards the location of the former drift ring; successive points then move in monotone fashion, either clockwise *or* anticlockwise (depending on starting condition) and settle to the attracting point.

The portrait at $A = 0.62$ shows a number of transients, each asymptotic to a saddle fixed point. Two of the transients are in the outset of the saddle, that is, they are asymptotic as $t \to -\infty$, while the others are in the inset of the saddle. The nearly vertical inset separates the starts leading to clockwise approach to the fixed point from starts leading to clockwise transients.

As the forcing amplitude A is decreased from $A = 0.62$ to a value near 0.611, the attracting fixed point of the Poincaré map and the saddle point move closer, approaching a saddle-node bifurcation. For lower forcing amplitudes, the fold catastrophe leaves an attracting set on a large drift ring. The location of this drift ring is essentially the same as the location of the saddle outset for larger A. Thus, across a threshold value of the control A, the attractor jumps discontinuously in size. The conditions for this to happen are a local catastrophic bifurcation in the context of a certain global configuration of invariant manifolds.

On the drift side of this bifurcation, the density of return points on the drift ring depends continuously on A; the closer A approaches the catastrophe border, the more densely points cluster near a single point. In an appropriate mathematical sense, the measure density of return points approaches a Dirac delta function situated at the saddle-node collision point, represented by a discrete measure. From the normal form of the saddle-node bifurcation, one can deduce a frequency scaling law for the flow and for the map (Stewart 1986); see Peinke *et al.* (1992) for application to semiconductors.

As we shall see below, the local/global bifurcation of mode locking is topologically related to the *intermittency* transitions to chaos studied by Pomeau and Manneville (1980). This *temporal intermittency* can be observed in a continuous time series signal, or more clearly in the Poincaré mapping. At drift very close to mode locking in the forced Van der Pol system (13.7), the measure density of return points is concentrated near a point (where the saddle node will occur); a time series consists of long intervals of nearly locked oscillations punctuated by time intervals during which there is a more rapid drift of the phase angle between forcing oscillation and response. As the bifurcation border at $A \simeq 0.611$ is approached by changing A, the intervals of drift become less frequent.

In the mode-locking bifurcation, although the measure density of trajectories on the attractor shifts continuously by varying a control, the attractor itself makes a discontinuous jump in size at the bifurcation border. Referring to our definition of continuous bifurcations in terms of attractrix paths, we find that mode locking is a discontinuous or catastrophic bifurcation. Some attractrix paths in control-phase space are interrupted, although some (passing through the saddle-node collision point) can be continued. This is prototypical of *intermittency catastrophes*; as we shall see, these may involve chaotic attractors as well as drift rings.

The mode-locking bifurcation illustrates why discontinuous bifurcations are called *unsafe boundaries* in the Russian literature, e.g. Shilnikov (1976); on the mode-locked side, as the saddle and node move closer, there is no indication of the impending jump in size to a larger attractor, unless one goes beyond the simple attractor diagram to consider the global topology of invariant manifolds.

The intermittency catastrophe illustrated in Figure 13.2 by the Poincaré mapping of a plane section has a counterpart for dynamical systems given by flows in a plane, in which a fold catastrophe for an equilibrium point substitutes for the fold of a period 1 fixed point. If the global configuration of outsets of the saddle equilibrium forms a closed loop, the intermittency catastrophe for the planar flow can cause discontinuous bifurcation from an equilibrium to a limit cycle attractor as illustrated in Figure 13.3. An example of this omega explosion for a planar flow is described by Zeeman (1982). The two intermittency catastrophes—equilibrium to limit cycle in planar flow and periodic fixed point to drift ring in planar mapping—achieve by discontinuous bifurcation what the supercritical Hopf and Neimark bifurcations respectively achieve by continuous bifurcation.

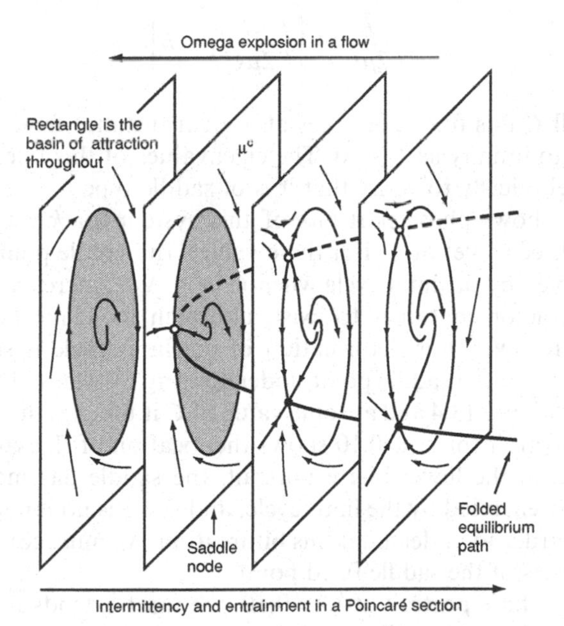

Figure 13.3 Diagrams of omega explosion in a flow, in which a folded equilibrium path generates a limit cycle. This illustrates the *flow explosion* and the *map explosion* of the Appendix

Hysteresis and blue sky catastrophe

The second type of global bifurcation in plane phase space is the discontinuous disappearance of a limit cycle from the phase portrait. This bifurcation event is not associated with any local bifurcation. However, it does have in common with the local catastrophic bifurcations that the attractor disappears by collision with a saddle, in this case a saddle equilibrium point. We shall illustrate this bifurcation by considering a Van der Pol system

$$\begin{aligned} \dot{x} &= ky + \mu x(b - y^2) \\ \dot{y} &= -x + C \end{aligned} \tag{13.8}$$

related to the velocity-forced system (6.7) of Shaw. Here we have simply replaced the sinusoidal AC forcing by a constant DC bias denoted by C, following Abraham and Simó (1986); see also Diener (1984). A different example of the same bifurcation can be found in Zeeman (1982).

With $C = 0$ equation (13.8) reduces to the familiar second-order equation (6.6) of Van der Pol, with its limit cycle surrounding a spiral repelling fixed point at the origin. For any $C \neq 0$, the phase portrait includes an additional fixed point of saddle type, whose coordinates are easily found (by setting $\dot{x} = \dot{y} = 0$ above) to be

$$\begin{aligned} x &= C \\ y &= \frac{k}{2\mu C} + \left[\left(\frac{k}{2\mu C}\right)^2 + b\right]^{1/2} \end{aligned} \tag{13.9}$$

Thus for small C this fixed point lies at a great distance from the limit cycle, moving away to infinity as $C \to 0$. The eigenvalues of this fixed point may be computed algebraically to verify that it is of saddle type.

Figure 13.4 shows phase portraits of this system for $k = 0.7$, $\mu = 10$ and $b = 0.1$, and three values of C. For these values, the saddle point is close to the location occupied by the limit cycle when $C = 0$. As C increases from zero, the limit cycle attractor continues to exist, although its shape becomes slightly asymmetric. However, this deformation of the limit cycle is small compared to the movement of the saddle point under varying C. The saddle point moves downwards in Figure 13.4 and at some value of C it touches the limit cycle. The upper phase portrait for $C = 0.10$ shows the location of the saddle still above the limit cycle; in the lower phase portrait, the saddle has moved inside the region formerly enclosed by the limit cycle, and there is no longer an attracting limit cycle. In order to understand this bifurcation, we must consider the global insets and outsets of the saddle fixed point.

In the upper phase portrait at $C = 0.10$, the inset extends roughly horizontally from the saddle point. The inset is a separator between two basins. Initial points below the inset generate trajectories that settle onto the limit cycle attractor. The downward branch of the outset appears to settle onto the limit cycle almost immediately, although we know that it is in fact only approaching

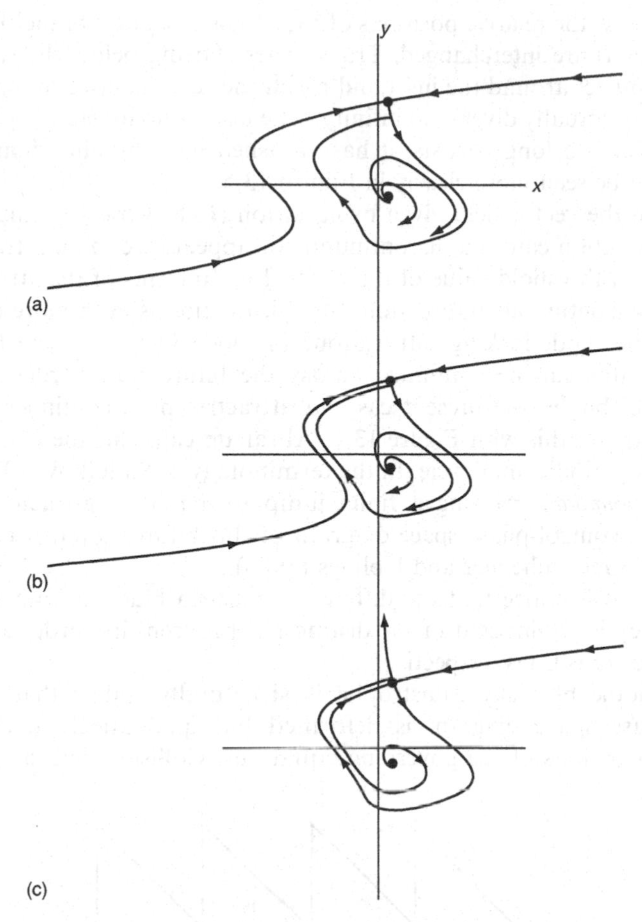

Figure 13.4 Phase portraits of blue sky disappearance of a limit cycle in acceleration-forced Van der Pol equations (13.8) with DC bias: (a) $C = 0.10$, (b) $C = 0.12$, (c) $C = 0.14$. This is the *saddle connection* of the Appendix

the limit cycle very rapidly. Initial points above the inset follow the upward branch of the outset and diverge to infinity.

As C increases, the saddle point and its inset move closer to the limit cycle, as shown in the middle phase portrait with $C = 0.12$. For some slightly larger value of C, the saddle point just touches the limit cycle. At that value of C, the left branch of the saddle inset and the lower branch of the outset both coincide with the location of the limit cycle; the saddle fixed point has a doubly asymptotic trajectory, or *homoclinic connection.* This homoclinic connection may be thought of as a limit cycle of infinite period; by increasing C, the attracting limit cycle oscillation has slowed down and ceased to oscillate.

The homoclinic connection exists only for a single value of C, and increasing C beyond this value gives a phase portrait like the bottom of Figure 13.4 at

$C = 0.14$. Now the relative positions of the global inset (left branch) and outset (lower branch) are interchanged. Trajectories starting below the saddle point follow the outset around the inset and reach the region above the saddle, from which they eventually diverge to infinity. The inset is no longer a separator, and the limit cycle no longer exists; it has vanished into the blue. Some of these features can be seen more clearly in Figure 13.5.

Although the vector field given by equation (13.8) varies continuously with C, this bifurcation causes a discontinuous disappearance of an attracting limit cycle across a threshold value of the control C. In terms of the attractrix path criterion for continuous bifurcation, this bifurcation is even more discontinuous than the mode-locking bifurcation: in mode locking, some but not all attractrix paths can be continued across the bifurcation border in control-phase space, but in the present case *no* attractrix path continues across the border; compare this with Figure 13.1. Abraham calls this the *blue sky catastrophe* for a periodic limit cycle. In the terminology of Shilnikov (1976) this is a *dangerous boundary*, causing a finite jump to a remote attractor. A three-dimensional control-phase space diagram of this bifurcation can be found on page 293 of Guckenheimer and Holmes (1983).

Physicists will notice that the difference between blue sky catastrophe and intermittency is reminiscent of the distinction between first-order and second-order phase transitions, respectively.

The periodic blue sky catastrophe is structurally stable; that is, the full control-phase space diagram is deformed but qualitatively unchanged by small perturbations of the governing equations. Collision with a saddle fixed

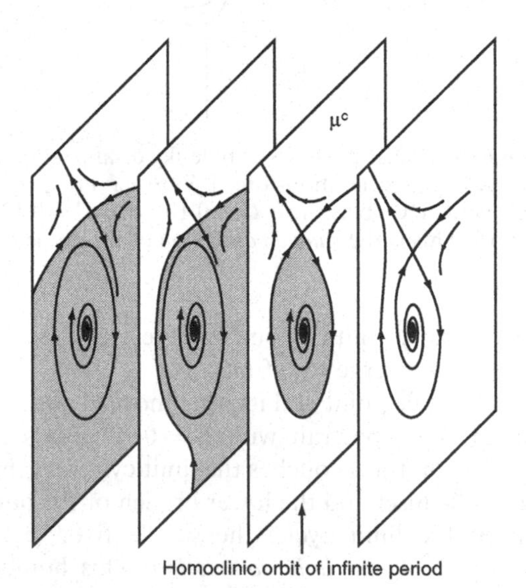

Figure 13.5 The blue sky catastrophe in which a homoclinic connection results in a limit cycle disappearing into the blue: control μ increases from left to right. This illustrates the *saddle connection* of the Appendix

point is the typical mechanism by which a limit cycle can abruptly vanish from a phase portrait. It is also useful to imagine the phase portraits of Figure 13.4 modified so as to include an additional attractor, say a fixed point, in the region above the saddle point. This could be easily accomplished for example by altering the equations (13.8) for large $y > 0$ only. Such a modified phase portrait could represent a dynamic hysteresis loop: starting in the oscillating state and increasing C, the limit cycle is destroyed and the system jumps to the equilibrium point at large y. Reversing the control and decreasing C, the saddle could be made to move back up in the phase plane: at a certain point the limit cycle would reappear in the phase portrait, but the system, now settled on the stable equilibrium, will not jump back to the oscillating state unless further decrease in C makes the equilibrium vanish, for example in a fold catastrophe. A diagram of this behaviour is given in Figure 13.6.

The blue sky catastrophe for an attracting limit cycle also has a complementary form in which a repelling limit cycle (in the phase plane) vanishes by collision with a saddle point. This would be illustrated by reversing all the arrows in Figure 13.4, e.g. the spiral point becomes attracting. Indeed, we have already encountered this complementary form, disguised by the addition of a third phase-space dimension, in the homoclinic connection of the Lorenz system at $R = 13.926\ldots$ This example shows that even global bifurcations can sometimes be at least partly understood by concentrating on an appropriate (global) centre manifold.

In summary, there are two common types of catastrophic bifurcation for limit cycles. One is exemplified by mode locking, with a discontinuous jump to a point attractor created inside the locus of the limit cycle. The other is discontinuous disappearance of the entire limit cycle. Both types need to be

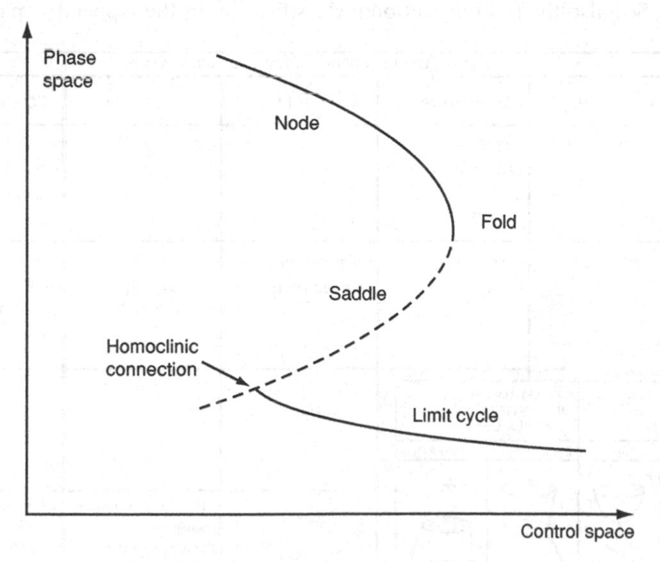

Figure 13.6 Control-phase diagram of a possible hysteresis loop involving a periodic blue sky catastrophe

understood in terms of outstructures, the inset and outsets, of saddle points, which are inevitably involved in such bifurcations. And both types serve as prototypes for the discontinuous bifurcations of chaotic attractors, as we shall see.

13.3 BIFURCATIONS OF CHAOTIC ATTRACTORS

Having studied the various types of bifurcations among point, limit cycle, and torus attractors, we now turn to the bifurcations of chaotic attractors. Again our aim is to identify topological types of bifurcation that occur as a single typical control is varied, i.e. the codimension 1 bifurcations.

It is convenient to organize the various bifurcations according to the topological dimension of the attractor on either side of the bifurcation point. Thus equilibrium points are zero-dimensional, limit cycles are one-dimensional, and the torus is a two-dimensional manifold. For simplicity, mappings are here treated as Poincaré mappings of flows, and attractor dimension refers to the flow rather than to the mapping. Assigning each attractor to a category by topological dimension, we may arrange the structurally stable bifurcations in a matrix with rows and columns of increasing attractor dimension. This is done in Table 13.1. Note that in addition to finite-dimensional attractors, it is possible for a phase portrait to have no attractor, that is, all trajectories eventually leave any bounded region of phase space.

Table 13.1 Bifurcations of codimension 1, arranged by topological dimension of the attracting sets before (row) and after (column) the bifurcation. Uppercase names refer to discontinuous bifurcations, while the continuous or subtle bifurcations appear in lower case. See also the full bifurcational classification in the Appendix, page 359

SEVERITY	ATTRACTOR DIMENSION AFTER THE BIFURCATION				
ATTRACTOR DIMENSION BEFORE THE BIFURCATION	0 POINT	1 CYCLE	2 TORUS	3 CHAOTIC	∞ NO ATTRACTOR
0 POINT	(Supercritical pitchfork) (Transcritical)	Supercritical Hopf [a] FLOW EXPLOSION [b]		?	STATIC FOLD SUBCRITICAL HOPF (SUBCRITICAL PITCHFORK)
1 CYCLE		Supercritical flip	Supercritical Neimark MAP EXPLOSION	(Flip cascade) INTERMITTENCY EXPLOSION MAP	CYCLIC FOLD SUBCRITICAL FLIP SUBCRITICAL NEIMARK SADDLE CONNECTION [c]
2 TORUS					
3 CHAOTIC				CHAOTIC EXPLOSION (namely the REGULAR-SADDLE and CHAOTIC-SADDLE EXPLOSIONS)	CHAOTIC BLUE SKY CATASTROPHE (namely the REGULAR-SADDLE and CHAOTIC-SADDLE CATASTROPHES)

Terminology

Subtle, i.e. continuous (lower case)	CATASTROPHIC i.e. DISCONTINUOUS (UPPER CASE)	
Safe	EXPLOSIVE	DANGEROUS
a (μ, x, 0, 1)	b (0, 1)	c (∞, 1)

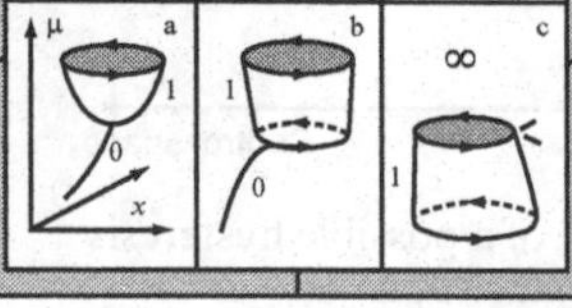

Each 'box' in the matrix of Table 13.1 contains the types of bifurcation from one attractor type to another. For example, the supercritical Hopf marks the change of a zero-dimensional to a one-dimensional attractor type; since it is always possible for a control parameter to cross the bifurcation value in either direction, the table must be symmetric about the diagonal. The table emphasizes bifurcations that are structurally stable under one control, so the supercritical pitchfork and transcritical bifurcations are in parentheses. The discontinuous bifurcations are identified by names all in capital letters.

Columns 0 through 2 of Table 13.1 are bifurcations we have already studied. The MAP EXPLOSION from one- to two-dimensional attractor is the local/global catastrophe associated with mode locking, as in equations (13.7). The FLOW EXPLOSION from zero- to one-dimensional attractor is the omega explosion catastrophe for planar flows. The SADDLE CONNECTION from one-dimensional limit cycle to no attractor is the planar homoclinic connection catastrophe exemplified by the velocity-biased Van der Pol system (13.8). The remaining bifurcations (excluding row 3 and column 3) are the subtle and catastrophic local bifurcations.

Row 3 and column 3 of Table 13.1 give the bifurcations of chaotic attractors. These are three-dimensional attractors in the sense of embedding dimension, that is, they require a phase space of at least three dimensions in order to be realized. However, the chaotic attractors themselves are not three-dimensional manifolds, and certain rules for computing their dimension such as the Hausdorff dimension yield non-integer values greater than 2.

As indicated in the table, only three basic topological types of codimension 1 bifurcation to (or from) a chaotic attractor occur generically, namely the flip or subharmonic cascade, intermittency, and blue sky catastrophe. In other words, transition to chaos in a typical dynamical system by varying one typical control will occur in one of these three ways. Any other type of transition to chaos would require specially coordinated adjustment of two or more controls simultaneously. We shall examine these three topological types of codimension 1 bifurcation in more detail, and conclude with some remarks about codimension 2 bifurcations.

Subharmonic cascade or period doubling

The period-doubling route to chaos is the most widely known and studied of the codimension 1 transitions. We have already studied this sequence of bifurcations in the logistic map and in Rössler's folded band attractor. As noted in Chapter 12, any chaotic attractor in which the fractal layers are rapidly compressed is a likely candidate for reduction to a one-dimensional mapping, as in Figure 12.3. If the one-dimensional map has the unimodal shape of the logistic map, then a typical control takes the dynamical system through a sequence of successive flip bifurcations. This flip cascade, whose geometric rate of accumulation was discovered by Feigenbaum, is the beginning of a complicated sequence of bifurcations for which the prototype is the logistic map diagram of Figures 9.8 or 9.10. See also Figure G.7 of the Glossary.

Although the Rössler band attractor is frequently observed beyond the accumulation point of a flip cascade, it is by no means the only possible outcome of a period-doubling sequence. Topologically different attractors resulting from period doubling have been observed by Marzec and Spiegel (1980) and by Thompson and Ghaffari (1983).

Furthermore, a dynamical system that undergoes one or more period doublings need not complete the entire infinite cascade; incomplete sequences of period doubling can be observed for example in simulation of a forced impact oscillator described in Chapter 15 and in a fluid dynamics experiment reported by Arnéodo *et al.* (1983). Such interrupted cascades become possible most notably when the Poincaré mapping cannot be summarized by a one-dimensional map; for example, we have seen in Figure 6.13 how the forced Duffing oscillator may have an attractor that appears very nearly one-dimensional or is essentially a two-dimensional structure, depending on the amount of damping. In dynamical systems that have interrupted period-doubling cascades, any transition to chaos while varying a single control will occur by intermittency or by blue sky catastrophe. There is mathematical evidence that interrupted cascades are in fact typical of dynamical systems not reducible to a one-dimensional mapping; see for example Holmes and Whitley (1984). This suggests an increasingly important role for global bifurcations in understanding more complicated attractors. And, as we shall see, it is by no means unlikely for global bifurcations to occur in simple attractors.

A completed period-doubling cascade is an infinite sequence of local bifurcations, each of which is a continuous bifurcation. Thus up to the Feigenbaum accumulation point the sequence of bifurcations is continuous in the sense defined above. This continuity is evident in the logistic map bifurcation diagram. We therefore refer to the period-doubling cascade as a *continuous transition to chaos*.

The objection might be raised that the bifurcations are of the continuous flip type only up to the Feigenbaum accumulation point $\mu = \mu_\infty$ whereas, strictly speaking, chaotic mixing is observed only by passing beyond the accumulation point to $\mu > \mu_\infty$. The theory of one-dimensional maps like the logistic map shows that the region $\mu > \mu_\infty$ is peppered with tiny windows of periodic behaviour associated with discontinuous bifurcations. Nevertheless, until μ goes well beyond μ_∞, these periodic attractors have very high periods and exist only for very narrow ranges of μ values, so they will be very difficult to observe in real dynamical systems subject to noise or uncertainty. Hence there is good reason to classify this transition to chaos as continuous. The corresponding term 'safe boundary' used in the Russian literature is also appropriate, as the approach to μ_∞ is clearly forewarned by the distinctive sequence of flip bifurcations.

The period-doubling cascade takes place over an interval of parameter values. We now turn to the remaining two types of bifurcation to chaos, which take place across an isolated threshold value of a control. These are both discontinuous bifurcations; subharmonic cascade is the only continuous path to chaos under one control.

Types of intermittency

A second commonly occurring type of transition to chaos involves intermittency. As in the example of the logistic map in Chapter 9, this can involve a periodic limit cycle destroyed by a catastrophic local bifurcation, for example by a fold. The local catastrophe by itself does not guarantee a transition to chaos; but if the global structure of phase space is such that transients settling to the periodic limit cycle were chaotic, then a chaotic attractor can be expected after the cycle vanishes.

An example is furnished by the nonlinear oscillator with potential $V(x) = x^4/4 - x^2/2$, which models the vibrating buckled beam. In Figure 13.7 Poincaré sections of the forced oscillator

$$\ddot{x} + 0.25\dot{x} - x + x^3 = A \sin t \tag{13.10}$$

are shown near a symmetric pair of cyclic fold bifurcations that occurred at forcing amplitude A just less than 0.265. In the upper part of the figure, the forcing amplitude $A = 0.265$ shows a chaotic attractor, with a high measure density of return points near the former locations of the two saddle-node collisions. In the lower part, the amplitude $A = 0.266$ is increased away from the bifurcation value, and the density of Poincaré section points is more evenly distributed. In Figure 13.8 time series of the same oscillations show nicely how in the second case the average time interval of nearly periodic behaviour has declined. Topologically the bifurcation is quite similar to Figure 13.2 with the drift ring replaced by a chaotic attractor.

In their analysis of intermittency, Pomeau and Manneville (1980) gave examples of equations in which the fold catastrophe is replaced by a subcritical Neimark bifurcation or by a subcritical flip bifurcation. Thus the local bifurcation event that triggers the transition to chaos may be any of the local catastrophes for limit cycle attractors. In all cases the chaotic attractor appears only if chaotic transients existed prior to the local catastrophe; that is, there was a non-attracting horseshoe (typically the result of transversely intersecting invariant manifolds) that becomes attracting when the limit cycle attractor inside vanishes or becomes non-attracting.

Both the intermittency by saddle node and by catastrophic flip have been observed in experiments, and their expected statistical behaviour (depending on the type of local catastrophe) confirmed; see for example Jeffries and Perez (1982), Yeh and Kao (1983) and Dubois *et al.* (1983). In principle, it is also possible that intermittent bifurcation to chaos might be caused by catastrophic disappearance of an equilibrium point attractor inside a phase-space region of chaotic transients. Pomeau (1983) has proposed a mechanism for this in four-dimensional phase space; it is an open question whether this can occur in three-dimensional phase space.

A basic topological feature of intermittency transitions is that the equilibrium or periodic attractor (before bifurcation) lies *inside* the chaotic attracting set (after bifurcation). If this were not the case, then the bifurcation would be hysteretic, as discussed below. When a small attractor bifurcates to a large

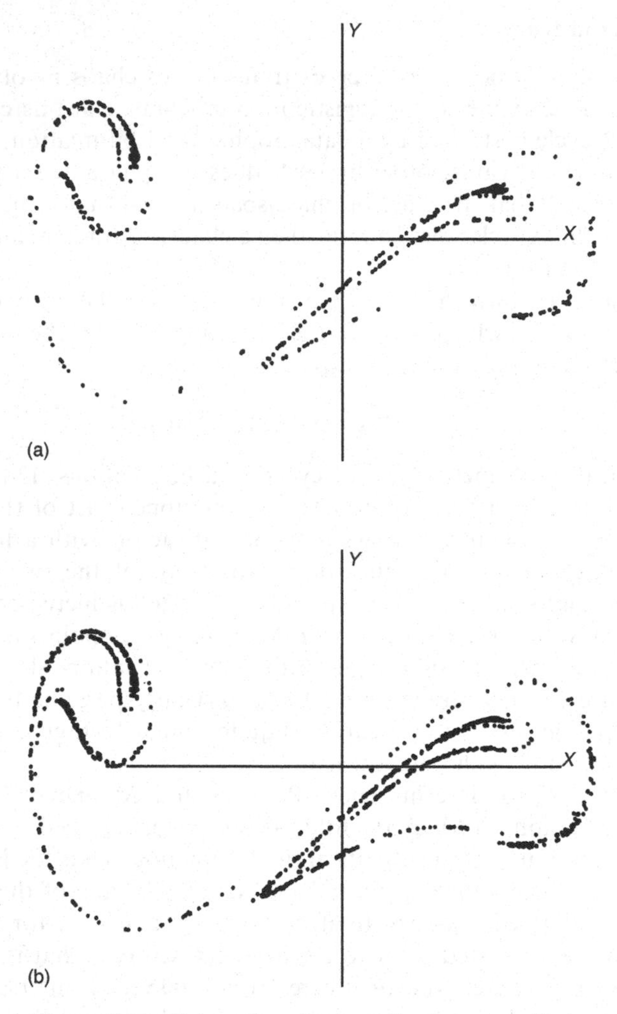

Figure 13.7 Poincaré sections of the forced oscillator equation (13.10) just past intermittency transition to chaos by cyclic fold catastrophe: (a) $A = 0.265$, (b) $A = 0.266$. This is the *intermittency explosion: map* of the Appendix

attractor that contains the locus of the small attractor, and in addition the measure density of return points spreads outwards continuously (in an appropriate topology on a space of measures, e.g. weak convergence), then the abstract topological conditions for intermittency are fulfilled. The topological inclusion of the small within the large attracting set is by itself sufficient to imply continuous shift of measure as the bifurcation parameter changes. Thus we may refer to intermittency bifurcations as *interior catastrophes*, also called *explosive bifurcations*, due to the sudden enlargement of the attractor.

We have already seen in the quadratic map that explosive bifurcations may start from a small attractor that is itself chaotic, for example at the end of the

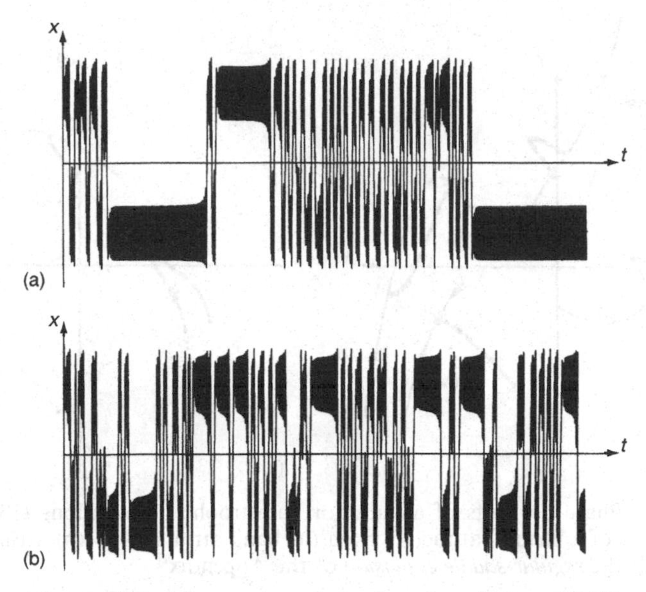

Figure 13.8 Time series of the forced oscillator (13.10) near intermittency transition: (a) $A = 0.265$, (b) $A = 0.266$

period 3 window in Figure 9.9. Typically these explosions happen when an unstable fixed point or periodic cycle of saddle type touches the small attractor. Grebogi *et al.* (1982) observed chaotic attractor explosions in several iterated mappings, and named them *interior crises*. In each case they found an unstable saddle fixed point or periodic point colliding with the small attractor. Before colliding, the unstable point may be attached to a fractal set of chaotic transients; these become part of the enlarged attractor.

Explosive bifurcations causing a jump in size in chaotic attractors occur commonly in differential equations; the first detailed picture of such an event was constructed by Ueda (1980b) using the forced Duffing oscillator with a DC bias. An example is illustrated in Figure 13.9 using the velocity-forced Van der Pol equation

$$\begin{aligned} \dot{x} &= y - B\sin 1.8t \\ \dot{y} &= -x + (1 - x^2)y \end{aligned} \tag{13.11}$$

In the left part of the figure, Poincaré section at $t = 0$ shows the attractors are chaotic Rössler bands. There are in fact only two attractors, since return points on one trajectory jump alternately between opposite pieces; that is, the four pieces of the Poincaré section are joined pairwise in three dimensions. Between each two adjacent pieces lies a saddle point where a period 2 saddle limit cycle cuts the Poincaré section plane. The insets of these saddle cycles are the boundaries between the basins, and are very close to the attractors. In the right part of Figure 13.9, increasing the forcing amplitude has caused the saddle insets to cut the attractors, so that a steady-state trajectory in one band

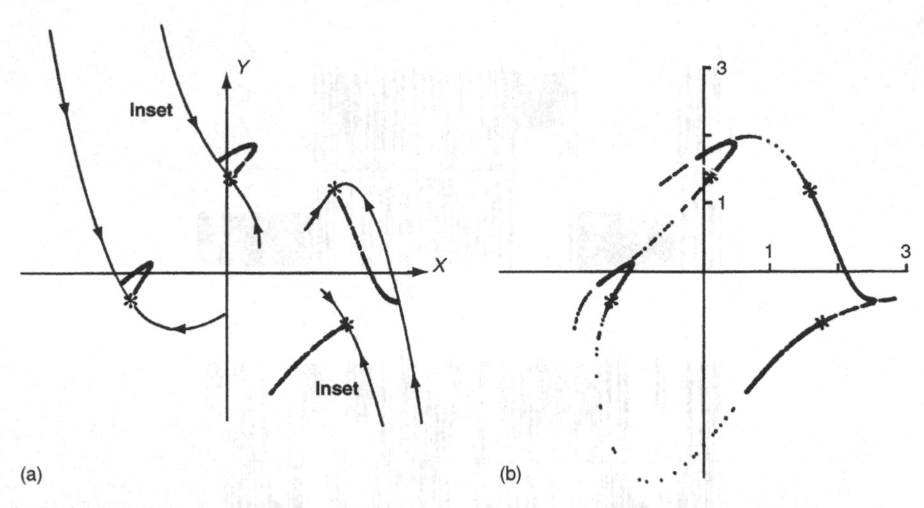

Figure 13.9 Phase portraits of an interior catastrophe in equations (13.11) causing a jump in size of a chaotic attractor, from (a) band attractors to (b) a Birkhoff–Shaw bagel. This is the *regular-saddle explosion* of the Appendix

pair will occasionally enter the region between bands, and subsequently wander across to the other pair of bands. In other words, there is now a single transitive attractor, the Birkhoff–Shaw attractor of Chapter 6. At the bifurcation value, the saddle insets are tangent to the band attractors (which are formed by outsets), so the bifurcation is a heteroclinic tangency. Further increase in forcing amplitude results in gradual increase of the density of return points in the regions between bands. But for *any* forcing amplitude even slightly past the bifurcation value, the entire region between bands is filled completely if enough return points are accumulated.

Explosive bifurcations thus have a discontinuous aspect and a continuous aspect, depending on whether one looks at the locus of the attracting set or at the measure density of trajectories. In terms of attractor locus, there is a threshold value of a typical control at which a discontinuous jump is size takes place; according to the basic definition, this is a discontinuous or catastrophic bifurcation. On the other hand, the shift in measure density lends a continuous aspect to this type of bifurcation. Explosive bifurcations are not safe boundaries because the bifurcation is discontinuous; nor are they dangerous boundaries, since there is no jump to a remote attractor.

In sum there are two known varieties of explosive bifurcations: intermittent transition from periodic to chaotic steady state, caused by a local catastrophic bifurcation in a global setting (chaotic transients); and jump in size of a chaotic attractor, associated with a global event (heteroclinic tangency) in an appropriate global setting. These explosive bifurcations are analogous to second-order phase transitions. We now turn to the first-order transitions, or blue sky catastrophes.

Hysteresis and blue sky catastrophe

If a chaotic attractor A bifurcates to another attractor B, and neither attractor is located inside the other, then the two attractors will typically be disjoint and separated in phase space by a finite distance. That is, of the three set relations shown in Figure 13.10, only the first two are generic under one control. Furthermore, in the disjoint case, it will not be generic for attractor B to appear suddenly at the *same* control threshold where attractor A vanishes. The attractor B will have existed previously, or it will not exist at all; in either case the bifurcation will consist of attractor A simply losing stability—it vanishes into the blue. Such *blue sky catastrophes* occur commonly for chaotic attractors of differential equations. And like the catastrophes we have examined so far, chaotic blue sky events involve collisions with saddle-type objects.

A topologically simple example is furnished by the velocity-forced Van der Pol system

$$\begin{aligned} \dot{x} &= 0.7y + 10x(0.1 - y^2) \\ \dot{y} &= -x + 0.25 \sin 1.5t + C \end{aligned} \tag{13.12}$$

where the constant C makes the sinusoidal forcing asymmetric. By increasing the value of C it is possible to make the Birkhoff–Shaw chaotic attractor of Chapter 6 vanish discontinously (Abraham and Stewart 1986). This bifurcation is a generalization of the blue sky disappearance of a limit cycle discussed above. Phase portraits of the periodically forced system (13.12) are illustrated in Figure 13.11.

In Figure 13.11(a) the Poincaré section at driving angle π is shown for $C = 0.08$. The Birkhoff–Shaw attractor is represented by 1000 return points computed from a single trajectory. The periodic forcing term makes the saddle equilibrium of Figure 13.4 into a saddle limit cycle, whose Poincaré section point is indicated in Figure 13.11 by an asterisk (*). Asymptotic to this saddle cycle as $t \to +\infty$ is an inset of infinitely many trajectories, represented in Figure 13.11(a) by points lying on a curve; in three-dimensional phase space the inset forms a smooth two-dimensional surface. This inset is a separator: initial conditions below the inset settle to the chaotic attractor, while starts above

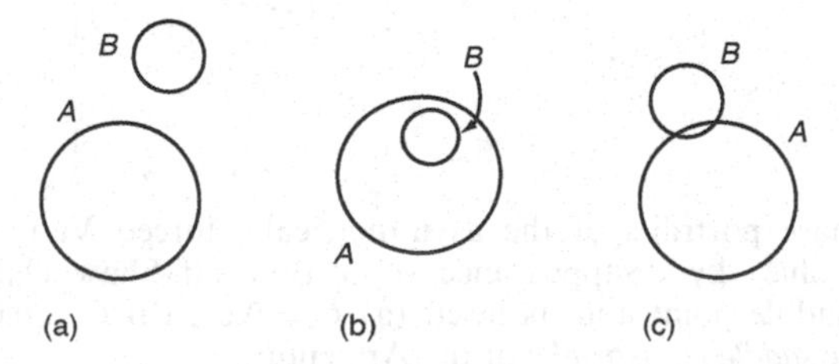

Figure 13.10 Possible set relations between attractors before and after a bifurcation: (a) hysteresis, (b) intermittency, (c) not generic

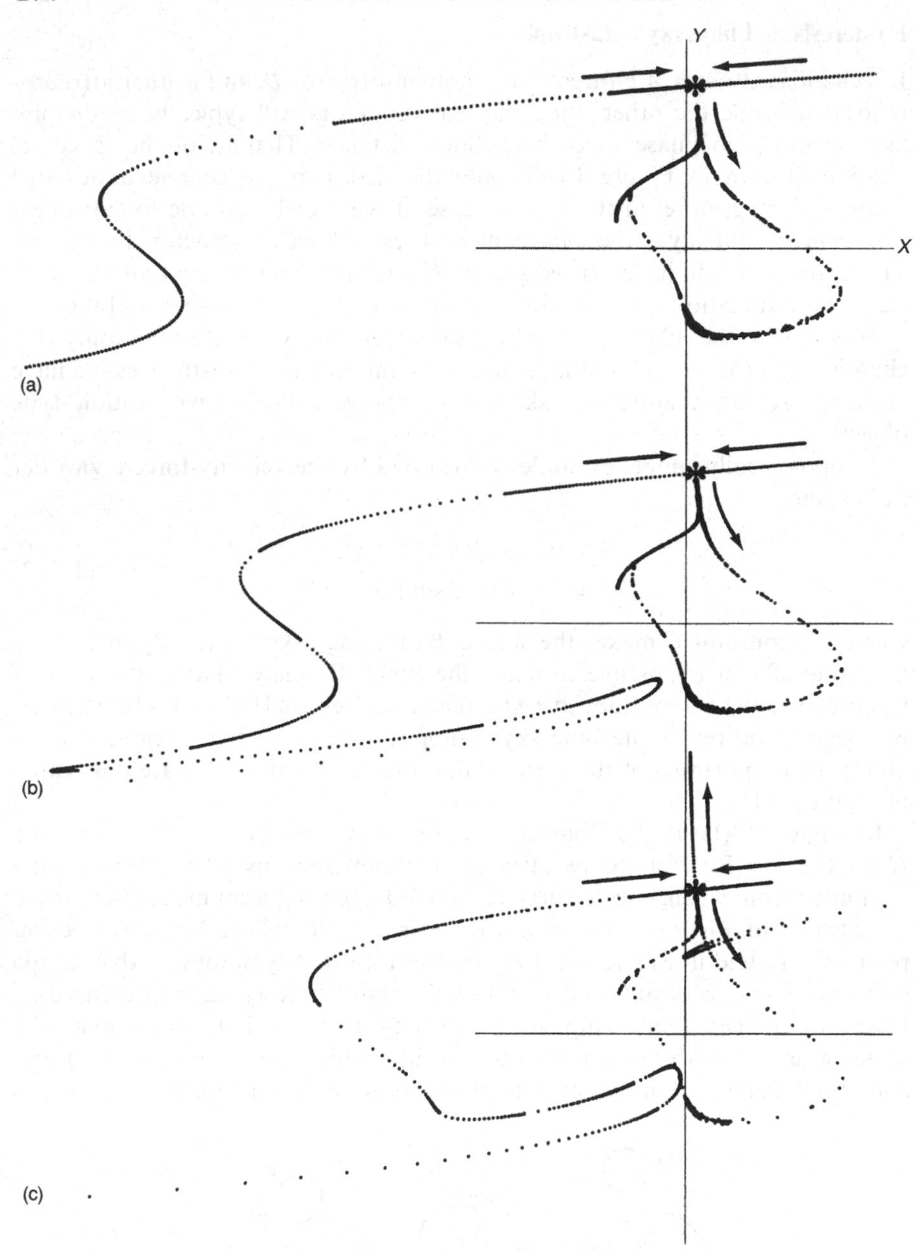

Figure 13.11 Phase portraits of the asymmetrically forced Van der Pol equations (13.12) showing blue sky disappearance of a Birkhoff–Shaw chaotic attractor by collision with a saddle point and its inset: (a) $C = 0.08$, (b) $C = 0.09$, (c) $C = 0.097$. This is the *regular saddle catastrophe* of the Appendix

the inset diverge to $y \to +\infty$ as $t \to +\infty$. The phase portrait for $C = 0.09$ is similar, but the saddle cycle is closer to the attractor, and the inset forms a finger that moves close to the attractor. As a necessary consequence of recurrence, there must be additional fingers moving in as C increases, but they have not yet entered the region of phase space shown.

The third phase portrait is at $C = 0.097$, just past the bifurcation threshold. The saddle cycle has moved very slightly into the region formerly occupied by the chaotic attractor. This region is indicated in Figure 13.11(c) by 500 points all started below the saddle on its outset, and followed a few forcing cycles with $C = 0.097$. As seen in the figure, some points have followed the mixing action to land above the saddle cycle, and are starting to diverge to $y \to +\infty$. Because the chaotic attractor at lower C values is a single transitive attractor (based on numerical observation, not mathematical proof), all points will eventually land above the saddle cycle and then diverge. Thus when the saddle cuts a small piece out of the attractor, by recurrence the entire transitive structure becomes non-attracting. There are still chaotic transients, but the attractor has vanished into the blue.

As an additional consequence of recurrence, the inset must cut the former attractor repeatedly in different locations. To understand this, we note that what we have been calling the former attractor is the outset of the saddle cycle (*). If the inset and the outset of the saddle intersect at a homoclinic point (in the Poincaré section) then the Poincaré mapped image of that point is also in both the inset and the outset (invariant manifolds). Figure 13.11(c) shows the first inset finger mapped backwards one forcing cycle to show a second inset finger. This process can be repeated to generate an infinite sequence of points on a *homoclinic trajectory*. Complete development of recurring inset–outset intersections generates a *homoclinic tangle*. Other examples of homoclinic tangles are illustrated in Hayashi (1975) and later in this book (e.g. chapter 16 page 341 and Glossary page 385).

Although the complete detailed phase portrait of this bifurcation is remarkably complex, the basic ingredients are a saddle cycle and its inset approaching and touching a chaotic attractor. The result is a discontinuous, catastrophic disappearance of the attractor. As with the blue sky disappearance of a limit cycle, no attractrix path in control-phase space can be continued beyond the bifurcation threshold, hence the name *chaotic blue sky catastrophe*.

The example above is conceptually simple because only a single attractor is involved. In other cases there may be additional attractors in the phase portrait, for example at large positive y in Figure 13.11. In such cases the blue sky catastrophe could be part of a hysteresis loop; that is, increasing C would cause steady-state chaotic motion on the Birkhoff–Shaw attractor to jump to another attractor. If that other attractor disappears at lower values of C, a full hysteresis loop involving a chaotic attractor exists, generalizing Figure 13.6. Typically, hysteresis involving chaotic attractors can be expected to involve a blue sky catastrophe associated with homoclinic or heteroclinic behaviour. In fact, we have already seen another example in the Lorenz system, as illustrated in Figure 11.13.

Other examples of chaotic blue sky catastrophes have been observed by Rössler (1976b) see also Gaspard and Nicolis (1983)—and in the Hénon map by Simó (1979), who also gives invariant manifold pictures. More examples are examined by Grebogi *et al.* (1983b), who named them *boundary crises* because the attractor touches its own basin boundary. In one-dimensional maps, the attractor touches its basin boundary at a saddle point having the lowest possible period (Stewart 1991). In forced oscillators, this is typical of cases (e.g. Figure 13.11) where the basin boundary is not tangled before the bifurcation threshold.

But it may be otherwise with forced oscillators. For example, Figure 13.12 shows a chaotic attractor in the twin-well Duffing oscillator equation (6.10), on the verge of touching a tangled inset. The fractal basin boundary (McDonald *et al.* 1985) contains a horseshoe-like *chaotic saddle*, but the fundamental period unstable saddle point S in the basin boundary does not touch the attractor (Stewart 1987).

In Figure 13.12 only parts of the basin of F are identified, by mapping the inset near S back only a few forcing cycles. The fractal structure of the true basin boundary is more clearly indicated in the phase portrait of Figure 6.19, obtained by exhaustive trials.

So discontinuous bifurcations are linked with definite geometric structures in control-phase space, namely saddles and their insets. These ideas may be used to forecast chaotic bifurcations (Stewart and Lansbury 1992).

A classification of bifurcations, included for easy reference in the Appendix, summarizes the generic attractor bifurcations under one control, relating topology in control-phase space to behaviour as in Figure 7.11.

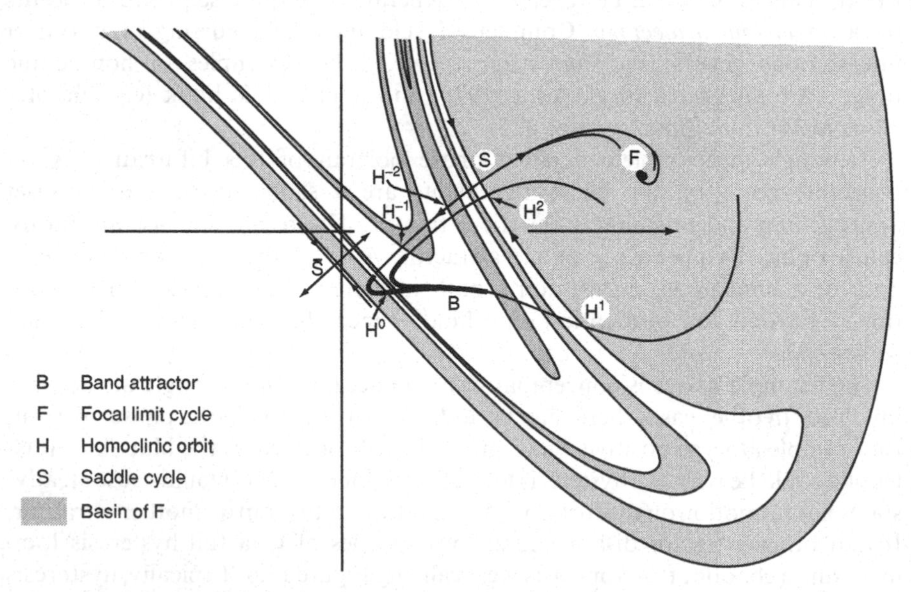

Figure 13.12 Poincaré section near a chaotic blue sky catastrophe, in which the inset of saddle S approaches attractor B while the saddle itself remains at a distance. This is the *chaotic-saddle catastrophe* of the Appendix

Transition to chaos via quasi-periodic motion

We have now examined the basic types of codimension 1 bifurcations that affect chaotic attractors. The larger terrain of *codimension 2* bifurcations includes many special types of qualitative change that can only be found when *two controls* of a dynamical system are adjusted.

One particularly important class of systems with two controls is the forced oscillator, considering the *amplitude* and the *frequency* of periodic forcing as controls. When the forced oscillator has two competing frequencies, as in the forced Van der Pol oscillator or the forced pendulum of Chapter 6, then it may be possible to observe a direct transition from attracting invariant torus to a chaotic attractor, e.g. the Birkhoff–Shaw bagel. Such a bifurcation occurs as the forcing amplitude is increased past the threshold for mixing (horseshoes) in phase space. But in coordination with the amplitude, the frequency must simultaneously be adjusted so as to avoid mode locking. In simple differential equations, this may or may not be possible.

A useful conceptual simplification of forced oscillators with two competing frequencies is represented by the circle map

$$\theta_{n+1} = \theta_n + \frac{K}{2\pi}\sin 2\pi\theta_n + \Omega \tag{13.13}$$

which maps the circumference of a circle (parametrized by $0 \leq \theta < 1$) to itself. Here the sinusoidal term represents the effect of periodic forcing with amplitude K. The ratio of the two frequencies (natural frequency and forcing frequency) is represented by Ω. It has long been known that for low to moderate forcing, two-frequency systems may either be mode locked or may appear to drift, with locking becoming increasingly likely as forcing amplitude increases. Arnold (1965, 1982) used the circle map to study mode-locking patterns in a prototypical two-frequency system.

The action of the circle map is illustrated in Figure 13.13 for two sets of parameters. The horizontal axis represents position along the circumference, so that $\theta = 0$ should be identified with $\theta = 1$. For $K < 1$ the mapping stretches some parts of the circumference and compresses other parts. For $K > 1$ the circumference is folded onto itself; the circle map is then non-invertible and chaotic behaviour may result. The picture for $K > 1$ may be compared with the Birkhoff–Shaw attractor (Figure 6.4).

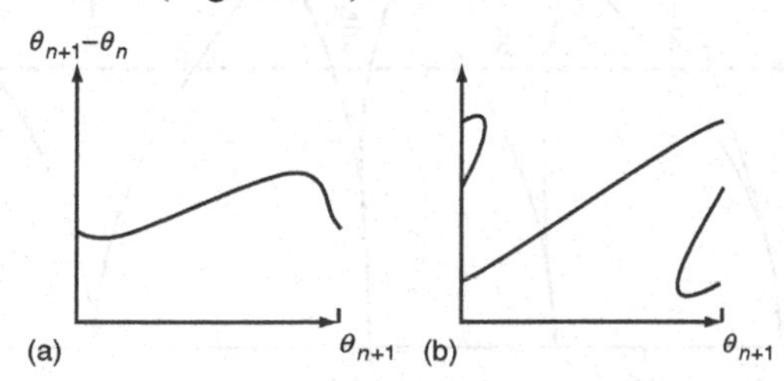

Figure 13.13 Graphs of circle map (13.13) for two different sets of parameters: (a) $\Omega = 0.46$, $K = 0.8$; (b) $\Omega = 0.46$, $K = 2.2$

For small K, the long-term behaviour of the iterated circle map tends to a unique attracting periodic fixed point whenever Ω is a rational number, $\Omega = p/q$. The denominator q then represents a subharmonic number for the periodic motion. The same periodic behaviour is also observed in a small range of Ω near p/q; the attracting periodic motion is structurally stable. At some Ω value on either side of p/q the periodic motion disappears by saddle-node bifurcation. These two saddle nodes mark the ends of a *window* of periodic behaviour.

It may seem paradoxical that drift is ever observed in the circle map, since every Ω is close to some rational p/q. However, Arnold noted that an equally proper question to ask is: what is the *probability* that a given Ω corresponds to mode locking? To answer this, one needs to add up the widths of all the periodic windows. Since there are a countable infinity of subharmonics, this is an infinite sum; the answer found by Arnold (1965) and Hermann (1979) is that the sum of window widths for fixed $K < 1$ and $0 < \Omega < 1$ is less than 1. Thus the probability of observing drift is greater than zero.

Some low-order periodic windows are shown in the (K, Ω) control plane in Figure 13.14. Each window increases in size as K increases; the stronger the coupling between the two frequencies, the more probable mode locking becomes. Each region of periodic motion, bounded by two arcs of saddle-node bifurcation, is called an *Arnold horn* or *tongue*. Although there are an infinity of resonant tongues, their width decreases so rapidly with increasing q that they do not fill the (K, Ω) control plane for $K < 1$. Thus if both K and Ω can be controlled simultaneously, it is possible to reach forcing amplitude $K = 1$ without encountering mode locking.

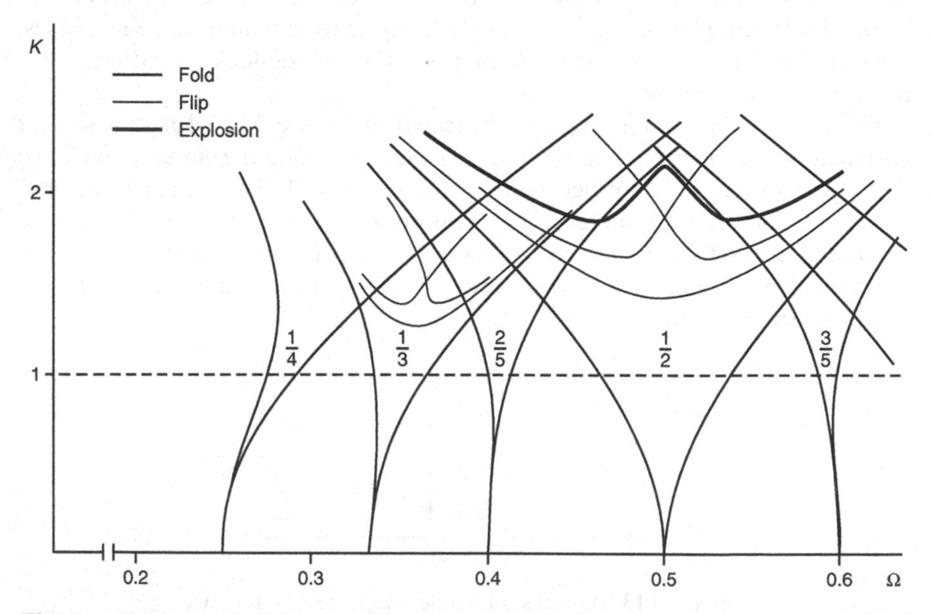

Figure 13.14 Control-space diagram of the circle map, $\theta_{n+1} = \theta_n + (K/2\pi) \sin 2\pi\theta_n + \Omega$, showing horn-like regions of locking for the lowest-period subharmonics

The line $K = 1$ in the (K, Ω) plane is the boundary between invertible and non-invertible circle mappings. For $K > 1$ there is folding of the phase space onto itself; a two-dimensional Poincaré mapping (of which the circle map is a one-dimensional idealization) would develop a topological horseshoe as the circle map becomes non-invertible. The line $K = 1$ is sometimes referred to as the transition to chaos in the circle map; but horseshoes need not be attracting and may correspond to transient chaos only, so we should say that $K = 1$ is the onset to transient chaos.

An interesting fact about the line $K = 1$ was discovered recently by Bak *et al.* (1984), who showed that the total width of all the tongues in $0 < \Omega < 1$ at $K = 1$ is precisely 1. That is, the probability of observing drift drops to zero as the onset of folding is reached. Furthermore, they demonstrated that for the circle map the set remaining at $K = 1$ when all periodic windows are removed is a fractal with dimension $D = 0.89\ldots$, which they conjectured to be universal for all smooth mappings of the circumference to itself. This conjecture is consistent with experiments with a forced electrical oscillator (Yeh and Kao 1983).

Above the line $K = 1$ there is always transient chaos, but chaotic attractors may not appear until K is increased well beyond 1. If only *one* control K is available, the work of Bak *et al.* (1984) shows that mode locking will always occur as K increases through the value 1. The transition to chaos will subsequently occur by one of the bifurcations in Table 13.1, i.e. subharmonic cascade, intermittency or blue sky catastrophe. These are all codimension 1 bifurcations, occurring across arcs in the (K, Ω) plane. The patterns of these arcs are described in much greater mathematical detail by MacKay and Tresser (1987). There is never a direct transition from drift ($K < 1$) to chaotic attractor when only one control is available, because mode locking is typically encountered at or before $K = 1$.

If both K and Ω are controlled simultaneously, it is possible to reach $K = 1$ along a path in the (K, Ω) plane without entering any of the tongues. In this way we reach a point in the fractal set identified by Bak *et al.* (1984). This point is accessible from $K < 1$ by a path containing only drift behaviour; it is unclear whether there is a continuous path from $K > 1$ to such a point via only chaotic attractor behaviour, although there are chaotic attractors nearby.

In a real physical system, mode locking to high-order subharmonics may be masked by noise. In this case the experimentally observable tongues will not fill $0 < \Omega < 1$ at $K = 1$. Thus it will be possible to pass $K = 1$ avoiding mode locking by approximate, but not infinitely fine, control of K and Ω. This probably explains the experimental observations of transition from quasi-periodic to chaotic behaviour, notably Fenstermacher *et al.* (1979).

The theorems proved by Ruelle and Takens (1971) and Newhouse *et al.* (1978) are sometimes said to establish the possibility of a transition from quasi-perodic to chaotic behaviour. In fact, this is a misinterpretation of their results; they showed only that a vector field having quasi-periodic flow can be approximated by vector fields having chaotic attractors; but they do not claim that these vector fields can be made to lie on any reasonable path in control space smoothly approaching the vector field of quasi-periodic flow. Thus there is no known quasi-periodic transition to chaos in mathematical dynamics (in the sense of an arc in control space); the only such transitions are observed in experimental dynamical systems.

It may be argued that the circle map is only an idealization of dynamical systems whose Poincaré maps involve two or more dimensions, so that mode locking might be avoided in, say, forced oscillators. However, as the Poincaré mapping becomes more two-dimensional, the resonant tongues typically broaden and in fact *more* than fill the line (or curve) in control space analogous to $K = 1$, $0 < \Omega < 1$ in the circle map. Thus it becomes even more difficult to avoid mode locking.

It appears that the presence of noise, and consequent masking of the narrowest resonant tongues, is responsible for experimentally observed transitions to chaos from quasi-periodic flow.

Furthermore, two independent controls, such as forcing amplitude and frequency, are needed to avoid mode locking before the onset of chaos. As the level of noise is reduced, finer control is required.

Figure 13.15 shows a control-space diagram for the velocity-forced Van der Pol equations, similar to equations (6.7) and (13.11), with forcing amplitude A and forcing frequency ω:

$$\begin{aligned} \dot{x} &= y + A \sin \omega t \\ \dot{y} &= (1 - x^2)y - x \end{aligned} \tag{13.14}$$

Comparison with Figure 13.14 illustrates the usefulness of the circle map as an archetype.

The subharmonic numbers in the tongues of Figure 13.15 correspond to the denominators in Figure 13.14. The dashed line in Figure 13.15 shows where the invariant torus of equations (13.14) develops folding. Moving upward within the period 2 subharmonic tongue, a mode-locked attractor of equations (13.14) doubles its period, crossing the bifurcation arc labelled d_2, and then again crossing the arc d_4; analagous bifurcation arcs are drawn in Figure 13.14 for the circle map. Moving further up toward the arc labelled e_2 in Figure 13.15, a period-doubling cascade leads to small chaotic attractors, as illustrated in the left phase portrait of Figure 13.9. As the bifurcation arc e_2 is crossed, the small chaotic attractors explode to a folded torus chaotic attractor. The result is as shown in the right phase portrait in Figure 13.9. Again there is a precisely analogous bifurcation in the circle map, shown as the thickly drawn curve in Figure 13.14.

Folded torus chaotic attractors predominate in the region of control space above the bifurcation arc e_2 up to the line labelled n, where harmonic entrainment occurs, typically involving a Neimark bifurcation.

Transition to chaos in two-frequency systems needs to be understood with reference to two control parameters, such as forcing amplitude and frequency. The codimension 1 bifurcations involved are none other than those we have previously encountered. But there are numerous different codimension

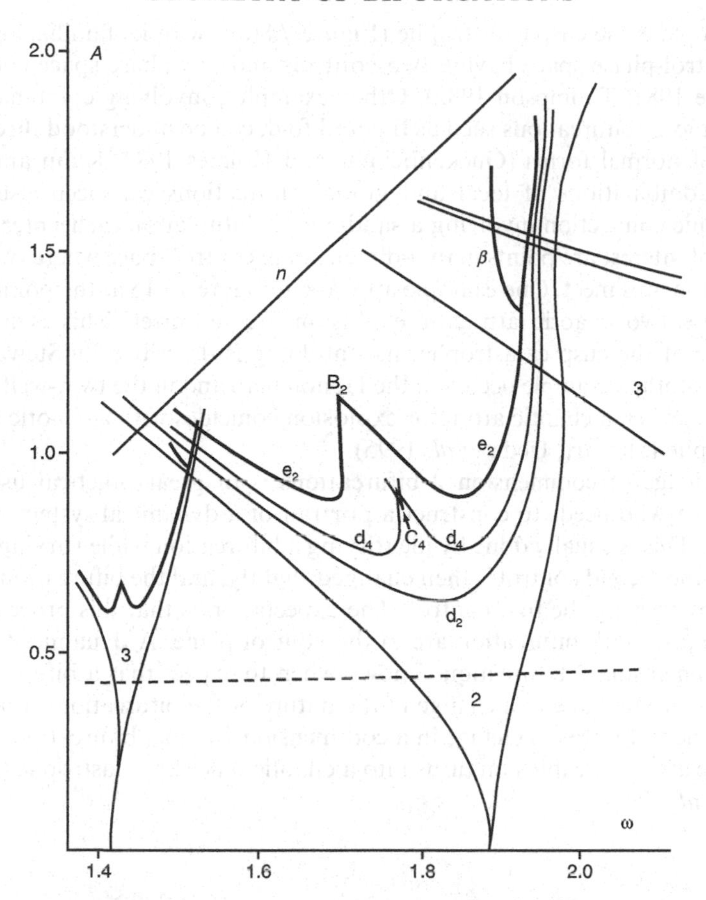

Figure 13.15 Diagram showing some bifurcations of the velocity-forced Van der Pol equations (13.14) in the two-dimensional control space of forcing frequency ω and forcing amplitude A

1 bifurcations, each represented by an arc in a control plane, and these form an overall pattern. Thus the transition from quasi-periodic to chaotic dynamics is not a bifurcation, but a scenario involving two controls, and different routes to chaos.

Codimension 2 bifurcation

Stepping back from this complex two-dimensional gestalt of bifurcation arcs, we may concentrate our attention on somewhat simpler phenomena that occur very commonly in dynamical systems with two controls. In a two-dimensional control plane, two arcs representing codimension 1 bifurcations may meet or intersect in a point. If these codimension 1 bifurcations interact with each other, then the point where the arcs intersect is a codimension 2 bifurcation.

An example is the cusp catastrophe (Figure 7.8(c)), with its familiar unfolding in a control-phase space having two controls and one phase space coordinate (Gilmore 1981; Thompson 1982). Other examples, involving combinations of different local bifurcations such as flip and fold, can be understood through the theory of normal forms (Guckenheimer and Holmes 1983; Kahn and Zarmi 1997). Combinations of local and global bifurcations can occur, such as a homoclinic connection involving a saddle-node bifurcation (Schechter 1987).

Also of interest are points in two-dimensional control space where two global bifurcation arcs meet. One example appears in Figure 13.15 at the point labeled B_2, where two chaotic attractor explosion arcs e_2 meet. This is a chaotic analogue of the cusp catastrophe; its unfolding is described by Stewart *et al.* (1990). Another example occurs in the Hénon map and in the twin-well Duffing oscillator, when a chaotic attractor explosion coincides with a chaotic blue sky catastrophe (Stewart, Ueda *et al.* 1995).

Knowledge of codimension 2 bifurcations is of great practical use to the dynamicist who needs to construct a portrait of a dynamical system with two controls. This is usually done by identifying a bifurcation while varying a single control; the second control is then changed slightly, and the bifurcation is again sought by varying the first control. The expectation is that this procedure will generate a smooth bifurcation arc in the control plane. A dynamicist familiar with codimension 2 bifurcations knows when to expect that a bifurcation arc will not continue, due to a change in the nature of the bifurcation. It may even be possible to forecast a change in a codimension 1 global bifurcation, as when a chaotic attractor explosion turns into a chaotic blue sky catastrophe (Stewart, Ueda *et al.* 1995).

Part IV

Applications in the Physical Sciences

14

Subharmonic Resonances of an Offshore Structure

The first two chapters of Part IV are devoted to a case history in marine technology in which subharmonic resonances and chaotic motions were recently identified in a simple deterministic model of an articulated mooring tower driven by steady ocean waves. The tower is modelled as a one-degree-of-freedom oscillator whose dynamics are nonlinear by virtue of a stiffness discontinuity. This simple yet practical resonance problem is used as an ideal vehicle with which to illustrate many of the general ideas that we have introduced in this volume.

We start then in this chapter by looking at the resonance of a system to be called the *bilinear* oscillator, namely a simple forced oscillator with different stiffnesses for positive and negative deflections. Although the behaviour of such an oscillator is linear in each half of the phase space, the discontinuity generates many features typical of nonlinear differential equations. Subharmonic resonances and multiple solutions dependent on the starting conditions are observed, so that questions of the existence and stability of periodic solutions are far from simple (Thompson, *et al.* 1983).

The motivation for this study came originally from potentially dangerous resonances reported in the naval and marine technology fields where a discontinuity of stiffness is generated by the periodic slackening of a mooring rope. A steady train of ocean waves typically provides a sinusoidal driving force in such a situation.

In particular, model tests on articulated mooring towers have revealed the existence of unexpected subharmonic resonances in steady waves. These buoys are essentially inverted pendulums, pinned to the seabed and standing vertically in still water due to their own internal buoyancy. They are increasingly used for loading oil products to tankers from the deep offshore installations currently being developed, as sketched for example in Figure 14.1. A massive tanker moored to such a tower is essentially a fixed object during tower oscillations, and the periodic slackening of its mooring line generates the stiffness discontinuity. The restoring force on the tower during its pendulum motion is thus due to 'buoyancy plus mooring line' in one direction, and just buoyancy in the other.

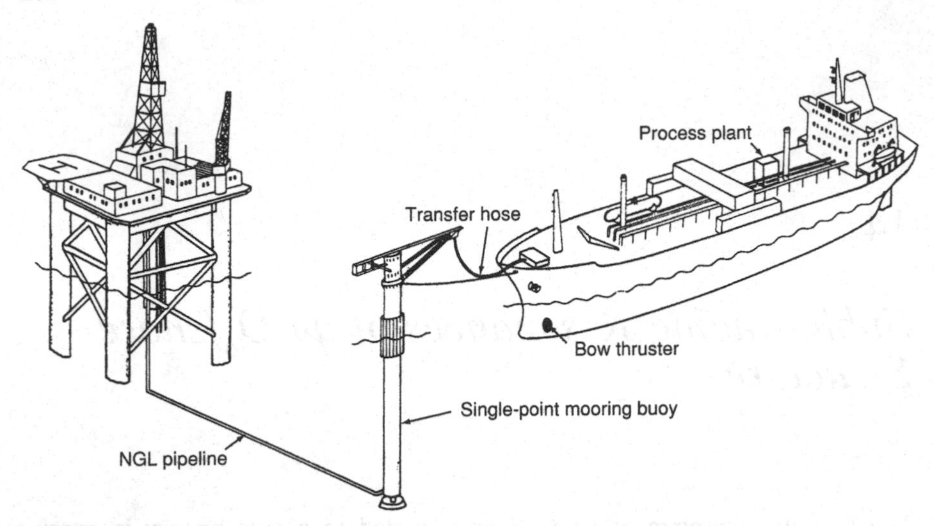

Figure 14.1 General layout of a proposed scheme for the offshore loading of natural gas liquids using an articulated tower as a single-point mooring buoy. Reproduced by permission of Three Quays Marine Services Ltd

The approach to the bilinear oscillator has been through the use of digital and analogue computations (the latter not reported here) guided by concepts of qualitative topological dynamics, making particular use of the Poincaré map. Analytical solutions were originally sought, but did not seem promising due to the unexpected complexity of the response. At each discontinuity, complementary functions are repeatedly generated, so *comprehensive* closed-form solution seemed to be ruled out by the successive transcendental equations involved at each matching point. We note however the useful analytical studies of Garry Tee (Figure 14.12) on the bilinear oscillator, and of Shaw and Holmes (1983a, b, c) on the bilinear and impact oscillators. A study of the infinite-stiffness *impact* oscillator is presented as a limiting case in the following chapter.

The discovery of continuous coexisting small-amplitude solutions under all the subharmonic resonances, and the chaotic non-periodic motions of the impact oscillator, seem to have important implications for offshore designers (Thompson 1983).

14.1 BASIC EQUATION AND NON-DIMENSIONAL FORM

The parameters representing our forced, damped, bilinear oscillator are the mass m, the damping coefficient c, the forcing amplitude F_0, the forcing frequency ω_f, and the two stiffnesses k_1 and k_2 for positive and negative displacements, respectively. Without loss of generality we assume that the stiffness k_1 is less than the stiffness k_2. Notice that we are here focusing attention on the just-tight mooring with the discontinuity at the origin: the more typical tensioned and slack moorings bring with them greater complexity, and will not be discussed at the present time.

The basic equation of motion for the displacement x at time t is then

$$mx'' + cx' + k_i x = F_0 \sin \omega_f t \tag{14.1}$$

where a prime denotes differentiation with respect to t, and $k_i = k_1$ or k_2 depending on the displacement region considered. Now, the free undamped vibrations of such a system with $c = F_0 = 0$ are composed of half sine waves, so we define the bilinear periodic time T as

$$T = \tfrac{1}{2}(T_1 + T_2) = \pi\sqrt{(m/k_1)} + \pi\sqrt{(m/k_2)} \tag{14.2}$$

and the bilinear circular frequency ω as

$$\omega = 2\pi/T = \sqrt{(K/m)} \tag{14.3}$$

where K is an equivalent stiffness defined as

$$K = \frac{4k_1k_2}{(\sqrt{k_1} + \sqrt{k_2})^2} \tag{14.4}$$

We can next define the bilinear critical damping as

$$c_c = 2m\omega \tag{14.5}$$

and the bilinear damping factor

$$\zeta = c/c_c = c/2m\omega \tag{14.6}$$

We introduce the frequency ratio

$$\eta = \omega_f/\omega \tag{14.7}$$

and dividing by F_0 we write the non-dimensional displacement as

$$X = \frac{x}{F_0/K} \tag{14.8}$$

Notice that the amplitude of forcing F_0 will now only influence the analysis as a scaling of the displacement, a linear feature that greatly simplifies the presentation of our results. Scaling the time we write

$$\tau = \omega_f t \tag{14.9}$$

and we shall now use a dot to denote differentiation with respect to τ. We finally write

$$\alpha = k_2/k_1 \tag{14.10}$$

and obtain the non-dimensional equation

$$\ddot{X} + 2\frac{\zeta}{\eta}\dot{X} + K_i X = \frac{1}{\eta^2}\sin\tau \tag{14.11}$$

where K_i is used to denote whichever of K_1 or K_2 is appropriate for the domain under consideration and

$$K_1 = (1 + \sqrt{\alpha})^2 / 4\alpha\eta^2 \tag{14.12}$$

$$K_2 = (1 + \sqrt{\alpha})^2 / 4\eta^2 \tag{14.13}$$

The problem now presents itself as follows. For a given level of damping ζ and a given stiffness ratio α, find the resonance response curve of the amplitude of X against the frequency ratio η. For $\alpha = 1$ this will degenerate to the well-known response curve of a linear oscillator.

14.2 ANALYTICAL SOLUTION FOR EACH DOMAIN

For each domain, the solution of the non-dimensional equation of motion (14.11) can be written in the form

$$X = X_{\mathrm{a}} + X_{\mathrm{b}} \tag{14.14}$$

where X_{a} and X_{b} are the complementary and the particular solutions, respectively. The particular solution is of the form

$$X_{\mathrm{b}} = M \sin \tau + N \cos \tau \tag{14.15}$$

where

$$M = \frac{K_i - 1}{\eta^2[(K_i - 1)^2 + (2\zeta/\eta)^2]} \tag{14.16}$$

and

$$N = \frac{-2\zeta/\eta}{\eta^2[(K_i - 1)^2 + (2\zeta/\eta)^2]} \tag{14.17}$$

The complementary function X_{a} can be written as

$$X_{\mathrm{a}} = \mathrm{e}^{-\zeta\tau/\eta}(A \sin \omega_{\mathrm{d}}\tau + B \cos \omega_{\mathrm{d}}\tau) \tag{14.18}$$

where the damped circular frequency, appropriate to either domain, is

$$\omega_{\mathrm{d}} = \sqrt{[K_i - (\zeta/\eta)^2]} \tag{14.19}$$

and the coefficients A and B, dependent on the initial conditions $(X_0, \dot{X}_0, \tau)$, are reset at every switchover from one domain to the other.

Substituting equations (14.15) and (14.18) into (14.14) and assuming the initial conditions to be $(X_0, \dot{X}_0, \tau)$, the two coefficients can be evaluated as

$$\begin{aligned} A = e^{\zeta\tau/\eta}\Big\{ & (X_0 - M \sin \tau - N \cos \tau) \sin \omega_{\mathrm{d}}\tau \\ & + \frac{1}{\omega_{\mathrm{d}}}\left[\dot{X}_0 + \frac{\zeta}{\eta} X_0 + \left(N - \frac{\zeta}{\eta} M\right) \sin \tau \right. \\ & \left. - \left(M + \frac{\zeta}{\eta} N\right) \cos \tau \right] \cos \omega_{\mathrm{d}}\tau \Big\} \end{aligned} \tag{14.20}$$

$$B = e^{\zeta\tau/\eta}\left\{(X_0 - M\sin\tau - N\cos\tau)\cos\omega_d\tau - \frac{1}{\omega_d}\left[\dot{X}_0 + \frac{\zeta}{\eta}X_0 + \left(N - \frac{\zeta}{\eta}M\right)\sin\tau - \left(M + \frac{\zeta}{\eta}N\right)\cos\tau\right]\sin\omega_d\tau\right\} \tag{14.21}$$

14.3 DIGITAL COMPUTER PROGRAM

The digital computer program was written to detect the stable steady-state solutions of the bilinear oscillator given the initial conditions $(X_0, \dot{X}_0, \tau)$, the frequency ratio η, the damping ratio ζ, and the stiffness ratio α.

The program uses the exact analytical solution $X(\tau)$ for each half of the phase space, and evaluates this at very small increments of time. It thus determines accurately the state at which the switch from one stiffness to the other occurs, say $X = 0$ for $\dot{X} = \dot{X}_i$, $\tau = \tau_i$. The new initial conditions for the next half of the phase space are then $(0, \dot{X}_i, \tau_i)$. This process is repeated until a periodic solution is detected.

This detection of the steady-state solutions of the oscillator is perhaps the most important part of the program, and makes use of the Poincaré map described earlier. The displacement X and its time derivative $\dot{X}$ are evaluated and recorded whenever τ is a multiple of 2π, and compared with all previous Poincaré points. So when, for example, the computer determines that

$$\{X(2m\pi), \dot{X}(2m\pi)\}$$

is equal, within a specified close tolerance, to

$$\{X(2(m-1)\pi), \dot{X}(2(m-1)\pi)\}$$

it knows that a steady fundamental solution ($n = 1$) has been reached. In the case of a subharmonic of order n the repetition will occur after a time delay of $2n\pi$.

Having detected such a repetition, the computer runs for a further time interval of 2π and makes sure that there is again a time-lag equality with the same subharmonic number as before. On receiving this verification, the program runs for a further response cycle making a record of the final waveform at very small time steps. This allows the amplitude to be evaluated and recorded.

So for each time integration, the following data are automatically recorded:

(a) the order of the subharmonic
(b) the approximate time taken to reach a steady state
(c) the number and coordinates of the switching points in a response cycle
(d) the Poincaré mapping points in a cycle
(e) the maximum and minimum values of the displacement, and hence the amplitude parameter

Keeping the system parameters the same, further runs are then made with different starting conditions, a scan of the two-dimensional $(X_0, \dot{X}_0)$ space at $\tau = 0$ being sufficient to fill the three-dimensional $(X, \dot{X}, \tau)$ phase space.

14.4 RESONANCE RESPONSE CURVES

The results of the digital computer time integrations are summarized in Figure 14.2, a second view of which is given by Thompson (1983). Basically these figures show plots of the amplitude parameter y, defined as half the positive peak to negative peak range of X, against η, the ratio of the forcing frequency to the bilinear frequency. These plots are repeated at integer values of $\alpha = k_2/k_1$ from 1 to 10: the limiting case of infinite α is discussed later as a special solution. All curves are here drawn for the fixed value of the damping ratio $\zeta = 0.1$, this being defined in terms of the bilinear circular frequency ω. The effect of varying the damping will be examined briefly in a later section.

We notice again that the magnitude of the forcing only appears indirectly through the definition of X; this is an important quasi-linear feature of the bilinear oscillator. The fact that the magnitude of the driving force has no effect on the response, other than magnifying x through the definition of X, greatly reduces the computational effort and the presentation of the resulting data.

The rear curve, for $\alpha = 1$, is simply the well-known resonance response curve of a *linear* oscillator with no discontinuity: it leaves the y axis at 1.0, rises to a maximum of 5.0 at approximately $\eta = 1$, and then tends asymptotically to zero as the frequency ratio increases.

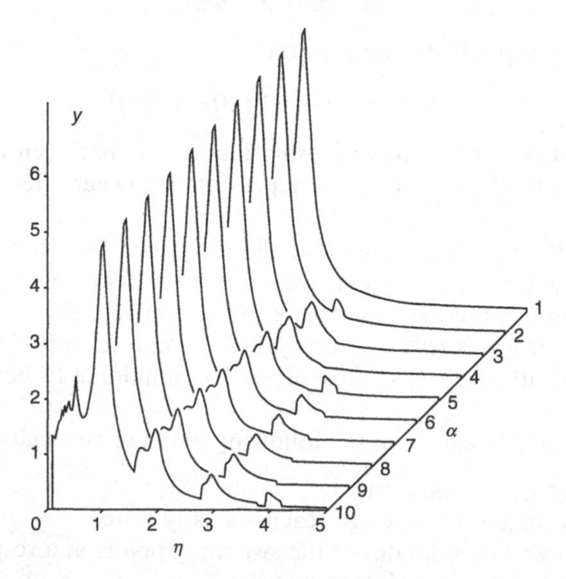

Figure 14.2 A three-dimensional view of the resonance response surface, showing the regular subharmonic resonant peaks ($\zeta = 0.1$)

The effect of increasing α, thereby increasing the strength of the stiffness discontinuity, is to generate new resonant peaks as shown. The main resonance at roughly $\eta = 1$ persists with slightly reduced magnitude for all values of α. The new peaks to the right of this main resonance will be called, perhaps rather arbitrarily, *regular* resonances; those to the left of the main resonance, with a less regular appearance, will be called *irregular* resonances.

The relationships between the three relevant circular frequencies should be noted. We remember that ω is the previously defined *bilinear* frequency, ω_f is the *forcing* frequency, and ω_R is the *response* frequency. Now a consideration of the phase trajectories shows that the wavenumber n defined as ω_f/ω_R can only take the integer values 1, 2, 3, ..., ∞. Fractional values are not possible, since the response cannot repeat itself under different conditions of forcing. In Figure 14.2 we see that the main resonance has $n = 1$ and $\eta = \simeq 1$, so $\omega_f = \omega_R$, both of which are *close* to ω. The *regular* resonances have $n = 2$, 3, 4, 5, ..., with $\eta \simeq n$, so they are typified by having ω_R approximately equal to ω with ω_f approximately equal to multiples of ω. Since these have ω_R a fraction of ω_f, they are normally called *subharmonic* resonances. The *irregular* resonances are enlarged for $\alpha = 10$ in Figure 14.3, and always have $n = 1$. In this plot they occur at values of η equal to 0.505, 0.360, 0.295 and 0.250. They are thus typified by $\omega_f = \omega_R$, both of which are non-simple fractions of ω. We shall show later how they correlate with the cyclic movement of the Poincaré point.

The magnitudes of the resonant peaks vary with α. We see that the strength of the main resonance decreases slightly with α, while the magnitudes of the other resonances increase with α, tending more or less asymptotically to their values at $\alpha = \infty$. The full response curve for infinite α will be presented in the next chapter.

We notice that the basic ($\alpha = 1$) linear response curve for $n = 1$ seems to *underlie* all the higher α curves, emerging particularly as the final response as η tends to infinity. This can be seen clearly in Figure 14.4 in which the $\alpha = 10$ curve is superimposed on the $\alpha = 1$ curve.

The response waveforms for the various resonance peaks are summarized in Figures 14.5 and 14.6 for $\alpha = 10$. Each waveform is set above its sinusoidal forcing function, so that the phase relationships can be observed. The waveform for the main resonance in Figure 14.5 is very similar to the free undamped vibrations of our bilinear system. Notice that the forcing becomes positive (τ is a multiple of 2π) at approximately the centre of the stiff halfwave. The waveforms for the regular subharmonic resonances are also shown in Figure 14.5. They all exhibit, as we would expect, large and long excursions into the low-stiffness domain and rather short, roughly sinusoidal, excursions into the high-stiffness domain.

Four irregular resonant waveforms are shown in Figure 14.6. As η decreases, these are increasingly typified by one relatively large excursion into the low-stiffness domain, followed by small damped oscillations predominantly in the high-stiffness domain. These damped oscillations can give rise to many switching points ($X = 0$) within a response cycle. It is the accommodation of an integral number of high-frequency oscillations in the high-stiffness domain that gives rise to the rather irregular resonant peaks, as we shall study later.

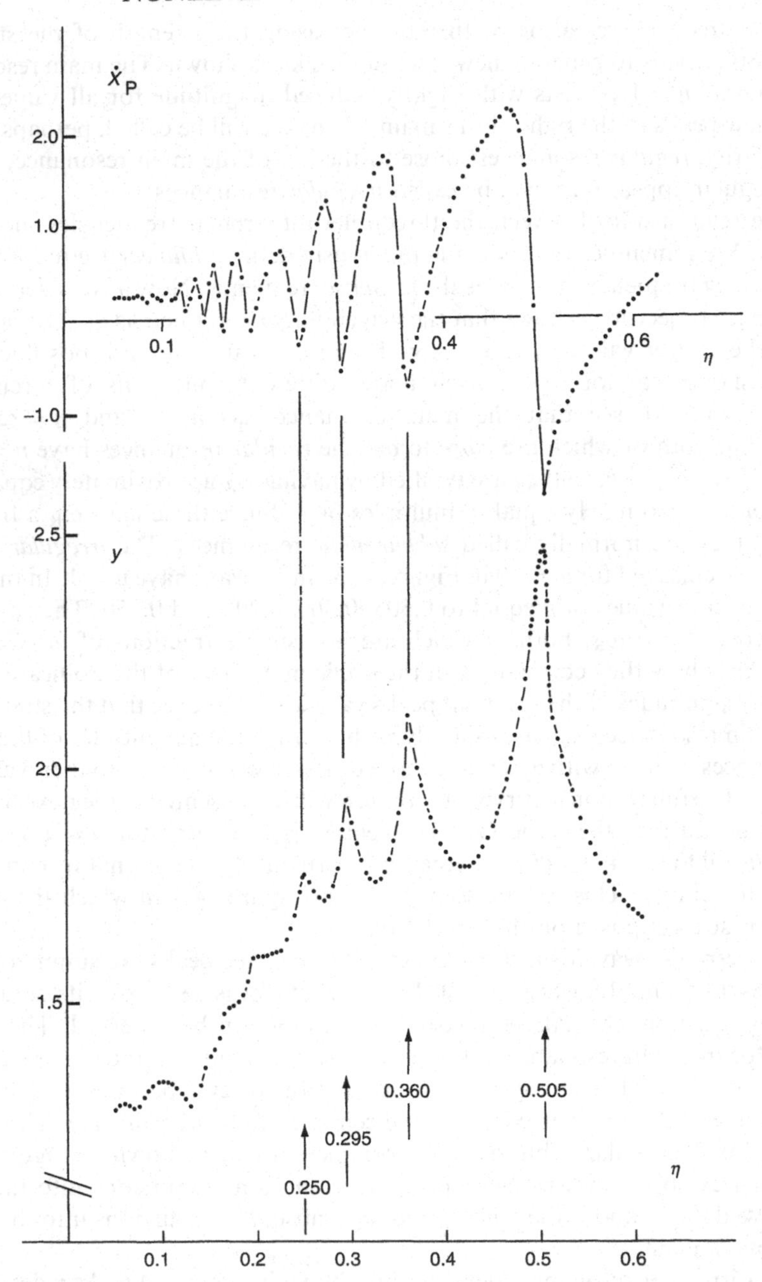

Figure 14.3 An enlarged view of the irregular resonances showing their correlation with the movement of the Poincaré point ($\alpha = 10$, $\zeta = 0.1$)

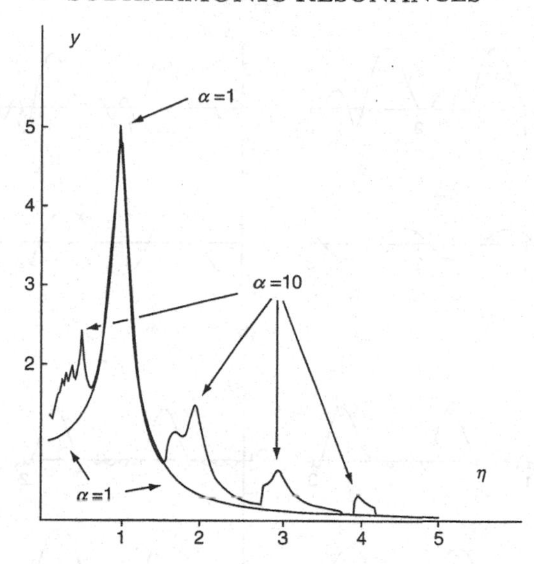

Figure 14.4 Superposition of the resonance response graphs for $\alpha = 1$ and $\alpha = 10$ ($\zeta = 0.1$)

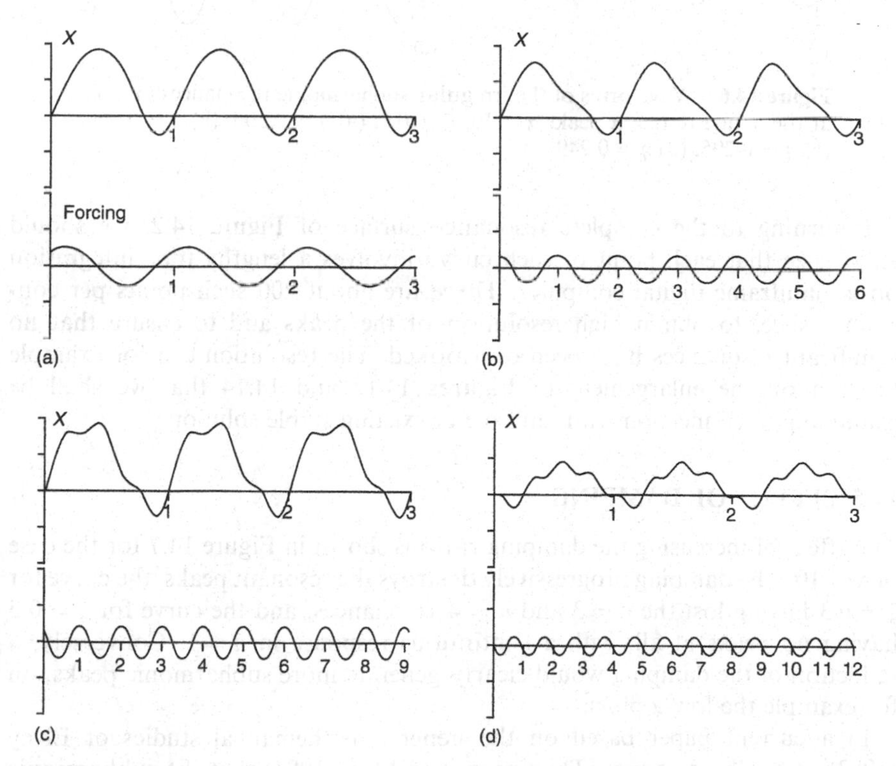

Figure 14.5 Waveforms of the regular subharmonic resonances, at the actual resonant peaks ($\alpha = 10$, $\zeta = 0.1$): (a) main $n = 1$, $\eta = 1.0$; (b) $n = 2$, $\eta = 1.95$; (c) $n = 3$, $\eta = 2.95$; (d) $n = 4$, $\eta = 3.95$

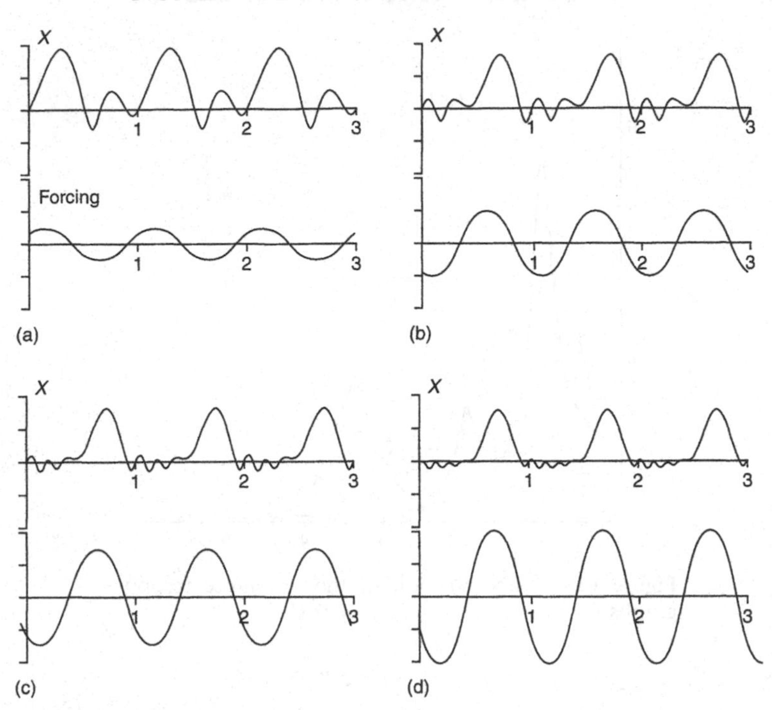

Figure 14.6 Waveforms of the irregular subharmonic resonances ($n = 1$), at the actual resonant peaks $\alpha = 10$, $\zeta = 0.1$: (a) $\eta = 0.505$, (b) $\eta = 0.360$, (c) $\eta = 0.295$, (d) $\eta = 0.249$

Returning to the complete resonance surface of Figure 14.2, we should emphasize that each point of each curve involves a lengthy time integration on a mainframe digital computer. There are about 200 such points per constant-α slice, to obtain high resolution of the peaks and to ensure that no significant resonances have been overlooked. The resolution can for example be seen on the enlargement of Figures 14.12 and 14.14 that we shall be examining in connection with multiple coexisting stable solutions.

14.5 EFFECT OF DAMPING

The effect of increasing the damping ratio is shown in Figure 14.7 for the case of $\alpha = 10$. The damping progressively destroys the resonant peaks, the curve for $\zeta = 0.3$ having lost the $n = 3$ and $n = 4$ resonances, and the curve for $\zeta = 0.5$ having no peaks at all, with a continuous response in $n = 1$. Conversely, a reduction of the damping would clearly generate more subharmonic peaks, on for example the low-α plots.

In a current paper based on the general mathematical studies of Elvey (1983), it has been shown (Thompson and Elvey 1984) that the subharmonic resonances can be designed out by increasing the damping to a prescribed level, or by varying other system parameters. Elvey's bound on the damping, above

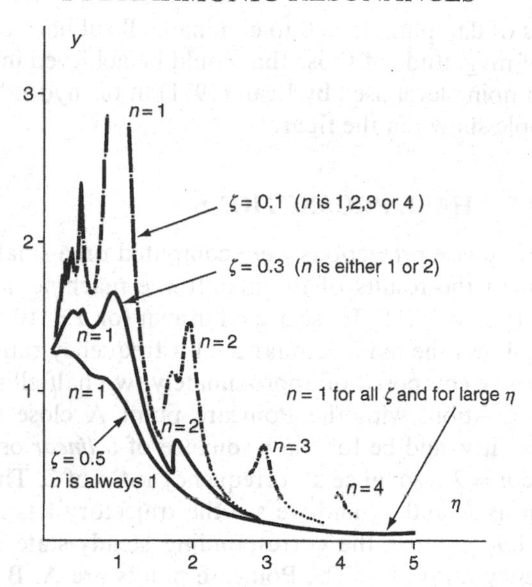

Figure 14.7 The effect of increasing the damping ratio for $\alpha = 10$

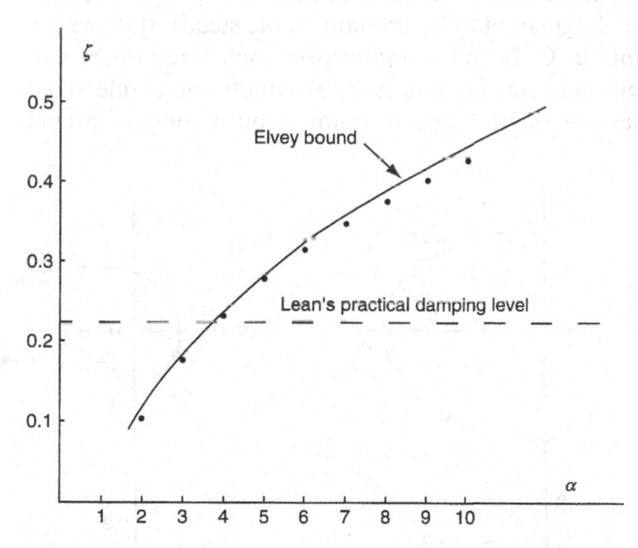

Figure 14.8 The theoretical bound of Elvey, compared with digital estimates of the corresponding $n = 2$ bifurcation. The dots show estimates of the bifurcation from $n = 2$ to $n = 1$ at $\eta = 2$ (with η free, slightly higher values of ζ would be expected)

which no steady subharmonics can exist for any η, is shown to be a very good one by comparison with special digital time integration on a desktop machine. The relevant curve of ζ against α is shown in Figure 14.8, where the digital verification corresponds to the vanishing of the critical $n = 2$ subharmonic at

$\eta = 2$. The values of damping needed to eliminate all subharmonics are seen to be of the order of magnitude of those that could be achieved in realistic design situations: the damping level used by Lean (1971) in his hydrodynamic simulations is for example shown in the figure.

14.6 COMPUTED PHASE PROJECTIONS

Some approximate phase projections were computed on a small desktop computer to supplement the results of the mainframe machine, and a sample is shown in Figure 14.9 to 14.11. These are all drawn for $\alpha = 10$ and $\zeta = 0.1$.

Figure 14.9(a) shows the main resonance at a frequency ratio of unity. The response is seen to be composed of approximately two half-ellipses akin to the undamped free vibration, with the Poincaré point A close to the state of minimum X where it would be for the resonance of a *linear* oscillator. Figure 14.9(b) shows the $n = 2$ resonance at a frequency ratio of 2. The two Poincaré points are shown as A and B, and we see the trajectory has no self-crossing points. Figure 14.9(c) shows the corresponding steady-state resonance with $n = 3$ at a frequency ratio of 3. The Poincaré points are A, B and C, and we notice that the steady-state trajectory crosses itself once in this phase *projection.* The fourth subharmonic resonance is shown in Figure 14.9(d) for a frequency ratio of 4. The large-amplitude resonant stable steady state with $n = 4$ has the Poincaré points B, C, D and E and crosses itself three times. Coexisting with this large-amplitude steady state is a low-amplitude stable steady state with $n = 1$: this has the single Poincaré point A on a roughly circular trajectory.

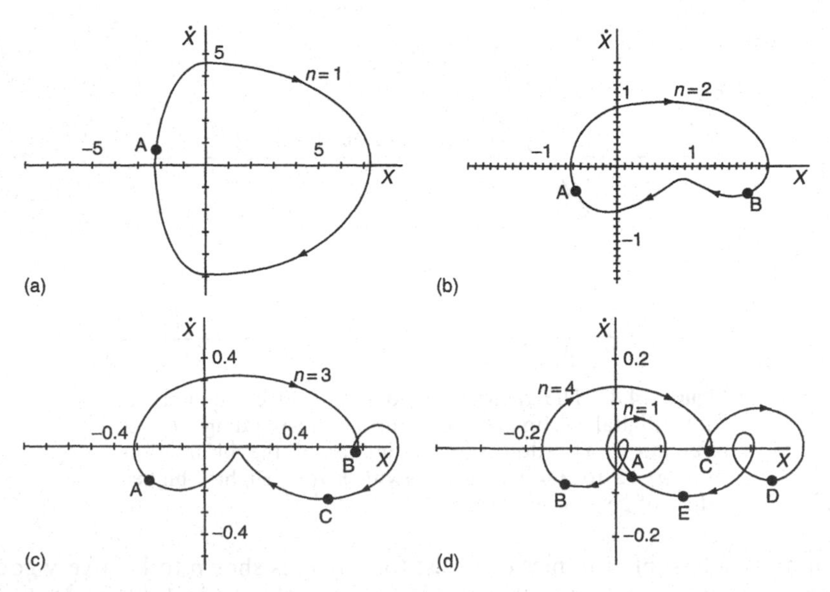

Figure 14.9 Approximate phase projections ($\alpha = 10$, $\zeta = 0.1$): (a) $\eta = 1$, (b) $\eta = 2$, (c) $\eta = 3$, (d) $\eta = 4$. In (d) the $n = 4$ resonant trajectory and the alternative coexisting $n = 1$ stable steady-state solutions are visible

The trajectory adopted is here dependent on the starting conditions, as we shall discuss at length later.

Finally, the first irregular resonance at a frequency ratio of 0.5 gives us the phase projection of Figure 14.10. This is a fundamental response with $n = 1$ and a single Poincaré point A, but with a single crossover point. Notice that we have here four switching points ($X = 0$) per cycle, in contrast to the two switching points observed above.

Phase projections such as these give us an explanation of the irregular resonances. For small frequency ratios the bilinear system experiences a number of relatively high-frequency damped vibrations in the stiff left-hand half-space while the forcing is negative and thus towards negative X. An extreme case of this for $\eta = 0.1$ is shown in Figure 14.11, with the single Poincaré point A near the origin. Essentially the size of the subsequent right-hand excursion depends on the phase relationship between the rebound from the stiff half-space and the driving force. In particular, a large-amplitude resonant excursion is generated when the phase is such that the Poincaré point (which coincides with the forcing becoming positive) is as low as possible, so that its $\dot{X}_P$ is a minimum.

The irregular $n = 1$ resonances do indeed correlate remarkably well with the minima of $\dot{X}_P$, the $\dot{X}$ coordinates of the single Poincaré points. This can be seen in Figure 14.3, where the four irregular resonances for $\alpha = 10$ lie accurately beneath the sharp minima of $\dot{X}_P$. We notice here the cyclic motion of $\dot{X}_P$ with the frequency ratio, a slow climb to a smooth maximum being followed by a rapid fall to a sharp minimum as η increases. This generation of resonance is strong while the rebound velocity is appreciable, but decreases as η tends to zero. The rebound in Figure 14.11 is, for example, too small to have an appreciable influence on the excursion amplitude.

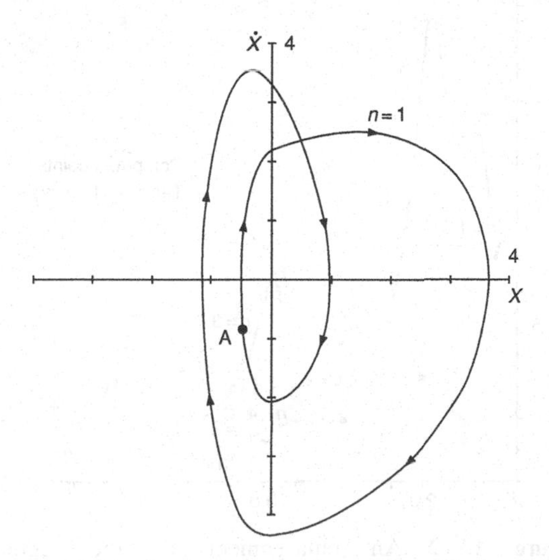

Figure 14.10 Approximate phase projection of the $n = 1$ irregular resonance at $\eta = 0.5$ ($\alpha = 10$, $\zeta = 0.1$)

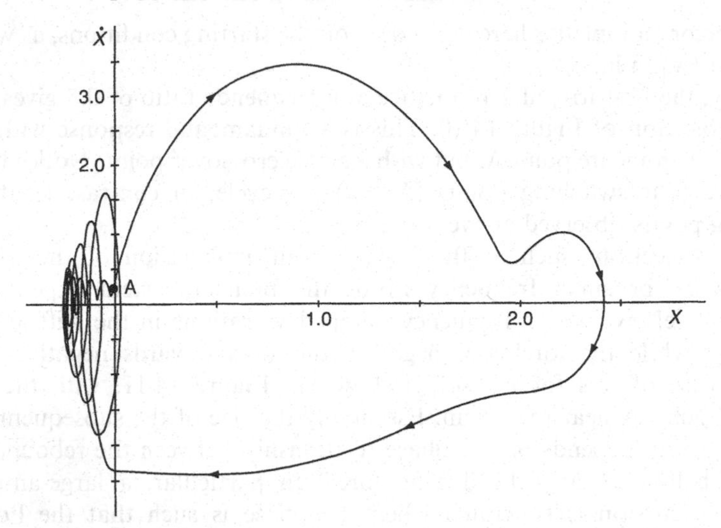

Figure 14.11 Approximate phase projection of the stable steady-state trajectory at $\eta = 0.1$ ($\alpha = 10$, $\zeta = 0.1$)

14.7 MULTIPLE SOLUTIONS AND DOMAINS OF ATTRACTION

As must be expected for any nonlinear system, our bilinear oscillator exhibits multiple stable steady states, depending only on the starting conditions. This can be seen clearly in Figure 14.12. For η in the region of 3.5, for example, there are two coexisting solutions a large-amplitude $n = 3$ motion and a

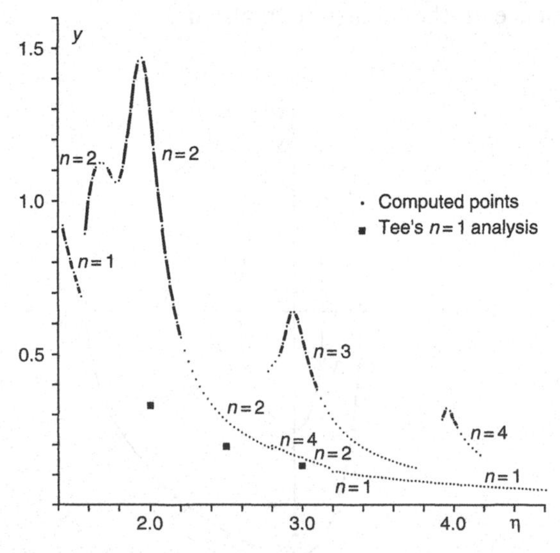

Figure 14.12 An enlargement of the regular subharmonic resonances, showing the coexistence of multiple stable steady states ($\alpha = 10$, $\zeta = 0.1$)

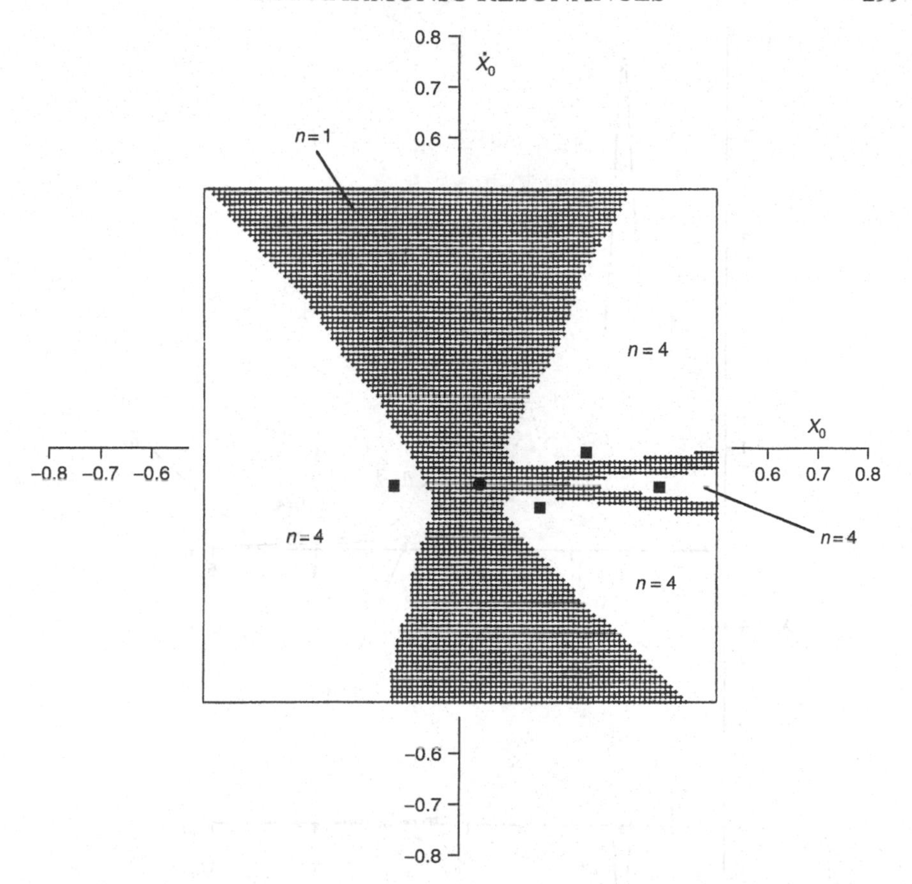

Figure 14.13 Catchment regions (basins of attraction) of the two coexisting $n = 1$ and $n = 4$ stable steady-state solutions ($\alpha = 10$, $\zeta = 0.1$, $\eta = 3.95$)

small-amplitude $n = 1$ vibration. More directly under the $n = 3$ resonance there are clearly bifurcations between $n = 1$, $n = 2$ and $n = 4$.

Under the $n = 2$ peak, no $n = 1$ solution was observed in the time integrations, but a very general result for a wide class of forced nonlinear oscillators (e.g. Thompson and Elvey 1984) assures that there will indeed be a fundamental $n = 1$ solution for all η. We conclude that there exists an *unstable* $n = 1$ solution under the $n = 2$ resonance, which would naturally not be picked up in the time integrations. In fact, a continuous smooth $n = 1$ curve presumably loses its stability at a flip bifurcation into $n = 2$ at about $\eta = 1.5$ and is restabilized at a second bifurcation at about $\eta = 3.2$. The existence of the (unstable) $n = 1$ steady state under the $n = 2$ peak is confirmed by semi-analytical work of our colleague Garry Tee, three of whose points are shown in the figure. Some similar analytical solutions using assumptions about the time of flight in each domain are made by Shaw and Holmes (1983a, b, c), who examine the appearance of supercritical flip bifurcations; they also make digital simulations similar to our own using the matching of the known analytical solutions.

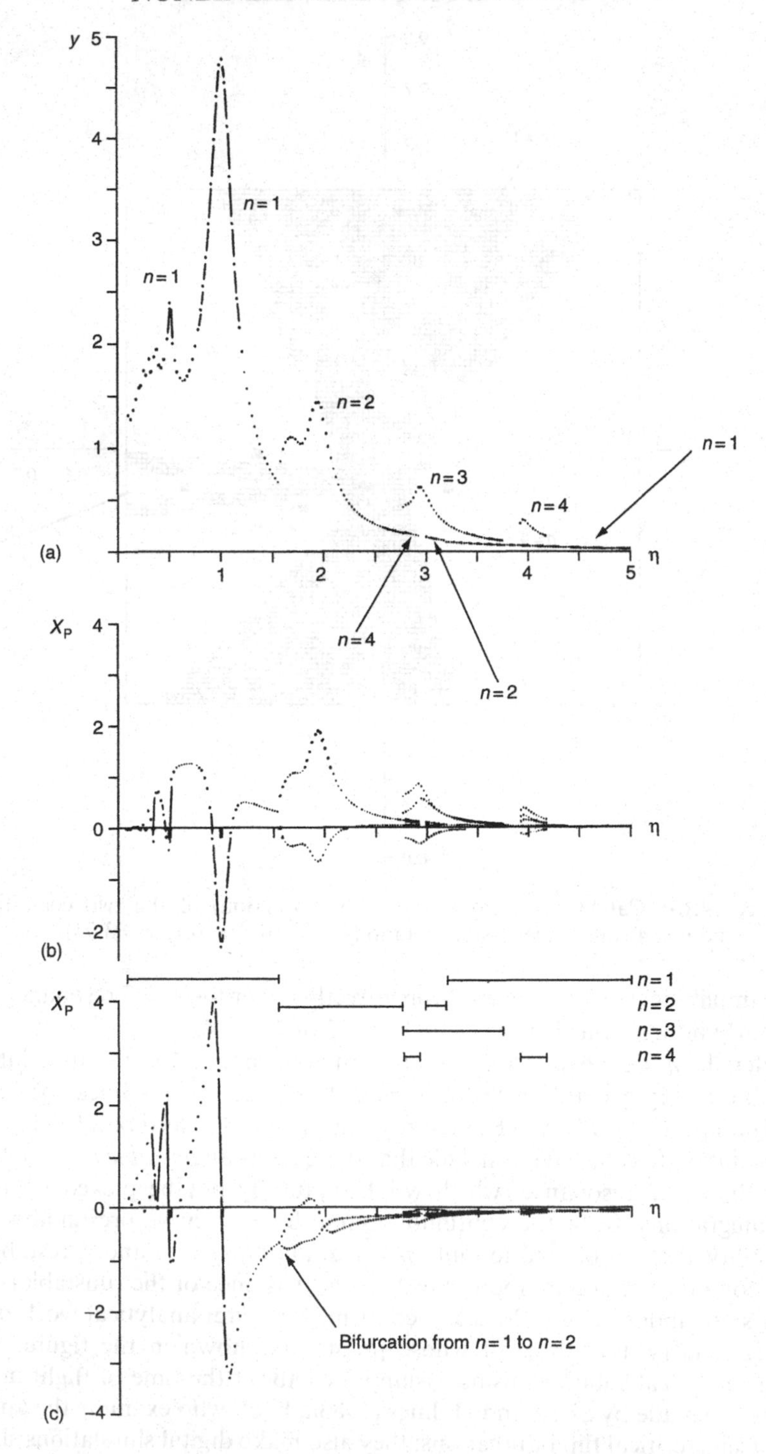

y
n=1
n=1
n=2
n=3
n=4
n=1
(a)
η
n=4
n=2
X_P
η
(b)
n=1
n=2
n=3
n=4
$\dot{X}_P$
η
Bifurcation from n=1 to n=2
(c)

Now in considering starting conditions, we can obviously generate all fibres of our three-dimensional phase space $(X, \dot{X}, \tau)$ by starting always at $\tau = 0$ but with a complete two-dimensional scan of X_0 and $\dot{X}_0$. We are then interested in the catchment regions (basins of attraction) in the $\tau = 0$ subspace, and a detailed study is shown in Figure 14.13.

Here, for $\alpha = 10$ and $\eta = 3.95$ at the $n = 4$ resonant peak, we have taken a mesh of 100×100 starting points. The hatched zone is really a set of small crosses corresponding to the mesh size, each cross denoting a start that leads to the small-amplitude $n = 1$ solution. The white area within the square has been studied at the same mesh size, and denotes starts that lead to the large-amplitude $n = 4$ resonant solution. The Poincaré points corresponding to these two coexisting solutions are shown. We see that the catchment areas are quite complex in shape, and contrary to any simple notion that an engineer might have, we observe that some large-amplitude starts lead to small-amplitude motions and vice versa.

A comprehensive, computer-plotted resonance curve for $\alpha = 10$ is finally shown in Figure 14.14. Here the two lower diagrams show the movements of the Poincaré coordinates X_P and $\dot{X}_P$.

This multiplicity of stable steady states for a given system means that care must be taken in determining the resonance response curve, since with unlucky starts a resonant peak could be entirely missed, the computer simply converging onto one of the coexisting low-amplitude solutions. Indeed, for computational efficiency, it is common to use the Poincaré point of one computation at $\eta = \eta_i$ as the starting condition for the next value of $\eta = \eta_{i+1}$. This means that the start is advantageous for the previously determined mode of vibration, so that there is an in-built tendency to follow a single resonance curve and not pick up any coexisting responses. Indeed, we have ourselves made use of this very technique to explore the full range of the subharmonic resonant peaks.

This danger of missing a resonant peak is particularly strong in the case of $n = 4$ (and presumable higher subharmonic resonances at lower damping) since it coexists over the whole of its range with an underlying stable $n = 1$ solution. As we have observed before, the close proximity of $n = 2$ and $n = 4$ solutions under the $n = 3$ resonance suggests a complex structure of bifurcations, which warrant further detailed study.

Figure 14.14 (a) Computer-plotted resonance response for $\alpha = 10$ showing multiple coexisting solutions. Graphs (b) and (c) show the movement of the X and $\dot{X}$ coordinates of the steady-state Poincaré points ($\alpha = 10$, $\zeta = 0.1$)

15

Chaotic Motions of an Impacting System

In this chapter we shall study the extreme case of the bilinear oscillator when the stiffness ratio α is infinite. This system has been modelled as an 'impact oscillator' that rebounds elastically whenever the displacement X drops to zero.

This system is highly relevant to marine offshore engineering where subharmonic motions and erratic oscillatory behaviour, due to discontinuity in stiffness, have been reported when vessels are moored in harbour against stiff fenders (Kilner 1961; Lean 1971). Here the stiffness of the fenders can often be regarded as infinite compared with the stiffness of the moorings, so that only an elastic rebound need be considered in which the impact velocity is exactly reversed. The number of impacts per cycle, which can be high, is an important factor to be considered in assessing possible damage to the vessel.

The impact oscillator is shown to exhibit complex dynamical behaviour. It exhibits a family of subharmonic resonant peaks between which we delineate cascades of period-doubling bifurcations leading to chaotic regimes typical of a strange attractor.

15.1 RESONANCE RESPONSE CURVE

The non-dimensionalized equation of motion of our impact oscillator can be written as follows (Thompson and Ghaffari 1982, 1983):

$$\ddot{X} + \frac{2\zeta}{\eta}\dot{X} + \frac{1}{4\eta^2}X = \frac{1}{\eta^2}\sin\tau \qquad \text{for } X > 0$$

where the parameters are as defined previously. The resonance response curve for this limiting oscillator has been explored using a modified digital computer program, which simply reflects the system with an equal and opposite elastic rebound velocity whenever X drops to zero. The curve for $\zeta = 0.1$ is shown in Figure 15.1, where for $\eta > 1$ only the subharmonic peaks have been located. The response between these peaks does indeed seem to be quite complex, and we shall discuss this point later.

We see that the strong and distinct subharmonic peaks and irregular resonances that appear on this diagram are analogous in all respects to our earlier $\alpha = 10$ slice. The wavenumber n takes the values indicated, and subharmonics

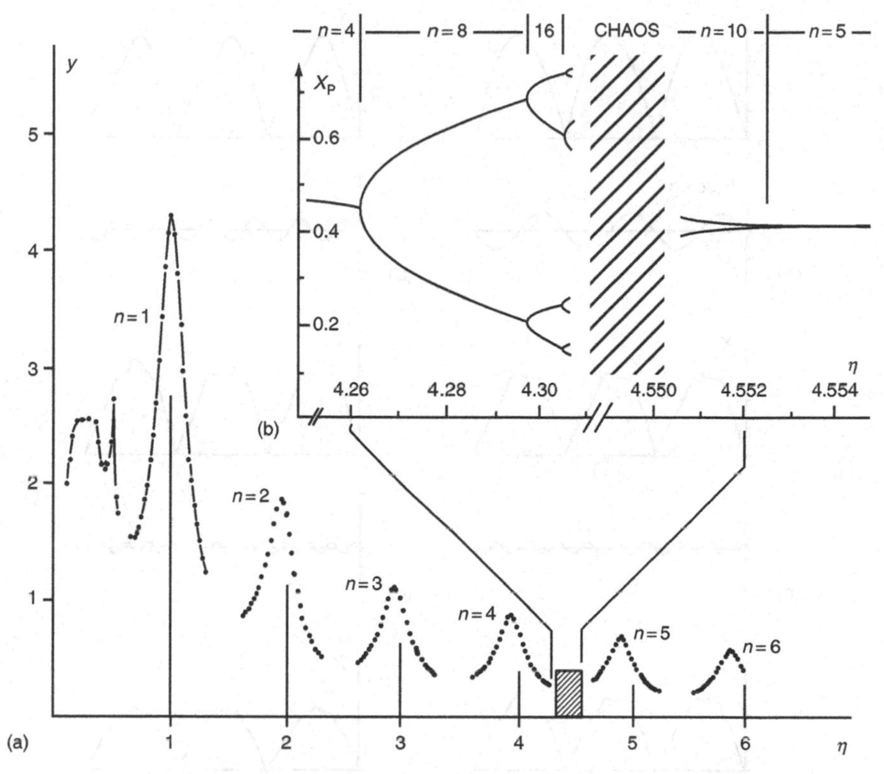

Figure 15.1 (a) The resonance response curve for the impact oscillator, showing the resonant peaks ($\alpha = \infty$, $\zeta = 0.1$). (b) A chaotic regime

of order 5 and 6 have now emerged. Clearly the number of observable subharmonic resonances depends crucially on the damping ratio ζ, the viscous damping being the key factor controlling the height of the peaks. The irregular resonances are similar in form to those of $\alpha = 10$, but the whole region of $\eta < 1$ is raised. This is the result of our particular definition of y, which makes the quasi-static response as η tends to zero correspond to a value of $y = 2$ for $\alpha = \infty$ as opposed to the value of $y = 1$ for $\alpha = 1$. An entirely analogous resonance response diagram for the impact oscillator is obtained analytically by Shaw and Holmes (1983a, b, c); this is for a closely related problem in which there is no viscous damping, energy dissipation being introduced via a coefficient of restitution for the impacts. The analysis is restricted to single impact responses, for which a stability analysis is also made. The bifurcations are shown to be supercritical flips.

The response waveforms for the regular resonances are shown in Figure 15.2. Notice that in these resonances there is only one impact per response cycle, each cycle being essentially composed of a large-amplitude complementary function half-wave plus a small-amplitude particular integral at the forcing frequency. Again, as with the earlier responses for finite α, there is a very precise phase correlation at resonance.

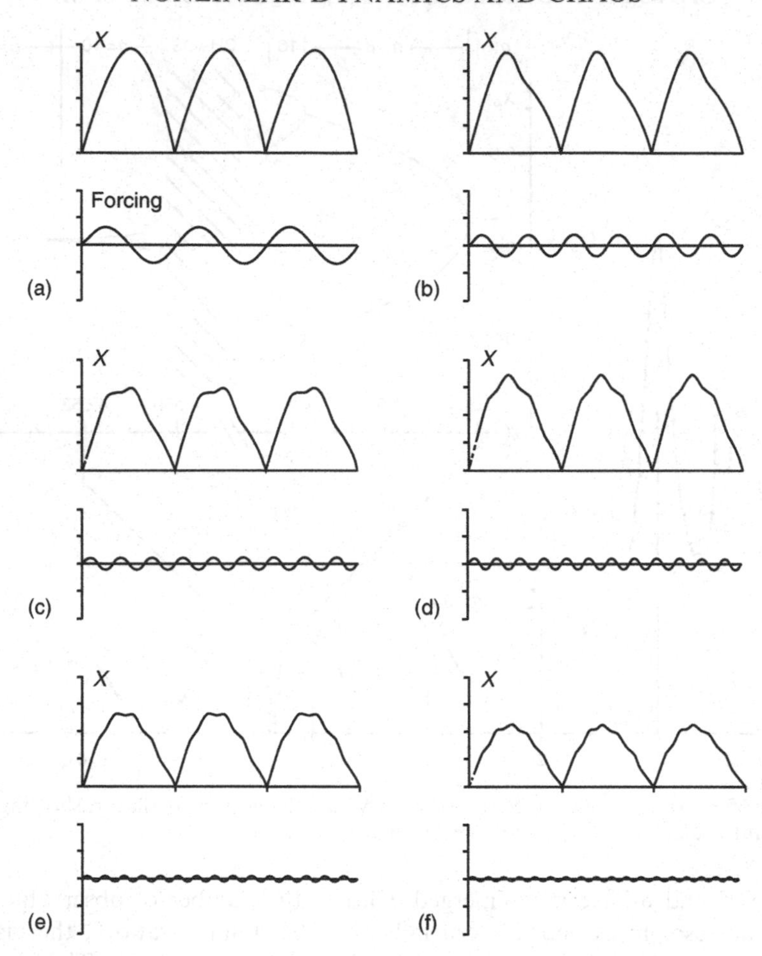

Figure 15.2 The regular subharmonic resonances for the impact oscillator showing the waveforms at the actual peaks ($\alpha = \infty$, $\zeta = 0.1$): (a) $\eta = 1.00$, $n = 1$; (b) $\eta = 1.95$, $n = 2$; (c) $\eta = 2.95$, $n = 3$; (d) $\eta = 3.925$, $n = 4$; (e) $\eta = 4.90$, $n = 5$; (f) $\eta = 5.875$, $n = 6$

Away from the regular resonant peaks for $\eta > 1$ complex solutions with multiple impacts are observed. The solutions from the digital computer seem to be very sensitive to initial conditions (X_0, $\dot{X}_0$) and we observe a complicated structure of period-doubling bifurcations as the frequency ratio is varied. We show later that a genuinely chaotic regime does indeed exist between two of the peaks.

Sample waveforms with two unequal impacts per response cycle are shown in Figure 15.3. Notice that we are here observing a subharmonic motion of order $n = 6$ at a value of $\eta = 3.30$.

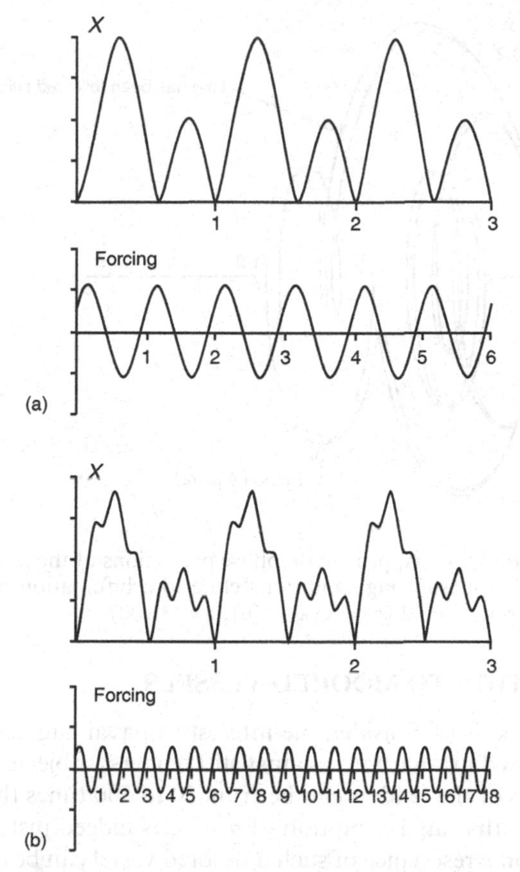

Figure 15.3 Double-impact waveforms for the impact oscillator between resonant peaks ($\alpha = \infty$, $\zeta = 0.1$): (a) $\eta = 1.35$, $n = 2$, (b) $\eta = 3.30$, $n = 6$

An approximate phase projection from a desktop machine is shown in Figure 15.4. This projection shows the between-resonance $\eta = 3.30$ solution and highlights the difficulties of establishing convergence. Here each line has been covered twice, as indicated by the double arrows. So using blindly a modest precision on the Poincaré points we would conclude that we have an $n = 12$ solution. However, viewing the whole phase projection we notice that we are quite close to an $n = 6$ solution. We are then left with the nagging doubt: have we correctly established an $n = 12$ solution, or is the system still converging slowly on to an $n = 6$ solution? In either case, we seem to be close to a bifurcation condition involving a 6–12 transition. From our earlier work at finite α, we might have thought that an $n = 12$ solution was most unlikely at such a low value of η.

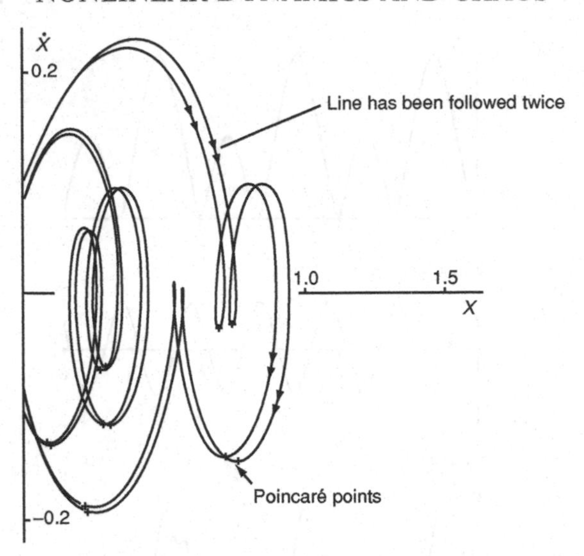

Figure 15.4 Approximate phase projections of the impact oscillator, showing conditions close to a bifurcation from $n = 6$ to $n = 12$ ($\alpha = \infty$, $\zeta = 0.1$, $\eta = 3.300$)

15.2 APPLICATION TO MOORED VESSELS

The case of $\alpha = \infty$ is of considerable interest to naval engineers when dealing with vessels moored in a harbour against stiff fenders, subjected to wave action. Here the stiffness of the fenders can be from 50 to 500 times the stiffness of the mooring lines, so that an assumption of $\alpha = \infty$ is indeed justified.

The subharmonic resonance of such a moored vessel can be quite severe, with damaging impacts against the fenders. Lean (1971) models the sway (lateral displacement without rotation) motions of a moored vessel precisely in terms of our impact oscillator. He solves the bilinear $\alpha = \infty$ equation numerically for an assigned low number of impacts per response cycle using a Fourier series technique, and his resonant peaks for the regular resonances compare well with those of the present study. For this comparison we have taken his value of $\zeta = 0.225$, which he chose to match his experimental test data from a model moored vessel. Agreement is found to be quite satisfactory.

A comparison of our theoretical computed waveform with an experimental waveform due to Lean has also been made (Thompson *et al.* 1984). The essential form is clearly reproduced, and in this application our theoretical waveform is in fact identical to that of Lean's Fourier study.

15.3 PERIOD-DOUBLING AND CHAOTIC SOLUTIONS

The automatic determination of stable subharmonics failed to give consistent solutions for the impact oscillator between the resonant peaks of Figures 15.1 and 15.5. To resolve this problem, we have looked at the interval between the $n = 4$ and $n = 5$ resonances by careful interactive computing.

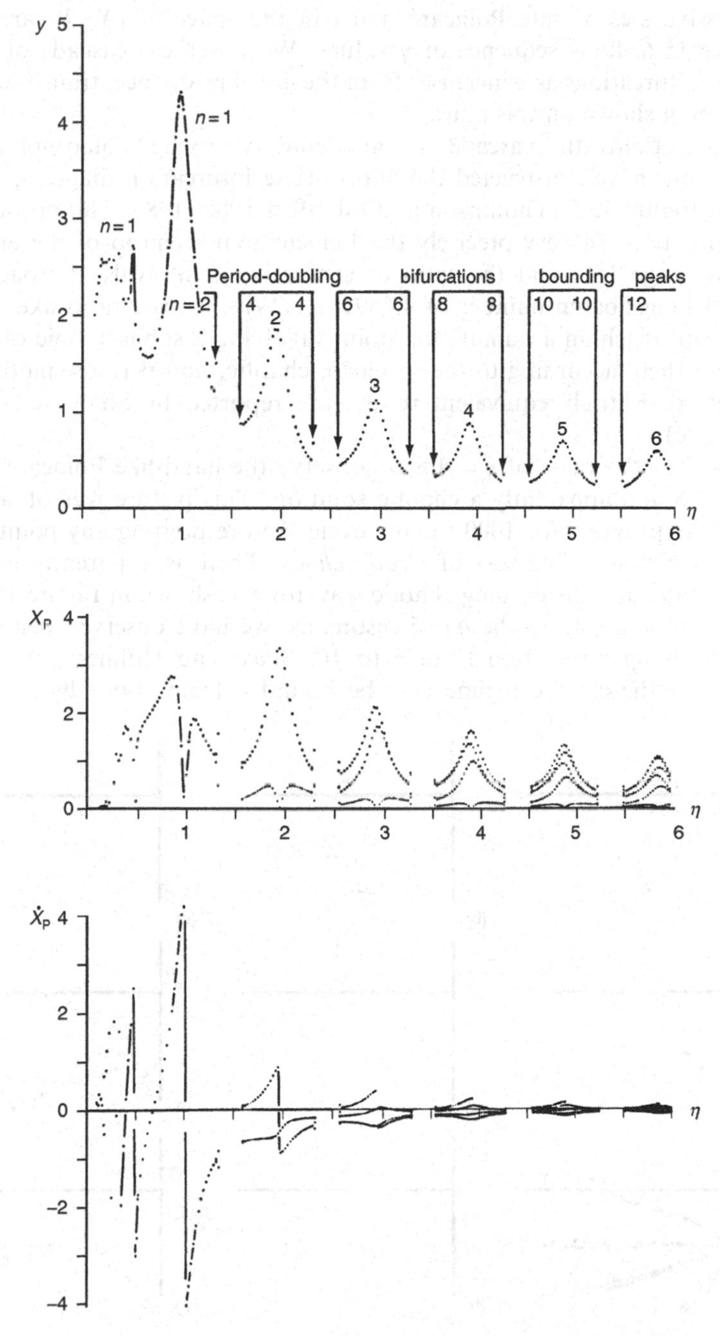

Figure 15.5 Resonance response diagram for the impact oscillator, showing all peaks bound by period-doubling bifurcations ($\zeta = 0.1$)

Successive steady-state Poincaré maps in the space of $(X, \dot{X})$ are shown in Figure 15.6, for a sequence of η values. We observe a cascade of period-doubling bifurcations as η increases from the $n = 4$ resonance, transitions up to $n = 32$ being shown on this figure.

We have studied this cascade in some detail, observing subharmonics up to $n = 128$, and have constructed the appropriate bifurcation diagram, as illustrated in Figure 15.7 (Thompson and Ghaffari 1982, 1983; Thompson 1983). The bifurcations fit very precisely the Feigenbaum scenario of the approach to chaos (page 168) and the ratio of successive η intervals approaches the universal Feigenbaum number of 4.66920.... With a fixed ratio like this, the cascade will reach an accumulation point involving a subharmonic of infinite order and then, according to the scenario, chaotic, non-periodic motions can be expected. Entirely equivalent results are reported by Shaw and Holmes (1983a, b, c).

Indeed, for the value of $\eta = 4.5$ we observe the hand-like Poincaré map of Figure 15.6 and apparently a chaotic solution. This picture was obtained by running the program for 1000 forcing cycles before plotting any points in the hope of achieving some sort of *steady chaos*. There is apparently no stable periodic state, and the ensuing chaotic waveform is shown in Figure 15.8.

On decreasing η from the $n = 5$ resonance we have observed just a single period-doubling bifurcation from 5 to 10. Shaw and Holmes (1983a, b, c) suggest that the chaotic regime may be bounded from above by a structure

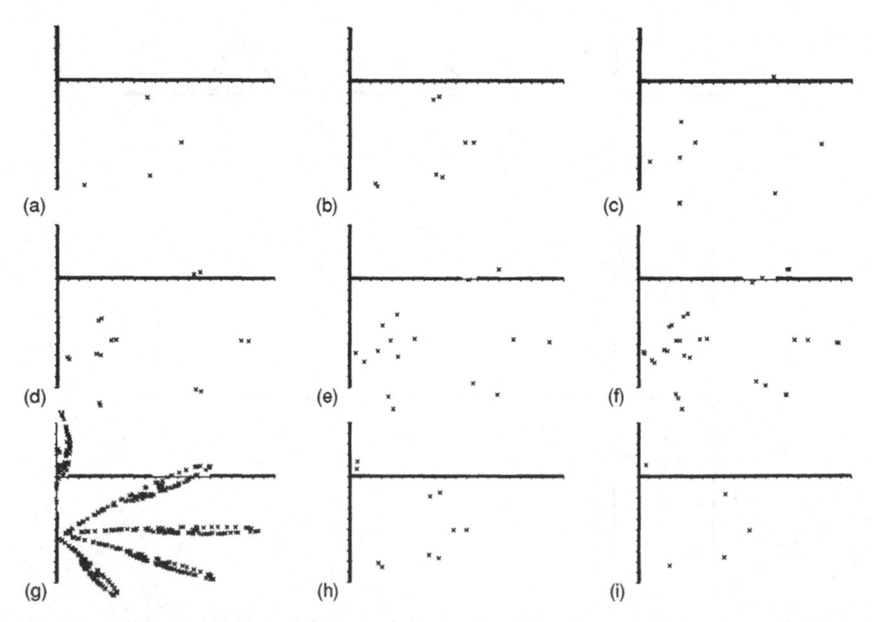

Figure 15.6 A sequence of steady-state Poincaré maps showing period-doubling bifurcations leading to a chaotic solution at $\eta = 4.5$, between the $n = 4$ and $n = 5$ resonance peaks ($\alpha = \infty$, $\zeta = 0.1$): (a) $n = 4$, $\eta = 4.260$; (b) $n = 8$, $\eta = 4.262$; (c) $n = 8$, $\eta = 4.295$; (d) $n = 16$, $\eta = 4.297$; (e) $n = 16$, $\eta = 4.304$; (f) $n = 32$, $\eta = 4.305$; (g) chaos $\eta = 4.500$; (h) $n = 10$, $\eta = 4.550$; (i) $n = 5$, $\eta = 4.555$

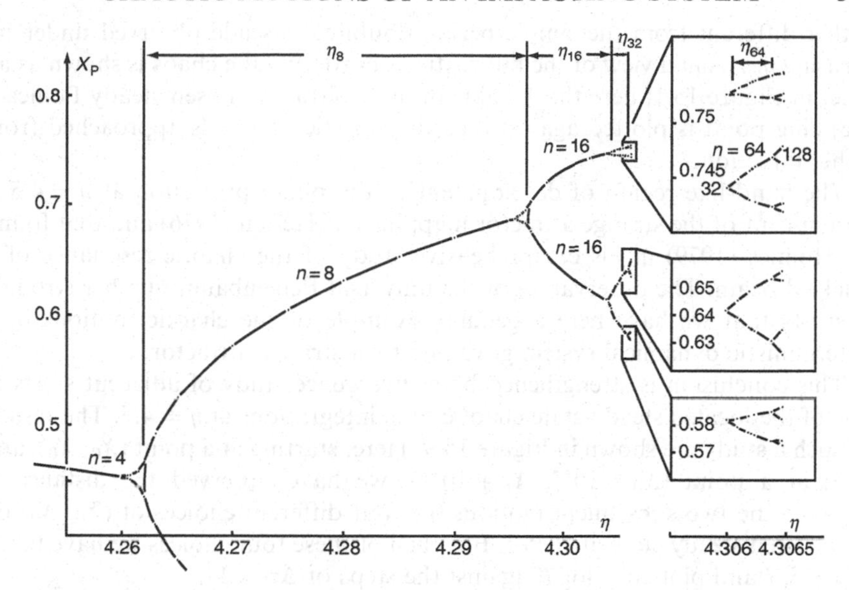

Figure 15.7 Period-doubling bifurcations leading to chaos, shown on a plot of the value of an arbitrarily chosen Poincaré coordinate, X_P, versus the frequency ratio $\eta : \eta_8/\eta_{16} = 4.56$, $\eta_{16}/\eta_{32} = 4.69$, $\eta_{32}/\eta_{64} = 4.64$. Compare these values with the Feigenbaum number 4.66920...

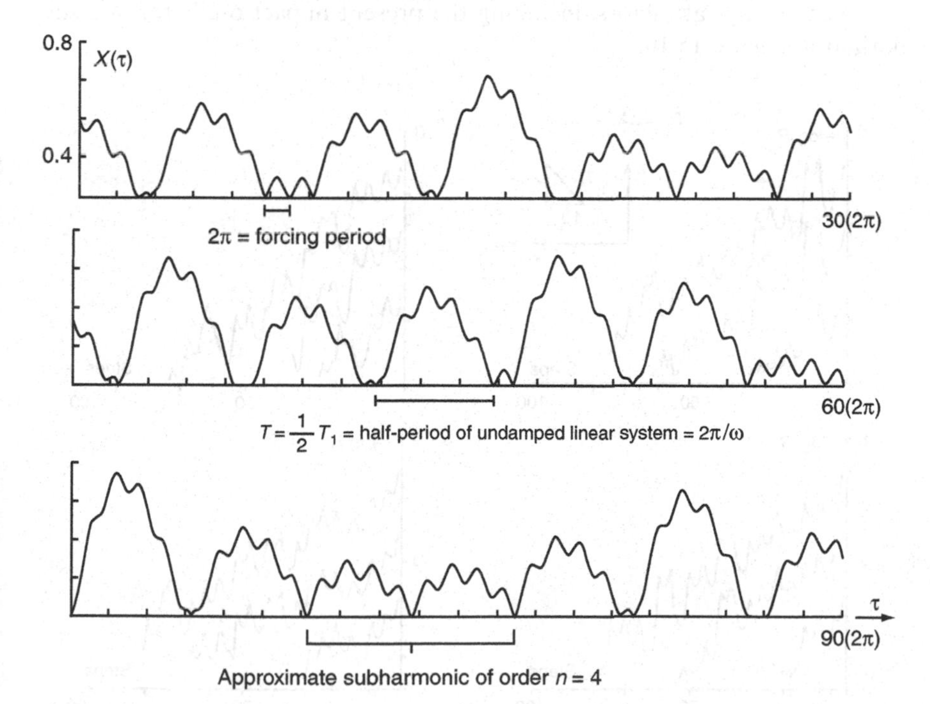

Figure 15.8 The steady-state chaotic waveform of the impact oscillator for $\eta = 4.5$ after 1480 forcing cycles ($\alpha = \infty$, $\zeta = 0.1$)

rather different from the simple period-doubling cascade observed under increasing η. A small view of the bifurcations bordering the chaos is shown as an inset in Figure 15.1; here the X value of an arbitrarily chosen steady Poincaré mapping point is plotted against η as the chaotic regime is approached from either direction.

The hand-like region of development in the phase projection at $\eta = 4.5$ is reminiscent of the strange attractor mappings of Hénon (1976) and that found by Holmes (1979) in his comprehensive study of the chaotic resonance of a buckled beam. The observation of the universal Feigenbaum number strongly suggests that we have here a genuine example of the chaotic motions of a deterministic dynamical system governed by a strange attractor.

This conclusion is strengthened by a divergence study of adjacent starts in one of the located steady-state chaotic time integrations at $\eta = 4.5$. The results of such a study are shown in Figure 15.9. Here, starting at a point $(X_0, \dot{X}_0)$, and then at a point $(X_0 + 10^{-r}, \dot{X}_0 + 10^{-r})$, we have observed the distance R between the two subsequent motions for four different choices of $(X_0, \dot{X}_0)$ on the located steady-state attractor. For each of these four choices we have taken $r = 3, 5, 7$ and plotted $-\log R$ against the steps of $\Delta\tau = 2\pi$.

The noisy straight lines on these logarithmic plots confirm that the adjacent solutions diverge exponentially (but noisily) before becoming completely uncorrelated. A similar result for the Hénon (1976) strange attractor is shown in Figure 9.11. The Liapunov numbers and characteristic exponents for three presumed strange attractors, including the present impact oscillator, are summarized in Figure 15.10.

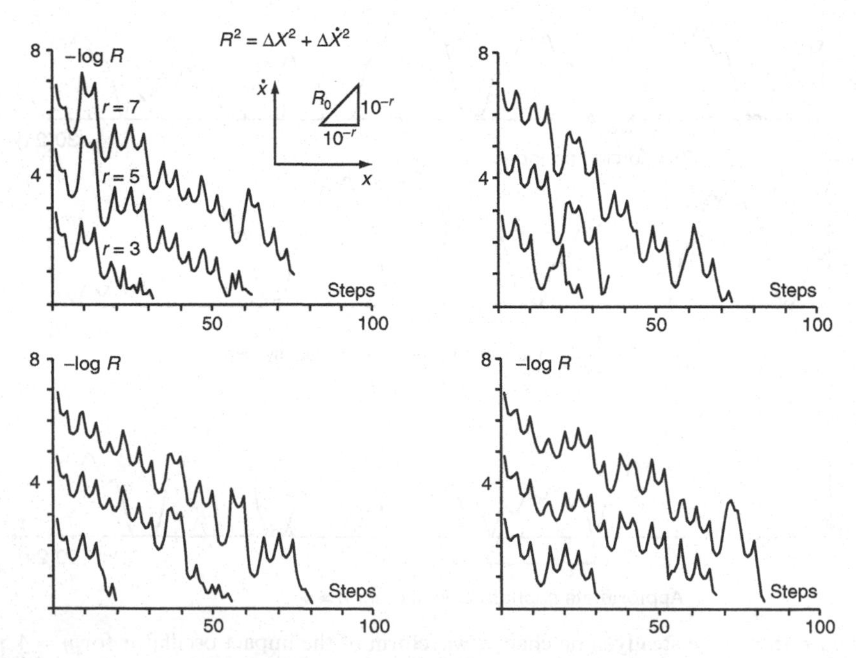

Figure 15.9 Four divergence studies showing a noisy exponential growth of the separation between adjacent starts ($\alpha = \infty$, $\eta = 4.5$, $\zeta = 0.1$)

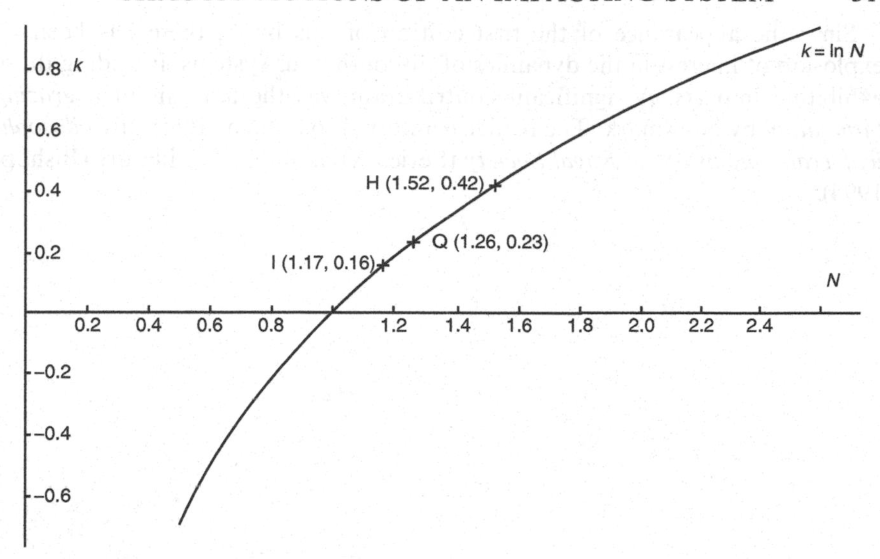

Figure 15.10 Characteristic exponent, k, versus Liapunov number, N, for three chaotic attractors ($R = R_0N^n = R_0e^{kn}$): (Q) quadratic map, $x_{i+1} = C - x_i^2$, $C = 1.790$, Grebogi *et al.*; (H) Hénon map, $x_{i+1} = y_i + 1 - Ax_i^2$, $y_{i+1} = Bx_i$, $A = 1.4$, $B = 0.3$, Feit and Thompson; (I) impact oscillator $\eta^2\ddot{x} + 2\eta\zeta\dot{x} + \beta x = \sin\tau$, $\beta = \frac{1}{4}$ or ∞, $\zeta = 0.1$, $\eta = 4.5$, Thompson and Ghaffari

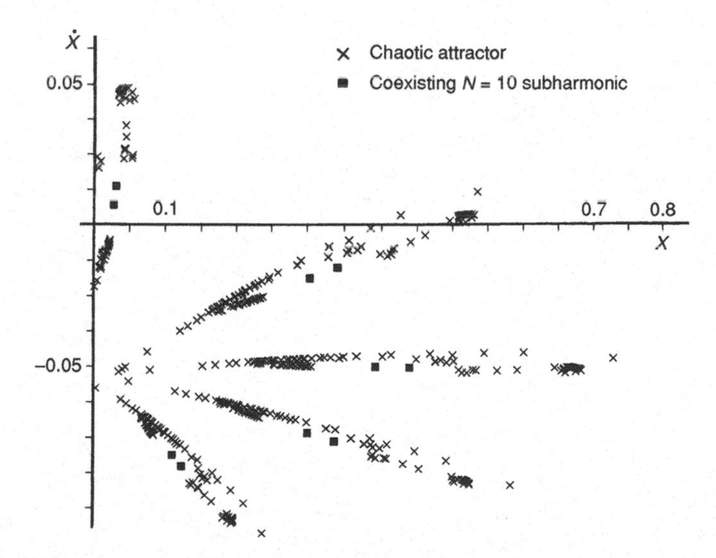

Figure 15.11 Attractors of two coexisting multiple solutions for a fixed system ($\eta = 4.55$). The attractor observed depends on the starting conditions

We would finally emphasize for marine engineers that the two phenomena that we have outlined, namely the coexistence of multiple steady states and the appearance of chaos, can arise at one and the same time. Thus a chaotic solution can coexist with a stable periodic solution, as shown for our impact oscillator in Figure 15.11.

Since the appearance of the first edition of this book, there has been an explosion of interest in the dynamics of discontinuous systems, including those exhibiting impacts. A significant contribution was the analysis of a *grazing bifurcation* by Nordmark. The reader is referred to a theme Issue of *Philosophical Transactions of the Royal Society* (Series A) on impact oscillators (Bishop, 1994).

16

Escape from a Potential Well

In this chapter we describe recent progress in understanding the complex behaviour of driven oscillators. Extensive references show where the interested reader can find more details. We illustrate in particular how the invariant manifolds of unstable saddle solutions structure the phase space and control the attractors, their basins and their bifurcations. We also illustrate features of some interesting indeterminate bifurcations.

We do this while studying escape from a potential well. In particular, we describe the manner in which a driven mechanical or electrical oscillator escapes from the cubic potential well typical of a metastable system close to a fold. This shows, in a well-defined framework, how the atoms of dissipative dynamics (saddle-node folds, period-doubling flips, cascades to chaos, boundary crises, etc.) can typically assemble to form molecules of overall response (hierarchies of cusps, incomplete Feigenbaum trees, etc.). Particular attention is given to the basin of attraction and the loss of engineering integrity that is triggered by a homoclinic tangle, the latter being accurately predicted by a Melnikov analysis.

16.1 INTRODUCTION

The escape from a potential well is a universal problem in the physical sciences, from activation energies of molecular dynamics to the gravitational collapse of massive stars. In applied physics, much recent work has centred on the response of Josephson junctions, modelled by the nonlinear pendulum equation with its sinusoidal potential function. Many failures in electrical systems are triggered when an underlying dynamical system escapes from a well: if power generators slip out of synchronization, the result could be the blackout of an entire city; in the phase-locked loop of a receiver, a slip from the locked configuration results in a loss of communication; if a synchronous motor slips under excessive load, timekeeping will be lost. In civil and aerospace engineering, a compressed structure is often required to operate in a metastable state, while in the field of naval architecture there is currently a flood of activity directed towards an improved understanding of the capsize of vessels (Thompson 1997; Spyrou and

Thompson 2000), including offshore oil production facilities and roll-on roll-off ferries.

Here we focus on the escape of a damped mechanical oscillator from a cubic potential well under the influence of a sinusoidal driving force. Such forcing might correspond to the RF-biased (radio frequency) driving of a Josephson junction, or more generally, the driving of charged particles by an electric field in a variety of condensed matter applications, including weakly pinned charge-density waves, and superionic conductors. In marine technology the hydrodynamic loads on a vessel in a train of regular ocean waves can give rise to a variety of sinusoidally driven capsize phenomena, including resonant and subharmonic rolling. A large flexible spacecraft with thin structural components prone to elastic buckling and excited by small imbalances in rotating electrical machinery is a nice example in which a structure in a noise-free environment is subjected to sinusoidal forcing at a very precise frequency.

As well as its wide applicability, we shall see that the escape from a potential well under sinusoidal forcing involves a multitude of chaotic phenomena due to the homoclinic tangling of the hilltop manifolds. It therefore serves as an excellent practical illustration of many phenomena described in this book. Comprehensive articles on the bifurcations of the escape process are Ueda *et al.* 1990), Thompson and McRobie (1993) and Stewart, Thompson *et al.* (1995).

16.2 ANALYTICAL FORMULATION

Escape equation and divergence characteristics

We consider the behaviour of a general mechanical oscillator having one degree of freedom and with inertia, linear viscous damping, linear stiffness and a quadratic stiffness nonlinearity. With suitable scaling of displacement and time, the equation of any such oscillator can be reduced to the standard form of our *escape equation*

$$\ddot{x} + \beta\dot{x} + x - x^2 = F \sin \omega t$$

where x is the dependent variable and a dot denotes differentiation with respect to the (scaled) time t. The positive coefficient β represents the magnitude of the damping, and the oscillator is driven by the sinusoidal force of magnitude F and circular frequency ω. This equation is worthy of detailed study because it represents motions in a potential well

$$V = \tfrac{1}{2}x^2 - \tfrac{1}{3}x^3$$

which is a universal form always encountered just before a system loses its stability at a saddle-node fold. This potential well is illustrated in Figure 16.1(a).

Despite this apparently narrow focus, the phenomena we shall describe are typical of a wide class of systems: systems with nonlinear damping, systems with different well shapes, systems with different direct and parametric forcing functions (time-periodic but not necessarily sinusoidal), and other systems as

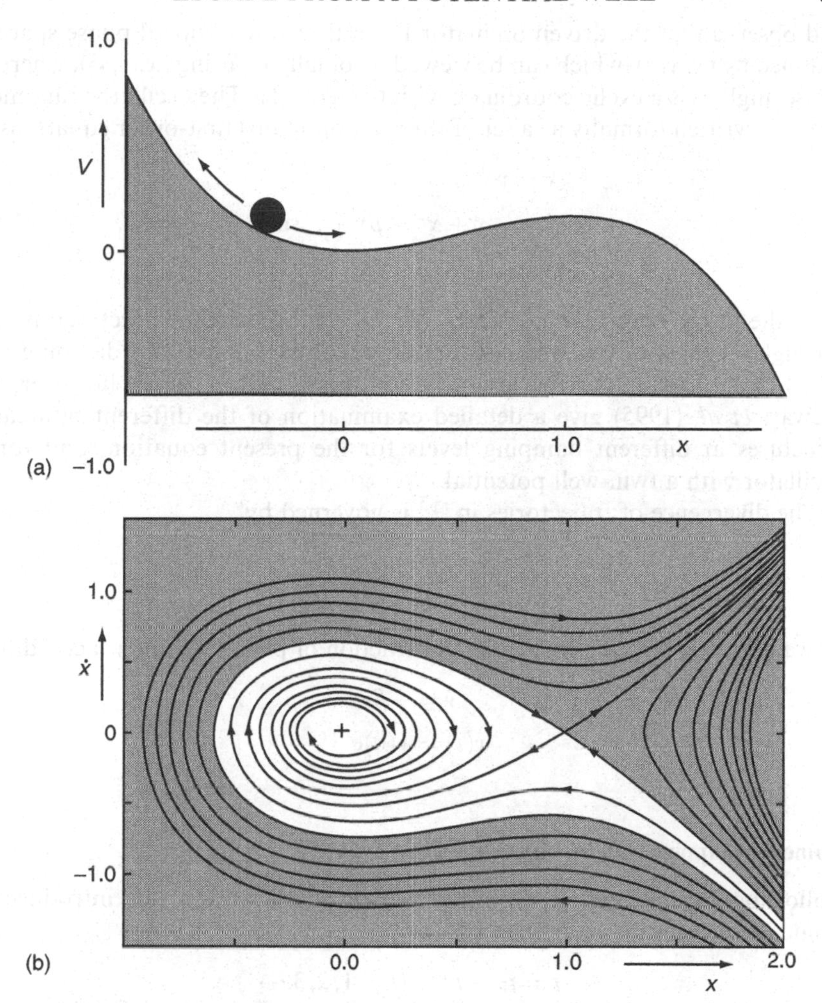

Figure 16.1 (a) Escape equation with its potential well, showing the two-dimensional phase portrait of the autonomous undriven system. All oscillators in an asymmetric well $V(x) = \int f(x)\mathrm{d}x$, with the general form $\ddot{x} + g(x, \dot{x}) + p_1(t)f(x) = p_2(t)$, $p_{1,2}(t) = p_{1,2}(t+T)$, exhibit universal features illustrated by the canonical escape equation, $\ddot{x} + \beta\dot{x} + x - x^2 = F \sin \omega t$, $T = 2\pi/\omega$. Here the potential well is given by $V = \frac{1}{2}x^2 - \frac{1}{3}x^3$. (b) Behaviour of autonomous system with $F = 0$, $\beta = 0.1$

well. Many of the features also arise in hardening systems, where there is no obvious escape mechanism at all (Soliman and Thompson 1996).

We write

$$\begin{aligned} y &= \dot{x} \\ \phi &= \omega t \quad (\text{mod } 2\pi) \\ T &= 2\pi/\omega \end{aligned}$$

and observe that the driven oscillator has a three-dimensional phase space $\mathbb{R}^3$ spanned by (x, y, t), which can be viewed toroidally by using (x, y, ϕ), where the phase angle ϕ is a cyclic coordinate with $0 \leq \phi < 2\pi$. The oscillator can, moreover, be written formally as a set of three autonomous first-order equations

$$\dot{x} = y$$
$$\dot{y} = -x + x^2 - \beta y + F \sin \phi$$
$$\dot{\phi} = \omega$$

with the three control parameters (F, ω, β). We shall in fact set $\beta = 0.1$ throughout most of our numerical studies, corresponding to a damping ratio of $\zeta = 0.05$, leaving the two primary controls F and ω. Note, however, that Stewart *et al.* (1995) give a detailed examination of the different bifurcation structures at different damping levels for the present equation, and for an oscillator with a twin-well potential.

The divergence of trajectories in $\mathbb{R}^3$ is governed by

$$\frac{\partial \dot{x}}{\partial x} + \frac{\partial \dot{y}}{\partial y} + \frac{\partial \dot{\phi}}{\partial \phi} = -\beta$$

so we have a constant exponential contraction of phase volume v, according to

$$\dot{v} = -\beta v$$
$$v(t) = v(0)\mathrm{e}^{-\beta t}$$

Poincaré mapping and its eigenvalues

Following the techniques described earlier in this book, we introduce the Poincaré sections (x, y) defined by

$$t = t_\mathrm{P} + iT \quad (i = 1, 2, 3, \cdots)$$
$$\phi = \phi_\mathrm{P} = \omega t_\mathrm{P}$$

and the associated map

$$P(\phi_\mathrm{P}) : [x(t_\mathrm{P}), y(t_\mathrm{P})] \rightarrow [x(t_\mathrm{P} + T), y(t_\mathrm{P} + T)]$$

This mapping depends on the prescribed phase, ϕ_P, and for most of our studies we use $\phi_\mathrm{P} = 0$. However, some phenomena (such as the final boundary crisis) are more clearly observed at a phase of $\phi_\mathrm{P} = \pi$. This map takes us iteratively from (x_i, y_i) to (x_{i+1}, y_{i+1}) according to an implied functional relationship

$$x_{i+1} = G(x_i, y_i)$$
$$y_{i+1} = H(x_i, y_i)$$

where G and H can be evaluated numerically for any (x_i, y_i) by making a Runge–Kutta numerical time integration through one forcing period.

Since there is no stretching action along the time axis, the exponential contraction of volume for the flow ensures that an area a in the Poincaré section contracts according to

$$a_{i+1} = \mathrm{e}^{-\beta T} a_i$$

which implies that the Jacobian determinant, D, of the map has the constant value

$$D = \begin{vmatrix} \partial G/\partial x & \partial G/\partial y \\ \partial H/\partial x & \partial H/\partial y \end{vmatrix} = \mathrm{e}^{-\beta T}$$

It follows that any fixed point, corresponding to a fundamental $n = 1$ oscillation with period T, will have mapping eigenvalues $\Lambda_i (i = 1, 2)$ constrained by the condition

$$\Lambda_1 \Lambda_2 = D = \mathrm{e}^{-\beta T} = \mathrm{e}^{-\beta 2\pi/\omega}$$

Similarly, the eigenvalues $\Lambda_i^{(n)}$ of the n-map P^n corresponding to a subharmonic of any order n are constrained by

$$\Lambda_1^{(n)} \Lambda_2^{(n)} = \mathrm{e}^{-n\beta T}$$

These constraints place restrictions on the sequences of folds ($\Lambda = +1$) and flips ($\Lambda = -1$) that can be observed. They also exclude a Neimark bifurcation (secondary Hopf bifurcation) in which a pair of complex conjugate eigenvalues leave the stable unit disc away from the real axis.

Linear solutions at low forcing

For zero forcing, $F = 0$, the behaviour of the autonomous system with $\beta = 0.1$ is shown in Figure 16.1(b). Here the (x, y) phase space has two equilibria at $x = 0$ and $x = 1$. The stable state S^0 of minimum energy at (0, 0) corresponds to an attracting focus, while the unstable state D^0 of maximum energy at (1, 0) corresponds to a saddle. It is the stable manifold, or inset, of this saddle, namely the set of all points which tend to the saddle as $t \to \infty$, that defines the boundary of the basin of attraction of S^0. Starting conditions outside this basin that lead to escape over the hilltop (with $x \to \infty$) are highlighted by the tint.

With light external forcing (small F), the 3D phase space has the form of Figure 16.2. The stable equilibrium in the well, S^0, has been transformed into a small stable periodic attractor (a fundamental $n = 1$ oscillation), and the unstable hilltop equilibrium state, D^0, has become a small unstable ($n = 1$) saddle cycle. The inset of this saddle is a sheet of solutions forming the boundary of the safe non-escaping basin. Figure 16.2 also illustrates how we sample the flow stroboscopically at the forcing period to obtain our Poincaré section and its associated 2D mapping.

We denote the path of cyclic attractors that emerges from S^0 by S^1. It is a simple matter to derive the complex mapping eigenvalues of the small

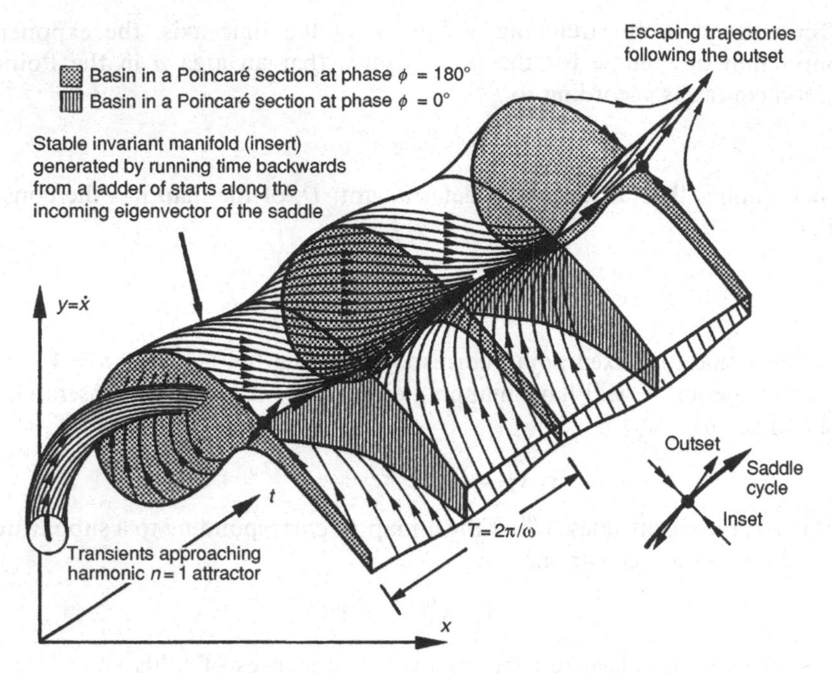

Figure 16.2 Sketch of the three-dimensional phase portrait of the non-autonomous, lightly driven system showing how the invariant manifold of the hilltop saddle cycle defines the safe basin boundary

oscillations, S^1, using the complementary function of the linearization about S^0. Similarly, we denote the path of saddle cycles emerging from D^0 by D^1 whose linearized equation is

$$\ddot{z} + \beta\dot{z} - z = F\sin\omega t$$

where $z = x - 1$. The complementary function, a variational solution about the particular integral, is written as $s = z - z_{\mathrm{PI}}$ with

$$s = Ae^{\lambda_1 t} + Be^{\lambda_2 t}$$

in terms of the real flow eigenvalues

$$\lambda_1 = \tfrac{1}{2}(-\beta + \sqrt{\beta^2 + 4})$$
$$\lambda_2 = \tfrac{1}{2}(-\beta - \sqrt{\beta^2 + 4})$$

with $\lambda_1 + \lambda_2 = -\beta$, and for small β we obtain

$$\lambda_1 \approx \tfrac{1}{2}(2 - \beta)$$
$$\lambda_2 \approx \tfrac{1}{2}(-2 - \beta)$$

More roughly, λ_1 is approximately $+1$ and λ_2 is approximately -1.

The linearized invariant manifolds, outset and inset, are given respectively by

$$\dot{s}/s = \lambda_1 \approx +1$$
$$\dot{s}/s = \lambda_2 \approx -1$$

so we see that they lie at about 45° to the x axis as will the manifolds of D^1 in the Poincaré section. The mapping eigenvalues of D^1 are $\Lambda_1 = e^{\lambda_1 T}$, $\Lambda_2 = e^{\lambda_2 T}$, and approximating λ_1 as $+1$ and λ_2 as -1, we have

$$\Lambda_1 \approx e^{+2\pi/\omega}$$
$$\Lambda_2 \approx e^{-2\pi/\omega}$$

So as $\omega \to 0$, when we are sampling the flow very infrequently, $\Lambda_1 \to \infty$ and $\Lambda_2 \to 0$.

16.3 OVERVIEW OF THE STEADY-STATE RESPONSE

A 3D response surface

Based on extensive numerical simulations, we have deduced the form of the response surface sketched in Figure 16.3. This is a simplified schematic diagram in which the 'response amplitude' might be the maximum value of the displacement, x_m, during a steady-state oscillation. The surface represents the steady-state solutions, with x_m and ω in the base plane, and the second control, F, plotted vertically for ease of visualization. The damping is constant at $\beta = 0.1$. Bifurcations on the surface project onto the (F, ω) control plane to give the

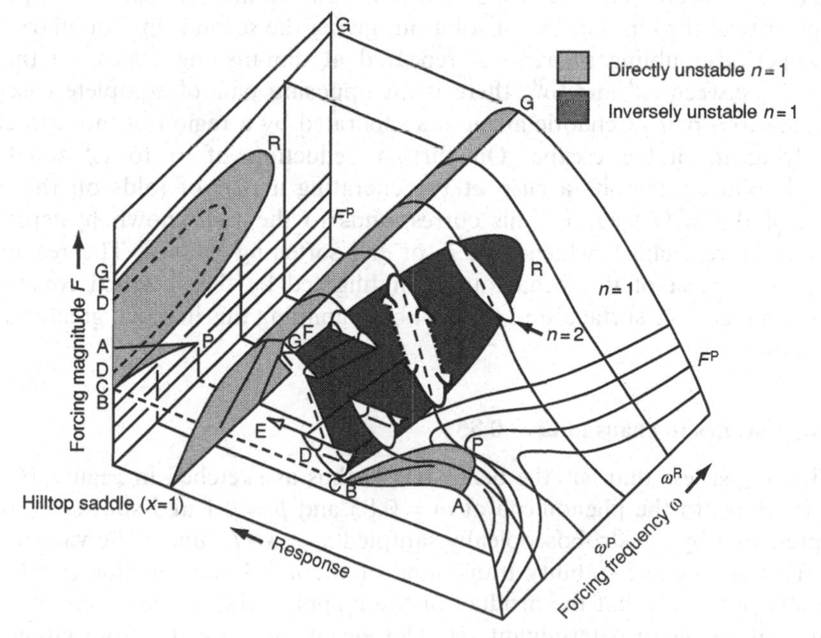

Figure 16.3 Schematic three-dimensional response surface for the escape equation at constant damping, showing folds, flips and period-doubling cascades

bifurcation diagram. The $n = 1$ surface cuts the $F = 0$ plane in two straight lines: $x_m = 0$ represents the stable equilibrium S^0, and $x_m = 1$ corresponds to the unstable hilltop state D^0.

Let us focus attention on the $x_m(F)$ paths at prescribed ω values. At the highest value of forcing frequency represented in the diagram ($\omega \approx 2.5$) the two equilibria are joined by a simple $n = 1$ path; the paths from the linearized solutions of S^1 and D^1 grow, merge and annihilate one another at the cyclic fold at G. Passing from S^1 to D^1 through this fold, one of the mapping eigenvalues, Λ_1 say, passes through $+1$, so that after G the $n = 1$ solution is directly unstable with both eigenvalues positive and $1 < \Lambda_1 < \infty$, $0 < \Lambda_2 < 1$. The physical system starting at S^0 is stable up to fold G, from which a fast dynamic jump might carry the system out of the well with $x \to \infty$. So at high ω (in this simplified overview) the fold line GG is the escape boundary.

This simple folding is preserved under decreasing ω until at $\omega^R \approx 2.2$ a flip into an $n = 2$ subharmonic is encountered. At the value of ω drawn between ω^P and ω^R the $n = 1$ path is cut by a closed $n = 2$ curve after which the $n = 1$ solution restabilizes, before finally losing its stability at G. Between the two opposing flips, the $n = 1$ solution is inversely unstable with both mapping eigenvalues real and negative, one inside the unit circle and the other outside, $-\infty < \Lambda_1 < -1$, $-1 < \Lambda_2 < 0$. These flips project into the boundary FRC in the control space. For the value of ω illustrated, the $n = 2$ solution is everywhere stable, so a physical system would experience a brief regime of stable $n = 2$ subharmonic oscillation between the two supercritical flips, before escape from the fold at G.

As ω is decreased, the $n = 2$ solution next exhibits a pair of opposing supercritical flips into an $n = 4$ solution, giving the second flip boundary DD. The period-doubling scenario is repeated at diminishing scales, so that at $\omega = 2$, between ω^P and ω^R, there is an opposing pair of complete cascades leading to a pair of chaotic attractors separated by a region of 'no attractor', implying inevitable escape. On further reduction of ω to $\omega^P \approx 0.9$ the $n = 1$ solution exhibits a cusp at P, generating a pair of folds on the early part of the $x_m(F)$ curve. This corresponds to the well-known hysteresis in nonlinear resonance, which here is of the softening variety. The resonance response aspect of this behaviour is highlighted by the sketched constant-F lines on the $n = 1$ surface, and we shall be examining this in much greater detail later on.

Two cascades to chaos at $\omega = 0.85$

With ω just less than ω^P the behaviour is thus as sketched in Figure 16.4(a). This relates to the phenomena at $\omega = 0.85$ and $\beta = 0.1$ and shows the paths represented by the stroboscopically sampled $x_i = x(iT)$ under the variation of F. To examine the stability transitions of the $n = 1$ path in this constant-ω section, we recall that the product of the mapping eigenvalues is equal to the constant Jacobian determinant, D. The eigenvalues are therefore either real with geometric mean

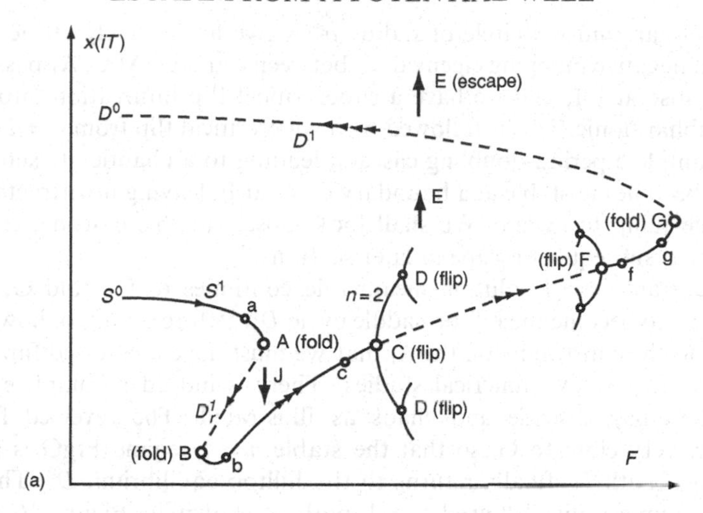

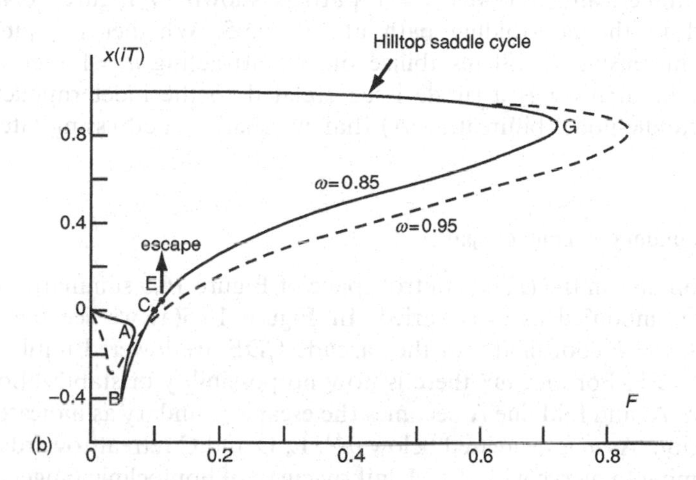

Figure 16.4 (a) Schematic sketch of the response curves at constant ω and constant β, showing the opposing flip cascades. (b) Steady-state $n = 1$ fundamental solution for $\beta = 0.1$ for two values of the forcing frequency

$$\sqrt{\Lambda_1 \Lambda_2} = e^{-\beta T/2} = \rho$$

or complex and constrained to lie on a circle of radius ρ centred on the origin of the Argand diagram.

From S^0 the path S^1 starts as an attracting focus, becomes a directly attracting node at point a, where the complex eigenvalues become real and positive, and folds at A as Λ_1 penetrates the unit circle at $+1$. From fold A to fold B, we have the directly unstable saddle D_r^1 (where subscript r, for resonant, distinguishes this from the hilltop saddle D^1), and the path restabilizes at fold B as Λ_1 re-enters the unit circle at $+1$. The Λ_i become complex at b, passing completely around the circle of radius ρ to give an inversely attracting node with real negative mapping eigenvalues between c and C. At C, Λ_1 passes out of the unit disc at -1, and we have a supercritical flip bifurcation into a stable $n = 2$ subharmonic. This is followed by a supercritical flip from $n = 2$ to $n = 4$, and a complete period-doubling cascade leading to a chaotic attractor, which quickly becomes unstable at a boundary crisis at E, leaving no attractor and an inevitable jump to escape. We shall look closely at this cascade, chaotic attractor and subsequent escape in later sections.

The unstable $n = 1$ solution meanwhile continues to the fold G, where it turns back to become the hilltop saddle cycle D^1. Before doing so, however, it is clear from the constraints on the Λ_i that we must have a reversed flip at F and this is confirmed by numerical studies. There is indeed a complete reversed period-doubling cascade and chaos as illustrated. The reversed flip F is, however, very close to G, so that the stable $n = 1$ regime FfgG is in reality very short. Path D^1 finally returns to the hilltop equilibrium D^0. The corresponding numerically followed $n = 1$ path is shown in Figure 16.4(b), and compared to the non-folding path at $\omega = 0.95$. Whether the jump from A under increasing F will restabilize on the attracting $n = 1$ focus as indicated by the arrow J is a subtle issue (related to the indeterminacy of the 'tangled saddle-node' bifurcation A) that we shall be addressing later in this chapter.

Escape boundary in control space

Bifurcation arcs in the (F, ω) control space of Figure 16.5 summarize how this response is modified as ω is varied. In Figure 16.5(a) we see that as ω is decreased, the F coordinates of the cascade CDE are lowered until at ω^{Q} we have $F^{\mathrm{A}} = F^{\mathrm{E}}$. For $\omega < \omega^{\mathrm{Q}}$ there is now no possibility of stabilization of the jump from A, and fold line A becomes the escape boundary as indicated by the tinted region. As ω is decreased below ω^{Q}, E, D and C retreat towards the fold B, all seeming to merge with the Melnikov curve of homoclinic tangency, M, at ω approximately 0.6. For frequencies above ω^{Q}, crisis arc E is the escape boundary, under increasing F, provided the earlier jump from fold A has restabilized onto the attracting focus. Notice that over a considerable range of ω, crisis E is initially quite close to the flip boundary C, so for practical purposes the flip into an $n = 2$ subharmonic can be used as a slightly conservative estimate of F^{E}.

The scenario of Figure 16.3, with its cusp and associated flip boundary, is repeated at diminishing scales at lower forcing frequencies (Figure 16.5(a)), two extra cusps of the fundamental $n = 1$ response surface mimicking the behaviour around the main cusp P. Each has a flip line passing transversely between the folds, which signals the onset of a period-doubling cascade to chaos and escape.

The continuation of the bifurcation arcs at higher values of ω is shown in Figure 16.5(b), where we see the flip arc C curving back at R to become the flip arc F, as sketched in Figure 16.3. Also shown in Figure 16.5(b) is the analytical

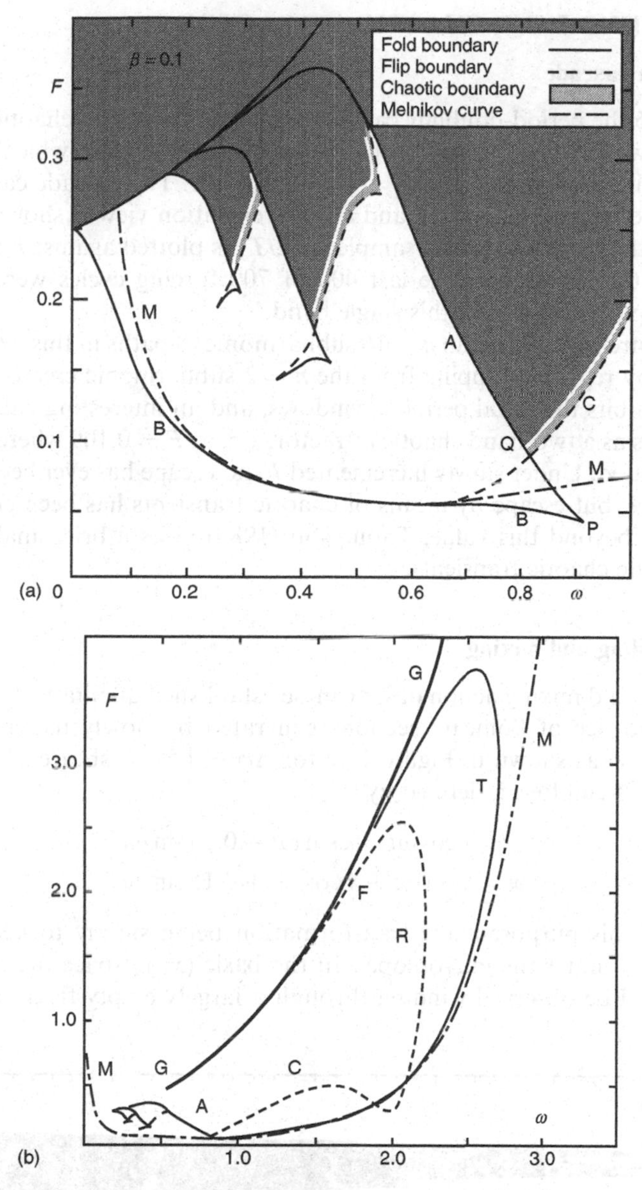

Figure 16.5 Two views of the (F, ω) control space showing the bifurcation arcs and the region of escape at $\beta = 0.1$

Melnikov curve M compared to the numerically determined arc of homoclinic tangency T. We shall be discussing these later, but we can note here that the two curves are in very close agreement for values of ω up to about 2.0. Beyond about 2.0 they diverge appreciably, and arc T actually turns back towards lower ω.

16.4 THE TWO-BAND CHAOTIC ATTRACTOR

Feigenbaum cascade

At $\omega = 0.85$ the period-doubling cascade leads to a two-band chaotic attractor, associated with two separate regions in the Poincaré section, which are visited alternately in the manner of an $n = 2$ subharmonic. The cascade can therefore be examined by sampling at $2T$ and a high-resolution view is shown in Figure 16.6. Here the stroboscopically sampled $x(2iT)$ is plotted against F for steps of $\Delta F = 5 \times 10^{-7}$: at each F, the last 400 of 700 forcing cycles were recorded, giving 200 plotted points in this single band.

This picture begins with an $n = 16$ subharmonic (8 paths in this $2T$ sampling) generated by repeated flipping from the $n = 2$ subharmonic created at C. The cascade exhibits the usual periodic windows, and an interesting chaotic explosion. It ends as a two-band chaotic attractor, C^2, at $F = 0.109$ where this figure was terminated. Under slowly incremented F, no escape has ever been observed before 0.109, but escape by means of chaotic transients has been encountered with F just beyond this value. Thompson (1989) gives a brief analysis of the scaling of the chaotic transients.

Chaotic folding and mixing

The folding and mixing action of C^2 can be established qualitatively by observing the sequence of Poincaré sections generated by slowly incrementing the phase angle ϕ as shown in Figure 16.7 for $\Delta\phi = 4.5°$. A sheared Van der Pol plane (u, v) is employed, defined by

$$u = x \cos \omega t - \dot{x} \sin \omega t + 0.15 \sin \omega t$$
$$v = x \sin \omega t + \dot{x} \cos \omega t + 0.15 \sin \omega t$$

the aim of this purpose-made transformation being simply to keep the two bands of C^2 'under the microscope'. In the basic (x, y) space the two narrow bands would be observed winding through a largely empty field of view. For

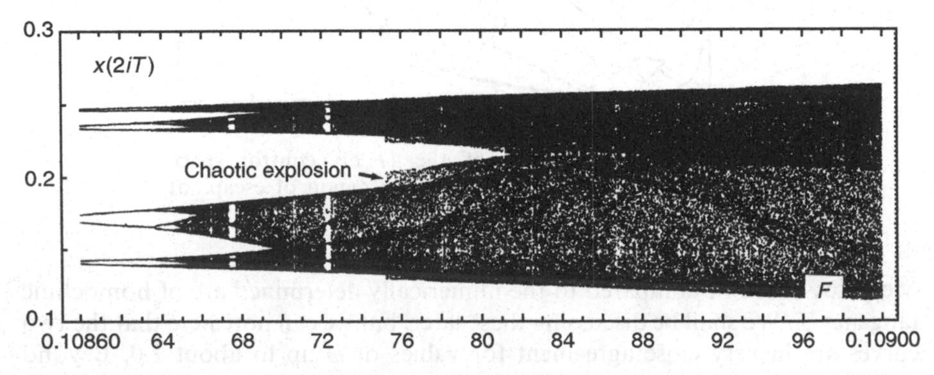

Figure 16.6 The period-doubling cascade leading to chaos and escape under increasing F at $\omega = 0.85$ and $\beta = 0.1$

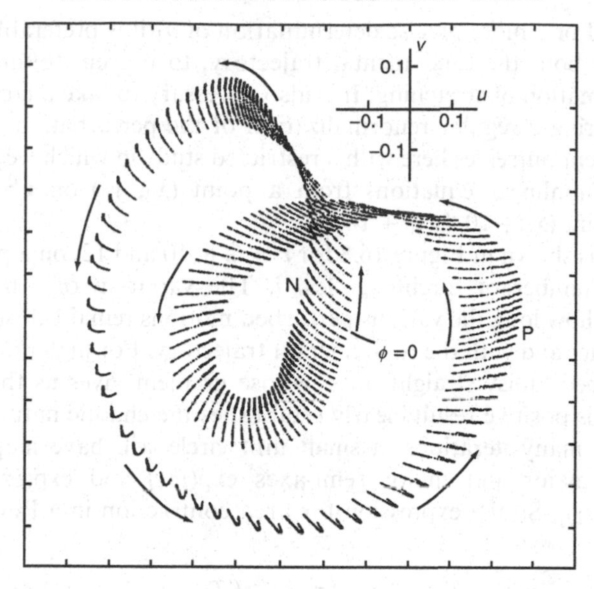

Figure 16.7 The chaotic attractor C^2 viewed in a sheared Van der Pol plane as a folding Möbius strip: $\beta = 0.1$, $\omega = 0.85$, $F = 0.109$

each of the 80 values of ϕ, a single trajectory is sampled at period T and the resulting two bands at $\phi = 0$ are indicated by NP. The chaotic attractor can be seen as a folding Möbius strip, the damping being such that one fold is almost complete, and lost to our view, before the next reversed folding begins. The fractal cross-section of the strip is therefore not observed significantly at this scale.

Positive Liapunov exponent

A readily computable quantitative measure of the stochasticity of a trajectory is provided by the Liapunov exponents, and we make here a rough estimate of the positive Liapunov exponent of C^2. In our Poincaré mapping, the Liapunov exponents σ of a trajectory can be defined in terms of the distance R_i between mapping points of the trajectory (x_i, y_i) and an infinitesimally perturbed trajectory $(x_i + \Delta x_i, y_i + \Delta y_i)$,

$$\sigma = \lim_{\substack{i \to \infty \\ R_0 \to 0}} \left(\frac{1}{i}\right) \ln \frac{R_i}{R_0}$$

where

$$R_i = \sqrt{\Delta x_i^2 + \Delta y_i^2}$$

This formula supplies σ_1, the greater of the two Liapunov exponents, for almost every initial perturbation, excluding only the zero-probability starts on the base vector of σ_2. For a more precise determination of σ_1 it is preferable to linearize the motions about the fundamental trajectory, to prevent folding interfering with our estimation of stretching. It is also necessary to take a large number of steps i, requiring a regular renormalization of the perturbation vector. However, we content ourselves here with a restricted study in which we just integrate the original nonlinear equations from a point (x_0, y_0) on C^2 and from a perturbed point $(x_0 + 10^{-r}, y_0 + 10^{-r})$.

The result is shown in Figure 16.8 for $r = 6, 8, 10$ and 12, on a plot of $-\ln R_i$ against the number of forcing cycles, i. The variation of r here serves to highlight for how long the various perturbed motions remain closely correlated with each other and with the fundamental trajectory. For high r and low i there are well-defined 'noisy straight lines' whose gradient gives us the estimate of $\sigma_1 \approx 1/4$. This positive result clearly establishes the chaotic nature of C^2.

Now after many iterations, a small unit circle will have mapped into an ellipse with major and minor semi-axes $\exp(i\sigma_1)$ and $\exp(i\sigma_2)$, and area $\pi \exp[i(\sigma_1 + \sigma_2)]$. So the expression for area contraction in a Poincaré section gives

$$\sigma_1 + \sigma_2 = -\beta T$$

We can thus calculate σ_2, giving us the full two-dimensional Liapunov spectrum of our mapping, which can be related to the fractal and information dimensions of the chaotic attractor.

Liapunov exponents can be viewed nicely in terms of information theory (Shaw 1981), measuring the rate of growth or decay of information. For this, they are expressed most naturally in terms of bits of information per iteration,

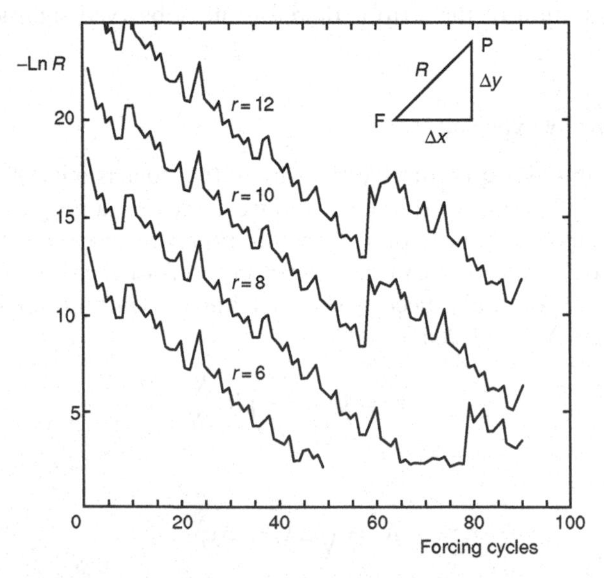

Figure 16.8 Estimation of the positive Liapunov exponent of the chaotic attractor C^2. Here $\Delta x_0 = \Delta y_0 = 10^{-r}$

and to get σ_1 into these units we must change our base e to base 2 by dividing by ln 2. Then in base 2 we have $\sigma_1 \approx 0.36$ bits per iteration. As we can see in Figure 16.8, this implies a predictability horizon of only about 100 forcing cycles if initial conditions are specified to 12 decimal places.

Collision with the D^6 subharmonic

The bifurcation at which C^2 finally loses its stability and the system jumps to infinity is at F^E, just above 0.109, where the attractor is in collision with D^6, as shown in Figure 16.9 for $\beta = 0.1$ and $\omega = 0.85$. This directly unstable $n = 6$ subharmonic is the saddle of a very recent saddle node that generates over a short F interval a complete $n = 6$ cascade; notice that sampling is at $2iT$ (at $\phi = 180°$), so that only half of the full picture is observed. At the end of the $n = 6$ cascade, when the $n = 6$ chaotic attractor collides with D^6 at $F \approx 0.1077$, a system evolving on the D-paths jumps back to the main sequence S^4 as indicated by arrow J.

The collision at F^E is shown in Figure 16.10 in the projection (x, y), sampling again at $2iT$ with phase $\phi = 180°$. The path of three points of D^6, in equal F steps, is shown relative to one band of C^2 at $F = 0.109$; this representation is useful because C^2 does not move significantly over the F range involved. The apparent crossing of the path and the attractor is illusory, because the chaotic attractor had not yet formed at the lower F values. High-resolution pictures of this event are given by Thompson and Ueda (1989).

We should remark here that the unexpected creation of a steady-state subharmonic (such as D^6) at a saddle-node fold is a typical event that happens all the time in heavily driven nonlinear oscillators. However, the events are not easily found numerically because the created stable orbits often have very small

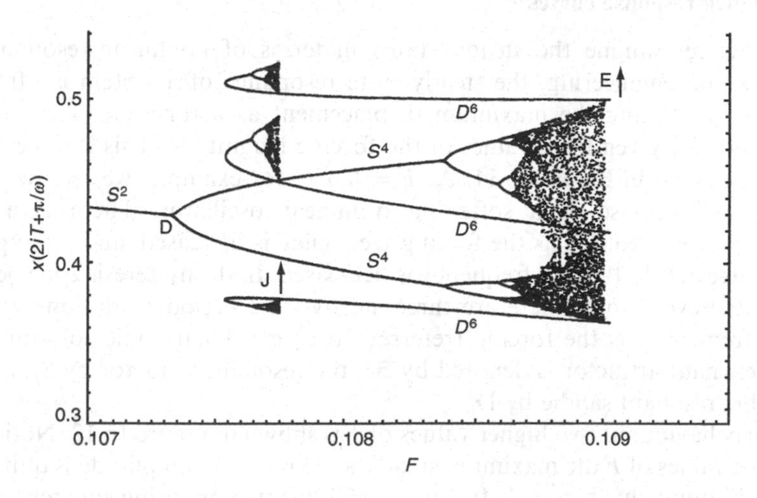

Figure 16.9 Collision of D^6 with the chaotic attractor at the final boundary crisis ($\beta = 0.1$, $\omega = 0.85$, $\phi = 180°$). Note that the mapping coordinate is here sampled at twice the forcing period

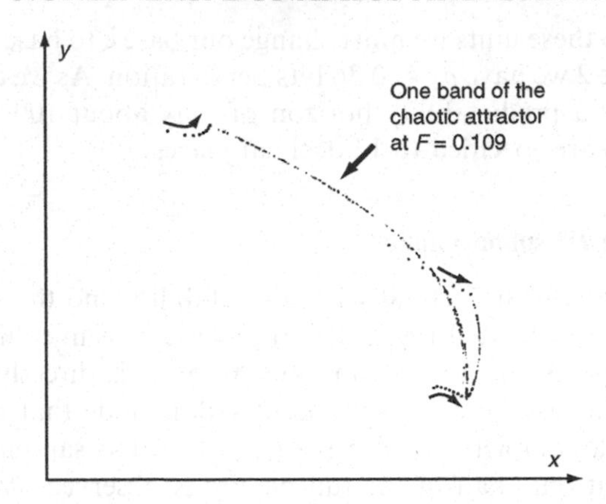

Figure 16.10 Boundary crisis at F^E viewed in the phase projection at $\phi = 180°$. The three curved arrows indicate the path of three points of the saddle D^6 for $F = 0.1076$ to 0.1090 in steps of 0.0002. The chaotic attractor does not move appreciably during this stepping

basins, and only exist in a very small parameter regime; they are appropriately called *fugitive* subharmonics.

To fully understand the significance of this collision event, we shall need the perspective provided by the concept of a chaotic saddle that we shall outline in Section 16.8.

16.5 RESONANCE OF THE STEADY STATES

Resonance response curves

We now re-examine the steady states in terms of nonlinear resonance. In physics and engineering, the steady-state resonance of a system is often displayed by plotting the maximum displacement as a function of the forcing frequency for given fixed values of the forcing magnitude. This is done for the present system in Figure 16.11. At $F = 0.056$, for example, we see the typical hysteresis response of a softening nonlinear oscillator. There is a jump to resonance at fold A as the forcing frequency is increased, and a jump from resonance at fold B as the frequency is decreased. In the hysteresis zone between these two cyclic folds there are three steady-state periodic solutions with the same frequency as the forcing (referred to as $n = 1$ harmonic solutions): the non-resonant attractor is denoted by S_n, the resonant attractor by S_r, and the unstable resonant saddle by D_r.

The behaviour at two higher values of F is shown in Figure 16.12. Notice that at these values of F the maximum steady-state response amplitude is quite close to the hilltop value of $x = 1$. In Figure 16.12(a) two opposing supercritical flip bifurcations generate a closed loop of $n = 2$ subharmonic attractors. In Figure16.12(b) this feature has developed into two opposing cascades to chaos.

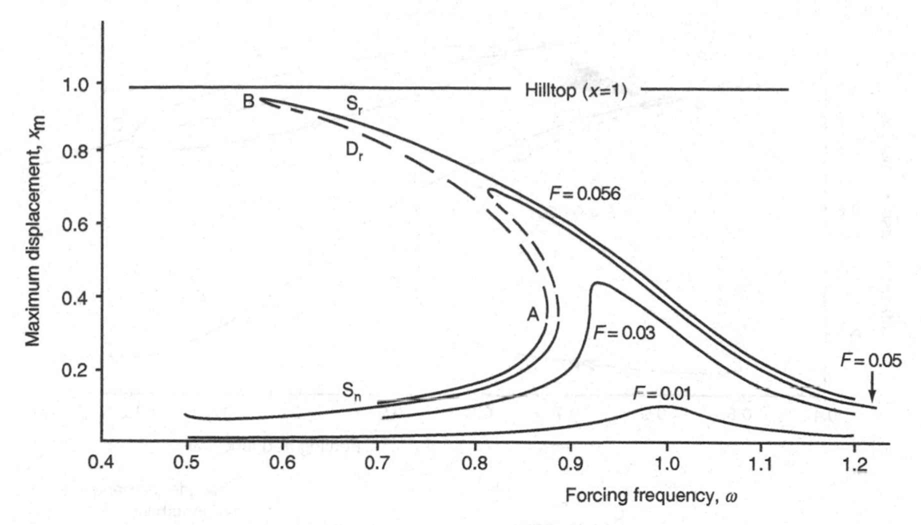

Figure 16.11 A nest of resonance response curves for the escape equation at relatively low forcing levels

A summary of the steady-state behaviour at three high values of F is finally given in Figure 16.13. In Figure 16.13(c) we observe a frequency interval with no main attractor. Remember, however, that the interval might contain high-order subharmonics with very small basins of attraction, existing over very small frequency ranges—the so-called fugitive subharmonics that we discussed earlier. Here, as we slowly increase the driving frequency, we come to fold A (with a relatively low response amplitude) from which the system would jump straight out of the well. If, conversely, the frequency is slowly decreased from an initially high value, then a period-doubling cascade to chaos occurs, followed by a chaotic crisis at which the system again jumps out of the well. In Figure 16.13(a), at lower F, the jump to resonance at fold A always restabilizes on the stable resonant state, R. Of more interest is the intermediate case (Figure 16.13(b)). Here the jump from fold A may or may not restabilize on R, the bifurcation being indeterminate with an outcome that depends sensitively on the way it is realized. We shall explain later how at point A the system is sitting precisely on a fractal basin boundary.

Note that the present idea of a safe jump should not be confused with the description of bifurcations as safe, explosive or dangerous (Thompson *et al.* 1994): in the bifurcational context fold A is always classified as a dangerous bifurcation.

Projection into the control space

The complete set of resonance response diagrams can be sketched as a surface, as in Figure 16.14, where the bifurcation sets have been projected into the base plane. Figure 16.15 shows more details of this (F, ω) control

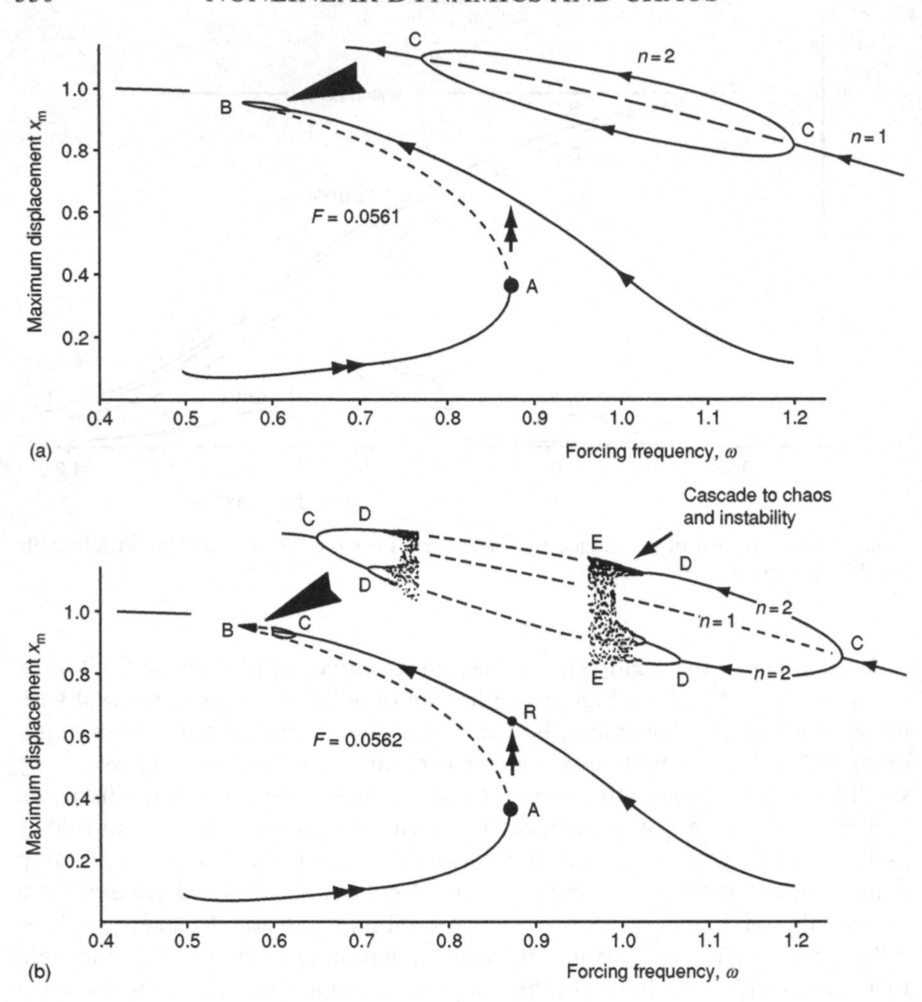

Figure 16.12 A pair of resonance response curves for the escape equation at two intermediate forcing levels

plane. Here the bifurcation arcs summarize the jump to resonance at fold A, the jump from resonance at fold B, the first period doubling to a stable subharmonic of order $n = 2$ at flip C, and the final loss of stability of the chaotic attractor at crisis E. As before, notice how the fold arcs A and B are generated by the cusp at P. On the crisis arc, the chaotic attractor is in collision with an unstable subharmonic of order $n = 6$, as we have seen; this arc can be approximately located by numerically following the Birkhoff signature change (see later) that occurs on arc S. The intersection of arcs A and E at Q marks the point of optimal steady-state escape.

If F is greater than F^Q, the resonance response curve will have a frequency interval with no (main sequence) attractor, as in Figure 16.13(c). Here, under increasing ω, the jump to resonance from fold A will inevitably carry the system

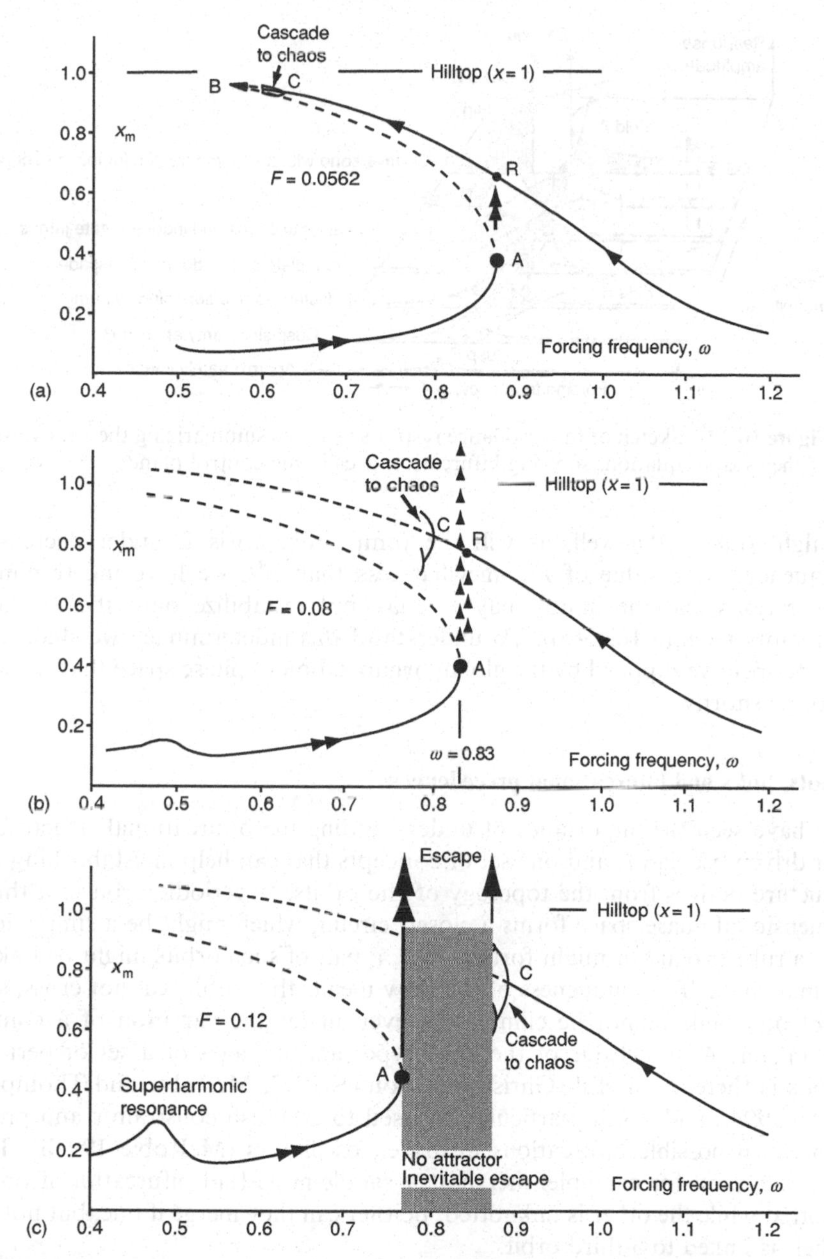

Figure 16.13 Three resonance response curves showing safe, indeterminate and unsafe jumps to resonance from the saddle-node fold A: (a) safe, a jump from A always restabilizes at R; (b) indeterminate, a jump from A may or may not restabilize at R; (c) unsafe, a jump from A always escapes to infinity

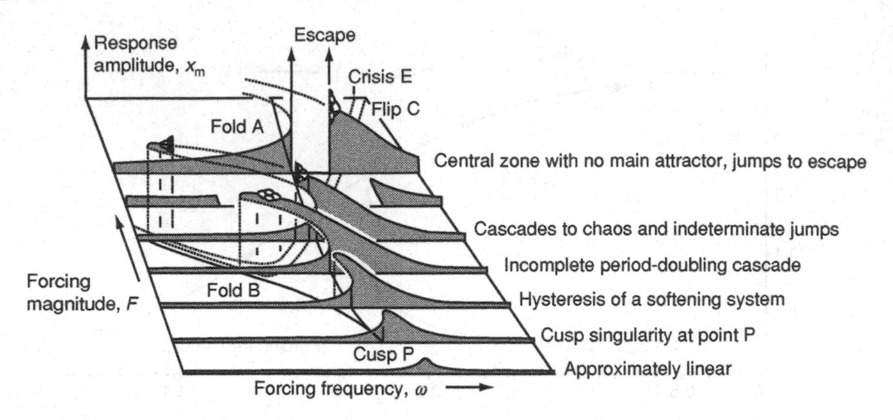

Figure 16.14 Sketch of the resonance response surface summarizing the behaviour of the escape equation, showing bifurcation arcs in the control plane

straight out of the well, as will the jump from crisis E under decreasing frequency. At a value of F somewhat less than F^Q, we have indeterminate bifurcations and the jumps may or may not restabilize onto the available attractors (Figure 16.13(b)). To understand this indeterminacy we shall need the perspective supplied by the global organization of phase space that we shall address shortly.

Knots, links and bifurcational precedences

We have seen the importance of understanding the bifurcational structure of our driven oscillator, and one set of concepts that can help in establishing this structure derives from the topology of the orbits. A periodic orbit in a three-dimensional phase space forms a closed circuit, which might be a simple loop like a rubber band or might form a *knot*; a pair of such orbits might be linked, as in a chain. The uniqueness of the flow means that orbits cannot cross, so a knot or a link cannot be eliminated, even under the variation of a control parameter. A knowledge of the knot types and linkages of a set of periodic orbits is therefore useful (Ghrist and Holmes 1993; McRobie and Thompson 1993, 1994), and can in particular be used to establish constraints and precedences on possible bifurcations in driven oscillators (McRobie 1992b). Two orbits cannot, for example, merge at a saddle-node fold bifurcation if one is knotted while the other is unknotted; neither can they merge if one, but not the other, is linked to a third orbit.

To establish the orbit topology for a driven oscillator, the time history $x(t)$ can be interpreted as a *braid* diagram, as in Figure 16.16. The orbit structure in the full 3D phase space of $(x, \dot{x}, t)$ is easily deduced from it because when two time histories cross it is obvious which has the greater $\dot{x}$. Similar deductions cannot be made from the $(x, \dot{x})$ phase projection.

A specific application is illustrated for the escape equation in Figure 16.17, which represents the $n = 1$ harmonic response shown earlier in Figure 16.4.

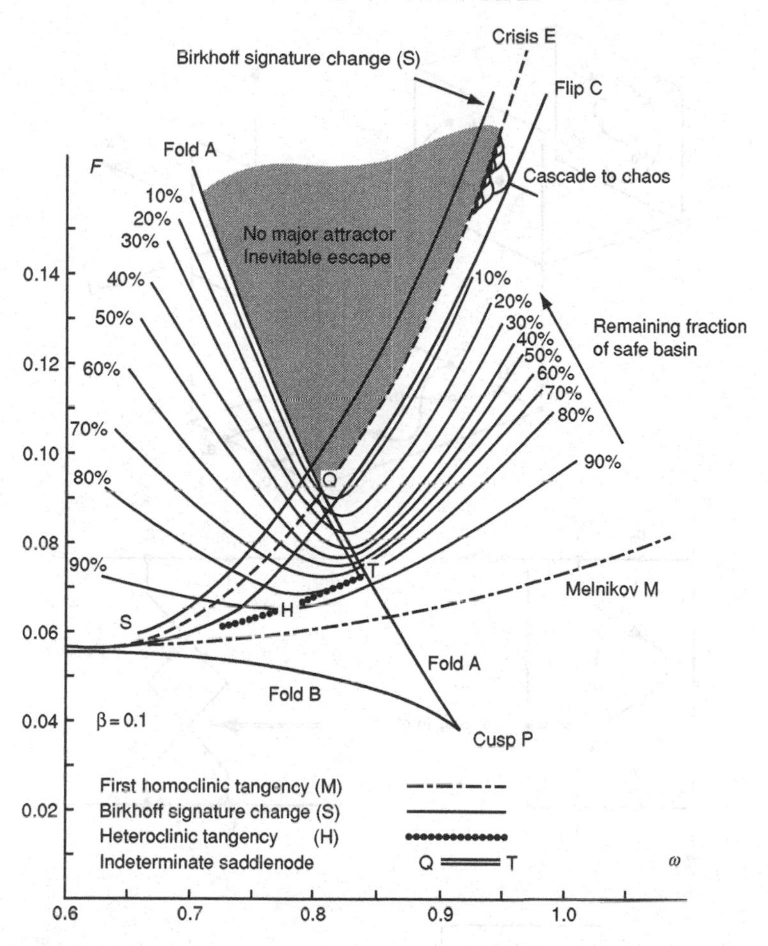

Figure 16.15 Bifurcation arcs of the escape equation in the control plane: superimposed contours show the remaining percentage of the safe basin

It can be shown that S_n, S_r and D_r are topologically linked as shown, while D_h is not. Hence we can establish that even under the continuous variation of another control parameter, such as ω or β, F^G must remain greater than F^A.

16.6 TRANSIENTS AND BASINS OF ATTRACTION

To understand fully the behaviour of a nonlinear dynamical system, we must pay attention not only to the steady states that we have discussed so far, but also to the transient motions that can arise from finite disturbances, sudden changes in excitation, etc. In fact, a global view of all possible transient motions supplies a view of the dynamics that is both more robust and more relevant than a detailed inspection of the steady states and their local bifurcations.

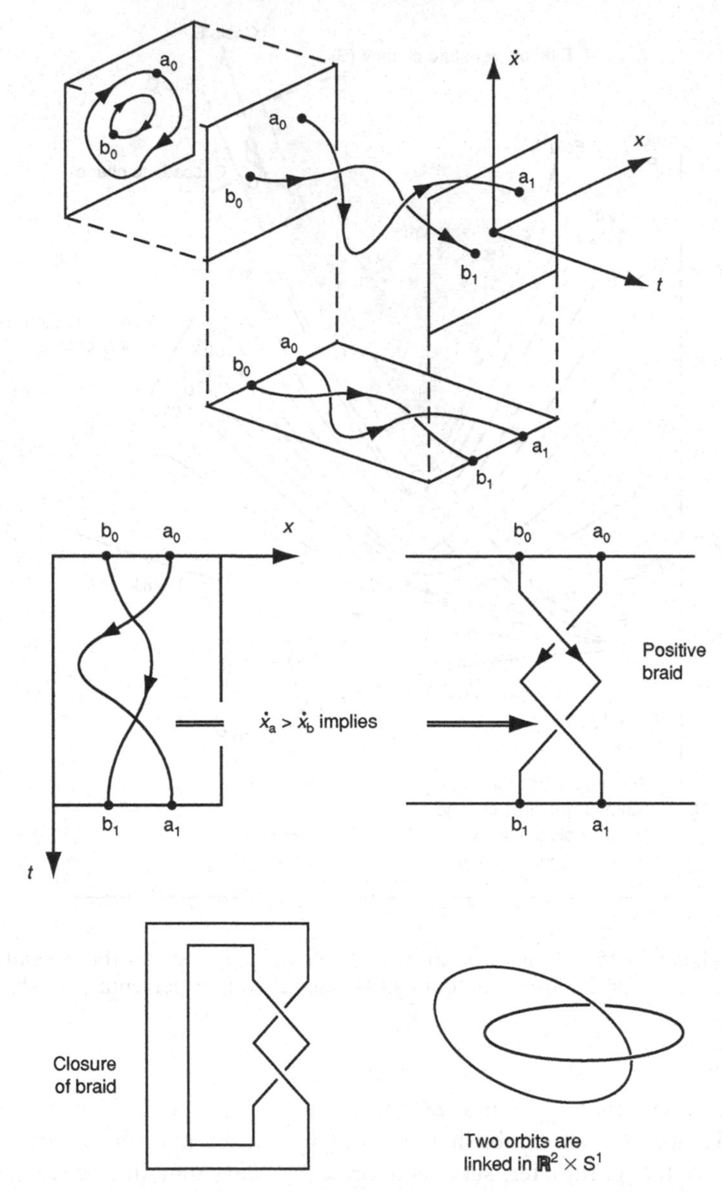

Figure 16.16 Imagining them to be steady-state responses of an oscillator driven at the period implied by the drawn Poincaré sections, the sketches show how a braid analysis of two simple $n = 1$ waveforms, $x(t)$, identifies them as linked orbits in the toroidal space $(x, \dot{x}, \phi)$, where ϕ is the phase angle

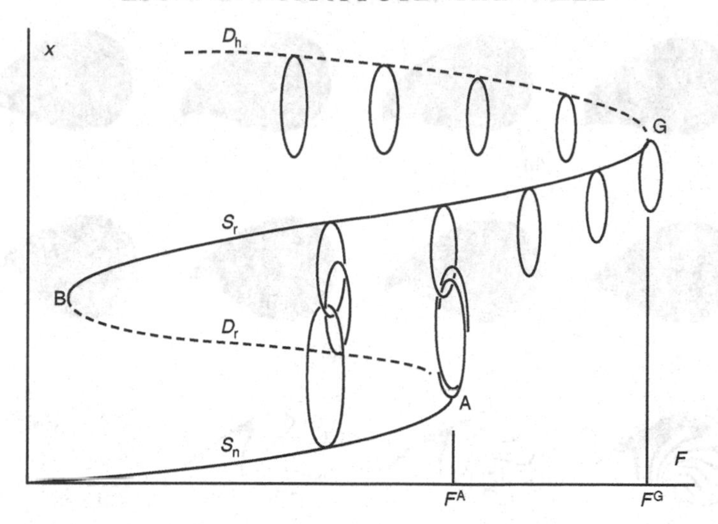

Figure 16.17 A schematic diagram showing how a linkage of $n = 1$ orbits in the escape equation generates a topological constraint, $F^G(\omega) > F^A(\omega)$, on the bifurcations

Erosion of the safe basin

In examining transients, it will suffice here to focus on the long-term behaviour by asking the basic question, Would the system escape or not from a given start if we were to wait for an infinite time? The more practical question concerning escape in a finite number (m) of forcing cycles can often be fairly accurately assessed once this basic question about the basins of attraction has been answered. Conversely, we have found that when studying the overall morphology of the safe basin, the escaping set for $m = \infty$ is approximated with good scientific accuracy by studies with m in the interval 8 to 64. We should notice, however, that while an $m = \infty$ safe basin can experience an instantaneous loss of area under increasing F, this will not be observed in finite m simulations.

In Figure 16.1 we have already examined the safe basin of attraction (comprising those starts that do not give escape over the hilltop with x going to $+\infty$) for the autonomous system with $F = 0$. Under increase of F, this basin is dramatically eroded by incursive fractal fingers, as illustrated in Figure 16.18. Here safe basins of attraction at increasing levels of F have been determined numerically using a grid in the $[x(0), y(0)]$ starting space. Starts leading to escape over the hilltop are denoted by white, while starts which settle onto finite motions within the well (harmonic, subharmonic or chaotic) are denoted by black. We observe that as F is increased up to 0.07, only a small reduction in the safe region occurs. But at about this value, a homoclinic tangency is formed (as described in a later section), creating fractal fingers which soon sweep dramatically across the safe basin; so at $F = 0.08$ most of the safe basin has been destroyed. This process is nicely shown in a computer-generated video made by Cusumano, who has also made ingenious experimental observations

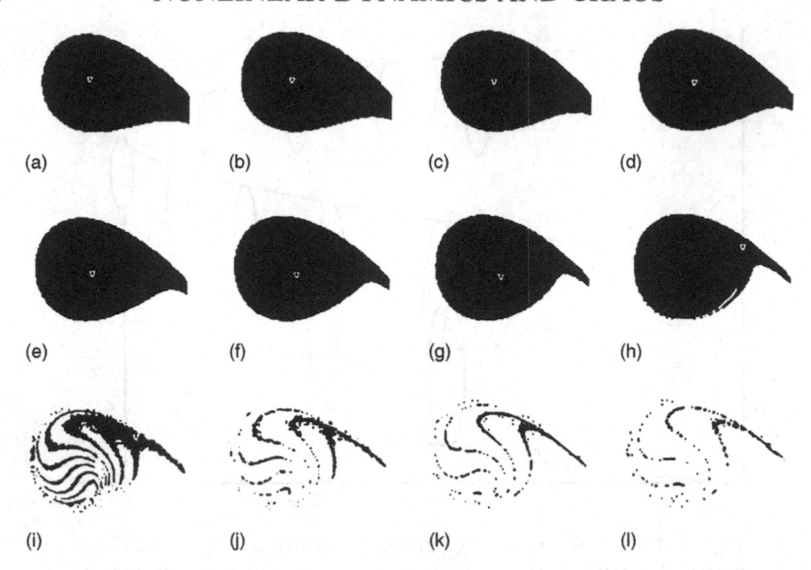

Figure 16.18 Basin erosion sequence of the escape equation, showing the black safe basin in the space of the starting conditions $[x(0), y(0)]$ as the forcing magnitude is increased: (a) $F = 0.000$, (b) $F = 0.010$, (c) $F = 0.020$, (d) $F = 0.030$, (e) $F = 0.040$, (f) $F = 0.050$, (g) $F = 0.060$, (h) $F = 0.070$, (i) $F = 0.080$, (j) $F = 0.090$, (k) $F = 0.100$, (l) $F = 0.110$

of the associated phenomena (Cusumano 1994). Finer details of the fractal incursion are shown in Figure 16.19, which was produced not by using a grid of starts, but by locating the fractal boundary defined by the inset of the hilltop saddle cycle. This was done numerically by running time backwards from a ladder of starts on the appropriate eigenvector of the saddle. When doing this, a complete picture is obtained if the short segment of the inset used as a building block is such that its endpoints are successive points on the same trajectory.

A characteristic feature of the (highly discontinuous) erosion process is the fold cascade crisis appearance of subharmonics at the tips of the fingers; an $n = 3$ subharmonic is seen in Figure 16.19(d). This picture also shows the Poincaré mapping sequence $0, 1, 2, \ldots, 7$ by which a trajectory steps from finger to finger before escaping to infinity.

The Dover cliff integrity curve

Perhaps the simplest way to quantify this basin erosion is to evaluate the area of the safe basin; this must be done within a suitable $[x(0), y(0)]$ window, because the safe basin has infinite area (due to the positive divergence of a damped system under time reversal). The area is one measure of the integrity of the system against disturbances; others are described in Soliman and Thompson (1989), and a plot of the area against F can be called an integrity diagram. Figure 16.20 plots the basin area (normalized with respect to the autonomous system) against the forcing magnitude. Up to the homoclinic tangency we have

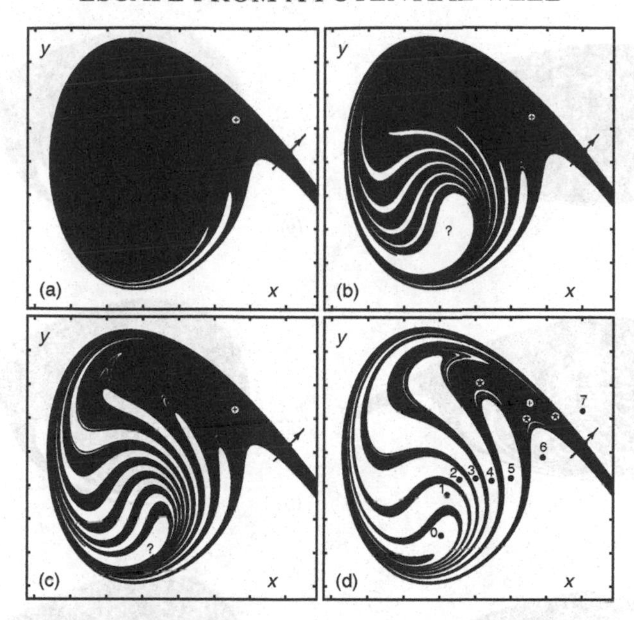

Figure 16.19 Development of the homoclinic tangle, showing how incursive fractals sweep across the centre of the safe basin: (a) $F = 0.0725$, (b) $F = 0.0750$, (c) $F = 0.0775$, (d) $F = 0.0872$. The cross denotes the main sequence harmonic (period 1) attractor. The stars denote a subharmonic attractor of period 3. A typical mapping sequence to escape is denoted by points 0, 1, 2, ..., 7

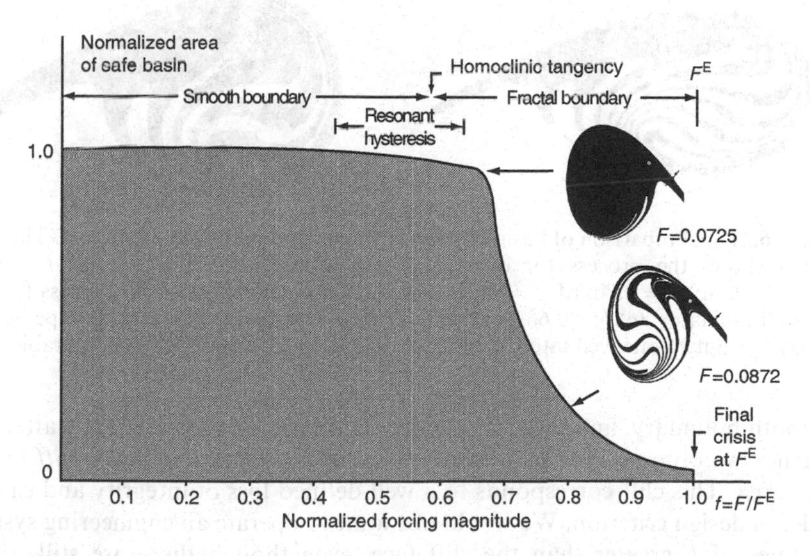

Figure 16.20 Dover cliff in the integrity curve of safe basin area versus forcing magnitude. Superimposed sections show the fractal incursion process

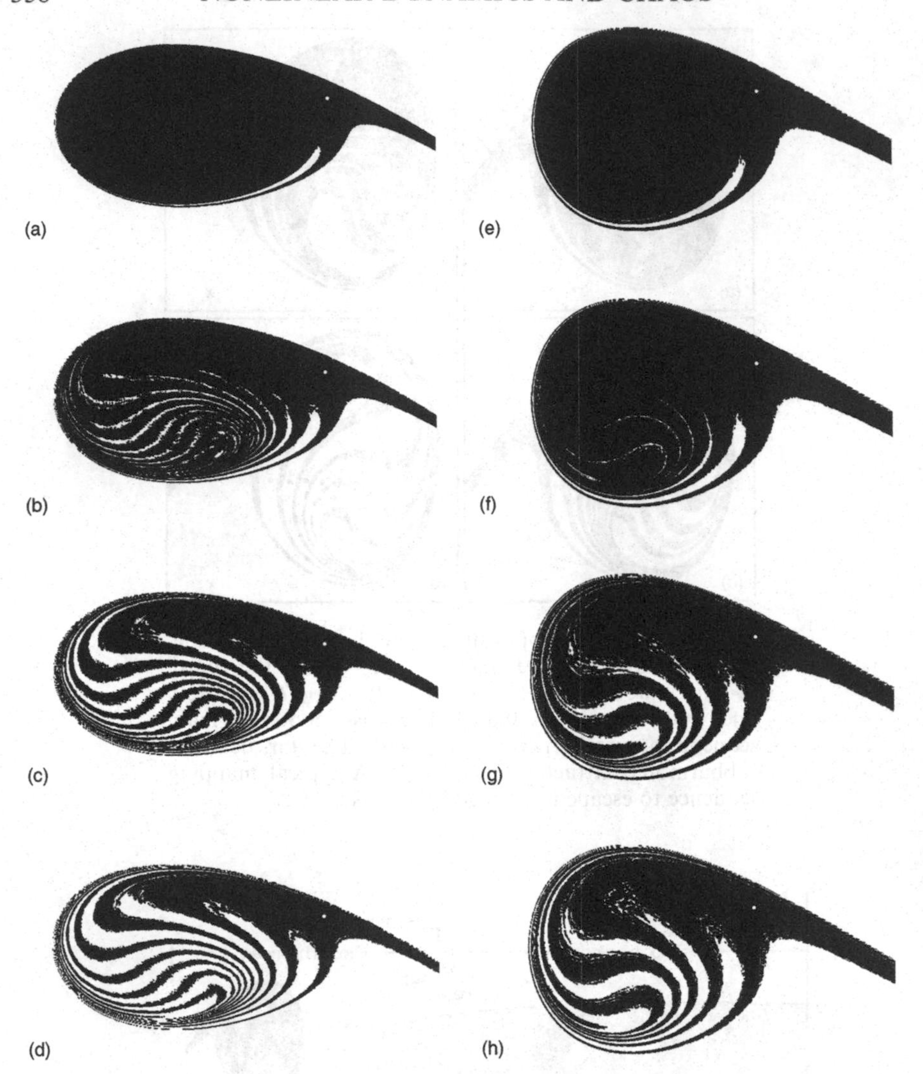

Figure 16.21 Comparison of basin erosion in single- and twin-well oscillators. The first column shows the process for our escape equation with $f \equiv F/F^{\mathrm{E}}$; (a) $f = 0.66$, (b) $f = 0.68$, (c) $f = 0.70$, (d) $f = 0.72$. The second column shows the process for the twin-well oscillator: (e) $f = 0.66$, (f) $f = 0.68$, (g) $f = 0.70$, (h) $f = 0.72$. Escape here is the passage out of one well into the other, with a defined in a directly comparable way

a smooth boundary, and there is little reduction in basin area. Shortly after the tangency we observe the rapid erosion process giving us the *Dover cliff* fall in basin area. This cliff corresponds to a well-defined loss of integrity and can be used as a design criterion. We would not want to operate an engineering system at values of F greater than the cliff face, even though there are still stable periodic motions within the well.

The Dover cliff is observed for a wide class of driven oscillators, and single- and twin-well potentials exhibit remarkably similar cliff profiles, as illustrated in the cell-to-cell mapping results of Figure 16.21 (Lansbury and Thompson 1990). Since the fractal fingers sweep rapidly across the centre of the section, the forcing magnitude of the cliff can be approximately located by running just one time simulation from the ambient state $[x(0), y(0)]$, rather than a complete grid. This gives a quick and economical way of locating the loss of engineering integrity. It has been proposed as a useful tool in the simulation and laboratory testing of model ships for capsize due to surface waves (Thompson *et al.* 1990, 1992; Rainey and Thompson 1991; MacMaster and Thompson 1994; Thompson 1997).

Fractal boundaries in control space

Transient trajectories from the ambient state, which we have just been discussing, have a deeper dynamical significance, because they define a fractal basin boundary in control space. At a fixed damping level, the safe basin of the escape equation can be thought of as a master basin (Thompson and Soliman 1990) in the 4D phase control space spanned by $x(0)$, $y(0)$, F and ω. The phase space basins of Figures 16.18 and 16.19 are cross-sections of this master basin corresponding to fixed values of F and ω. The control space basins that we are now considering are likewise just cross-sections of the master basin, this time at the fixed values of $x(0) = 0$, $y(0) = 0$.

Figure 16.22(a) shows this control space cross-section of the safe basin of the escape equation; here safe is denoted by white, unsafe by black. Also shown on

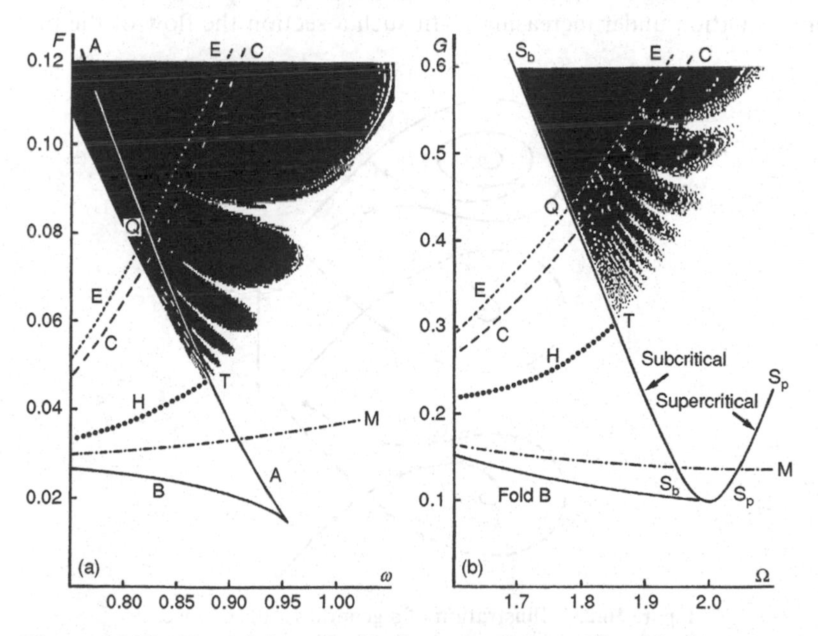

Figure 16.22 Fractal boundaries in control space for (a) the escape equation, $\ddot{x} + \beta\dot{x} + x - x^2 = F \sin \omega t$ $(x(0) = \dot{x}(0) = 0)$ and (b) for a similar parametrically excited system, $\ddot{x} + \beta\dot{x} + (1 + G \sin \Omega t)(x - x^2) = 0$ $(x(0) = 0.01,\ \dot{x}(0) = 0)$. The damping level is $\beta = 0.05$ in each. These pictures are based on the work of Soliman (1994)

the figure are the steady-state bifurcation arcs. Notice that thin black fingers extend down to point T, which is the lower limit for indeterminate jumps to resonance governed by a heteroclinic connection between the hilltop saddle and the resonant saddle. We shall say more about this in a later section, and further details can be found in Soliman (1994), from which this diagram has been reproduced. An experimental study of this control space cross-section is reported by (Gottwald *et al.* 1995). To emphasize just how universal these features are, Figure 16.22(b) is an equivalent picture for a parametrically excited system, also from Soliman (1994).

16.7 HOMOCLINIC PHENOMENA

Global bifurcations are the key to understanding the many complex issues that we have encountered so far in the escape story. The first bifurcation we must address is the homoclinic tangency.

Homoclinic tangency

We recall that in a 2D flow, an outset of a saddle can return as an inset of the same saddle (Figure 16.23). In this example the saddle collides with an expanding stable cycle, which vanishes into the blue at the global bifurcation. The equivalent event in a 2D mapping is the homoclinic tangency (Figure 16.24), which might represent three stroboscopic Poincaré sections of the escape equation under increasing F. In such a section the flow of the previous

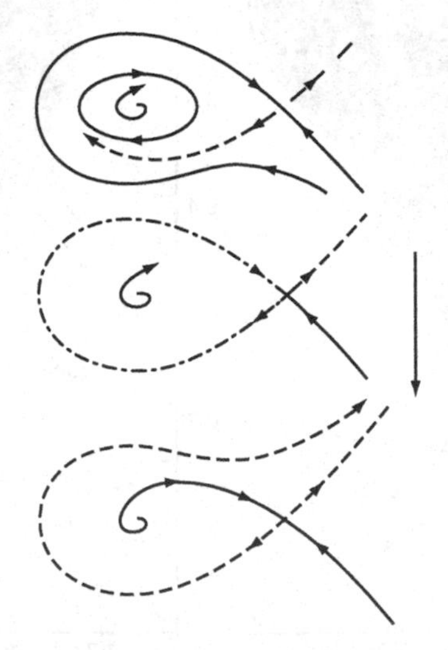

Figure 16.23 Illustration of a generic saddle connection in the two-dimensional flow of an autonomous oscillator. The stiffness function is $x - x^2$; the linear damping coefficient is negative and decreasing. The cubic damping coefficient is a positive constant

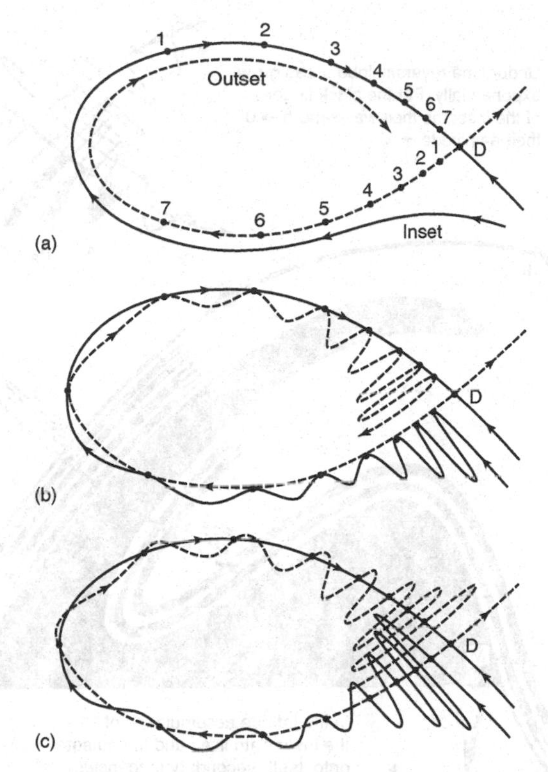

Figure 16.24 Schematic illustration of a homoclinic tangency in a Poincaré section. Here D is a directly unstable saddle cycle, and invariant manifolds are sketched (a) before the tangency, (b) at the tangency and (c) after the tangency

figure is replaced by a differentiable mapping (a diffeomorphism), and by considering the forward and backward iterations of the map it is clear that if the inset touches (or intersects) the outset once, it will touch (or intersect) it an infinite number of times. This fact, together with the asymptotically slow dynamics near the saddle D, implies that after the tangency there will be an infinite number of intersection points accumulating onto D, as emphasized in Figure 16.25. After a homoclinic tangency we say that we have a homoclinic tangle. The 3D flow manifolds of such a tangle are shown schematically in Figure 16.26, which emphasizes how the flow surfaces determine the capture properties of the mapping.

Tangles and fractal boundaries

We have seen that at $F = 0$ the basin of attraction of the stable equilibrium at the bottom of our cubic well is bounded by the trajectory that flows asymptotically into the unstable hilltop equilibrium. As F is increased from zero, this

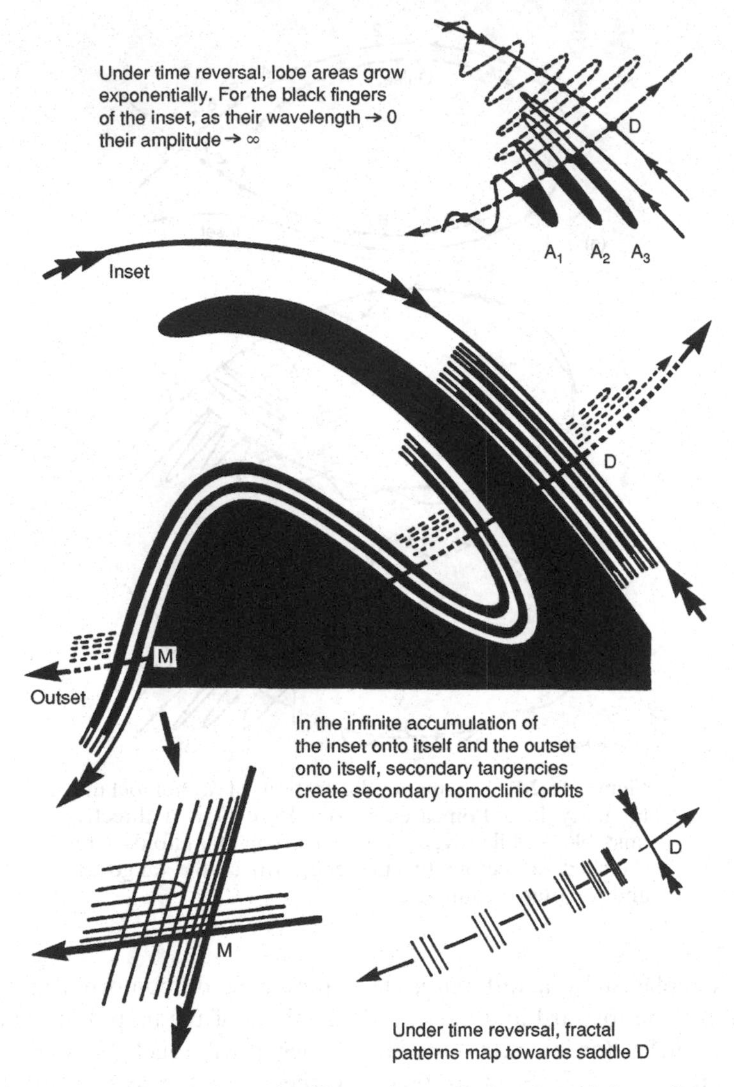

Under time reversal, lobe areas grow exponentially. For the black fingers of the inset, as their wavelength → 0 their amplitude → ∞
D
A1 A2 A3
Inset
D
M
Outset
In the infinite accumulation of the inset onto itself and the outset onto itself, secondary tangencies create secondary homoclinic orbits
D
M
Under time reversal, fractal patterns map towards saddle D

Figure 16.25 Schematic illustration of the fractal boundary accumulation generated by a homoclinic tangency. Notice how secondary tangencies will create secondary homoclinic orbits

hilltop equilibrium is transformed into a small unstable oscillation across the hilltop, denoted here by D_h, and its inset continues to delineate the boundary between escaping and non-escaping starts. Under further increase of F, this boundary undergoes a series of metamorphoses at global bifurcations, the most significant of which are the homoclinic tangency, M, and the heteroclinic tangency, H, which occur on the arcs shown in Figures 16.15 and 16.22 in the (F, ω) control space. At the homoclinic tangency, M, the inset and outset of the hilltop saddle cycle, D_h, first touch one another. For $F < F^M$ any small inwards

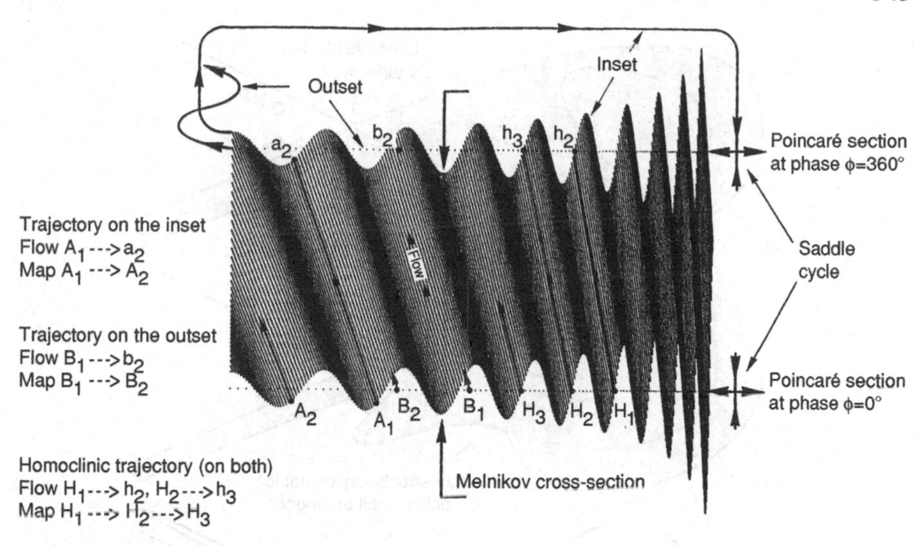

Figure 16.26 Schematic illustration of the invariant manifolds in a homoclinic tangle, showing the flow between two Poincaré sections

disturbance of D_h gives a trajectory that falls onto one of the safe, bounded attractors within the well (Figure 16.27(a)). At the homoclinic tangency (Figure 16.27(b)) the forcing is just strong enough to allow one (disturbed) trajectory to depart asymptotically from D_h along an outgoing eigenvector, pass across the floor of the well and return asymptotically to D_h along an incoming eigenvector. At the higher F of Figure 16.27(c), some starts from the outgoing eigenvector will be safe, while others will give escape.

The inset of D_h (strictly its closure) remains the boundary of the safe basin throughout this sequence of events, but beyond M the homoclinic tangling gives it a fractal nature (McDonald *et al.* 1985). An infinite number of infinitely thin tails of the escaping and non-escaping basins are intertwined around the edge of the safe basin, making prediction impossible within this fractal fringe where we have *final state sensitivity* (see Glossary). The structure of the invariant manifolds within a homoclinic tangle is amenable to analysis by the techniques of lobe dynamics (McRobie and Thompson 1991, 1992). The contributions of Allan McRobie show how the manifolds of the three-striped Smale horseshoe can be used to locate orbits and establish important bifurcational precedences (McRobie 1992b; McRobie and Thompson 1994). Similar techniques using manifold tangencies have been applied to the estimation of symmetry breaking and escape by Clifford and Bishop (1995).

At the first homoclinic tangency, at F^M, the union of the basins of the non-escaping attractors acquires a fractal boundary. So for F marginally above F^M, there will be a thin fractal fringe around the safe non-escaping basin. Note, however, that this fringe has initially (before the subsequent incursion) very little effect on the overall dynamics of the oscillator. At the centre of the safe basin there will often still be a regular period 1 attractor whose local transients

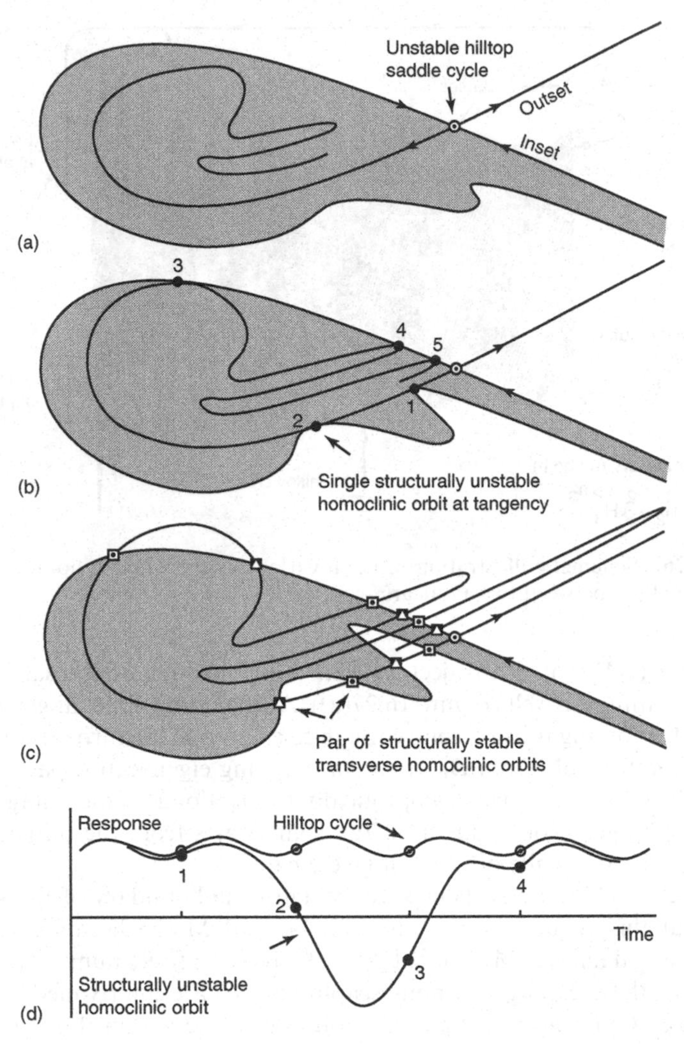

Figure 16.27 Initial homoclinic tangency of the hilltop saddle cycle. Poincaré sections (a) before, (b) at and (c) after the tangency show how the safe basin acquires a fractal boundary. Part (d) shows the time history of the homoclinic orbit

are entirely normal. The period doubling to chaos of the central, main sequence attractor typically occurs long after the first tangency, ($F^{C} \gg F^{M}$) and is a totally unrelated event. This being the case, we must ask what meaning can be attributed to the often heard, and usually misunderstood, statement that a homoclinic tangency implies chaos. The answer is that associated with the Smale horseshoe created at the tangency there will be chaotic transients and highly localized fold cascade crisis scenarios, but all of them located within the thin fractal zone around the edge of the basin.

Melnikov analysis

The homoclinic tangency at F^{M} can be estimated analytically using the Melnikov perturbation method. This can be interpreted as an energy balance approach (e.g. Thompson 1996b) in which the energy input due to the forcing is equated to the energy lost due to dissipation. These energies are calculated by integrating the work done as the real system is imagined to move around the saddle loop of the unforced and undamped Hamiltonian system. Applied to the escape equation, the Melnikov result is

$$F^{\mathrm{M}} = \beta \sinh(\pi\omega)/5\pi\omega^2$$

The corresponding arc in the (F, ω) control space at the damping level $\beta = 0.1$ is almost coincident with the true arc of homoclinic tangency over the frequency regime of interest as we have seen in Figure 16.5(b); see also Foale and Thompson (1991).

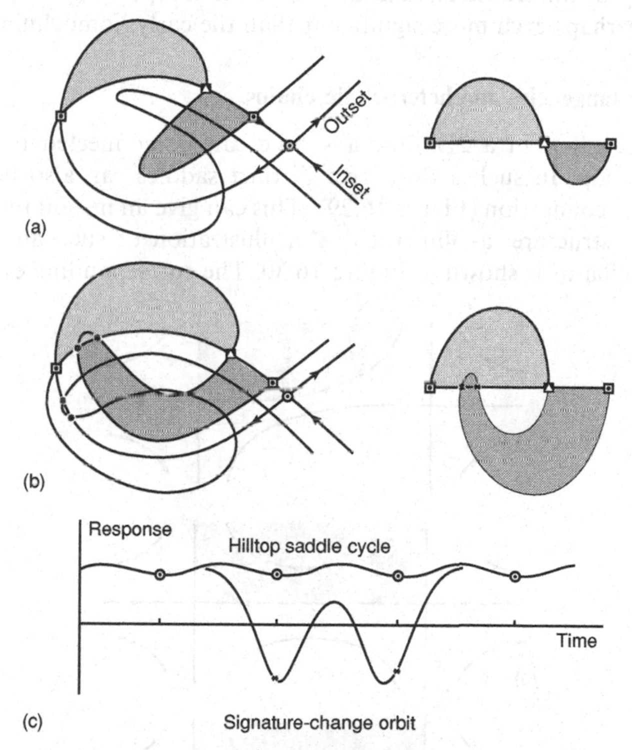

Figure 16.28 Schematic illustration of two Poincaré sections showing the first period 1 Birkhoff signature change. (a) The basic signature after the initial homoclinic tangency: pips and associated lobes; the pips are created at initial tangency. (b) The new signature after the internal homoclinic tangency: two adjacent homoclinic orbits created at signature change (*–*). (c) A time history of the signature-change orbit

Birkhoff signature change

Under further increase of F, the inset and outset of the hilltop saddle, D_h, become increasingly tangled. Further secondary homoclinic tangencies arise; a key tangency is where there is a change in the period 1 Birkhoff signature (McRobie 1992a). This is illustrated in Figure 16.28, and loosely corresponds to the lowest F at which a trajectory can depart from the outgoing eigenvector of D_h, make one complete oscillation within the well during one period of the applied forcing, and return to the incoming eigenvector of D_h. This Birkhoff signature change occurs on the global bifurcation arc S of Figure 16.15, lying just above the crisis arc, E. Thus arc S can be used as a guide to the location of arc E, as discussed by McRobie (1992a).

16.8 HETEROCLINIC PHENOMENA

In our study of the transients and basins of the escape equation, heteroclinic events are perhaps even more significant than the early homoclinic tangency.

Heteroclinic tangencies and heteroclinic chains

We have seen that in a 2D flow a saddle can be connected to itself by a homoclinic loop. In such a flow, two distinct saddles can also be linked by a heteroclinic connection (Figure 16.29). This can give an instantaneous change to the basin structure, as illustrated. An illustration of such an event in an undriven oscillator is shown in Figure 16.30. The corresponding event in a 2D

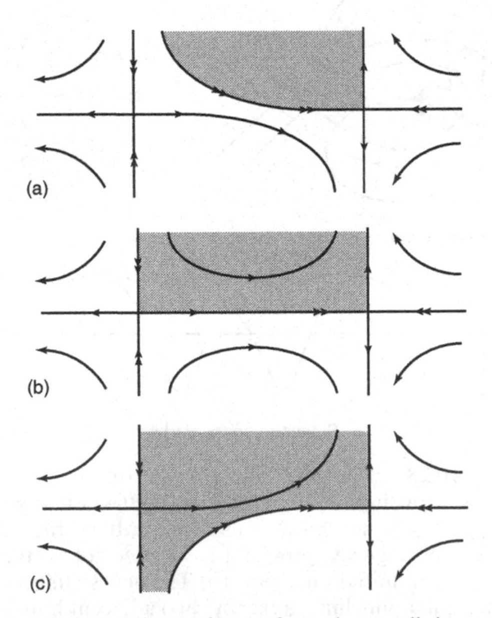

Figure 16.29 Illustration of a heteroclinic saddle connection in a 2D flow, showing the corresponding basin metamorphosis: (a) before, (b) at, (c) after

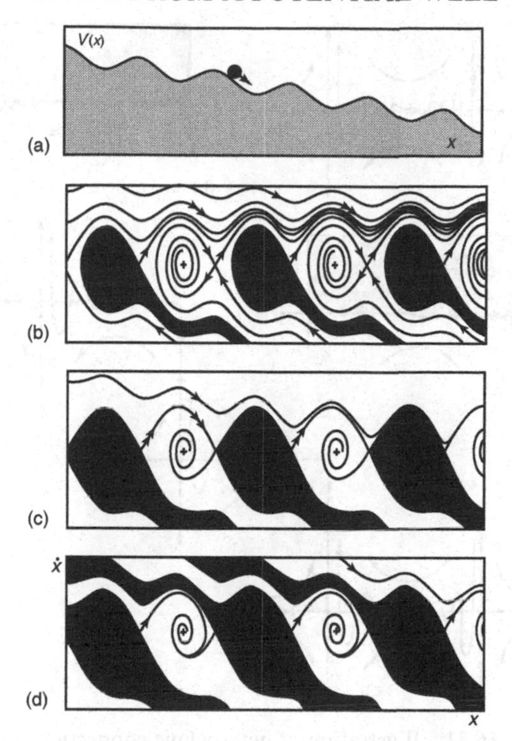

Figure 16.30 An example of a heteroclinic saddle connection for a ball rolling on a corrugated surface. This is the same as a pendulum subjected to an applied torque ($\ddot{x} + \beta\dot{x} + \sin x = \mu$, $\mu = 0.24$), though for the pendulum it could be regarded as a homoclinic connection: (a) potential energy curve; (b) light damping, $\beta = 0.1$, the two-headed arrows indicate steady motion; (c) critical damping, $\beta = 0.19$, saddle connection, $\mu/\beta \approx 4/\pi$; (d) heavy damping, $\beta = 0.25$

mapping is the heteroclinic tangency in Figure 16.31. Like the homoclinic tangency, this has an infinite number of tangencies (or intersections) that accumulate onto the saddles. If the heteroclinically tangled inset forms a basin boundary, an infinite number of infinitely thin fingers of the basin will accumulate onto the saddle as illustrated. Notice, however, that this is a smooth accumulation (sometimes called a mosquito coil) and does not generate any fractal structure. Heteroclinic connections in driven oscillators often form chains (Figure 16.31(d)). Notice that, in this chain, a small perturbation of the lower left saddle can give a trajectory that ends up close to the upper right saddle.

Accessible orbits and the chaotic saddle

An engineer contemplating a fractal basin boundary will immediately be concerned about the extent of the fractal fringe. Can his or her dynamics team be

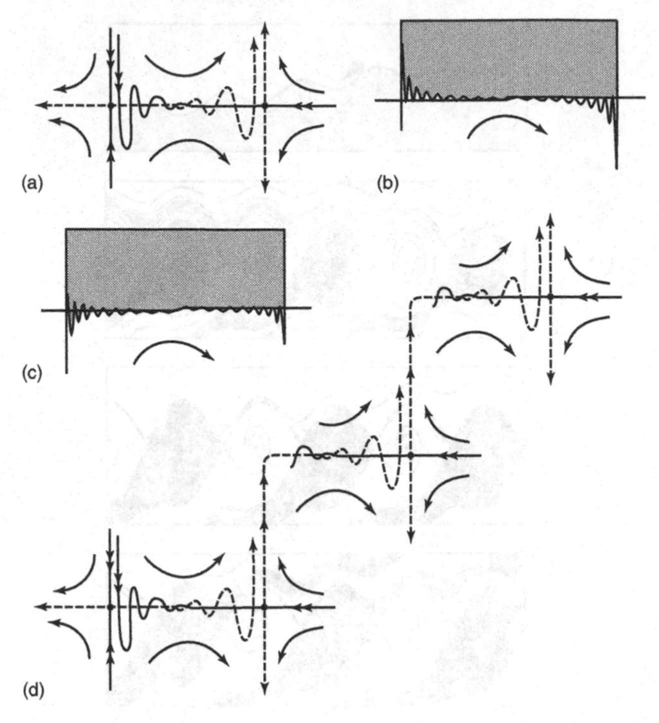

Figure 16.31 Illustration of heteroclinic connections in a 2D map: (a) connection between two saddles, after tangency, (b) basin accumulation at tangency at a heteroclinic connection, (c) basin accumulation after tangency at a heteroclinic connection, (d) a chain of heteroclinic connections

sure that no thin escaping tails penetrate into the centre of the safe basin? The good news is that the penetration of the tails is always blocked by their accumulation onto an unstable saddle solution, heteroclinically tangled with D_h, that is called the *accessible orbit* (Grebogi *et al.* 1987). The bad news is that this accessible orbit is usually very hard to locate numerically. Under increase of F, the role of accessible orbit is taken by a sequence of different subharmonic saddles, allowing the tails to make a corresponding sequence of implosions into the safe regime (Figure 16.32). Here we see how a heteroclinic connection from A to B allows the fingers to implode: originally blocked by the inset of saddle A, they are next blocked by the inset of saddle B. The very bad news is that to determine the region that is unpermeated by the escaping fingers, we need to locate not only the accessible orbit, but also its entire inset manifold!

As we progress down the Dover cliff, the erosion process involves a chain of such events, shown schematically in Figure 16.33; this figure is based on the fine manifold studies of Lansbury *et al.* (1992). Just before the homoclinic tangency at F^M, where the basin boundary becomes fractal, we see an $n = 3$ fold cascade crisis scenario which leaves behind a directly unstable period 3 orbit. At the first homoclinic tangency of the hilltop saddle, D_h, there is a simultaneous

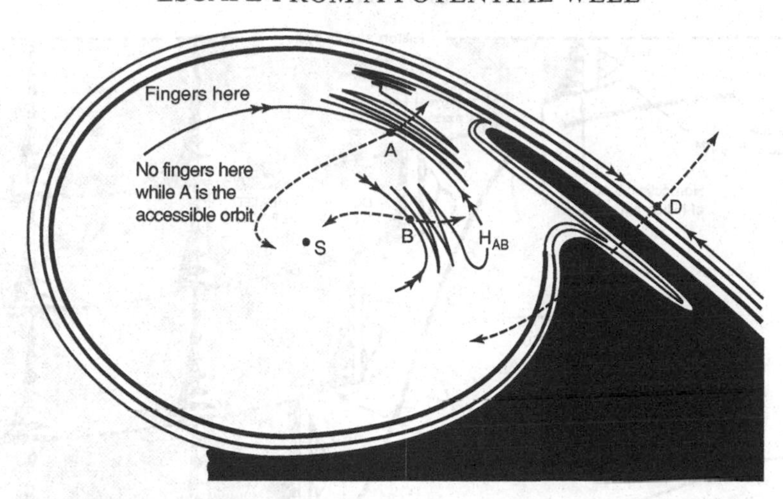

Figure 16.32 A sketch of a Poincaré section showing how the accessible orbits block the penetration of fractal fingers: D = the main governing (hilltop) saddle, whose homoclinic tangle has generated the black/white fractal basin boundary; A = the first accessible saddle orbit, whose inset blocks the penetration of the fractal fingers; B = the second accessible saddle orbit, which takes over the blocking role from A when the second heteroclinic connection H_{AB} is made; S = the central periodic attractor

heteroclinic connection between this orbit and D_h, which makes it the first accessible orbit. The fractal escaping fingers thus accumulate onto this orbit, and are thereby blocked from further encroachment into the safe basin. Subsequent progress down the Dover cliff is characterized by discontinuous changes in the accessible orbit. At the top of the cliff we see, for example, an $n = 5$ fold cascade crisis scenario, the fold of which creates instantaneously a small basin, for the short-lived $n = 5$ subharmonic attractor and its progeny, in the outer region of the safe basin.

When the subsequent period 5 chaotic attractor is destroyed at a crisis, this basin reverts temporarily to the main safe basin, but shortly afterwards there is a heteroclinic connection between the unstable $n = 5$ orbit created at the fold and the current accessible orbit. At this heteroclinic tangency the $n = 5$ subharmonic becomes the new accessible orbit onto which escaping fingers are accumulated and thereby blocked. This mechanism of implosion is repeated down the cliff face as shown. At each stage, the inset of the current accessible orbit blocks the encroaching escape fingers, dividing the safe basin into an outer domain, permeated by (though not completely filled by) escaping fingers, and an inner domain which is unpermeated and completely free from escape fingers. The subharmonics of this process are often located at the tips of the incursive fractal fingers, like the $n = 3$ subharmonic denoted by stars in Figure 16.19. This is nicely seen in figure 8 of Lansbury *et al.* (1992).

We should notice here that the heteroclinic chain that exists between the hilltop saddle and the current accessible orbit forms a chaotic saddle. It is the

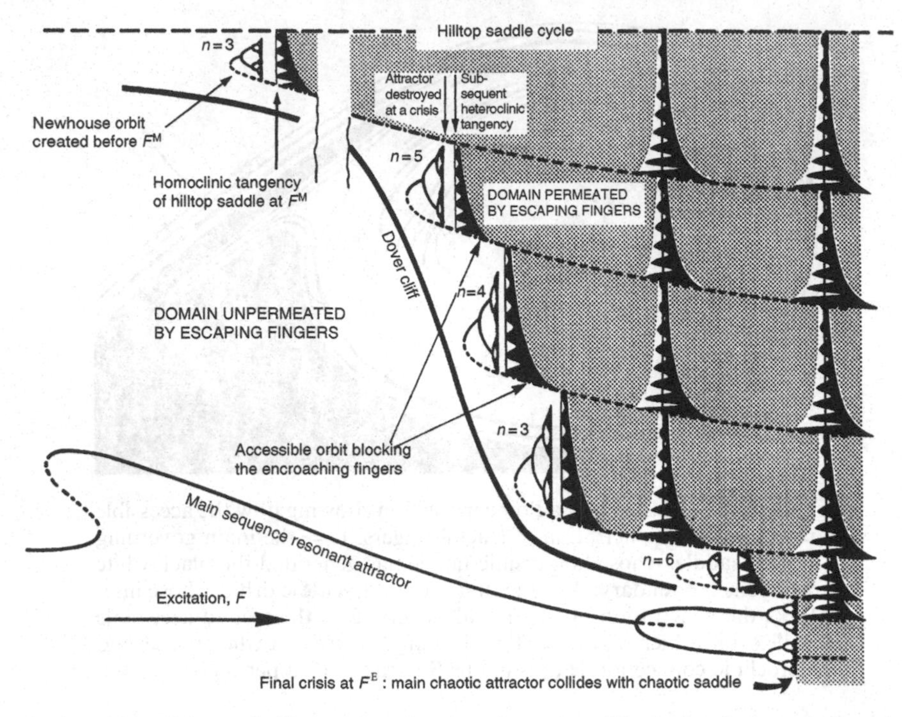

Figure 16.33 Schematic illustration showing the series of basin implosions during the Dover cliff erosion process; these implosions are associated with discontinuous changes of the accessible orbit. The diagram is based on numerical results for the twin-well oscillator

collision with the current accessible orbit (the collision with the D^6 subharmonic that we described in Section 16.4), and hence with this chaotic saddle (Stewart 1987), that finally destroys the main sequence chaotic attractor at the boundary crisis at F^E. This explains why the final main sequence chaotic oscillation is observed to be remote from the unstable hilltop saddle oscillation at the instant of its final crisis; see for example figure 12 of Thompson (1989). Poincaré sections just before and just after the cliff implosions are given by Lansbury *et al.* (1992). The straddle orbit location of a chaotic saddle is described by Mitsui *et al.* (1994).

Heteroclinic connection to the resonant saddle

We have just described the basin erosion process that corresponds to a value of the forcing frequency above that of the cusp P (Figure 16.15), where there is no regime of hysteresis. At frequencies below that of P, the existence of the resonant saddle, D_r, can complicate the story. In particular, the heteroclinic chain can reach D_r, giving it a heteroclinic link to the hilltop saddle, D_h. The heteroclinic tangency that creates this major heteroclinic connection is shown as arc H in Figure 16.15. This arc ends at point T, where it hits the line of fold A on which D_r is annihilated at a saddle-node bifurcation. Arc H, and its imagined extension beyond the fold line, gives a good estimate of the steepest region of the Dover cliff, which can be identified on the figure by the clustering of the contour lines. A heuristic explanation of this is outlined by Soliman and Thompson (1992a).

This heteroclinic tangency is illustrated in Figure 16.34. It corresponds to the first value of F for which a perturbation of D_r can reach D_h. Beyond this value of F, escaping fingers accumulate onto the resonant saddle, D_r, and some outwards perturbations of D_r lead to escape, while others still lead to the

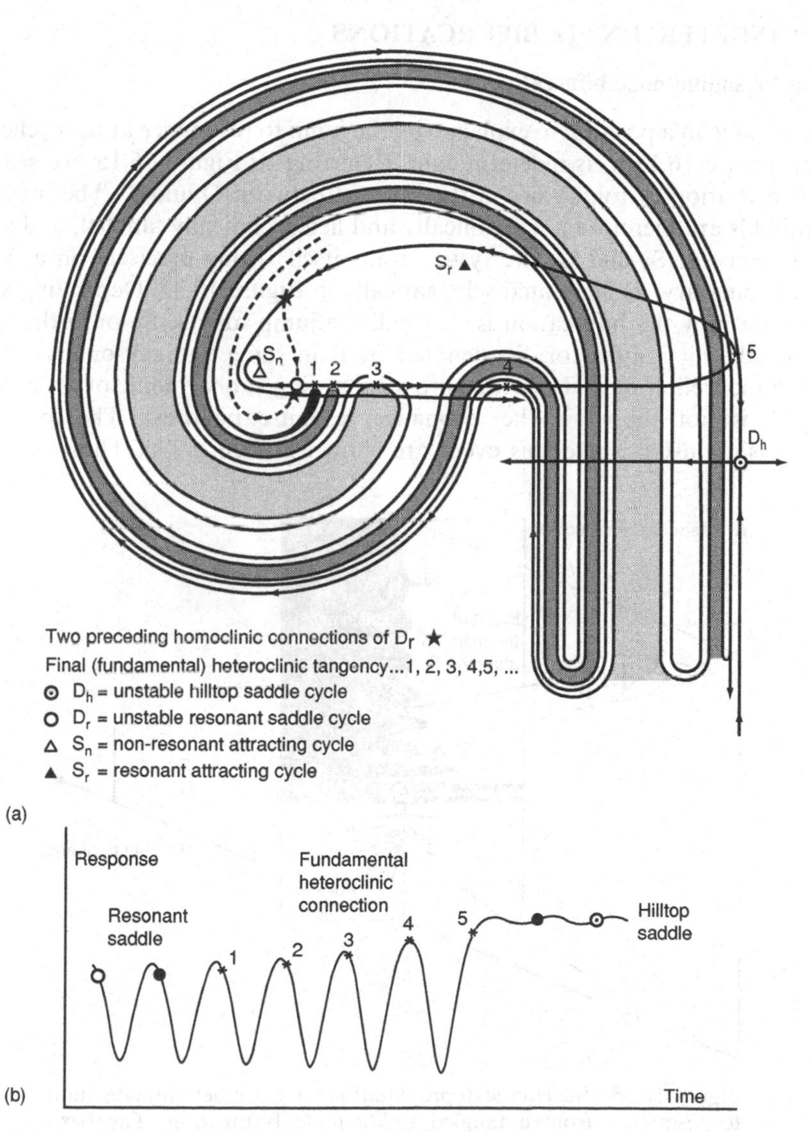

Figure 16.34 Sketch of the heteroclinic tangency between the resonant saddle and the hilltop saddle. The upper diagram shows the Poincaré section, while the lower diagram shows the corresponding time history

resonant attractor S_r. Once we are above the arc H in Figure 16.15, D_r is chained to D_h, and when D_r is subsequently annihilated at fold A, two important phenomena are observed. On the one hand, there is an instantaneous loss of safe basin area as fractal fingers sweep through the residual basin of the annihilated non-resonant attractor, S_n (as we shall see later in Figures 16.38 and 16.39). On the other hand, the jump from fold A, where the non-resonant attractor, S_n, collides with D_r, is indeterminate, as we shall now describe.

16.9 INDETERMINATE BIFURCATIONS

Tangled saddle-node bifurcation

We are now in a position to explain why the jump to resonance at the cyclic fold A in Figure 16.13(b) is indeterminate. Referring to Figure 16.15, we see that this realization of fold A lies in the interval between Q and T. The invariant manifolds are therefore homoclinically and heteroclinically tangled, and at the coalescence of S_n and D_r the system finds itself sitting precisely on a fractal basin boundary, as illustrated schematically in Figure 16.35. Depending sensitively on how the bifurcation is realized, the jump may settle onto the large-amplitude $n = 1$ attractor, S_r (denoted by R in some figures), or may escape out of the well; it may also settle onto a coexisting subharmonic of order $n = 3$ (which is not shown in the resonance response pictures). The manifolds and basins just prior to this event are shown in figure 7 of Thompson and

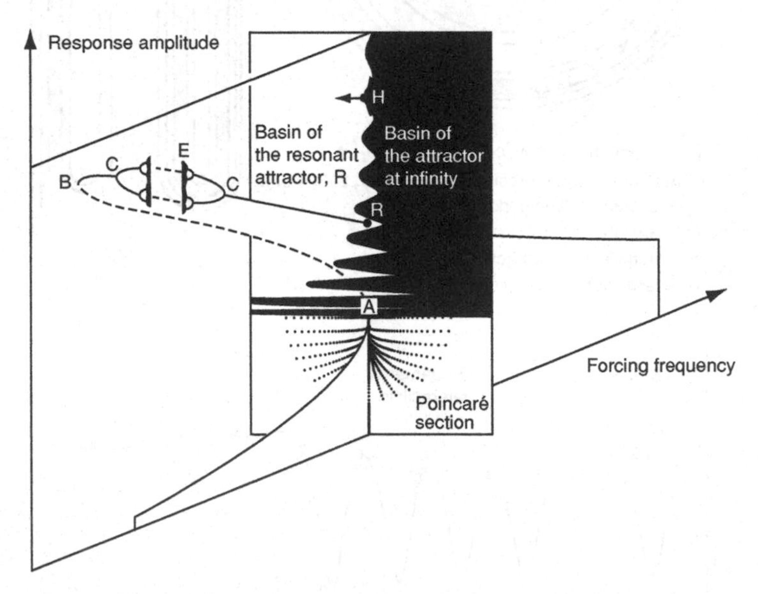

Figure 16.35 Schematic representation of an indeterminate jump to resonance from a tangled saddle-node bifurcation. The basin boundary is the tangled inset of the hilltop saddle, H, which accumulates on the saddle-node bifurcation, A. The jump to resonance from A may either stabilize at R or diverge to infinity

Soliman (1991), and the $x(t)$ time histories of three different jumps from fold A are shown in Figure 16.36.

As far as any local analysis is concerned, the saddle node that we are examining here is a perfectly regular one. The high degree of indeterminacy that arises when the system, under a slow sweep of either ω or F to the fold, finds itself sitting on a fractal basin boundary, can only be predicted or understood by an examination, as above, of the global manifold structures. In view of the homoclinic and heteroclinic tangling involved, we have called the event a tangled saddle-node bifurcation (Thompson and Soliman 1991; Soliman and Thompson 1991). It is a prime example of *final state sensitivity* (see Glossary).

This has implications for escape as illustrated in Figure 16.37. If the control parameters F and ω are swept slowly forward from $F = 0$ along the general loading arcs shown, there will be a zone of indeterminate response as indicated. In this zone we cannot predict whether escape will be on arc QT or on arc QT′; this is because of the indeterminate outcomes as we sweep slowly across the arc of the tangled fold A.

Striation of the residual basin

It is instructive to see what happens to the basins at this tangled saddle-node bifurcation, and the basins just before and just after fold A are shown in Poincaré stroboscopic section in Figure 16.38. In Figure 16.38(a) we see the

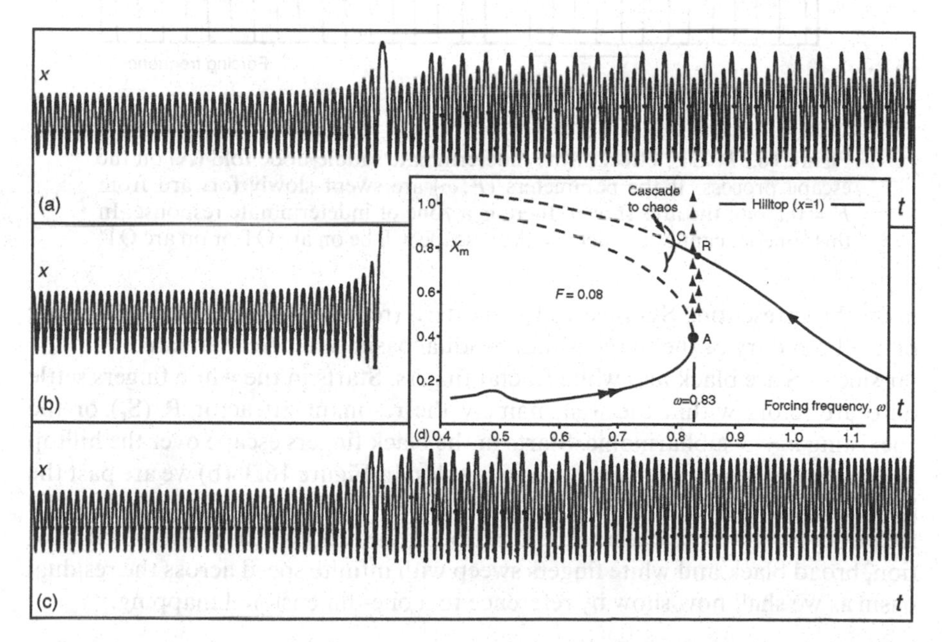

Figure 16.36 Three close realizations of the jump from the tangled saddle node, A: (a) a jump to a coexisting $n = 3$ attractor (a fugitive subharmonic), F is incremented from 0.08004 to 0.080068; (b) an escape to infinity, F is incremented from 0.08004 to 0.080069; (c) a jump to the $n = 1$ resonant attractor, F is incremented from 0.08004 to 0.080070; (d) maximum displacement versus forcing frequency. Here $F^{\mathrm{A}} \approx 0.08004$

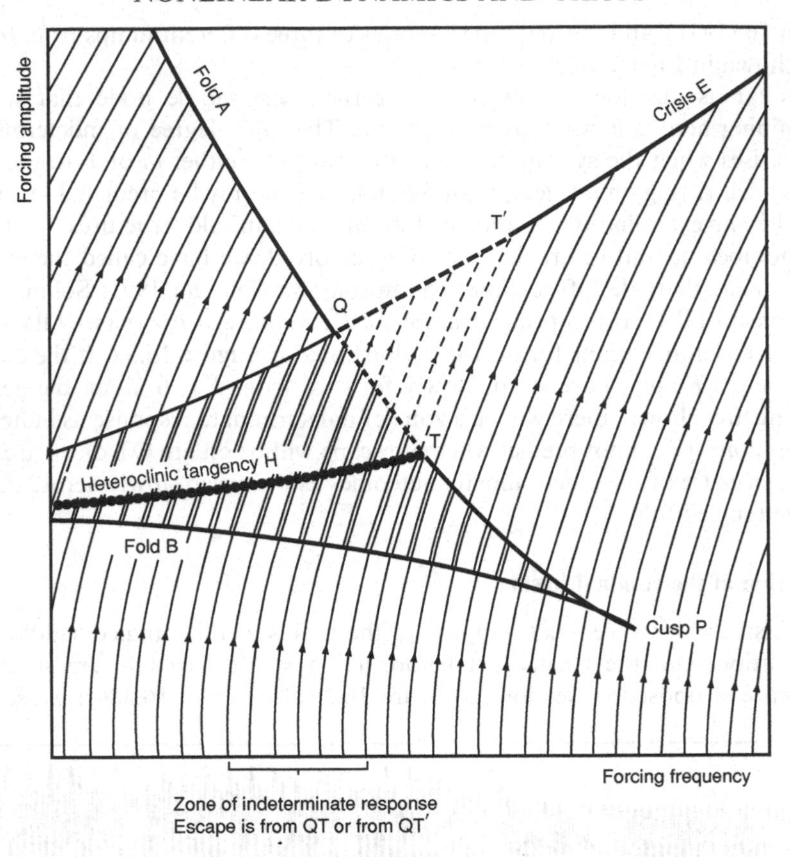

Figure 16.37 The effect of the indeterminate saddle-node fold (A) on the escape process. If the parameters (F, ω) are swept slowly forward from $F = 0$ along the arcs shown, there is a zone of indeterminate response. In this zone we cannot predict whether escape will be on arc QT or on arc QT′

node N (representing S_n) close to the saddle S (representing D_r), the latter lying on the boundary of the (large white) residual basin of N. Accumulated onto the far side of S are black and white fractal fingers. Starts in the white fingers settle onto attractors within the well, namely the resonant attractor R (S_r) or the coexisting $n = 3$ subharmonic. Starts in the black fingers escape over the hilltop within the displayed number of forcing cycles. In Figure 16.38(b) we are past the bifurcation (having made here a small change of F, rather than varying the frequency) and we can observe the striation of the residual basin. At the bifurcation, broad black and white fingers sweep with infinite speed across the residual basin as we shall now show by reference to a one-dimensional mapping.

Vertical cliff at the tangled saddle node

Now, bifurcation theory tells us that close to the saddle-node fold, the Poincaré mapping is essentially one-dimensional, so the mechanism of striation can be

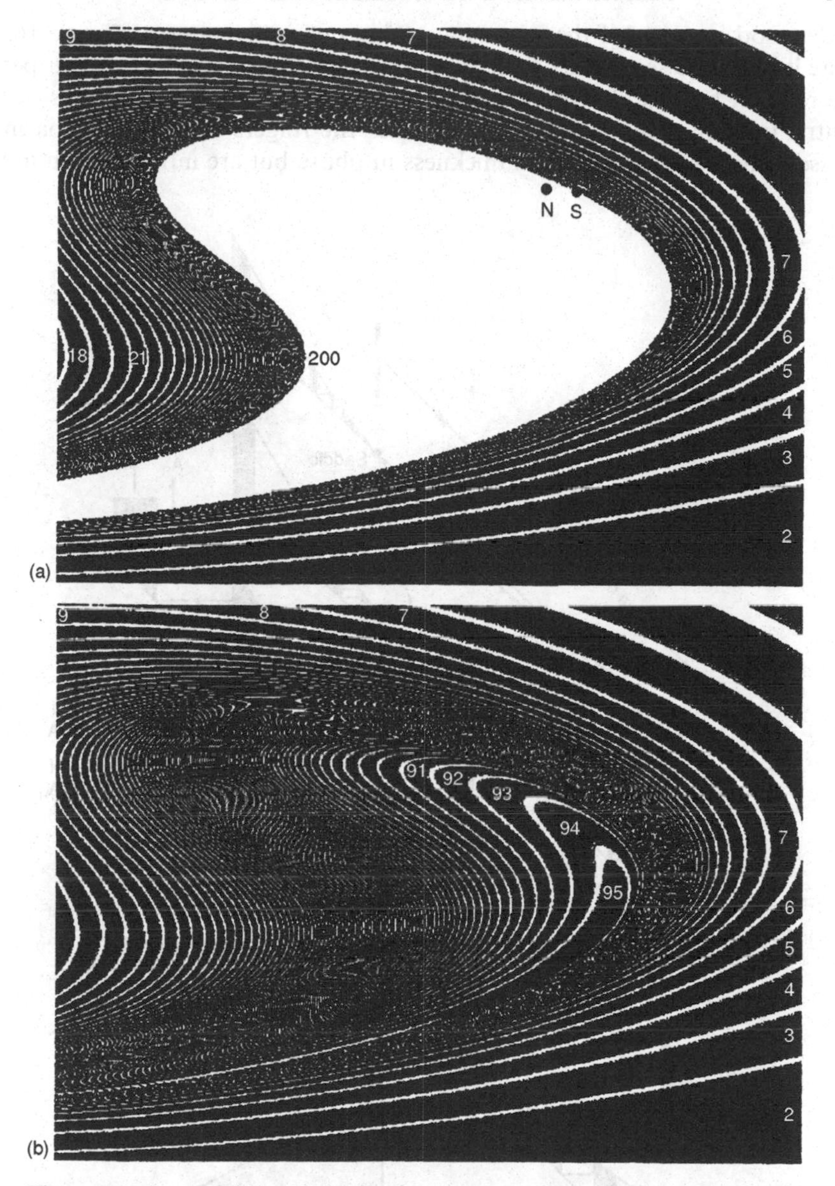

Figure 16.38 Striation of the residual basin of the tangled saddle-node bifurcation: the Poincaré section (a) just before the bifurcation ($F = 0.08$) and (b) just after the bifurcation ($F = 0.08008$). Here black denotes escape within the displayed number of forcing cycles. The main central white area in (a) is the residual basin of the node N; starts in the thin white stripes settle onto either the resonant harmonic attractor, R, or the coexisting $n = 3$ subharmonic

understood in terms of the schematic 1D mapping diagrams of Figure 16.39. Here it is easy to see that as F decreases towards its critical value, the stripes in the residual basin will acquire infinite speed due to the slow dynamics of the central tunnelling motion. This implies that the fingers of the master basin in phase-control space have finite thickness in phase but are infinitely thin in the

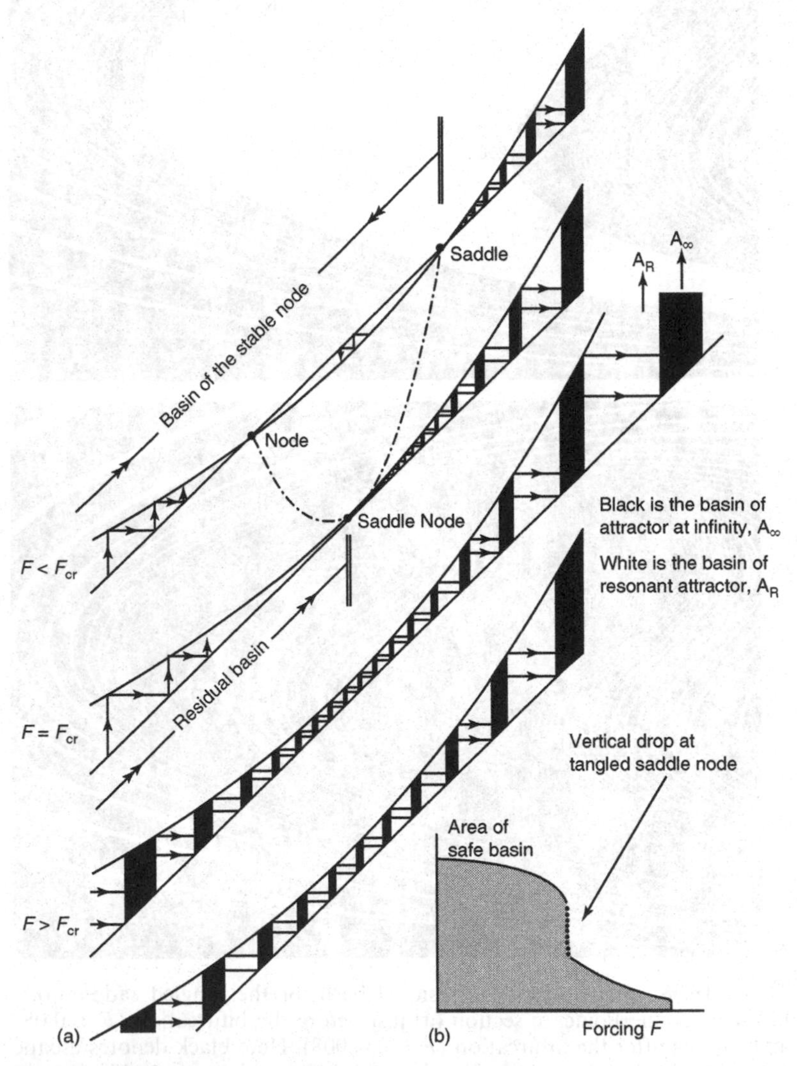

Figure 16.39 (a) The basin striation at a tangled saddle-node bifurcation using a one-dimensional mapping. As F decreases towards F_{cr}, the stripes in the residual basin acquire infinite speed. The stripes of the phase-control basins are then thick in phase but infinitely thin in control. The well-defined black fraction of the residual basin is instantaneously lost from the safe basin at the saddle node. (b) A well-defined fraction of the residual basin is lost from the safe basin, giving a vertical drop to an integrity curve

control direction. An analysis of this for a cyclic fold in an autonomous 2D flow is given by Thompson (1992).

The well-defined black fraction of the residual basin is instantaneously lost from the safe basin of the saddle node. This implies a vertical drop in the integrity diagram of safe basin area versus F at the tangled saddle-node bifurcation, as shown. Notice that such an instantaneous drop would only be observed if transient motions were allowed to run for an infinite length of time. Integrity curves plotted on the basis of escape within a finite number of forcing cycles would show a steep but not a vertical drop.

An indeterminate boundary crisis

In our discussion so far we have focused on the indeterminacy that exists in the jump to resonance from fold A. An entirely similar indeterminacy arises in the jump from resonance from the crisis E, as illustrated in Figure 16.40. Now, as we have shown, the jump from fold A is indeterminate over the interval from Q to T, namely once we are above arc TH where the resonant saddle, D_r, is chained to the hilltop saddle, D_h. In a similar way, the jump from crisis E is indeterminate over the interval from Q to S, once we are above arc SJ where the destroyer governing E is chained to D_h. This destroyer is the colliding unstable subharmonic saddle described earlier.

Thus the resonance response diagram at $F = 0.08$, shown in Figures 16.13(b) and 16.40(b), has two indeterminate jumps, at A and E respectively. Approaching E, under decreasing frequency, the system period-doubles to chaos and the jump from resonance at crisis E (a chaotic saddle catastrophe) is indeterminate by virtue of a fractal accumulation (Stewart and Ueda 1991; Stewart, Thompson *et al.* 1995). The jump from E either stabilizes onto the non-resonant attractor N (namely S_n) or escapes to infinity.

Indeterminacy in other oscillators

An extensive survey of parametrically excited systems by Mohamed Soliman has identified a number of new indeterminate bifurcations. Of these the subcritical and transcritical forms lie strictly outside the list of codimension 1 events, being rather special bifurcations generated by parametric excitation.

The indeterminate subcritical bifurcation (Soliman and Thompson 1992b) arises in parametrically excited oscillations within a cubic well when the driving frequency is approximately twice the linear natural frequency of the system; this is the regime of the main (principal) parametric resonance that is of greatest interest to engineers. Bifurcations in control space are shown in Figure 16.22(b), where the governing equation is displayed. Here we see a rather similar sequence of events to that of our escape equation in Figure 16.22(a). The regime of the indeterminate subcritical bifurcation extends from Q to T, where T is the intersection with an arc of heteroclinic tangency. Notice that the fractal fingers of transient escape extend down to T in both diagrams; this gives us a quick way to locate T, as discussed by Soliman (1994). Indeterminate

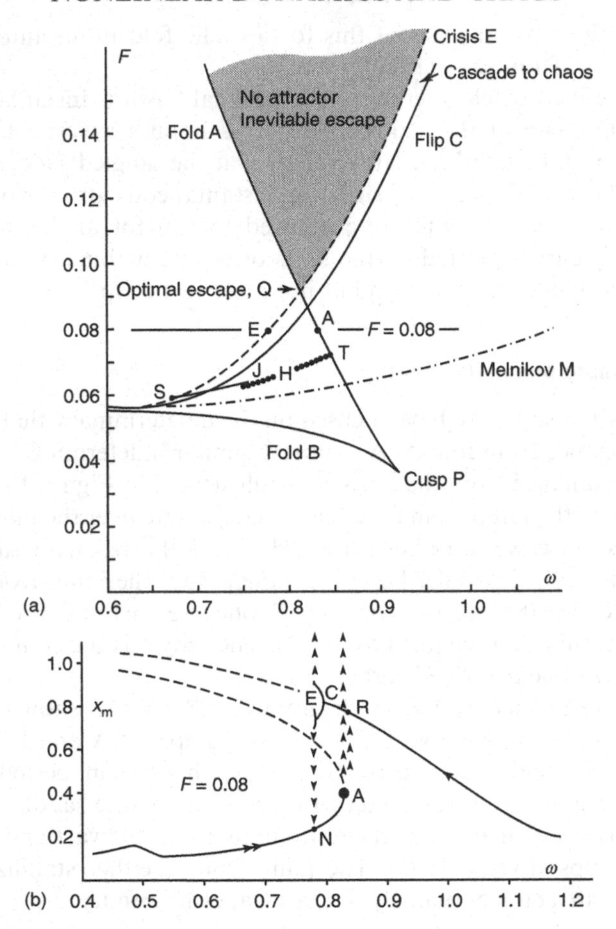

Figure 16.40 Indeterminate jumps to and from resonance in the escape equation. Arc TH: the resonant saddle, D_r, is chained to D_h. Arc SJ: destroyer governing E is chained to D_h. Fold A is indeterminate from T to Q; crisis E is indeterminate from S to Q. The indeterminacy at the crisis E is described in Stewart *et al.* (1995)

subcritical bifurcations have also been found in parametrically excited hardening systems (Soliman and Thompson 1996) with a restoring force equal to $x + x^3$; here there is no hilltop in the continuously hardening potential function.

Transcritical indeterminacies (Soliman and Thompson 1992c) have been identified in softening oscillators (governed again by the equation of Figure 16.22(b)) in the smaller resonance observed when the driving frequency is approximately equal to the linear natural frequency of the system. An example of an indeterminate subcritical flip has recently been observed by Mohamed Soliman for an oscillator in a cubic well subjected simultaneously to direct and parametric excitation.

Appendix

Codimension 1 Bifurcations

INTRODUCTION

In this phenomenological classification of the generic codimension 1 attractor bifurcations of dissipative dynamics we speak of a *forward* control sweep of a parameter, μ, as one which generates instability or increased complexity. Attention is focused on flows of continuous systems and their mappings generated by a Poincaré section. The minimum flow dimension necessary to observe a bifurcation in a Euclidean space is denoted by D, and descriptions assume that we are in this dimension. So a fold will be described on its centre manifold, as a repellor–attractor transition; embedding in a higher-dimensional phase space, with attraction onto this manifold, would give the more familiar terminology of a saddle-node bifurcation. Notice that although a flip can occur on a 2D Möbius strip, our Euclidean restriction necessitates $D = 3$.

Indeterminate bifurcations

Within the *dangerous* bifurcations of our classification we give some emphasis to the concept of bifurcational indeterminacy. Now, the most unpredictable response of a nonlinear dynamical system is one that starts nominally on a basin boundary. External noise and random perturbations can then nudge the trajectories to alternative attractors with totally different characteristics: one might be safe, while another might represent the collapse of an engineering or ecological system. Such a start might be thought unlikely, but under the slow sweep of a control there are generic bifurcations that bring a system precisely onto a basin boundary. These are the indeterminate dangerous events that give an unpredictable dynamic jump to one of two or more alternative coexisting attractors.

The simplest examples of potentially indeterminate bifurcations are the cyclic fold and the subcritical Hopf bifurcation. These both have a 2D outset, which allows generic connections with any number of saddles giving the possibility of an indeterminate jump to any number of coexisting attractors (Figure G.6 in the glossary). An analytical example of such a generic indeterminacy is given in Thompson (1992). In driven oscillators, cyclic folds can occur on fractal basin boundaries, giving the tangled saddle-node bifurcation that we examined in Chapter 16. Similar indeterminacies can also arise in the global dangerous bifurcations. When a chaotic attractor expands to hit a saddle cycle whose smooth, untangled inset forms its basin boundary, we have the regular saddle catastrophe (regular boundary crisis) for which indeterminacy is generically possible. Similarly, when a chaotic attractor expands to hit the accessible orbit within a fractal basin boundary, we have the chaotic saddle catastrophe (chaotic boundary crisis). Here generic indeterminacy with a fractal accumulation has been observed in the twin-well Duffing oscillator (Stewart and Ueda 1991), and in the escape equation of Chapter 16. Mohamed Soliman has made an extensive study of indeterminate bifurcations, finding examples of many of the generically possible forms (Soliman 1994; Soliman and Thompson 1992b, 1992c, 1996).

Abbreviations for books

Before starting on the classification, the reader should notice that under 'Pictures' we make abbreviated reference to four books: TB denotes this book; A&M denotes Abraham and Marsden (1978); A&S denotes Abraham and Shaw (1992); and G&H denotes Guckenheimer and Holmes (1983).

A SAFE BIFURCATIONS

- **Subtle** (i.e. continuous) bifurcations with the continuous supercritical growth of a new attractor path.
- **Safe** with no fast dynamic jump or instantaneous enlargement of the attracting set.
- **Determinate** with a single outcome even under small noise excitation.
- **No hysteresis** with attractor paths retraced on reversal of the control sweep.
- **No basin change**, with basin boundary remote from the bifurcating attractors.
- **No intermittency** in the steady-state responses of the attractors.

A.1 Local Supercritical Bifurcations

Supercritical Hopf ($D = 2$)

Point to cycle

A spiral point attractor becomes a spiral repellor as a complex conjugate pair of flow eigenvalues, $\lambda = \alpha \pm i\beta$, leaves the left-hand stable half-space. A stable supercritical periodic attractor expands parabolically around the primary path of point repellors. Here are two examples:

$$x' = -y + x[\mu - (x^2 + y^2)] \qquad y' = x + y[\mu - (x^2 + y^2)]$$

$$r' = r(\mu - r^2) \qquad \theta' = 1$$

- **Precursor**: local transients have the form $e^{\alpha t}\sin\beta t$ with the negative α increasing linearly through zero at C.
- **Other names**: in aeroelasticity, galloping or flutter; the mapping equivalent is the Neimark bifurcation.
- **Applications**: the galloping and flutter of elastic solids in a fluid flow, chemical oscillations.
- **Pictures**: TB fig. 7.5; A&M fig. 7.7.2; A&S (first excitation) fig. 17.1.8; G&H fig. 3.4.4.

Supercritical Neimark ($D = 3$)

Cycle to torus

A spirally attracting cycle becomes repelling as a complex conjugate pair of mapping eigenvalues, $\Lambda = \alpha \pm i\beta = \rho e^{\pm i\phi}$, leaves the stable unit disc. A stable supercritical toroidal attractor grows parabolically around the primary path of repellors. Special resonances occur when the eigenvalues satisfy $\Lambda^3 = 1$ or $\Lambda^4 = 1$.

- **Precursor**: local transients in a suitable polar map are $r_i = \rho^i r_0$, $\theta_i = \theta_0 + i\phi$ with ρ increasing linearly through $+1$.
- **Other names**: supercritical secondary Hopf bifurcation; the flow equivalent is the Hopf bifurcation.
- **Applications**: Taylor–Couette flow, internal autoparametric resonance in coupled driven oscillators.
- **Pictures**: TB fig. 8.18; A&M fig. 7.7.6; A&S (second excitation) fig. 17.2.6.

Supercritical flip ($D = 3$)

Cycle to cycle

An inversely attracting nodal cycle becomes an inverting saddle as a real mapping eigenvalue, Λ, leaves the stable unit disc at -1. A stable supercritical periodic attractor, with twice the period of the fundamental cycle, grows parabolically away from the saddles of the primary path. An example in a 1D map is

$$x_{i+1} = -(1 + \mu)x_i + x_i^3$$

- **Precursor**: Local mapping transients separate as Λ^i with Λ decreasing linearly through -1 at the bifurcation.
- **Other names**: supercritical period-doubling bifurcation, subharmonic resonance; the flip has no flow equivalent.
- **Applications**: subharmonic resonances in driven oscillators, and is a building brick of the Feigenbaum cascade.
- **Pictures**: TB figs 7.10, 8.6, 8.7 and 9.4; A&M fig. 7.7.4; A&S (octave jump) figs 17.3.7 and 17.4.5; G&H fig. 3.5.1.

A.2 Global Bifurcations

Band merging ($D = 3$)

Chaos to chaos

A chaotic attractor with noisy 2^n periodicity becomes a chaotic attractor with noisy 2^{n-1} periodicity on absorbing a period 2^{n-1} inverting saddle. These bifurcations, discussed by Lorenz, form the noisy reverse cascade in the logistic map which follows the more familiar Feigenbaum cascade.

- **Precursor**: the separation between adjacent chaotic bands decreases in a locally linear fashion.
- **Other names**: Lorenz reverse cascade.
- **Applications**: an ingredient of the universal period-doubling route to chaos, encountered in many fields.
- **Pictures**: TB (logistic map) fig. 9.8 and G.7 A&S (chaotic octave jump) fig. 22.2.2.

B EXPLOSIVE BIFURCATIONS

- **Catastrophic** (i.e. discontinuous) global bifurcations with an abrupt enlargement of the attracting set.
- **Explosive** enlargement, but no jump to remote disconnected attractor.
- **Determinate** with a single outcome even under small noise excitation.
- **No hysteresis** with attractor paths retraced on reversal of control sweep.
- **No basin change**, with basin boundary remote from the bifurcating attractors.
- **Intermittency**: supercritical lingering in old domain, flashes through the new domain.

Flow explosion ($D = 2$)

Point to cycle

A path of equilibrium fixed points exhibits locally a regular saddle-node fold; meanwhile the global dynamics are such that the saddle outset flows around a closed loop to the node; when the saddle and node on this loop annihilate one another at the fold, a stable limit cycle is created; initial period is infinite, due to critical slowing.

- **Precursor**: the local subcritical behaviour is identical to that of the static fold.
- **Other names**: omega explosion; the mapping equivalent is the map explosion.
- **Applications**: chemical oscillations, low-temperature semiconductors.
- **Pictures**: TB fig. 13.3; A&S (blue loop in 2D) fig. 21.1.6; G&H fig. 3.4.5.

Map explosion ($D = 3$)

Cycle to torus

This mapping equivalent of the flow explosion has a saddle-node annihilation of two cycles; these lie on a drift ring in the Poincaré section, so that a toroidal attractor is created after the collision; critical slowing is observed in the toroidal flow as the mapping point passes through the region of the recent cyclic fold.

- **Precursor**: the local subcritical behaviour is identical to that of the cyclic fold.
- **Other names**: omega explosion, mode locking, mode unlocking; the flow equivalent is the flow explosion.
- **Applications**: arises in the mode locking and unlocking of oscillators with two intrinsic frequencies.
- **Pictures**: TB figs 13.2 and 13.3; A&S (blue loop in 3D) fig. 21.2.5.

Intermittency explosion: flow ($D = 4$)

Point to chaos

According to a geometrical model that has been proposed in 4D, a subcritical Hopf bifurcation can give not a *dangerous* jump to a remote attractor, but an *explosive* enlargement of the attracting set from a point to a chaotic attractor containing the unstable focus. It is not clear whether this is possible in 3D.

- **Precursor**: the subcritical transient behaviour is identical to that of the Hopf bifurcation.
- **Other names**: none.
- **Applications**: none.
- **Pictures**: none.

Intermittency explosion: map ($D = 3$)

Cycle to chaos

As in the map explosion, a cyclic fold can give not a *dangerous* jump, but an *explosive* enlargement of the attracting set; when this is to a chaotic attractor (as in the opening of a periodic window), critical slowing gives an irregular intermittency in the chaotic motion. Subcritical Neimark and flip bifurcations can replace the fold.

- **Precursor**: the subcritical transient behaviour is identical to that of the constituent local bifurcation.
- **Other names**: temporal intermittency, Pomeau–Manneville (types I, II, III) intermittency.

- **Applications**: arises as the opening of a periodic window in the ubiquitous period-doubling route to chaos.
- **Pictures**: TB figs 13.7 and 13.8, (opening of window in logistic map) figs 9.7 and 9.9; A&S (Zeeman's blue tangle) fig. 21.3.4.

Regular-saddle explosion ($D = 3$)

Chaos to chaos

A chaotic attractor has an abrupt enlargement to a larger chaotic attractor which includes the smaller as a subset, on colliding with a regular saddle cycle. The larger attractor has a global annular structure, as in the bagel or folded torus. The regular-saddle and chaotic-saddle explosion pair has an analogy in the catastrophes.

- **Precursor**: subcritical lingering near the impinging saddle, significant when the saddle is only weakly repelling.
- **Other names**: (regular) interior crisis (Grebogi *et al.* 1983b).
- **Applications**: explosion from band attractors to bagel (folded torus) in the velocity-forced Van der Pol equation.
- **Pictures**: TB fig. 13.9.

Chaotic-saddle explosion ($D = 3$)

Chaos to chaos

A chaotic attractor undergoes an abrupt instantaneous enlargement to a larger chaotic attractor which includes the original attractor as a subset, on colliding with a chaotic saddle (Ueda 1980b; Grebogi *et al.* 1983b; Lai *et al.* 1992). A simple example of this is seen in the closing of a periodic window of the logistic map.

- **Precursor**: subcritical lingering near the impinging saddle, significant when the saddle is only weakly repelling.
- **Other names**: (chaotic) interior crisis (Grebogi *et al.*1983b).
- **Applications**: escape from chaotic one-well to chaotic cross-well motions in the twin-well Duffing oscillator.
- **Pictures**: TB (closing of periodic window in logistic map) fig. 9.9; A&S (Ueda's chaotic explosion) fig. 21.4.9.

C DANGEROUS BIFURCATIONS

- **Catastrophic** (i.e. discontinuous) bifurcations with the blue sky disappearance of the attractor.
- **Dangerous** with a sudden fast dynamic jump to a distant unrelated attractor of any type.

- **Determinate or indeterminate** in outcome, depending on the global topology.
- **Hysteresis** with original attractor not reinstated on reversal of the control sweep.
- **Basin** shrinks to zero (C.2) or attractor hits boundary of a residual basin (C.1 and C.3).
- **No intermittency** but note the critical slowing in the global bifurcations.

C.1 Local Saddle-Node Bifurcations

Static fold ($D = 1$)

From point

A path of point attractors folds back parabolically as a path of repellors on reaching an extreme value of the control. Moving around the path, a real flow eigenvalue λ leaves the left-hand stable half-space at the node–repellor transition. An example is

$$x' = \mu + x^2$$

The related pitchfork (cusp) and transcritical bifurcations are not directly generic.

- **Precursor**: local transients have the form $e^{\lambda t}$, with the real negative λ increasing linearly through zero at C.
- **Other names**: saddle-node bifurcation; in elasticity, limit point; mapping equivalent is below.
- **Applications**: snap-buckling of shallow elastic structures, including arches and shells; runaway of reactors.
- **Indeterminacy**: in $D = 2$ the 1D flow outset needs a non-generic coincident saddle connection for indeterminacy.
- **Pictures**: TB figs 5.14 and 7.4; A&M fig. 7.7.1; A&S (static fold) figs 18.1.10 and 18.2.6; G&H fig. 3.4.1.

Cyclic fold ($D = 2$)

From cycle

A path of periodic attractors folds back parabolically as a path of periodic repellors on reaching an extreme value of the control. Moving around the path a real mapping eigenvalue, Λ, leaves the stable unit disc at $+1$ as the stable cycle becomes unstable at the extremum. An example in a one-dimensional map is

$$x_{i+1} = x_i + \mu + x_i^2$$

- **Precursor**: local mapping transients separate as Λ^i with Λ increasing linearly through $+1$ at the extremum.

- **Other names**: saddle-node bifurcation, dynamic fold, periodic fold; flow equivalent is previous entry.
- **Applications**: jumps to and from resonance in driven oscillators under direct and parametric excitation.
- **Indeterminacy**: smooth mosquito coils (Stewart and Ueda 1991; Thompson 1992) or fractal basins (Thompson and Soliman 1991) can accumulate onto saddle's remote outset.
- **Pictures**: TB figs 7.9 and 8.3 (indeterminate) G.6, (tangled) 16.35; A&M fig. 7.7.3; A&S (periodic fold) figs 18.3.10 and 18.4.9.

C.2 Local Subcritical Bifurcations

Subcritical Hopf ($D = 2$)

From point

A spiral point attractor becomes a spiral repellor as complex conjugate flow eigenvalues, $\lambda = \alpha \pm i\beta$, leave the left-hand stable half-space. An unstable subcritical periodic repellor shrinks parabolically around the attractor, pinching its basin of attraction to zero. Two examples are

$$x' = -y + x[\mu + (x^2 + y^2)] \qquad y' = x + y[\mu + (x^2 + y^2)]$$

$$r' = r(\mu + r^2) \qquad \theta' = 1$$

- **Precursor**: local transients have the form $e^{\alpha t} \sin \beta t$ with the negative α increasing linearly through zero at C.
- **Other names**: in aeroelasticity, galloping or flutter; the mapping equivalent is the Neimark bifurcation.
- **Applications**: the galloping and flutter of elastic solids in a fluid flow, onset of turbulence.
- **Indeterminacy**: an example is aeroelastic galloping in an asymmetric well (Thompson 1992).
- **Pictures**: TB fig. 7.5 (indeterminate) G.6; A&S (spiral pinch) fig. 19.1.4.

Subcritical Neimark ($D = 3$)

From cycle

A spirally attracting cycle becomes repelling as a complex conjugate pair of mapping eigenvalues, $\Lambda = \alpha \pm i\beta = \rho e^{\pm i\phi}$, leaves the stable unit disc. An unstable subcritical toroidal repellor shrinks parabolically around the attractor, pinching its basin to zero. Special resonances occur when $\Lambda^3 = 1$ or $\Lambda^4 = 1$.

- **Precursor**: local transients in a suitable polar map are $r_i = \rho^i r_0$, $\theta_i = \theta_0 + i\phi$ with ρ increasing linearly through $+1$.
- **Other names**: subcritical secondary Hopf bifurcation; the flow equivalent is the Hopf bifurcation.

- **Applications**: electrical circuits.
- **Indeterminacy**: generically possible but no specific example is known to the authors.
- **Pictures**: TB (reversed time) fig. 8.18; A&S (vortical pinch) fig. 19.2.6.

Subcritical flip ($D = 3$)

From cycle

An inversely attracting nodal cycle becomes an inverting saddle as a real mapping eigenvalue, Λ, leaves the stable unit disc at -1. An unstable subcritical periodic saddle, with twice the period of the fundamental cycle, shrinks parabolically onto the attractor, pinching its basin of attraction to zero. An example is

$$x_{i+1} = -(1+\mu)x_i - x_i^3$$

- **Precursor**: local mapping transients separate as Λ^i with Λ decreasing linearly through -1 at the bifurcation.
- **Other names**: subcritical period-doubling bifurcation; the flip has no flow equivalent.
- **Applications**: Rayleigh–Bénard convection, subharmonic resonance.
- **Indeterminacy**: escape from a cubic potential well under simultaneous direct and parametric excitation.
- **Pictures**: TB fig. 8.8; A&M fig. 7.7.5; A&S (octave pinch) figs 19.3.6 and 19.4.6.

C.3 Global Bifurcations

Saddle connection ($D = 2$)

From cycle

A stable cycle expands towards a saddle fixed point whose inset forms the boundary of its basin of attraction; the period of the cycle goes to infinity as it touches the saddle in a homoclinic saddle connection; at the connection the cycle is its own basin boundary, and beyond it the cycle and basin vanish into the blue.

- **Precursor**: the period of the cyclic attractor tends to infinity due to slow dynamics near the approaching saddle.
- **Other names**: homoclinic connection, separatrix loop; no mapping equivalent because connection becomes a tangle.
- **Applications**: chemical oscillations.
- **Indeterminacy**: in $D = 2$ the 1D flow outset needs a non-generic coincident saddle connection for indeterminacy.
- **Pictures**: TB figs 13.4 and 13.5; A&S (periodic blue sky bifurcation) fig. 20.2.7; G&H fig. 4.4.2.

Regular-saddle catastrophe ($D = 3$)

From chaos

A chaotic attractor expands to hit a saddle cycle whose smooth, untangled inset forms its basin boundary; at the collision the saddle simultaneously becomes homoclinic and the chaotic attractor and its residual basin vanish into the blue. A simple example is seen at the end of the logistic map.

- **Precursor**: subcritical lingering near the impinging saddle, significant when the saddle is only weakly repelling.
- **Other names**: (regular) boundary crisis (Grebogi *et al.* 1983b), chaotic blue sky catastrophe.
- **Applications**: blue sky instability of Birkhoff–Shaw folded torus in asymmetrically forced Van der Pol equation.
- **Indeterminacy**: generically possible with a smooth mosquito-coil accumulation, but no example known.
- **Pictures**: TB fig. 13.11, (end of logistic map) fig. 9.8 and G.7; A&S (chaotic blue sky bifurcation) fig. 20.3.11.

Chaotic-saddle catastrophe ($D = 3$)

From chaos

A chaotic attractor expands to hit the accessible saddle orbit within a fractal basin boundary, but remaining at a distance from the main saddle cycle whose prior homoclinic tangling generated the fractal boundary. At the collision the chaotic attractor and its residual basin vanish into the blue (Stewart 1987).

- **Precursor**: subcritical lingering near the impinging saddle, significant when the saddle is only weakly repelling.
- **Other names**: (chaotic) boundary crisis (Grebogi et al. 1983b), chaotic blue sky catastrophe.
- **Applications**: escape from a cubic potential well where bounded chaotic motions suddenly jump out of the well (as described in chapter 16).
- **Indeterminacy**: determinate and indeterminate (fractal accumulation) examples in the twin-well Duffing equation (Stewart and Ueda 1991) (see also Figure 16.40).
- **Pictures**: TB fig. 13.12; A&S (Rössler's blue sky bifurcation) fig. 20.4.8.

Illustrated Glossary

We give here an informal, encyclopedic glossary of terms and formulae covering the new geometrical concepts of nonlinear dynamics and chaos. It is arranged as an alphabetical dictionary, which can also be read as a connected introduction to the subject by following the *go to* instructions starting with the entry *dynamical system* and ending with *archetypal maps*. The reader who follows this *guided tour* through the main entries, and pursues all the *see also* instructions will thereby cover all (substantive) entries. Boldface is used to draw particular attention to a related entry, while italics are used in the normal manner for emphasis, etc.

The guided tour starts with the types of dynamical system, and their phase spaces: flows and maps are introduced, the latter often arising from a Poincaré section. Emphasis is given to the negative phase-space divergence generated by energy dissipation in macroscopic real-world applications: this gives rise to the first important concept of an *attractor*. Liapunov stability theory is outlined, leading to an examination of the fixed points of flows and maps including the regular attractors, repellors and saddles. The chaotic post-transient states are next introduced, together with their characterisation using Liapunov exponents and fractal dimensions. The second major concept is that of a *basin* of attraction, and the manifolds commonly forming its boundary are described. The fractal basin boundary, subject of much recent research, is described.

The third important concept is that of a *bifurcation*, typically encountered at a structurally unstable phase portrait under the slow sweep of a control parameter. The physical outcome of such a sweep can form the basis of a useful phenomenological classification into subtle, catastrophic-explosive and catastrophic-dangerous forms. The local bifurcations, and their eigenvalue characterisations, are presented. In the discourse on global bifurcations, emphasis is given to the homoclinic and heteroclinic tangles that are so important in driven oscillators where they govern the fractal basin boundaries and indeterminate bifurcations such as the tangled saddle-node bifurcation. The guided tour through the glossary ends with a listing of the main *archetypal equations* that have played such an important role in the development of the subject. More information about many of the entries can be found in *A Dictionary of Nonlinear Dynamics* (Kapitaniak and Bishop 1999).

TOUR OUTLINE

Dynamical system autonomous, vector field, phase space and portrait, flow, invariants (two types)
Time integration Euler time integration, Runge–Kutta
Poincaré section and map map and mapping, diffeomorphism, stroboscopic sampling
Divergence
Dissipation Hamiltonian and conservative systems, Liouville's theorem, volume preserving
Recurrent state and behaviour steady-state, transient, non-wandering state, wandering
Stability Liapunov stability, hyperbolic point
Attractor
Point attractor fixed point of a flow, equilibrium point
Periodic attractor limit cycle, harmonic and subharmonic oscillations, fixed point of a map
Quasi-periodic attractor incommensurate, rotation number, winding number, Arnold horns
Instability index of instability, saddle, repellor, index (Poincaré)
Chaotic attractor strange attractor, chaos, period doubling, Feigenbaum, cascade, embedding dimension
Chaotic saddle
Chaotic transient
Chaotic repellor
Sensitive dependence on initial conditions Liapunov exponent
Fractal Cantor set, fractal, capacity, Hausdorff and information dimensions
Symbolic dynamics horseshoe
Basin
Basin boundary outstructures, inset, outset, separatrix
Invariant manifolds stable manifold, unstable manifold
Fractal basin boundary final state sensitivity
Control parameters and space phase-control space
Structural stability coarse system
Bifurcation codimension, unfolding
Bifurcational classification subtle bifurcation, catastrophic bifurcations (two forms), hysteresis, intermittency
Local bifurcation centre manifold
Saddle-node bifurcation fold
Supercritical bifurcation
Subcritical bifurcation
Transcritical bifurcation
Hopf bifurcation
Flip bifurcation
Neimark bifurcation secondary Hopf bifurcation
Global bifurcation crisis

Homoclinic connection saddle connection, homoclinic tangency, homoclinic tangle, tangle
Melnikov theory
Shilnikov homoclinic connection
Newhouse sinks
Heteroclinic connection heteroclinic tangency, heteroclinic tangle
Determinate bifurcation indeterminate bifurcation
Archetypal differential equations Duffing, Van der Pol, Lorenz
Archetypal maps logistic, quadratic, Hénon, circle, standard

THE GLOSSARY

Archetypal differential equations

These are carefully chosen differential equations with simple nonlinearities, representative of a wide class, that can be studied in detail to reveal characteristics typical of the class. *See also* Duffing's equation, Van der Pol equation, Lorenz equations. *Go to* archetypal maps.

Archetypal maps

These are carefully chosen iterative mappings with simple nonlinearities, representative of a wide class, that can be studied in detail to reveal characteristics typical of the class. *See also* logistic map, quadratic map, Hénon map, circle map, standard map.

Arnold horns or tongues

Arnold horns (tongues) are tongue-like regimes in the control space of generic two-frequency oscillators—as modelled for example by the **circle map**—in which the two periodic components of the response are mode-locked into periodic motion. Outside these regimes there is drift, corresponding for example to **quasi-periodic** motion. The tongue boundaries are cyclic folds. *See* quasi-periodic attractor.

Attractor

In a system with **dissipation**, the negative **divergence** ensures that a cloud or ensemble of starts will shrink onto an attracting set of zero phase volume. Such a post-transient set can be a point attractor, a periodic or quasi-periodic attractor, or a **chaotic attractor**. Generically each attractor is entirely surrounded in phase space by its own **basin** of attraction. Consequently, all transient motions initialized in a small neighbourhood around the attractor move asymptotically back to it, making it *asymptotically stable* in the sense of Liapunov. *Go to* point attractor.

Autonomous

A continuous dynamical system described by a set of ordinary differential equations is autonomous if time does not appear explicitly in the equations. Non-autonomous equations can be rendered autonomous by identifying the time as an extra phase coordinate governed by the dummy equation $t' = 1$. *See* dynamical system.

Basin

The **phase space** of a dissipative dynamical system decomposes into basins of attraction bordered and separated from each other by basin boundaries. Trajectories initialized within a basin tend asymptotically, as time goes to plus infinity, to the **attractor** lying within the basin. *Go to* basin boundary.

Basin boundary

Trajectories initialized precisely on a basin boundary often flow towards a **saddle** solution which attracts within the boundary but repels across it (Figure G.1). The boundary is then formed by the **inset** (stable manifold) of the saddle and can be located computationally by running time backwards from one or more starts close to the saddle. *See also* outstructures, separatrix. *Go to* invariant manifolds.

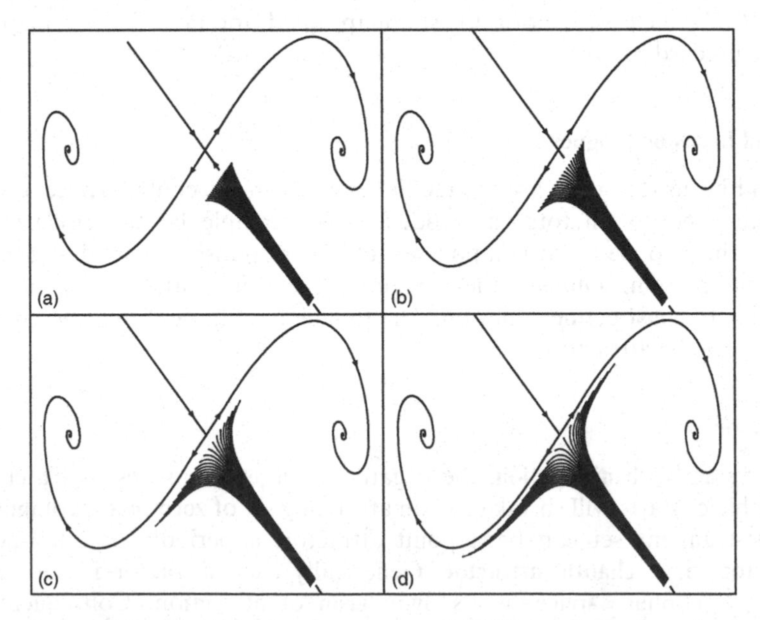

Figure G.1 The flow in phase space straddling a basin boundary. Varying the duration of the time integrations emphasizes the continuity of end-states in finite time: (a) $T = 300$, (b) $T = 400$, (c) $T = 500$, (d) $T = 600$

Bifurcation

A bifurcation is a qualitative change in the topology of the attractor-basin *phase portrait*, realisable under the quasi-static variation of a control parameter μ across a critical value of $\mu = \mu_c$. The phase portrait at μ_c is **structurally unstable**. **Local bifurcations**, restricted to a small neighbourhood of phase space, involve the creation, destruction and splitting of attractors. **Global bifurcations** involve connections between the **outstructures** of distant saddles, often producing abrupt changes in the basin structure. *See also* codimension, unfolding. *Go to* bifurcational classification.

Bifurcational classification (see also Appendix)

We classify the **codimension 1** bifurcations of attractors according to the response that would be observed physically if a control parameter were slowly varied to sweep the system through the bifurcation in the direction that generates instability or increased complexity (Figure G.2). *Subtle* (i.e. continuous) bifurcations are the local **supercritical** bifurcations with the continuous growth of a new attractor path. They are *safe* with no fast dynamic jump or instantaneous enlargement of the attracting set. They are *determinate* with a single outcome, and generate no **hysteresis** on reversal of the control sweep. The *catastrophic-explosive* bifurcations (our first subdivision of the discontinuous bifurcations) involve a sudden, instantaneous enlargement of the attracting set, but with no jump to a remote disconnected attractor. They are *determinate* in outcome and generate no *hysteresis*. The *catastrophic-dangerous* bifurcations (our second subdivision) have a blue sky disappearance of the initial attractor, giving a dangerous fast dynamic jump to a distant unrelated attractor;

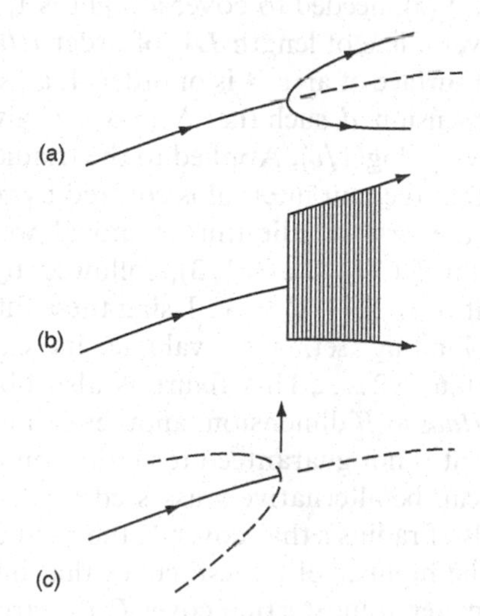

Figure G.2 Bifurcations can usefully be classified according to their outcome. Here we show examples of (a) safe, (b) explosive and (c) dangerous bifurcations

examples are the **subcritical** local bifurcations. They can be *determinate* or *indeterminate* in outcome, depending on the global topology. The original attractor is not reinstated on reversal of the control sweep, often generating a *hysteresis* cycle. *See also* subtle bifurcation, catastrophic-explosive bifurcation, catastrophic-dangerous bifurcation, hysteresis, intermittency. *Go to* local bifurcation.

Cantor set

A Cantor set is an infinitely desiccated set such as the triadic Cantor set. The latter is generated by removing the middle third from the unit interval, and then progressively removing the middle thirds from the remaining intervals until in the limit we are left with just an (uncountable) infinity of disconnected points. Since the total length of line removed is $(1/3) + 2(1/3)^2 + 4(1/3)^3 + 8(1/3)^4 + \ldots = 1$, the length or *measure* of the remaining points is zero; on the other hand, the points remaining can be put in one-to-one correspondence with all points in the original interval, so the cardinality is not reduced. Because of their distribution in space, the points can, however, be assigned a dimension; their **capacity dimension** can for example be evaluated as $(\log 2)/(\log 3) = 0.63092\ldots$, a non-integer result between that of a point (0) and a line (1). The final set of points has the attributes of a **fractal**, with every magnification revealing more detail, which in this simple example is always self-similar. *See* fractal.

Capacity dimension

The capacity or box dimension is based on the covering of a set, as we shall illustrate using cubes of side ε to cover an object in a three-dimensional space. The number of cubes, $N(\varepsilon)$, needed to cover a point is 1, so $N(\varepsilon)$ here scales as ε^0. The number to cover a line of length L is of order L/ε, scaling as ε^{-1}, while the number to cover a surface of area A is of order A/ε^2, scaling as ε^{-2}. We thus define the capacity dimension, d, such that $N(\varepsilon) \propto \varepsilon^{-d}$, giving d as the limit (as ε tends to zero) of $\log N(\varepsilon)/\log(1/\varepsilon)$. Applied to the triadic **Cantor set** in a plane (Figure G.3), we see that the unit interval is covered by one square of unit side ($N = 1, \varepsilon = 1$): after the first middle-third removal we can cover with two squares of side one-third ($N = 2, \varepsilon = 1/3$), followed by $N = 2^2$, $\varepsilon = (1/3)^2$, with the general result $N = 2^i$, $(1/\varepsilon) = 3^i$. Using these differently sized squares to cover the *final* limiting set, we evaluate its capacity dimension as $d = (\log 2)/(\log 3) = 0.63092\ldots$. This figure is also obtained for the triadic set if we evaluate its *Hausdorff* dimension, another commonly employed measure, but such agreement is not guaranteed for more complex sets. The *capacity dimension* of a set F can be alternatively assessed by determining the smallest number of closed balls of radius ε that cover F; the smallest number of cubes of side ε that cover F; the number of ε-mesh cubes that intersect F; the smallest number of sets of diameter at most ε that cover F; the largest number of disjoint balls of radius ε with centres in F (Falconer 1990, p. 41). *See* fractal.

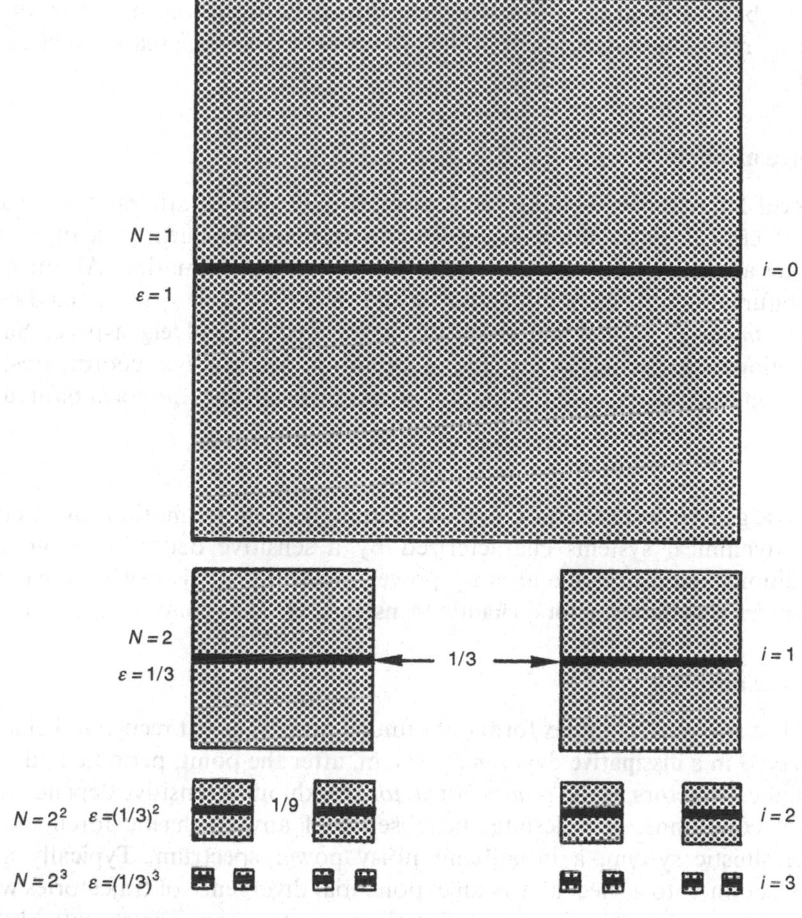

Figure G.3 Method to determine the fractal (capacity) dimension of the triadic Cantor set

Cascade

The name given to the infinite sequence of period-doubling bifurcations that is the most familiar precursor to chaos. *See* chaotic attractor.

Catastrophic-dangerous bifurcation

In our bifurcational classification these bifurcations are those that give rise to a fast dynamic jump to a remote attractor. *See* bifurcational classification.

Catastrophic-explosive bifurcation

In our bifurcational classification these bifurcations are those that involve an abrupt, reversible enlargement of the attracting set. *See* bifurcational classification.

Centre manifold

A **local bifurcation** involving the loss of stability of an attractor will have a set of critical eigenvalues for which the response is linearly *neutral*, being neither attracting nor repelling within the linear approximation. All the essential bifurcational behaviour can be viewed in the reduced space defined by the *centre manifold* originating from the associated critical eigenspace. Similar reductions are achieved by the elimination of passive coordinates, the Fredholm alternative and the Liapunov–Schmidt scheme. *See* local bifurcation.

Chaos

A loose generic term for the complex, seemingly irregular motions of deterministic dynamical systems characterized by a sensitive dependence on initial conditions and a broadband noisy power spectrum. In **dissipative** dynamics it embraces **chaotic attractors**, **chaotic transients**, etc. *See* chaotic attractor.

Chaotic attractor

The last and most complex form of bounded post-transient recurrent behaviour observed in a **dissipative** dynamical system, after the point, periodic and quasi-periodic attractors, is the *chaotic attractor*. It exhibits a sensitive dependence on initial conditions, and despite the absence of any stochastic forcing of the deterministic system, a broadband noisy power spectrum. Typically a fast contraction onto a sheet allows an exponential divergence of trajectories within the sheet, consistent with an overall volume contraction. The repeated folding of the sheet, as in the making of flaky pastry, gives the attractor a bounded **fractal** structure and induces a complex mixing of trajectories. The chaotic nature of an attractor can be established by examining the divergence of trajectories, quantified by **Liapunov exponents**. Other signatures of a chaotic attractor include its bifurcational precursors, such as a **period-doubling** cascade; and its **fractal dimension** as assessed for example by the **capacity dimension** or the **information dimension**. In the case of time series data, where the existence of a low-dimensional phase space is in doubt, a crucial property is the **embedding dimension**. *See also* strange attractor, chaos, period doubling cascade, embedding dimension. *Go to* chaotic saddle.

Chaotic repellor

Chaotic transients may be governed by a recurrent structure which repels nearby trajectories in all directions. Such a structure, which would necessarily become an attractor under reversal of the sense of time, is a chaotic repellor. *Go to* sensitive dependence on initial conditions.

Chaotic saddle

Just as point and periodic steady states can be stable **attractors** or unstable **saddles**, so can the recurrent chaotic solutions. Chaotic saddles, which attract nearby trajectories in at least one direction and repel trajectories in at least one complementary direction, play a particularly important role in **fractal basin boundaries** of driven oscillators. *Go to* chaotic transient.

Chaotic transient

A chaotic transient is an irregular transient motion, observed for example in the vicinity of a **chaotic saddle** forming a **fractal basin boundary**. It can be of arbitrarily long duration, and may lead eventually to an attractor of any type (point, periodic, etc). *Go to* chaotic repellor.

Circle map

An archetypal one-dimensional mapping of the circumference of a circle onto itself, used to illustrate the response of oscillators with two competing frequencies. Regimes of mode locking are called **Arnold tongues**. *See* archetypal maps.

Coarse system

The term employed in the Russian literature to describe a **structurally stable** system. *See* structural stability.

Codimension

Generic bifurcations observed under variation of one control lie on surfaces of dimension one less than the dimension of control space, and are said to have codimension 1. A bifurcational event in **phase-control** space is declared *structurally stable* if the whole event is topologically robust against additional perturbations of the system. In the manner popularized by catastrophe theory, bifurcations that are not structurally stable under one control can sometimes be made structurally stable, or **unfolded**, by embedding in a higher-dimensional control space. The codimension of a bifurcation is the number of control parameters needed to achieve this **structural stability**. *See* bifurcation.

Conservative system

Same as a **Hamiltonian system**. *See* dissipation.

Control parameters and space

Control parameters in the functions governing a **dynamical system** are understood to remain constant during the dynamical motions of the system. They are, however, imagined to be under the control of an external agent who can give them very slow quasi-static variation, thereby driving the system slowly from one parameter regime to another; in this way **bifurcations** can be realized. An example would be a human controller slowly varying a throttle or rheostat setting. The control space is the conceptual multidimensional space whose coordinates are the control parameters. *See also* phase-control space. *Go to* structural stability.

Crisis

A crisis is a **global bifurcation** of a chaotic attractor in which the attracting set changes abruptly and discontinuously. An *interior crisis* is a **catastrophic-explosive bifurcation** of a chaotic attractor, while a *boundary crisis* is a **catastrophic-dangerous bifurcation** of a chaotic attractor. *See* global bifurcation.

Determinate bifurcation

A bifurcation whose outcome is insensitive to small details such as the rate of the control sweep, small perturbations, and noise; see **bifurcational classification**. The outcome is therefore predictable and determinate. *See also* indeterminate bifurcation. *Go to* archetypal differential equations.

Diffeomorphism

A smooth *differentiable* one-to-one **mapping** with a unique and smooth differentiable inverse. The Poincaré mapping of a smooth continuous dynamical system will typically be a diffeomorphism. *See* Poincaré section.

Dissipation

A mechanical system without dissipation, namely a **Hamiltonian system**, has an identically zero **divergence** in phase space: this result is known as **Liouville's theorem**. The phase flow is then akin to that of an incompressible fluid. Dissipation of energy gives a negative divergence to the flow; for example, the oscillator $x'' + bx' + cx = 0$ is reduced to the first-order form $x' = y$, $y' = -cx - by$ giving $\text{div} = -b$. In the wider context of non-mechanical systems, we can refer to **volume-preserving** systems, etc. Similar results and subdivisions apply to iterated **mappings**. We use the adjective *dissipative* to describe any system that does not have an identically zero divergence. Very often a system so described will be *totally dissipative*, in the sense that the divergence function is everywhere negative. But we also encounter systems (of the Van der Pol type, for example, with an energy source) in which the phase space might have regimes of positive divergence, containing for example a **repellor**. *See also* Hamiltonian system, Liouville's theorem, volume preserving. *Go to* recurrent state.

Divergence

The non-crossing trajectories of a continuous dynamical system give a fluid-like **flow** in the phase space. Writing the set of n first-order differential equations as $x'_i = f_i(x_j)$, we have the important scalar divergence, $\text{div}(x_i) = \partial f_1/\partial x_1 + \partial f_2/\partial x_2 + \ldots + \partial f_n/\partial x_n$. The rate of change of a small volume V of the phase 'fluid' is given by $V'/V = \text{div}\, x_i$. The analogous result for the mapping $x_{i+1} = G(x_i, y_i), y_{i+1} = H(x_i, y_i)$ gives us the ratio of small areas $A_{i+1}/A_i = D$, where D is the Jacobian determinant $(\partial G/\partial x)(\partial H/\partial y) - (\partial G/\partial y)(\partial H/\partial x)$. *Go to* dissipation.

Duffing's equation

The equation of a driven oscillator with a cubic, or more generally a polynomial, nonlinear restoring force. *See* archetypal differential equations.

Dynamical system

A continuous dynamical system can be described by an autonomous set of first-order differential equations, $\mathbf{x}' = f(\mathbf{x})$, giving a stationary vector field in the phase space spanned by the components of vector $\mathbf{x}$. The coordinates of the phase space describe the state of the system at any time. Solutions are seen as trajectories that trace out in phase space the evolution of the system over time. A driven mechanical oscillator can, for example, be put into this form by identifying the velocity as a second phase coordinate and the time as a third. A discrete dynamical system is described by the iterated **mapping**, $\mathbf{x}_{i+1} = F(\mathbf{x}_i)$. The **Euler** time integration of a continuous system would, for example, generate a discrete system of this type. More globally, continuous dynamical systems of dimension n are formally reduced to a mapping of dimension $n - 1$ by the use of a **Poincaré section**. *See also* autonomous, vector field, phase space, flow, invariants. *Go to* time integration.

Embedding dimension

An integer number giving the dimension of a **phase space** in which the trajectories of a dynamical system do not cross through each other in finite time, and the behaviour is deterministic and predictable for short times into the future. *See* chaotic attractor.

Equilibrium point

This means the same as the **fixed point of a flow**. *See* point attractor.

Euler time integration

The vector field of a continuous dynamical system $\mathbf{x}' = f(\mathbf{x})$ leads naturally to the Euler time integration scheme. For a small time step Δt we can write $\Delta \mathbf{x} = f(\mathbf{x})\Delta t$, allowing us to make a small finite step (Figure G.4) from point k to the next point $k+1$ using $\mathbf{x}^{k+1} = \mathbf{x}^k + f(\mathbf{x}^k)\Delta t$. An improvement of this basic scheme is the **Runge–Kutta** method. *See* time integration.

Feigenbaum

The name often associated with the period-doubling cascade and its universal number. *See* chaotic attractor.

Final state sensitivity

In a dynamical system with **dissipation**, start-up transients can involve **recurrent behaviour** that hesitates for a finite but arbitrarily long time before settling to one of several attractors. A cloud of nearby initial conditions can separate at a rate that increases geometrically over time, and the basin boundary has **fractal** structure. As a result, even the most minute variations of initial condition may change the attractor ultimately chosen by the system. This is one of three closely related forms of unpredictability; the other two are **sensitive dependence on initial conditions** and **indeterminate bifurcation**. *See* fractal basin boundary.

Fixed point of a flow

An *equilibrium* or *fixed point* of a **flow**, $\mathbf{x}' = f(\mathbf{x})$, is characterized by $\mathbf{x}' = f = 0$. It can be stable or unstable, and is the first (trivial) form of steady-state

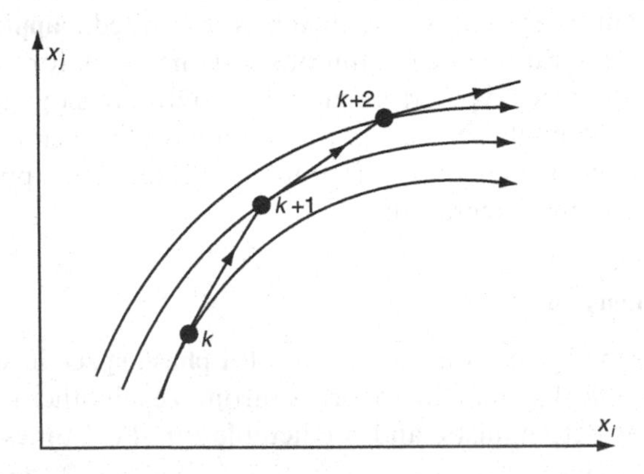

Figure G.4 Illustration of the Euler time integration scheme. The Runge-Kutta method is an improvement on this basic scheme which prevents the gross slippage from one trajectory to another that is apparent in the sketch

recurrent behaviour. If it is asymptotically stable it is a **point attractor**. *See* point attractor.

Fixed point of a map

A fixed point of a map $\mathbf{x}_{i+1} = F(\mathbf{x}_i)$ is characterized by $\mathbf{x}_{i+1} = \mathbf{x}_i$. It can be stable or unstable, and is the trivial form of **recurrent** behaviour for an iterated mapping. *See* periodic attractor.

Flip bifurcation

This is a **local bifurcation** of a map in which a continuous fundamental control-phase path of fixed points loses its stability as it intersects a secondary path of period-doubled fixed points. When the secondary path is stable and exists at supercritical values of the control parameter, we have a **supercritical bifurcation**. Conversely, when the secondary path is unstable and exists at subcritical values of the control parameter, we have a **subcritical bifurcation**. When the map is the Poincaré mapping of a flow, the flip bifurcation generates secondary cycles with twice the period of the fundamental cycle. The location of a flip on a fundamental path is characterized by a real linear mapping eigenvalue passing through -1; the nature of the bifurcation, subcritical or supercritical, depends on the sign of a nonlinear coefficient. *Go to* Neimark bifurcation.

Flow

Trajectories in the phase space of a continuous dynamical system are said to form a flow, in contrast to the **mapping** of a discrete system. *See* dynamical system.

Fold

The typical form of a **saddle-node bifurcation**. *See* saddle-node bifurcation.

Fractal

A fractal, or fractal set, has fine detail on all scales, the term being coined by Mandelbrot from the Latin *fractus* 'broken'. There is as yet no precise, agreed mathematical definition. Falconer (1990) suggests that the term is best used loosely for sets, often defined by a simple recursion, that have properties such as fine structure, with detail on arbitrarily small scales; local and global irregularity which cannot be described by traditional geometry; self-similarity, perhaps approximate or statistical; a non-integer **fractal dimension**. A simple example is the **Cantor set**. *See also* Cantor set, fractal dimension, capacity dimension, Hausdorff dimension, information dimension. *Go to* symbolic dynamics.

Fractal basin boundary

In a two-dimensional dissipative map, a basin boundary formed by a saddle inset will become fractal when, after a **homoclinic tangency**, the saddle develops a **homoclinic tangle**. A non-integer **fractal dimension** can then be estimated for the boundary. The system exhibits **final state sensitivity**: minute variations of initial condition may change the attractor ultimately chosen by the system. *See also* final state sensitivity. *Go to* control parameters.

Fractal dimension

The *fractal dimension* of an infinite set of points in an n-dimensional space is a measure, non-integer and less than n, of the extent to which the points fill the space. Of the numerous definitions for determining such a measure, one of the most useful is the *capacity* dimension; a more mathematical measure is the *Hausdorff* dimension. *See* fractal.

Global bifurcation

The adjective *global* is used to describe **bifurcations** and other phenomena that are not essentially described in a **local** region of phase space. Global bifurcations often involve **homoclinic connections** and **heteroclinic connections** between the **outstructures** of saddles, producing changes in the basin structure of the phase portrait. *See also* crisis. *Go to* homoclinic connection.

Hamiltonian system

A **conservative**, **autonomous** mechanical system with no energy **dissipation**, whose **divergence** function is therefore identically zero, by **Liouville's theorem**, is called a *Hamiltonian system*. The equations of motion can be written in terms of the Hamiltonian function H (numerically equal to the sum of the kinetic and potential energies) as $q_i' = \partial H / \partial p_i, p_i' = -\partial H / \partial q_i$, where q_i are the n generalized coordinates and p_i the generalized momenta. Notice that not all **volume-preserving** systems can be reduced to this classical canonical form. *See* dissipation.

Harmonic oscillation

A periodic oscillation (not normally sinusoidal) with the same period as the driving or sampling is called a harmonic oscillation. The more restrictive expression *simple harmonic motion* is used for sinusoidal behaviour. See for comparison **subharmonic oscillation**. *See* periodic attractor .

Hausdorff dimension

One of the non-integer dimensions that can be evaluated for a fractal set, giving results sometimes identical to the **capacity dimension**. *See* fractal.

Hénon map

A **mapping** of the plane onto itself, governed by two coupled difference equations with one quadratic nonlinearity. The Hénon map can be seen as an embedding of the **quadratic map**, to which it degenerates as one of its parameters drops to zero. *See* archetypal maps.

Heteroclinic connection

In a two-dimensional **flow** the **outset** of a saddle can be the **inset** of a second remote saddle forming a *heteroclinic* (saddle) connection; compare this with **homoclinic connection**. In a **dissipative system** such a connection is **structurally unstable**; it is typically encountered under the sweep of a single control parameter, identifying it as a **codimension 1** global bifurcation. The introduction of periodic forcing typically thickens this single event into a train of events in a two-dimensional **map** that starts with a **heteroclinic tangency** and gives rise to a **heteroclinic tangle**. On the other hand, higher-dimensional flows also commonly exhibit heteroclinic connections, in which an outset lies inside an inset, or vice versa, at a single parameter value. *See also* heteroclinic tangency, heteroclinic tangle. *Go to* determinate bifurcation.

Heteroclinic tangency

In a two-dimensional **dissipative map**—such as the **Poincaré map** of a driven damped oscillator, generated by **stroboscopic sampling**—the **outset** of one saddle can touch the **inset** of a second remote saddle in a *heteroclinic tangency*. This **structurally unstable** event is typically encountered during the sweep of a single control parameter, identifying it as a global bifurcation of **codimension 1**. As a simple consequence of the forward and backward mapping along the manifolds, when they touch once they will simultaneously touch an infinite number of times. Sweeping the control parameter beyond the first tangency generates a **heteroclinic tangle**. *See* heteroclinic connection.

Heteroclinic tangle

After a **heteroclinic tangency** the inset and outset will have an infinite number of structurally stable transverse intersections and will form a **heteroclinic tangle**. Unlike a **homoclinic tangle**, this is not by itself sufficient to generate fractal geometry. *See* heteroclinic connection.

Homoclinic connection

In a two-dimensional **flow** the **outset** of a saddle can return as its **inset**, forming a *homoclinic* (saddle) connection. In a **dissipative system** such a connection is **structurally unstable**; it is typically encountered under the sweep of a single control parameter, identifying it as a **codimension 1** global bifurcation. The introduction of periodic forcing typically thickens this single event into a train of events in a two-dimensional **map** that starts with a **homoclinic tangency** and gives rise to a **homoclinic tangle**. On the other hand, higher-dimensional flows also commonly exhibit homoclinic connections, in which an outset lies inside an inset, or vice versa, at a single parameter value. A significant event is the **Shilnikov homoclinic connection**. Associated periodic orbits are the **Newhouse sinks**. *See also* homoclinic tangency, homoclinic tangle. *Go to* Melnikov theory.

Homoclinic tangency

In a two-dimensional **dissipative map**—such as the Poincaré map of a driven damped oscillator, generated by stroboscopic sampling—the **outset** of a saddle can touch its **inset** in a homoclinic tangency. This **structurally unstable** event is typically encountered during the sweep of a single control parameter, identifying it as a global bifurcation of **codimension 1**. As a simple consequence of the forward and backward mapping along the manifolds, when they touch once, they will simultaneously touch an infinite number of times. Sweeping the control parameter beyond the first tangency generates a **homoclinic tangle**. *See* homoclinic connection.

Homoclinic tangle

After a **homoclinic tangency** the inset and outset of a saddle will have an infinite number of structurally stable transverse intersections and will form a *homoclinic tangle*. This generates very complex dynamics, associated with the formation of a **horseshoe** (Figure G.5). The inset, which may often represent a basin boundary, will now have a **fractal** nature. *See* homoclinic connection.

Hopf bifurcation

This is a **local bifurcation** of a flow in which a continuous fundamental control-phase path of fixed points loses its stability as it intersects a secondary path of cycles. When the secondary path is stable and exists at supercritical values of the control parameter, we have a **supercritical bifurcation**. Conversely, when the secondary path is unstable and exists at subcritical values of the control parameter, we have a **subcritical bifurcation**. The location of a Hopf bifurcation on the fundamental path is characterized by a complex conjugate pair of linear eigenvalues whose real part passes through zero; the nature of the bifurcation, subcritical or supercritical, depends on the sign of a nonlinear coefficient. *Go to* flip bifurcation.

Horseshoe

A two-dimensional **mapping** introduced by Smale to explore the folding and mixing actions of chaotic motion. *See* symbolic dynamics.

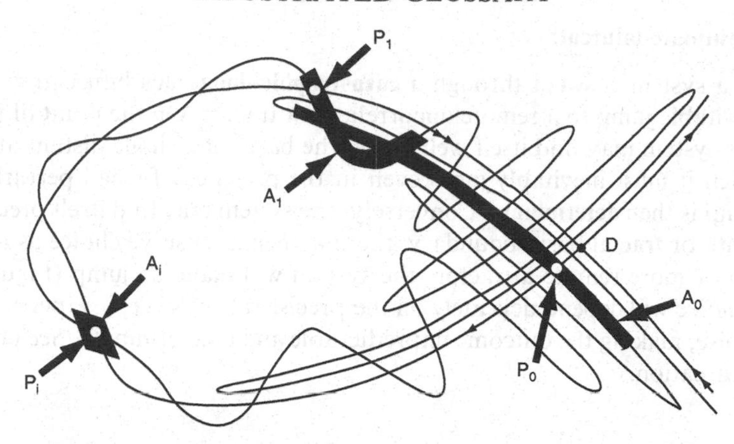

Figure G.5 Horseshoe dynamics in a homoclinic tangle Here the inset and outset manifolds of a saddle, D, form a homoclinic tangle. State (0) maps to the intermediate state (i) in M iterates. Then state (i) maps to the final state (1) in N further iterates. The mapping of homoclinic points, P, and small areas, A, is indicated. The horseshoe intersection of A_0 and A_1 thus occurs after $M+N$ iterates, giving dynamics that is conjugate to a left shift on two symbols. There are infinitely many periodic orbits close to the homoclinic orbits. Reproduced, with permission, from Arrowsmith and Place (1990)

Hyperbolic point

A fixed point (equilibrium) of a flow or map that has all its linear eigenvalues in the stable or unstable domains is called *hyperbolic*. There are then no critical eigenvalues corresponding to neutral stability, and the phase portrait around the fixed point is **structurally stable** against perturbations of the system. We should note carefully, however, that the term *hyperbolic point* is used differently in the literature on Hamiltonian systems to mean a saddle, near which trajectories follow a roughly hyperbolic shape. *See* stability.

Hysteresis

The phenomenon, generated by a **catastrophic-dangerous** bifurcation, in which an attractor path is not (immediately) reinstated on the reversal of the control sweep. Two such bifurcations typically generate a *hysteresis loop*, observed for example during frequency sweeps through nonlinear resonance. *See* bifurcational classification.

Incommensurate

If the ratio of two periodic times is irrational, the periodic times are *incommensurate* and a composite motion will be quasi-periodic. *See* quasi-periodic attractor.

Indeterminate bifurcation

When a system is swept through a **catastrophic-dangerous** bifurcation there is an inevitable jump to a remote uncorrelated attractor. On the point of instability, the system may find itself well inside the basin of a single distant attractor, to which it must inevitably jump even in the presence of small perturbations; the jump is then **determinate**. Conversely, the system may find itself precisely on a **smooth** or **fractal** basin boundary; there is then a sensitive choice as to which of two or more remote attractors the system will make its jump (Figure G.6). This choice will depend delicately on the precise rate of sweeping, perturbations and noise, making the outcome unpredictable and indeterminate. *See* determinate bifurcation.

Index of instability

The *instability index* of a **hyperbolic** steady state is the degree of instability based on the linear eigenvalues. For a fixed point of a flow, it is the number of eigenvalues with a positive real part; for a fixed point of a map (perhaps representing a cyclic flow) it is the number of eigenvalues with modulus greater than one. *See* instability.

Index (Poincaré)

For a plane vector field, the Poincaré index of a fixed point is the net rotation of the vector field direction along a nearby closed path encircling the point. The index of an attractor or repellor is $+1$, the index of a saddle is -1. This index concept can be generalized to mappings and to higher-dimensional phase space. Topological theorems give global constraints on the number and type of fixed points based on the Poincaré index. *See* instability.

Information dimension

A kind of non-integer dimension in which points in a dynamical invariant set are weighted according to their relative probability of occurrence in a typical long trajectory. *See* fractal.

Inset

The **outstructure** coming in to a saddle solution. *See* basin boundary.

Instability

A steady state is unstable in the sense of *Liapunov* if any single adjacent motion moves out of the immediate phase-space neighbourhood. A totally **dissipative** system may have unstable steady-state **saddle** solutions (equilibria, cycles,

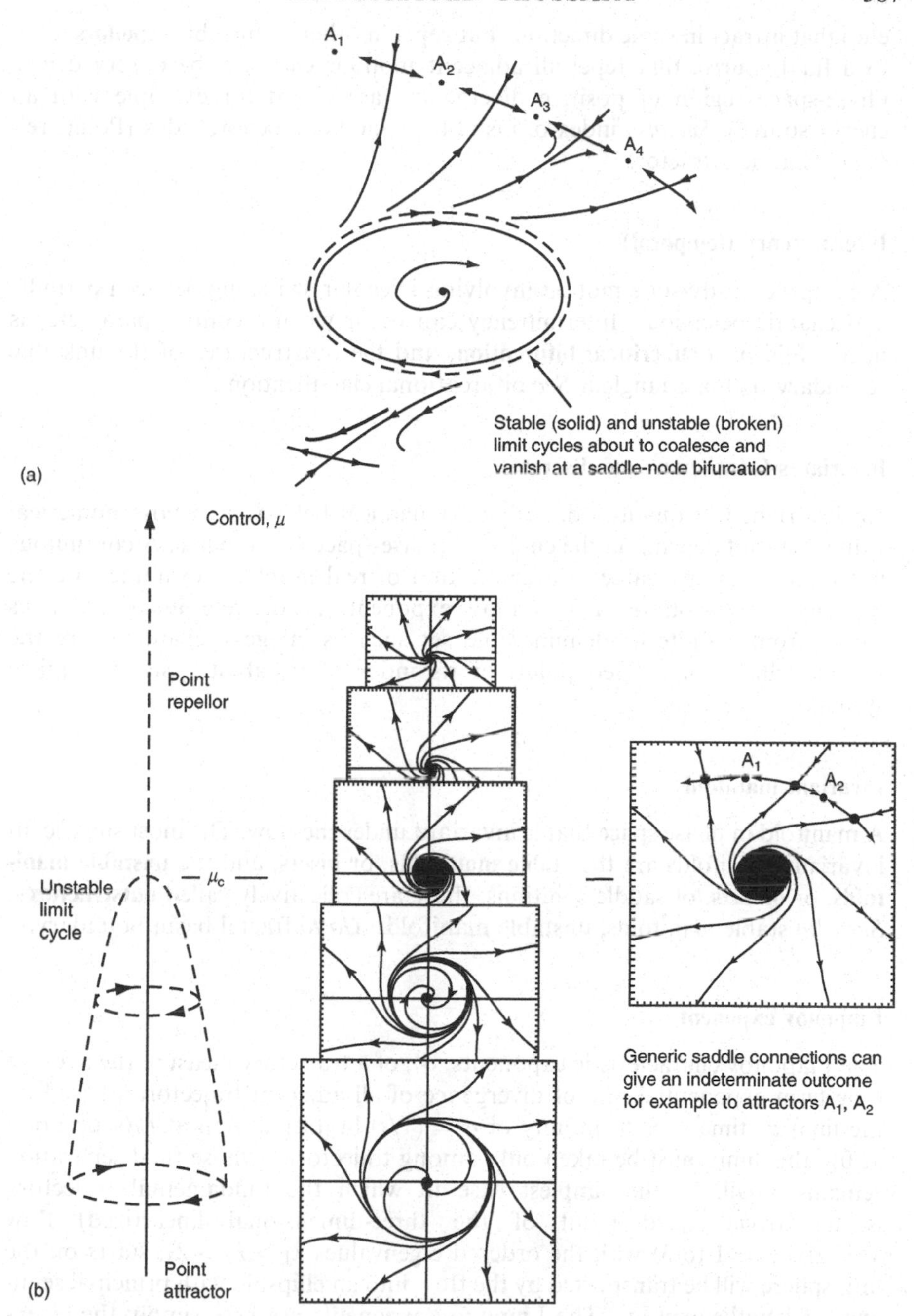

Figure G.6 Two examples of an indeterminate bifurcation. (a) The two-dimensional outflow from a cyclic saddle-node fold allows any number of generic saddle connections: the jump from the bifurcation can lead to any of the available attractors, $A_1, A_2, A_3, \ldots, A_N$. (b) An entirely similar situation, with unpredictable outcomes, can be observed in the subcritical Hopf bifurcation

etc.) that attract in some directions but repel in others. Unstable **repellors**, akin to a fluid source, that repel all adjacent motions can only be observed in a phase-space region of positive divergence (associated for example with an energy source). *See also* index of instability, saddle, repellor, index (Poincaré). *Go to* chaotic attractor.

Intermittency (temporal)

A complex steady-state motion involving irregular switching between periodic and chaotic behaviour. Intermittency can occur when a control parameter is near a **fold** or a **subcritical bifurcation**, and the outstructures of the unstable secondary path are tangled. *See* bifurcational classification.

Invariants (continuous and discrete)

An invariant is a quantity describing dynamical behaviour, whose numerical value does not depend on the choice of phase-space coordinates. A continuous invariant takes any value in a continuum of real numbers; examples are the capacity dimension or a Liapunov exponent. A discrete invariant takes values from a finite or denumerable set such as integers; examples are the Poincaré index of a fixed point, or its index of instability. *See* dynamical system.

Invariant manifold

A manifold in phase space that is invariant under the flow. The most significant invariant manifolds are the **stable manifolds**, or **insets**, and the **unstable manifolds**, or **outsets**, of saddle solutions; these are collectively called **outstructures**. *See also* stable manifolds, unstable manifolds. *Go to* fractal basin boundary.

Liapunov exponent

The Liapunov characteristic exponents, σ_i, of a trajectory measure the average long-term exponential rate of divergence of all adjacent trajectories based on the limit as time goes to infinity of $\sigma = (1/t)$ ln [(separation at t)/(separation at 0)]; this limit must be taken only among trajectories whose final separation remains small. In the simplest case in which the fundamental trajectory is the trivial fixed point of the three-dimensional linear(ized) flow $x_i' = \lambda_i x_i$ ($i = 1$ to 3) with the ordered eigenvalues $\lambda_1 > \lambda_2 > \lambda_3$, starts on the unit sphere will be transported by the flow into an ellipsoid with principal semi-axes of lengths $\exp(\lambda_i t)$. The Liapunov exponents are here simply the eigenvalues, $\sigma_i = \lambda_i$, with the sum, $\sigma_1 + \sigma_2 + \sigma_3$, equal to the **divergence** of the flow.

This result holds for more general situations, so in a totally dissipative system for which the (volume) divergence is negative, the sum of the Liapunov exponents of any trajectory will be negative, but with no such restriction on any single exponent. If the maximum exponent, σ_1, is positive then some adjacent trajectories will diverge from the fundamental, and there will be a sensitive dependence on initial conditions (albeit a trivial one for the above fixed point). A positive Liapunov exponent in a bounded attractor is a sign of chaotic motion. For trajectories on a bounded chaotic attractor, the numerical values of the Liapunov exponents are the same, no matter where within the attractor the bundle of nearby trajectories initiates, and no matter what coordinate system is used—they are **invariants**. *See* sensitive dependence on initial conditions.

Liapunov stability

The classical notion of **stability** and, correspondingly, **instability**. *See* stability.

Limit cycle

A periodic oscillation (very often implying stability). *See* periodic attractor.

Liouville's theorem

This states that the phase space **divergence** of a **Hamiltonian system** is identically zero. *See* dissipation.

Local bifurcation

The adjective *local* is used to describe **bifurcations** and other phenomena whose effects are restricted to a neighbourhood of a point or cycle in phase space; bifurcations which are not so restricted are termed **global bifurcations**.

A local bifurcation will be observed on an equilibrium path of a **flow** as the real part of a linear eigenvalue passes through zero; see **stability**. If a real eigenvalue passes through zero, we have a **saddle-node bifurcation**. Typically this will be observed as a fold, at which the path folds smoothly back at a maximum (or minimum) value of the control parameter. But in the presence of symmetry, or other constraints, the saddle-node bifurcation can manifest itself as a sub-, super- or transcritical bifurcation in which a secondary equilibrium path bifurcates from the monotonically increasing fundamental path. If the real part of a complex conjugate pair of flow eigenvalues passes through zero, we have the **Hopf bifurcation** in which stable (unstable) cycles emerge from the monotonic fundamental path at supercritical (subcritical) values of the control parameter.

Similarly, a local bifurcation can be observed on a path of fixed points of a **map**, representing for example a path of cycles in a stroboscopically sampled flow. A loss of **stability** will now correspond to a mapping eigenvalue passing through the unit circle. If a real eigenvalue passes through $+1$ we have a saddle-node bifurcation, which typically manifests itself as a (cyclic) fold. If a real eigenvalue passes through -1 we have a **flip bifurcation** at which a secondary stable (unstable) path of period-doubled fixed points bifurcates off the monotonic fundamental path at supercritical (subcritical) values of the control parameter. If a complex conjugate pair of eigenvalues penetrates the unit circle, we have the **Neimark bifurcation**. *See also* centre manifold. *Go to* saddle-node bifurcation.

Logistic map

A nonlinear one-dimensional **mapping**, typically arising in population dynamics (or indeed in any growth phenomenon with a 'pollution' effect), that forms a classical introduction to the **period-doubling** cascade to chaos (Figure G.7). It is entirely equivalent to the quadratic map, to which it can be transformed by a change of variable. *See* archetypal maps.

Lorenz equations

An autonomous set of three first-order ordinary differential equations with simple nonlinearities. Devised by Lorenz as a model of thermal convection relating to his atmospheric studies, they can exhibit chaotic motions and played a seminal role in the development of chaos theory. *See* archetypal differential equations.

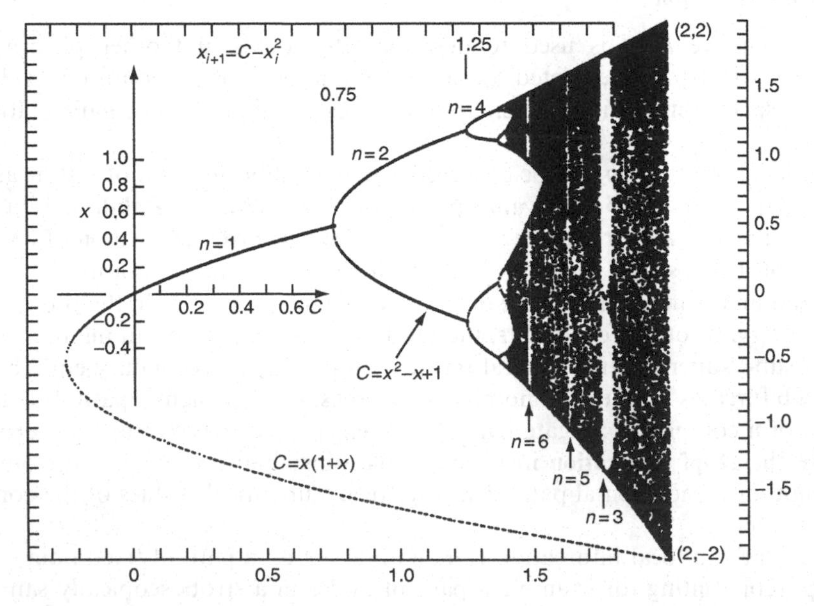

Figure G.7 The logistic map (in one of its equivalent forms) showing the period-doubling cascade to chaos

Map and mapping

An iteration of the form $\mathbf{x}_{i+1} = F(\mathbf{x}_i)$ which defines a discrete-time **dynamical system** is called a *mapping* or *map*. *See* Poincaré section.

Melnikov theory

Melnikov methods allow an estimation of the parameter values at which **homoclinic** and **heteroclinic tangencies** will occur. Using a perturbation from an underlying integrable system (often a Hamiltonian flow) for which an analytical closed-form solution is available, the Melnikov function provides a measure of the distance between the relevant **inset** and **outset**. The Melnikov result for a damped driven oscillator (such as the escape equation) can be quickly and easily derived by an energy balance approach (Thompson 1996b). *Go to* Shilnikov homoclinic connection.

Neimark bifurcation

The mapping equivalent of the **Hopf bifurcation**, corresponding to two complex conjugate mapping eigenvalues passing out of the unit circle. It can appear in supercritical and subcritical forms, and is sometimes called a **secondary Hopf bifurcation**. *Go to* global bifurcation.

Newhouse sinks

Infinite sets of stable periodic orbits shown by Newhouse to exist at certain parameter values close to those at which a **homoclinic tangency** occurs. *Go to* heteroclinic connection.

Non-wandering state

A *non-wandering state* is one that has arbitrarily close states that return arbitrarily close. This is a generalisation of a **recurrent state**; any recurrent state is non-wandering, but not vice versa. This generalisation is needed to embrace a homoclinic orbit. *See* recurrent state.

Outset

The **outstructure** departing from a saddle solution. *See* basin boundary.

Outstructures

Trajectories that flow or map towards a **saddle** solution as time goes to plus infinity, approaching it tangentially along the stable incoming eigenvectors, are called the *stable manifolds* or *insets*. Similarly, trajectories asymptotic to a saddle as time goes to minus infinity are called the *unstable manifolds* or *outsets*.

Insets and outsets are known collectively as outstructures, and being invariant with time they are *invariant manifolds*. The insets are particularly important in organising a phase portrait, and typically form basin boundaries. They can be located computationally by running time backwards from close to the saddle. In a two-dimensional flow, one start displaced a small distance along an incoming eigenvector will be enough to locate the corresponding inset. In a two-dimensional map, a *ladder of starts* along the eigenvector will be needed to fill out the whole manifold; a complete ladder (also called a fundamental neighbourhood), in which the bottom rung maps to the top rung, ensures that the inset is properly filled with computed points. *See* basin boundary.

Period doubling

Under the slow variation of a control parameter, a supercritical **flip bifurcation** generates a stable oscillation with twice the period of its precursor. A cascade of such period-doubling bifurcations is a common route to chaos. In this classical cascade the bifurcations become more and more closely spaced, the ratio of successive parameter intervals tending in the limit to the **Feigenbaum number** $\delta_\infty = 4.66920\ldots$. This number is universal (generic) in the sense that it arises in a very wide class of problems. This limiting ratio ensures that in such a *Feigenbaum cascade* the repeatedly doubled period quickly reaches infinity at an accumulation point in a finite parameter interval. After the accumulation point, a complex pattern of chaotic motions and periodic windows is observed under further increase of the control parameter; see for example Figure G.7 showing the logistic map. *See* chaotic attractor.

Periodic attractor

In the phase space of a **flow**, a closed orbit satisfying **recurrence** by returning precisely to its starting point after its periodic time, T, is called a *periodic motion*. In a **dissipative** system, such an orbit is also called a *limit cycle*. If the motion is asymptotically stable, it is called a *periodic attractor*. The stability of periodic motion is best assessed using a **Poincaré section** in which it will appear as a **fixed point** of the Poincaré map. This section serves to assess the integrated contractions and expansions around the cycle (Figure G.8). *See also* limit cycle, harmonic oscillation, subharmonic oscillation, fixed point of a map. *Go to* quasi-periodic attractor.

Phase space and portrait

A trajectory of a continuous **dynamical system** traces out its evolution over time in an abstract *phase space*, whose coordinates describe the dynamical state of the system at any particular time. Ensembles of trajectories fill the *phase space* to form a *phase portrait*. In a dissipative system this portrait will show the structure of the attractors and basins; to emphasize this, it is sometimes called the attractor-basin phase portrait. *See* dynamical system.

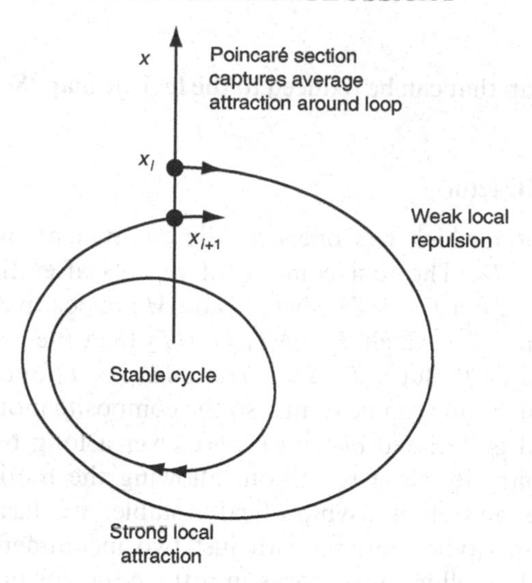

Figure G.8 Picture illustrating how a Poincaré section assesses the average attraction or repulsion around a cycle

Phase-control space

The conceptual $n + m$ dimensional space spanned by the n phase coordinates and the m control parameters is called the phase-control space. Paths of attractors will be observable in this space. *See* control parameters.

Poincaré section and map

A section transverse to the flow in the phase space of an n-dimensional continuous dynamical system which generates an $(n - 1)$-dimensional *Poincaré mapping*, taking a point on the surface of the section to its image upon first return to the section, is called a *Poincaré section*. The mapping has the same general stability properties as the flow. For a mechanical oscillator driven by a periodic excitation of period T, Poincaré sections can be defined by the planes $t = iT$ where $i = 1, 2, 3, \ldots$. This corresponds to the **stroboscopic sampling** of the velocity and displacement. The Poincaré mapping of a smooth continuous dynamical system will typically be a diffeomorphism. *See also* map, diffeomorphism, stroboscopic sampling. *Go to* divergence.

Point attractor

A stable equilibrium or **fixed point** of a **flow** is called a *point attractor*. It is the simplest (trivial) form of **recurrent** behaviour. *See also* fixed point of a flow. *Go to* periodic attractor.

Quadratic map

An archetypal map that can be reduced to the **logistic map**. *See* archetypal maps.

Quasi-periodic attractor

Consider a motion which has one periodic component of period T_1 and a second of period T_2. The first component repeats after time $t = NT_1$ while the second repeats after $t = MT_2$ where N and M are positive integers. If we can find a (lowest) time for which $T = NT_1 = MT_2$ then the composite motion is *periodic* with period T. But if T_1/T_2 is *irrational*, i.e. T_1 and T_2 are *incommensurate*, then no such time can be found, so the composite motion never precisely repeats itself and is declared *quasi-periodic*. Over a long timescale there will, however, be arbitrarily close repetition, allowing the motion to be declared **recurrent**. If the motion is asymptotically stable, we have a *quasi-periodic attractor*. Quasi-periodic motions with just two incommensurate frequencies can be visualized as filling out a torus in a three-dimensional phase space. In strongly coupled systems, quasi-periodic attractors with more than two incommensurate frequencies are unlikely to be observed experimentally because they are easily perturbed into **chaotic attractors**. *See also* incommensurate, rotation number, Arnold horns. *Go to* instability.

Recurrent state and behaviour

In a dynamical system with **dissipation**, the negative **divergence** in phase space ensures that a cloud of starts will shrink onto an attracting set of zero volume. To distinguish final-time steady-state behaviour from start-up transients, geometrical dynamics uses the concept of a *recurrent state*. A particular state of a dynamical system is deemed recurrent if, after sufficient time, the system returns arbitrarily close to the state. The relaxation of the definition away from precise repetition is here used to embrace quasi-periodic and chaotic motion as recurrent. A further relaxation is to the **non-wandering** state to embrace homoclinic trajectories. An ensemble of recurrent states linked together by a single trajectory constitutes *recurrent behaviour*. *See also* steady state, non-wandering state. *Go to* stability.

Repellor

An unstable steady-state solution (equilibrium, cycle, etc.) that repels all adjacent motions in the manner of a fluid source is called a repellor. Such a state can only be observed in a phase regime of positive divergence. Note that saddles in higher-dimensional phase spaces may become repellors if the dynamics is projected down to a lower dimension; on the centre manifold of a saddle-node bifurcation, for example, the saddle will appear as a one-dimensional repellor. *See* instability.

Rotation number

If a motion has two periodic components with frequencies ω_1 and ω_2 then in terms of the basic frequency ω_1, say, the rotation number, or **winding number**, measures the average number of orbits of frequency ω_2 during one orbit of frequency ω_1. *See* quasi-periodic attractor.

Runge–Kutta

The Runga–Kutta time integration scheme is an elaboration of the basic **Euler** scheme using an Euler-type prediction followed by corrections to achieve higher-order accuracy. *See* time integration.

Saddle

An unstable steady-state solution (point, cycle, etc.) that repels in some phase directions but attracts in others is called a *saddle*. Such saddles are typical ingredients of *dissipative* and *totally dissipative* phase spaces. Although, being unstable, they are not directly observable in a physical system, they play a key role in structuring the phase space; in particular, saddles with one unstable direction frequently organize basin boundaries. We can note in particular the existence of a **chaotic saddle**. *See* instability.

Saddle connection

A connection between two saddles, a **heteroclinic connection**, or between a saddle and itself, a **homoclinic connection**. *See* homoclinic connection.

Saddle-node bifurcation

The simplest **local bifurcation** in which a control-phase path of fixed points (representing equilibria in a flow, or cycles in a Poincaré section) reaches an extreme value, say a maximum μ_c of a control parameter μ. As μ is increased towards μ_c a saddle and a node coalesce; there is locally no fixed point for $\mu > \mu_c$, forcing the system to jump dynamically to a distant, unrelated attractor. Alternative names are the *fold* of catastrophe theory, the *limit point* of elastic stability. In a flow the saddle-node bifurcation is characterized by a real linear eigenvalue passing through zero; in a map it is characterized by a real eigenvalue passing through $+1$. *See also* fold. *Go to* supercritical bifurcation.

Secondary Hopf bifurcation

A common name for the **Neimark bifurcation**. *See* Neimark bifurcation.

Sensitive dependence on initial conditions

Chaos is unpredictable over long times because any two trajectories starting close to one another on a chaotic attractor will separate as they both advance forward in time. Using recurrence to measure time, a small initial separation grows to a multiple of itself at the first recurrence, and typically grows by a similar multiple at the next recurrence (so long as the rules governing the dynamics are continuous functions). Thus the two trajectories will separate at a rate which, on average, increases geometrically with time, and is described by an exponent—the largest **Liapunov exponent**. This can be confirmed by plotting the logarithms of separations against time, as in Figure 9.11 and Figure 16.8. Exponential growth means that, for any given uncertainty in initial conditions, there is a well-defined time horizon beyond which prediction is impossible, even if the dynamical system is a perfect model of real behaviour. Two other forms of unpredictability are closely related to sensitive dependence on initial conditions: **final state sensitivity**, associated with **fractal** basin boundaries, and **indeterminate bifurcation**. *See also* Liapunov exponent. *Go to* fractal.

Separatrix

In an n-dimensional phase space, an invariant manifold of dimension $n-1$ that separates regions of phase space is termed a *separatrix*. For dissipative systems the separatrix of major interest will be the basin boundary. *See* basin boundary.

Shilnikov homoclinic connection

In three-dimensional flows a structurally unstable homoclinic connection may involve a fixed point with one-dimensional outset lying inside a two-dimensional inset with complex conjugate eigenvalues Shilnikov showed that if the real eigenvalue is larger in magnitude than the real part of the complex eigenvalues, the existence of (persistent) horseshoes can be inferred (for a nearby interval of parameters). *Go to* Newhouse sinks.

Stability

An **attractor** of a dissipative dynamical system is *asymptotically stable* in the sense of *Liapunov* because all local trajectories flow back to the attractor. Fixed points of a **Hamiltonian system** can be at most *neutrally stable* in the sense of Liapunov, with all local trajectories staying close to the point, though not returning to it. *Orbital stability* relates to a phase-space criterion in which divergence of the unseen time coordinate is deemed unimportant.

A (Liapunov) stability analysis of a fixed point starts with the linearized equations describing small variations about the point. For a fixed point of a flow, stability hinges on the signs of the real parts of the eigenvalues. Typically the point will be **hyperbolic** (non-critical) with no zero real parts; then the necessary and sufficient condition for stability is that all signs be negative; and the necessary and sufficient condition for instability is that at least one sign be positive.

For a fixed point of a map (perhaps representing the Poincaré mapping of a cycle), stability hinges on the moduli of the eigenvalues. If the point is hyperbolic with no modulus equal to unity, then the necessary and sufficient condition for stability is that all moduli be less than unity; and the necessary and sufficient condition for instability is that at least one modulus is greater than unity. Note the rather different concept of **structural stability**. *See also* Liapunov stability, hyperbolic point. *Go to* attractor.

Stable manifold

One of the invariant manifolds, namely the **inset**, of a saddle solution. *See* invariant manifolds.

Standard map

An archetypal two-dimensional area-preserving mapping of the surface of a torus onto itself, used to illustrate the transition from regular to chaotic motion in conservative Hamiltonian systems. *See* archetypal maps.

Steady state

A typical start of a dissipative system will experience a *transient* motion before settling asymptotically onto a stable *steady-state* solution, called an **attractor**. Other steady states are the **saddles** and **repellors**. *See* recurrent state.

Strange attractor

In experimental and computational dynamics this term is often used loosely and interchangeably with **chaotic attractor**. Strict mathematical definitions of a strange attractor are not always mutually consistent, but often demand that it should contain a transversal homoclinic orbit. Sometimes the term *strange* is taken to refer only to a fractal structure, which does not always imply sensitive dependence on initial conditions. *See* chaotic attractor.

Stroboscopic sampling

Sampling of the position and velocity of a mechanical oscillator, for example, at the period of the driving excitation. *See* Poincaré section.

Structural stability

Because the parameters of a physical system are never known precisely, and may be subject to small variation, it can be argued that a good mathematical model should have a phase portrait that is robust against small changes in the model itself. Small changes of the parameters and functions of the model should not qualitatively change the topology of the phase portrait. A phase portrait that is robust in this way is declared *structurally stable*. Structurally unstable portraits encountered under the variation of a control parameter signal a bifurcation. For an extension of the concept to phase-control space, see **codimension**. A precise definition of structural stability appropriate to chaotic and homoclinic structures is not yet agreed upon. *See also* coarse system. *Go to* bifurcation.

Subcritical bifurcation

This is a local bifurcation in which a continuous fundamental control-phase path of fixed points loses its stability as it intersects an unstable secondary path that only exists at subcritical values of the control parameter; compare with **supercritical bifurcation** and see **bifurcational classification**. When the secondary path is a trace of cycles, we have the **codimension 1** subcritical **Hopf bifurcation**. When the secondary path is a trace of fixed points, we have the *subcritical pitchfork bifurcation* of elastic stability theory; this bifurcation is rendered **structurally stable** by the addition of a second symmetry-breaking control parameter to give the codimension 2 unstable *cusp* of catastrophe theory (Figure 7.8). *Go to* transcritical bifurcation.

Subharmonic oscillation

A subharmonic oscillation of order n is a steady-state periodic oscillation with period n times that of the driving or sampling. A **harmonic** oscillation can be identified as a subharmonic of order unity. *See* periodic attractor.

Subtle bifurcation

In our bifurcation classification the subtle (i.e. continuous) bifurcations are the local **supercritical** bifurcations that generate the smooth continuous growth of a new attracting path. *See* bifurcational classification.

Supercritical bifurcation

This is a local bifurcation in which a continuous fundamental path of fixed points loses its stability as it intersects a stable secondary path that only exists at supercritical values of the control parameter; see **bifurcational classification**. When the secondary path is a trace of cycles, we have the **codimension 1** supercritical **Hopf bifurcation**. When the secondary path is a trace of fixed points, we have the *supercritical pitchfork* bifurcation of elastic stability theory; this bifurcation is rendered **structurally stable** by the addition of a second symmetry-breaking control parameter to give the codimension 2 stable *cusp* of catastrophe theory (Figure 7.8). *Go to* subcritical bifurcation.

Symbolic dynamics

The theory and manipulation of sequences of symbols, used for example in exploring the invariant sets of the Smale **horseshoe**. A simple example is the shifting of the decimal point in the bi-infinite sequence ...010010.1110101011...to display the characteristics of chaotic motion (Figure G.9). *Symbolic dynamics* should not be confused with similar expressions describing automated algebraic manipulations on a computer. *See also* horseshoe. *Go to* basin.

Tangle

A shorthand for **homoclinic tangle** or **heteroclinic tangle**. *See* homoclinic connection.

Time integration

Time integration of the equations describing a dynamical system is readily performed on a digital computer. Many standard integration routines are available. *See also* Euler time integration, Runge–Kutta. *Go to* Poincaré section.

Transcritical bifurcation

This is a local bifurcation in which a continuous fundamental path of fixed points loses its stability as it intersects a secondary path of fixed points; see **bifurcational classification**. The continuous secondary path exists at subcritical and supercritical values of the control parameter, being unstable below and stable above. This *exchange of stability* was discussed by Poincaré, and the transcritical bifurcation is the *asymmetric* bifurcation of elastic stability theory; it is essentially a pathological form of the **saddle-node bifurcation**. *Go to* Hopf bifurcation.

Transient

Motions leading to a **steady state**. *See* recurrent state and behaviour.

Unfolding

A structurally unstable bifurcation can be unfolded into a **structurally stable** form by the introduction of suitable control parameters. The number of control

Transparent Chaos in Deterministic Dynamics

Consider the iterated map, $x_{i+1} = 10x_i$, where x is mod 1, with $0 < x < 1$

Successive iterates move the decimal point one place to the right, and remove the leading digit, D. Notice that D identifies a sector in the circular state-space.

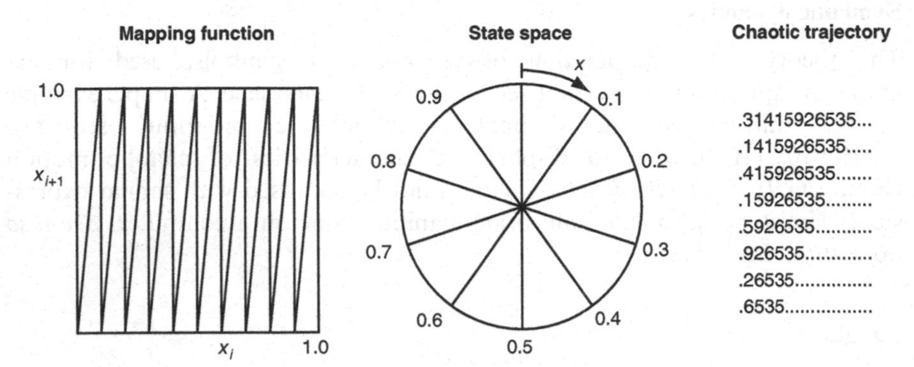

Chaotic trajectories emerge from irrational (non-repeating) starts such as $x_0 = \pi/10$.

Periodic trajectories emerge from rational starts (such as $x_0 = 1/7 = 0.142857142...$).

Adjacent starts, identical to ***any finite accuracy*** (say 8 figures) become uncorrelated:

.3141592677777777...	steady high response after 8 iterates
.3141592618181818...	period 2 response after 8 iterates
.3141592600000000...	steady zero response after 8 iterates

Initial exponential divergence of adjacent starts

Fundamental start	.555555550
Perturbed start	.555555551

Separation after n iterates ($n < 9$) is 10^{n-9}. Divergence is by a factor of 10 on each iteration.

Typical starts give chaos: since almost all numbers are irrational, all typical (experimental) starts give chaos.

Simulations on an idealized computer settle to zero: a simulation, from a finite decimal, will (in the absence of spurious round-offs) settle onto zero.

Chaos wanders almost everywhere, since any value of x that we care to specify to N figures (N finite) will certainly appear somewhere in a random sequence of digits.

Chaos as random as a coin toss: any sequence of digits (generated by a coin or die) can be replicated as a sector sequence.

Symbolic dynamics of logistic map: the binary map $x_{i+1} = 2x_i$ (mod 1) describes, under change of variable, the fully developed chaos at the end of the logistic map, $x_{i+1} = 4x_i(1 - x_i)$.

Figure G.9 Illustration of how symbolic dynamics can make transparent the complexities of chaotic motion. In this scenario the reader should realize that we can devise any (tricky) outcome we like, because we are free to choose any sequence of digits as our starting condition. Mathematicians, of course, usually use base 2, but the use of base 10 makes the illustration more colourful

parameters needed to achieve this is the **codimension**. A typical example is the symmetry-breaking imperfection needed to unfold a symmetric bifurcation into the codimension 2 *cusp* of catastrophe theory. *See* bifurcation.

Unstable manifold

One of the invariant manifolds, namely the **outset** of a saddle solution. *See* invariant manifolds.

Van der Pol equation

The equation of a driven or undriven oscillator with a nonlinear damping characteristic such that the autonomous system is capable of sustained self-excited oscillation in a limit cycle. *See* archetypal differential equations.

Vector field

A multidimensional space having a vector associated with each point. In a typical phase space, the vectors vary smoothly with position, and trajectories are everywhere tangent to them. *See* dynamical system.

Volume preserving and contracting

Volume preserving (volume contracting) is used as an adjective to describe a system with zero (negative) phase-space divergence. *See* dissipation.

Wandering

The converse of **non-wandering**. *See* recurrent state and behaviour.

Winding number

Means the same as **rotation number**. *See* quasi-periodic attractor.

Bibliography

Abarbanel, H. D. I. (1983). Universality and strange attractors in internal-wave dynamics. *J. Fluid Mech.*, **135**, 407–434.

Abarbanel, H. D. I. (1996). *Analysis of Observed Chaotic Data*. Springer-Verlag: New York.

Abraham, F. (1990). *A Visual Introduction to Dynamical Systems Theory for Psychology*. Aerial Press: Santa Cruz.

Abraham, N. B. (1983). A new focus on laser instabilities and chaos. *Laser Focus*, **19**, no. 5 (May), 73–81.

Abraham, N. B., Gollub, J. P. and Swinney, H. L. (1984). Meeting Report: Testing nonlinear dynamics. *Physica*, **11D**, 252–264.

Abraham, R. A., Corliss, J. B. and Dorband, J. E. (1991). Order and chaos in the toral logistic lattice. *Int. J. Bifurcation and Chaos*, **1**, 227–234.

Abraham, R. H. (1979). Dynasim: exploratory research in bifurcations using interactive computer graphics. In Bifurcation Theory and Applications in Scientific Disciplines, O. Gurel and O. E. Rössler (eds), pp. 676–683. New York Academy of Sciences: New York.

Abraham, R. H. (1985). Chaostrophes, intermittency, and noise. In *Chaos, Fractals, and Dynamics*, P. Fischer and W. R. Smith (eds). Dekker: New York.

Abraham, R. H. (1986). *Complex Dynamical Systems: Selected Papers*. Aerial Press: Santa Cruz, CA.

Abraham, R. H. and Marsden, J. E. (1978). *Foundations of Mechanics*. Benjamin/ Cummings: Reading, MA.

Abraham, R. H. and Robbin, J. (1967). *Transversal Mappings and Flow*. Benjamin: Reading, MA.

Abraham, R. H. and Shaw, C. D. (1982). *Dynamics: The Geometry of Behavior*, Part One, *Periodic Behaviour*. Aerial Press: Santa Cruz, CA.

Abraham, R. H. and Shaw, C. D. (1983). *Dynamics: The Geometry of Behavior*, Part Two, *Periodic Behaviour*. Aerial Press: Santa Cruz, CA.

Abraham, R. H. and Shaw, C. D. (1985). *Dynamics: The Geometry of Behavior*, Part Three, *Global Behavior*. Aerial Press: Santa Cruz, CA.

Abraham, R. H. and Shaw, C. D. (1988). *Dynamics: The Geometry of Behavior*, Part Four, *Bifurcation Behavior*. Aerial Press: Santa Cruz, CA.

Abraham, R. H. and Shaw, C. D. (1992). *Dynamics: The Geometry of Behaviour*. Addison-Wesley: Redwood City.

Abraham, R. H. and Simó, C. (1986). Bifurcations and chaos in forced Van der Pol systems. In Dynamical Systems and Singularities, S. Pnevmatikos (ed.), pp. 313–323. North-Holland: Amsterdam.

Abraham, R. H. and Stewart, H. B. (1986). A chaotic blue sky catastrophe in forced relaxation oscillations. *Physica*, 21D, 394–400.

Abraham, R. H., Gardini, L. and Mira, C. (1997). *Chaos in Discrete Dynamical Systems*. Springer-Verlag: New York.

Abraham, R. H., Kocak, H. and Smith, W. R. (1985). Chaos and intermittency in an endocrine system model. In *Chaos, Fractals, and Dynamics*, P. Fischer and W. R. Smith (eds). Dekker: New York.

Afraimovich, V. S., Bykov, V. V. and Shilnikov, L. P. (1977). On the origin and structure of the Lorenz attractor. *Sov. Phys. Dokl.*, **22**, 253–255.

Aihara, K., Numajiri, T., Matsumoto, G. and Kotani, M. (1986). Structure of attractors in periodically forced neural oscillators. *Phys. Lett.*, **116A**, 313–317.

Aizawa, Y. and Ueza, T. (1982). Global aspects of the dissipative dynamical systems. II. Periodic and chaotic responses in the forced Lorenz system. *Prog. Theor. Phys.*, **68**, 1864–1879.

Alligood, K. T., Sauer, T. D. and Yorke, J. A. (1997). *Chaos – An Introduction to Dynamical Systems*, Springer-Verlag: New York.

Ananthakrishna, G. and Valsakumar, M. C. (1983). Chaotic flow in a model for repeated yielding. *Phys. Lett.*, **95A**, 69–71.

Andereck, C. D., Dickman, R. and Swinney, H. L. (1983). New flows in a circular Couette system with co-rotating cylinders. *Phys. Fluids*, **26**, 1395–1401.

Andronov, A. A. and Pontryagin, L. (1937). Systèmes Grossiers. *Dokl. Adad. Nauk. SSSR*, **14**, 247–251.

Andronov, A. A., Vitt, E. A. and Khaiken, S. E. (1966). *Theory of Oscillators*. Pergamon Press: Oxford.

Arecchi, F. T. and Califano, A. (1984). Low-frequency hopping phenomena in nonlinear system with many attractors. *Phys. Lett.*, **101A**, 443–446.

Arecchi, F. T., Badii, R. and Politi, A. (1984). Scaling of first passage times for noise induced crises. *Phys. Lett.*, **103A**, 3–7.

Arecchi, F. T., Meucci, R., Puccioni, G. and Tredicce, J. (1982). Experimental evidence of subharmonic bifurcations, multistability, and turbulence in a Q-switched gas laser. *Phys. Rev. Lett.*, **49**, 1217–1220.

Aref, H. (1983). Integrable, chaotic, and turbulent vortex motion in two-dimensional flows. *Annu. Rev. Fluid Mech.*, **15**, 345–389.

Armstrong, M. A. (1979). *Basic Topology*. McGraw-Hill: Maidenhead.

Arnéodo, A., Coullet, P. and Tresser, C. (1981a). A possible new mechanism for the onset of turbulence. *Phys. Lett.*, **81A**, 197–201.

Arnéodo, A., Coullet, P. and Tresser, C. (1981b). Possible new strange attractors with spiral structure. *Commun. Math. Phys.*, **79**, 573–579.

Arnéodo, A., Coullet, P. and Tresser, C. (1982). Oscillators with chaotic behavior: an illustration of a theorem by Shilnikov. *J. Stat. Phys.*, **27**, 171–182.

Arnéodo, A., Coullet, P., Tresser, C., Libchaber, A., Maurer, J. and d'Humières, D. (1983). On the observation of an uncompleted cascade in a Rayleigh-Bénard experiment. *Physica*, **6D**, 385–392.

Arnold, V. I. (1965). Small denominators. I. Mappings of the circumference onto itself. *Am. Math. Soc. Transl., Ser. 2*, **46**, 213–284.

Arnold, V. I. (1977). Loss of stability of self-oscillations close to resonance and versal deformations of equivalent vector fields. *Functional Anal. Appl.*, **11**, 85–92.

Arnold, V. I. (1978). *Mathematical Methods of Classical Mechanics*. Springer-Verlag: New York, Heidelberg and Berlin.

Arnold, V. I. (1983). *Geometrical methods in the Theory of Ordinary Differential Equations*. Springer-Verlag: New York, Heidelberg and Berlin.

Aronson, D. G., Chory, M. A., Hall, G. R. and McGeehee, R. P. (1982). Bifurcations from an invariant circle for two-parameter families of maps of the plane: a computer assisted study. *Commun. Math. Phys.*, **83**, 303–354.

Arrowsmith, D. K. and Place, C. M. (1990). *An Introduction to Dynamical Systems*. Cambridge University Press: Cambridge.

Baesens, C. and Nicolis, G. (1983). Complex bifurcations in a periodically forced normal form. *Z. Phys. B*, **52**, 345–354.

Bak, P., Bohr, T. and Jensen, M. H. (1984). Mode-locking and the transition to chaos in dissipative systems. *Phys. Scr.*, **T9**, 50–58.

Baker, G., McRobie, F. A. and Thompson, J. M. T. (1997) Implications of chaos theory for engineering science. *Proc. Instn Mech. Engrs*, **211C**, 349–363.

Baker, G. L. and Gollub, J. P. (1990). *Chaotic Dynamics: An Introduction*. Cambridge. University Press: Cambridge.

Barenblatt, G. I., Iooss, G. and Joseph, D. D. (eds) (1983). *Nonlinear Dynamics and Turbulence*. Pitman: London.

Behringer, R. P. (1985). Rayleigh–Bénard convection and turbulence in liquid helium. *Rev. Mod. Phys.*, **57**, 657–687.

Belair, J. and Glass, L. (1985). Universality and self-similarity in the bifurcations of circle maps. *Physica*, **16D**, 143–154.

Benedicks, M. and Carleson, L. (1991). The dynamics of the Hénon map. *Annals of Mathematics*, **133**, 73–169.

Ben-Jacob, E., Goldhirsch, I., Imry, Y. and Fishman, S. (1982). Intermittent chaos in Josephson junctions. *Phys. Rev. Lett.*, **49**, 1599–1602.

Benjamin, T. B. (1978). Bifurcation phenomena in steady flows of viscous fluid. *Proc. R. Soc. Lond. A*, **359**, 1–26, 27–43.

Benjamin, T. B. and Mullin, T. (1981). Anomalous modes in the Taylor experiment. *Proc. R. Soc. Lond. A*, **377**, 221–249.

Bergé, P., Dubois, M., Manneville, P. and Pomeau, Y. (1980). Intermittency in Rayleigh–Bénard convection. *J. Phys. Lett.*, **41**, L341–L345.

Bergé, P., Pomeau, Y. and Vidal, Ch. (1984). *L'Ordre dans le Chaos*. Hermann: Paris.

Berry, M. V. (1976). Waves and Thom's theorem. *Adv. Phys.*, **25**, 1–26.

Berry, M. V. (1981). Regularity and chaos in classical mechanics, illustrated by three deformations of a circular 'billiard'. *Eur. J. Phys.*, **2**, 91–102.

Bier, M. and Bountis, C. (1984). Remerging Feigenbaum trees in dynamical systems. *Phys. Lett.*, **104A**, 239–244.

Birkhoff, G. D. (1913). Proof of Poincaré's geometric theorem. *Trans. Am. Math. Soc.*, **14**, 14–22; *Collected Mathematical Papers*, vol. 1, pp. 673–681.

Birkhoff, G. D. (1927). *Dynamical Systems*. American Mathematical Society: Providence, RI.

Birkhoff, G. D. (1932). Sur quelques courbes fermées remarquables. *Bull. Soc. Math. Fr.*, **60**, 1–26; *Collected Mathematical Papers*, vol. 2, pp. 418–443.

Birkhoff, G. D. (1935). Sur le problème restrient des trois corps. *Ann. Sc. Norm. Sup. Pisa (2)*, **4**, 267–306; *Collected Mathematical Papers*, vol. 2, pp. 466–505.

Birkhoff, G. D. (1950). *Collected Mathematical Papers*, vols 1–3. American Mathematical Society: Providence, RI.

Birkhoff, G. and Rota, G.-C. (1978). *Ordinary Differential Equations*, 3rd edn. Blaisdell: Waltham, MA.

Bishop, S. R. (ed.) (1994) *Impact Oscillators*, theme issue. *Phil. Trans. R. Soc. Lond.*, **347A**, 345–448.

Bishop, S. R., Leung, L. M. and Virgin, L. N. (1986). Predicting incipient jumps to resonance of compliant marine structures in an evolving sea-state. *5th Intl. Conf. Offshore Mechanics and Arctic Engineering*, ASME, Tokyo.

Bowen, R. (1978). *On Axiom A Diffeomorphisms* (*CBMS Regional Conference Series in Mathematics*, vol. 35). American Mathematical Society: Providence, RI.

Bowen, R. and Ruelle, D. (1975). The ergodic theory of Axiom A flows. *Invent. Math.*, **79**, 181–202.

Brandstäter, A., Swift, J., Swinney, H. L., Wolf, A., Farmer, J. D., Jen, E. and Crutchfield, P. J. (1983). Low-dimensional chaos in a hydrodynamic system. *Phys. Rev. Lett.*, **51**, 1442–1445.

Brorson, S. D., Dewey, D. and Linsay, P. S. (1983). Self-replicating attractor of a driven semiconductor oscillator. *Phys. Rev.*, **28**, 1201–1203.

Buffoni, B., Champneys, A. R. and Toland, J. F. (1996). Bifurcation and coalescence of a plethora of homoclinic orbits for a Hamiltonian system. *J. Dyn. Diff. Eq.*, **8**, 221–281.

Bullard, E. (1978). The disk dynamo. In *Topics in Nonlinear Mechanics*, S. Jorna (ed.), pp. 373–389. American Institute of Physics: New York.

Campbell, D. K. and Rose, H. A. (eds) (1983). Order in chaos. *Physica*, **7D**, no. 1.

Carr, J. (1981). *Applications of Center Manifold Theory*. Springer-Verlag: New York, Heidelberg and Berlin.

Carr, J. and Eilbeck, J. C. (1984). One-dimensional approximations for a quadratic Ikeda map. *Phys. Lett.*, **104A**, 59–62.

Cartwright, M. L. (1948). Forced oscillations in nearly sinusoidal systems. *J. Inst. Electr. Eng.*, **95**, 88–96.

Cartwright, M. L. and Littlewood, J. E. (1945). On nonlinear differential equations of the second order. I. The equation $\ddot{y} + k(1 - y^2)\dot{y} + y = b\lambda k \cos(\lambda t + a), k$ large. *J. Lond. Math. Soc.*, **20**, 180–189.

Casati, G., Ford, J., Vivaldi, F. and Visscher, W. M. (1984). One-dimensional classical many-body system having a normal thermal conductivity. *Phys. Rev. Lett.*, **52**, 1861–1864.

Champneys, A. R. and Thompson, J. M. T. (1996). A multiplicity of localized buckling modes for twisted rod equations. *Proc. R. Soc. Lond.*, **A452**, 2467–2491.

Champneys, A. R., Hunt, G. W. and Thompson, J. M. T. (eds) (1999). *Localization and Solitary Waves in Solid Mechanics*. World Scientific, Singapore.

Champneys, A. R., van der Heijden, G. H. M. and Thompson, J. M. T. (1997). Spatially complex localization after one-twist-per-wave equilibria in twisted circular rods with initial curvature. *Phil. Trans. R. Soc. Lond.*, **A355**, 2151–2174.

Chenciner, A. and Iooss, G. (1979). Bifurcations de tores invariants. *Arch. Rat. Mech. Anal.*, **69**, 108–198.

Chillingworth, D. R. J. (1976). *Differentiable Topology with a View to Applications*. Pitman: London.

Chirikov, B. V. (1979). A universal instability of many dimensional oscillator systems. *Phys. Rep.*, **52**, 263–379.

Chow, S. N. and Hale, J. K. (1982). *Methods of Bifurcation Theory*. Springer-Verlag: New York, Heidelberg and Berlin.

Ciliberto, S. and Gollub, J. P. (1984). Pattern competition leads to chaos. *Phys. Rev. Lett.*, **52**, 922–925.

Clifford, M. J. and Bishop, S. R. (1995). Estimation of symmetry breaking and escape by observation of manifold tangencies. *Int. J. Bifurcation and Chaos*, **5**, 883–890.

Coddington, E. A. and Levinson, N. (1955). *Theory of Ordinary Differential Equations*. McGraw-Hill: New York.

Collet, P. and Eckmann, J.-P. (1980a). On the abundance of aperiodic behaviour for maps on the interval. *Commun. Math. Phys.*, **73**, 115–160.

Collet, P. and Eckmann, J.-P. (1980b). *Iterated Maps on the Interval as Dynamical Systems* (*Progress in Physics*, vol. 1). Birkhäuser: Boston.

Collet, P., Eckmann, J.-P. and Koch, H. (1981). Period doubling bifurcations for families of maps on R^n. *J. Stat. Phys.*, **25**, 1–14.

Collet, P., Eckmann, J.-P. and Lanford, O. E. (1980). Universal properties of maps on an interval. *Commun. Math. Phys.*, **76**, 211–254.

Cooperrider, N. K. (1980). Nonlinear behavior in rail vehicle dynamics. In *New Approaches to Nonlinear Problems in Dynamics*, P. J. Holmes (ed.) pp. 173–194. SIAM: Philadelphia.

Cornfeld, I. P., Fomin, S. V. and Sinai, Ya. G. (1982). *Ergodic Theory*. Springer-Verlag: New York, Heidelberg and Berlin.

Coullet, P., Tresser, C. and Arnéodo, A. (1979). Transition to stochasticity for a class of forced oscillators. *Phys. Lett.*, **72A**, 268–270.

Crutchfield, J. P. (1984). *Chaotic Attractors of Driven Oscillators* (16 mm film). Aerial Press: Santa Cruz, CA.

Crutchfield, J. P., Farmer, J. D. and Huberman, B. A. (1982). Fluctuations and simple chaotic dynamics. *Phys. Rep.*, **92**, 45–82.

Curry, J. (1978). A generalized Lorenz system. *Commun. Math. Phys.*, **60**, 193–204.

Curry, J. (1979). On the Hénon transformation. *Commun. Math. Phys.*, **68**, 129–140.

Curry, J. H. and Johnson, J. R. (1982). On the rate of approach to homoclinic tangency. *Phys. Lett.*, **92**, 217–220.

Cusumano, J. P. and Kimble, B. W. (1994). Experimental observation of basins of attraction and homoclinic bifurcation in a magneto-mechanical oscillator. In *Nonlinearity and Chaos in Engineering Dynamics*, J. M. T. Thompson and S. R. Bishop (eds), pp. 71–85. John Wiley & Sons: Chichester.

Cvitanović, P. (ed.) (1984). *Universality in Chaos*. Adam Hilger: Bristol.

Davis, P. and Ikeda, K. (1984). T^3 in a model of a nonlinear optical resonator. *Phys. Lett*, **100A**, 455–459.

Day, R. H. (1982). Irregular growth cycles. *Am. Econ. Rev.*, **72**, 406–414.

Day, R. H. (1983). The emergence of chaos from classical economic growth. *Q. J. Econ.*, 201–213.

Denjoy, A. (1932). Sur les courbes définies par les équations différentielles à la surface du tore. *J. Math.*, **17**, 333–375.

Derrida, B., Gervois, A. and Pomeau, Y. (1979). Universal metric properties of bifurcations of endomorphisms. *J. Phys. A*, **12**, 269–296.

Devaney, R. L. (1977). Blue sky catastrophes in reversible and Hamiltonian systems. *Indiana Univ. Math. J.*, **26**, 247–263.

Devaney, R. and Nitecki, Z. (1979). Shift automorphisms in the Hénon mapping. *Commun. Math. Phys.*, **67**, 137–148.

D'Humières, D., Beasley, M. R., Huberman, B. A. and Libchaber, A. (1982). Chaotic states and routes to chaos in the forced pendulum. *Phys. Rev. A*, **26**, 3483–3496.

Diener, M. (1984). The canard unchained, or how fast/slow dynamical systems bifurcate. *Math. Intell.*, **6**, no. 3, 38–49.

Donnelly, R. J., Park, K., Shaw, R. and Walden, R. W. (1980). Early nonperiodic transitions in Couette flow. *Phys. Rev. Lett.*, **44**, 987–989.

Dowell, E. H. (1975). *Aeroelasticity of Plates and Shells*. Noordhoff: Leyden.

Dowell, E. H. (1980). Nonlinear aeroelasticity. In *New Approaches to Nonlinear Problems in Dynamics*, P. J. Holmes (ed.), pp. 147–172b. SIAM: Philadelphia.

Dowell, E. H. (1982). Flutter of a buckled plate as an example of chaotic motion of a deterministic autonomous system. *J. Sound Vib.*, **85**, 333–334.

Dowell, E. H. (1984). Observations and evolution of chaos for an autonomous system. *ASME J. Appl. Mech.*, **51**, 664–673.

Drazin, P. G. (1992). *Nonlinear Systems*. Cambridge University Press: Cambridge.

Dubois, M., Bergé, P. and Croquette, V. (1982). Study of non-steady convective regimes using Poincaré sections. *J. Phys. Lett.*, **43**, L295–L298.

Dubois, M., Rubio, M. A. and Bergé, P. (1983). Experimental evidence of intermittencies associated with a subharmonic bifurcation. *Phys. Rev. Lett.*, **51**, 1446–1449.

Duffing, G. (1918). *Erzwungene Schwingungen bei Veränderlicher Eigenfrequenz*. Vieweg: Braunschweig.

Eckmann, J.-P. and Ruelle, D. (1985). Ergodic theory of chaos and strange attractors. *Rev. Mod. Phys.*, **57**, 617–656.

Eilbeck, J. C. (1984). The sine-Gordon equation—from solitons to chaos. *Bull. Inst. Math. Appl.*, **20**, 77–81.

Elvey, J. S. N. (1983). On the elimination of de-stabilizing motions of articulated mooring towers under steady sea conditions. *IMA J. Appl. Math.*, **31**, 235–252.

Epstein, I. R. (1983). Oscillations and chaos in chemical systems. *Physica*, **7D**, 47–56.

Epstein, I. R. and Pojman, J. A. (1998). *An Introduction to Nonlinear Chemical Dynamics: Oscillations, Waves, Patterns*, Oxford University Press: Oxford

Falconer, K. (1990). *Fractal Geometry*. John Wiley & Sons: Chichester.

Farmer, J. D. (1982). Chaotic attractors of an infinite-dimensional dynamical system. *Physica*, **4D**, 366–393.

Farmer, J. D., Crutchfield, J., Froehling, H., Packard, N. and Shaw, R. (1980). Power spectra and mixing properties of strange attractors. In *Nonlinear Dynamics*, R. H. G. Helleman (ed.). New York Academy of Sciences: New York.

Farmer, J. D., Hart, J. and Weidman, P. (1982). A phase space analysis of baroclinic flow. *Phys. Lett.*, **91A**, 22–24.

Farmer, J. D., Ott, E. and Yorke, J. A. (1983). The dimension of chaotic attractors. *Physica*, **7D**, 153–180.

Fatou, P. (1919). Sur les équations fonctionnelles. *Bull. Soc. Math. Fr.*, **47**, 161–270; **48**, 33–95; **48**, 208–314.

Fauve, F., Laroche, C. and Perrin, B. (1985). Competing instabilities in a rotating layer of mercury heated from below. *Phys. Rev. Lett.*, **55**, 208–210.

Fauve, S., Laroche, C., Libchaber, A. and Perrin, B. (1984). Chaotic phases and magnetic order in a convective fluid. *Phys. Rev. Lett.*, **52**, 1774–1777.

Feigenbaum, M. J. (1978). Quantitative universality for a class of nonlinear transformations. *J. Stat. Phys.*, **19**, 25–52.

Feigenbaum, M. J. (1979). The onset spectrum of turbulence. *Phys. Lett.* **74A**, 375–378.

Feigenbaum, M. J. (1983). Universal behavior in nonlinear systems. *Physica*, **7D**, 16–39.

Feigenbaum, M. J., Kadanoff, L. P. and Shenker, S. J. (1982). Quasiperiodicity in dissipative systems: a renormalization group analysis. *Physica*, **5D**, 370–386.

Feingold, M. and Peres, A. (1983). Regular and chaotic motion of coupled rotators. *Physica*, **9D**, 433–438.

Feit, S. D. (1978). Characteristic exponents and strange attractors. *Commun. Math. Phys.*, **61**, 249–260.

Fenstermacher, P. R., Swinney, H. L. and Gollub, J. P. (1979). Dynamical instabilities and the transition to chaotic Taylor vortex flow. *J. Fluid Mech.*, **94**, 103–128.

Feroe, J. A. (1982). Existence and stability of multiple impulse solutions of a nerve equation. *SIAM J. Appl. Math.*, **42**, 235–246.

Field, R. J. and Burger, M. (1984). *Oscillations and Traveling Waves in Chemical Systems*. John Wiley & Sons: New York.

FitzHugh, R. (1961). Impulses and physiological states in theoretical models of nerve membrane. *Biophys. J.*, **1**, 445–466.

Flaherty, J. E. and Hoppensteadt, F. C. (1978). Frequency entrainment of a forced Van der Pol oscillator. *Stud. Appl. Math.*, **18**, 5–15.

Flashner, H. and Hsu, C. S. (1983). A study of nonlinear periodic systems via the point mapping methods. *Int. J. Numer. Meth. Eng.*, **19**, 185–215.

Foale, S. and Thompson, J. M. T. (1991). Geometrical concepts and computational techniques of nonlinear dynamics. *Computer Methods in Appl. Mechs and Engng*, **89**, 381–394.

Foale, S., Thompson, J. M. T. and McRobie, F. A. (1998) Numerical dimension-reduction methods for nonlinear shell vibrations. *J. Sound and Vibration*, **215**, 527–545.

Ford, J. (1978). A picture book of stochasticity. In *Topics in Nonlinear Dynamics*, S. Jorna (ed.), pp. 121–146. American Institute of Physics: New York.

Ford, J. (1983). How random is a coin toss? *Phys. Today*, **36**, no. 6 (July), 40–48.

Fowler, A. C. and McGuinness, M. J. (1982a). A description of the Lorenz attractor at high Prandtl number. *Physica*, **5D**, 149–182.

Fowler, A. C. and McGuinness, M. J. (1982b). Hysteresis in the Lorenz equations. *Phys. Lett.*, **92A**, 103–106.

Fraser, A. M. and Swinney, H. L. (1986). Using mutual information to find independent coordinates for strange attractors. *Phys. Rev. A*, **33**, 1134–1140.

Frauenthal, J. C. (1984). Population dynamics and demography. In *Proc. Symp. on Applied Mathematics*, vol. 30, pp. 9–18. American Mathematical Society: Providence, RI.

Frederickson, P., Kaplan, J. L., Yorke, E. D. and Yorke, J. A. (1983). The Liapunov dimension of strange attractors. *J. Differ. Eq.*, **49**, 185–207.

Fujisaka, H. and Yamada, T. (1983). Stability theory of synchronized motion in coupled-oscillator systems. *Prog. Theor. Phys.*, **69**, 32–47; **70**, 1240–1248; **72**, 885–894.

Garrido, L. (ed.) (1983). *Dynamical Systems and Chaos* (*Springer Lecture Notes in Physics*, vol. 179). Springer-Verlag: Berlin and New York.

Gaspard, P. and Nicolis, G. (1983). What can we learn from homoclinic orbits in chaotic dynamics? *J. Stat. Phys.*, **31**, 499–518.

Gavrilov, N. K. and Shilnikov, L. P. (1972–73). On three-dimensional dynamical systems close to systems with a structurally unstable homoclinic curve. *Math. USSR Sb.*, **17**, 467–485; **19**, 139–156.

Ghrist, R. and Holmes, P. (1993). Knots and orbit genealogies in three dimensional flows. In *Bifurcations and Periodic Orbits of Vector Fields*, NATO ASI Series, Kluwer.

Gibson, G. and Jeffries, C. (1984). Observation of period doubling and chaos in spinwave instabilities in yttrium iron garnet. *Phys. Rev.*, **29**, 811–818.

Giglio, M., Musazzi, S. and Perini, U. (1981). Transition to chaotic behavior via a reproducible sequence of period-doubling bifurcations. *Phys. Rev. Lett.*, **47**, 243–246.

Gilmore, R. (1981). *Catastrophe Theory for Scientists and Engineers*. John Wiley & Sons: New York.

Gilpin, M. E. (1979). Spiral chaos in a predator-prey model. *Am. Naturalist*, **113**, 306–308.

Gioggia, R. S. and Abraham, N. H. (1983). Routes to chaotic output from a singlemode, DC-excited laser. *Phys. Rev. Lett.*, **51**, 650–653.

Glass, L. and Mackey, M. C. (1988). *From Clocks to Chaos: The Rhythms of Life*. Princeton University Press: Princeton.

Glass, L. and Perez, R. (1982). The fine structure of phase locking. *Phys. Rev. Lett.*, **48**, 1772–1775.

Glass, L., Guevara, M. R., Shrier, A. and Perez, R. (1983). Bifurcation and chaos in a periodically stimulated cardiac oscillator. *Physica*, **7D**, 89–101.

Glendinning, P. (1994). *Stability, Instability and Chaos*. Cambridge University Press: Cambridge.

Gollub, J. P. and Benson, S. V. (1980). Many routes to turbulent convection. *J. Fluid Mech.*, **100**, 449–470.

Gollub, J. P. and Swinney, H. L. (1975). Onset of turbulence in a rotating fluid. *Phys. Rev. Lett.*, **35**, 927–930.

Golubitsky, M. and Stewart, I. (1985). Hopf bifurcation in the presence of symmetry. *Arch. Rat. Mech. Anal.*, **87**, 107–165.

Golubitsky, M. and Langford, W. (1981). Classification and unfoldings of degenerate Hopf bifurcations. *J. Differ. Eq.*, **41**, 375–415.

Golubitsky, M. and Schaeffer, D. (1979). A theory for imperfect bifurcation via singularity theory. *Commun. Pure Appl. Math.*, **32**, 21–98.

Golubitsky, M. and Schaeffer, D. (1985). *Singularity and Groups in Bifurcation Theory*. Springer-Verlag: New York.

Gorman, M., Reith, L. A. and Swinney, H. L. (1980). Modulation patterns, multiple frequencies and other phenomena in circular Couette flow. In *Nonlinear Dynamics*, R. H. G. Helleman (ed.). New York Academy of Sciences: New York.

Gorman, M., Widmann, P. J. and Robbins, K. A. (1984). Chaotic flow regimes in a convection loop. *Phys. Rev. Lett.*, **52**, 2241–2244.

Gottwald, J. A., Virgin, L. N. and Dowell, E. H. (1995). Routes to escape from an energy well. *J. Sound and Vibration*, **187**, 133–144.

Graham, R. (1976). Onset of self-pulsing in lasers and the Lorenz model. *Phys. Lett.*, **58A**, 440–442.

Grassberger, P. and Procaccia, I. (1983). Measuring the strangeness of strange attractors. *Physica*, **9D**, 189–208.

Grebogi, C. and Yorke, J. A. (eds) (1997). *The Impact of Chaos on Science and Society*. United Nations University Press: New York.

Grebogi, C., Ott, E. and Yorke, J. A. (1982). Chaotic attractors in crisis. *Phys. Rev. Lett.*, **48**, 1507–1510.

Grebogi, C., Ott, E. and Yorke, J. A. (1983a). Are three-frequency quasiperiodic orbits to be expected in typical nonlinear dynamical systems? *Phys. Rev. Lett.*, **51**, 339–342.

Grebogi, C., Ott, E. and Yorke, J. A. (1983b). Crises, sudden changes in chaotic attractors, and transient chaos. *Physica*, **7D**, 181–200.

Grebogi, C., Ott, E. and Yorke, J. A. (1987). Basin boundary metamorphoses: changes in accessible boundary orbits. *Physica*, **24D**, 243–262.

Greene, J. M. and Percival, I. C. (1981). Hamiltonian maps in the complex plane. *Physica*, **3D**, 530–548.

Greenspan, B. and Holmes, P. (1984). Repeated resonance and homoclinic bifurcation in a periodically forced family of oscillators. *SIAM J. Math. Anal.*, **15**, 69–97. Guckenheimer, J. (1973). Bifurcation and catastrophe. In *Dynamical Systems*, M. M. Pexoto (ed.). Academic Press: New York.

Greenspan, B. D. and Holmes, P. J. (1982). Homoclinic orbits, subharmonics and global bifurcations in forced oscillations. In *Nonlinear Dynamics and Turbulence*, G. Barcnblatt, G. Iooss and D. D. Joseph (eds). Pitman: London.

Guckenheimer J. (1977). On the bifurcation of maps of the interval. *Invent. Math.*, **39**, 165–178.

Guckenheimer, J. (1976). A strange strange attractor. In *The Hopf Bifurcation and Its Applications*, J. E. Marsden and M. McCracken (eds), pp. 368–381. Springer-Verlag: New York, Heidelberg and Berlin.

Guckenheimer, J. (1979). Sensitive dependence on initial conditions for one-dimensional maps. *Commun. Math. Phys.*, **70**, 133–160.

Guckenheimer, J. (1980a). Symbolic dynamics and relaxation oscillations. *Physica*, **1D**, 227–235.

Guckenheimer, J. (1980b). Bifurcations of dynamical systems. In *Dynamical Systems*, CIME Lectures, Bressanone, Italy, June 1978, pp. 115–231 (*Progress in Mathematics*, vol. 8) Birkhauser: Boston.

Guckenheimer, J. (1981). On a codimension two bifurcation. In *Dynamical Systems and Turbulence*, D. A. Rand and L. S. Young (eds), pp. 99–142 (*Springer Lecture Notes in Mathematics*, vol. 898). Springer-Verlag: New York, Heidelberg and Berlin.

Guckenheimer, J. and Buzyna, G. (1983). Dimension measurements for geostrophic turbulence. *Phys. Rev. Lett.*, **51**, 1438–1441.

Guckenheimer, J. and Holmes, P. (1983). *Nonlinear Oscillations, Dynamical Systems, and Bifurcations of Vector Fields*. Springer-Verlag: New York, Berlin and Heidelberg.

Guckenheimer, J. and Williams, R. F. (1979). Structural stability of Lorenz attractors. *Publ. Math. IHES*, **50**, 59–72.

Guevara, M. R., Glass, L. and Shrier, A. (1981). Phase locking, period-doubling bifurcations, and irregular dynamics in periodically stimulated cardiac cells. *Science*, **214**, 1350–1352.

Gumowski, I. and Mira, C. (1980a). *Dynamique Chaotique*. Cepadues: Toulouse.

Gumowski, I. and Mira, C. (1980b). *Recurrences and Discrete Dynamical Systems* (*Springer Lecture Notes in Mathematics*, vol. 809). Springer-Verlag: New York, Heidelberg and Berlin.

Gurel, O. and Rössler, O. E. (eds) (1979). *Bifurcation Theory and Applications in Scientific Disciplines* (*Annals of the New York Academy of Sciences*, vol. 316). New York Academy of Sciences: New York.

Gutzwiller, M. (1990). *Chaos in Classical and Quantum Mechanics*. Springer-Verlag: New York.

Haken, H. (1975). Analogy between higher instabilities in fluids and lasers. *Phys. Lett.*, **53A**, 77–78.

Haken, H. (ed.) (1982). *Evolution of Order and Chaos in Physics, Chemistry, and Biology* (*Springer Series in Synergetics*, vol. 17). Springer-Verlag: Berlin, Heidelberg and New York.

Hale, J. K. (1969). *Ordinary Differential Equations*. John Wiley & Sons: New York.

Hammel, S. M., Yorke, J. A. and Grebogi, C. (1987). Do numerical orbits of chaotic dynamical processes represent true orbits? *J. Complexity*, **3**, 136–145.

Hao, B.-L. (1984). *Chaos*. World Scientific: Singapore.

Hao, B.-L. (1985). Bifurcations and chaos in a periodically forced limit cycle oscillator. In *Advances of Science in China: Physics*, vol. 1, Zhu Hong-yuan *et al.* (eds), p. 113. Science Press: Beijing.

Hao, B.-L. and Zhang, S. Y. (1982). Subharmonic stroboscopy as a method to study period-doubling bifurcations. *Phys. Lett.*, **87A**, 267–270.

Hao, B.-L. and Zheng, W.-M. (1998). *Applied Symbolic Dynamics and Chaos*. World Scientific: Singapore.

Harrison, R. G., Firth, W. J., Emshary, C. A. and Al-Saidi, I. A. (1983). Observation of period doubling in an all-optical resonator containing NH_3 gas. *Phys. Rev. Lett.*, **51**, 562–565.

Hart, J. E. (1984). A new analysis of the closed loop thermosyphon. *Int. J. Heat Mass Transfer*, **27**, 125–136.

Harth, E. (1983). Order and chaos in neural systems: an approach to the dynamics of higher brain functions. *IEEE Trans. Syst. Man, Cybern.*, **SMC-13**, 782–789.

Hartman, P. (1964). *Ordinary Differential Equations*. John Wiley & Sons: New York.

Hassard, B. D., Kazarinoff, N. D. and Wan, Y.-H. (1981). *Theory and Applications of Hopf Bifurcation*. Cambridge University Press: Cambridge.

Hastings, S. P. (1982). Single and multiple pulse waves for the FitzHugh–Nagumo equations. *SIAM J. Appl. Math.*, **42**, 247–260.

Hayashi, C. (1964). *Nonlinear Oscillations in Physical Systems*. McGraw-Hill: New York.

Hayashi, C. (1975). *Selected Papers on Nonlinear Oscillations*. Nippon Printing and Publishing: Osaka.

Hayashi, C. (1980). The method of mapping with reference to the doubly asymptotic structure of invariant curves. *Int. J. Nonlinear Mech.*, **15**, 341–348.

Hayashi, H., Ishizuka, S. and Hirakawa, K. (1983). Transition to chaos via intermittency in the Onchidium pacemaker neuron. *Phys. Lett.*, **98A**, 474–476.

Hénon, M. (1976). A two-dimensional mapping with a strange attractor. *Commun. Math. Phys.*, **50**, 69–77.

Hénon, M. and Pomeau, Y. (1976). Two strange attractors with a smiple structure. In *Turbulence and Navier–Stokes Equations* (*Springer Lecture Notes in Mathematics*, vol. 565). Springer-Verlag: New York.

Held, G. A., Jeffries, C. and Haller, E. E. (1984). Observation of chaotic behavior in an electron–hole plasma in Ge. *Phys. Rev. Lett.*, **52**, 1037–1040.

Helleman, R. H. G. (1979). Exact results for some linear and nonlinear beam–beam effects. In *Nonlinear Dynamics and the Beam–Beam Interaction*, M. Month and J. C. Herrara (eds), pp. 236–256. Amer. Inst. Phys: New York.

Helleman, R. H. G. (ed.) (1980). *Nonlinear Dynamics* (*Annals of the New York Academy of Sciences*, vol. 357). New York Academy of Sciences: New York.

Herman, M. R. (1977). Mesure de Lebesgue et nombre de rotation. In *Geometry and Topology*, J. Palis and M. deCarmo (eds), pp. 271–293 (*Springer Lecture Notes in Mathematics*, vol. 597). Springer-Verlag: New York, Heidelberg and Berlin.

Herman, M. R. (1979). Sur la conjugaison différentiable des diffeomorphismes du cercle à des rotations. *Publ. Math. IHES*, **49**, 5–233.

Hirsch, J. E., Huberman, B. A. and Scalapino, D. J. (1982). Theory of intermittency. *Phys. Rev. A*, **25**, 519–532.

Hirsch, M. W. (1976). *Differential Topology*. Springer-Verlag: New York, Heidelberg and Berlin.

Hirsch, M. W. (1984). The dynamical systems approach to differential equations. *Bull. Am. Math. Soc.*, **11**, 1–64.

Hirsch, M. W. and Smale, S. (1974). *Differential Equations, Dynamical Systems and Linear Algebra*. Academic Press: New York.

Hirsch, M. W., Pugh, C. C. and Shub, M. (1977). *Invariant Manifolds* (*Springer Lectures Notes in Mathematics*, vol. 583). Springer-Verlag: New York, Heidelberg and Berlin.

Hodgkin, A. L. and Huxley, A. F. (1952). A quantitative description of membrane current and its application to conduction and excitation in nerve. *J. Physiol.*, **117**, 500–544.

Holden, A. V. and Winlow, W. (1983). Neuronal activity as the behavior of a differential system. *IEEE Trans. Syst. Man, Cybern.*, **SMC-13**, 711–719.

Holmes, P. J. (1977). Bifurcations to divergence and flutter in flow-induced oscillations: a finite-dimensional analysis, *J. Sound Vib.*, **53**, 471–503.

Holmes, P. J. (1979). A nonlinear oscillator with a strange attractor. *Phil. Trans. R. Soc. Lond.*, **A292**, 419–448.

Holmes, P. J. (1980). Unfolding a degenerate nonlinear oscillator. In *Nonlinear Dynamics*, R. H. G. Helleman (ed.), pp. 473–488. New York Academy of Sciences: New York.

Holmes, P. J. (1982). The dynamics of repeated impacts with a sinusoidally vibrating table. *J. Sound. Vib.*, **84**, 173–189.

Holmes, P. J. and Marsden, J. E. (1978). Bifurcations to divergence and flutter in flowinduced oscillations: an infinite-dimensional analysis. *Automatica*, **14**, 367–384.

Holmes, P. J. and Marsden, J. E. (1981). A partial differential equation with infinitely many periodic orbits: chaotic oscillations of a forced beam. *Arch. Rat. Mech. Anal.*, **76**, 135–166.

Holmes, P. J. and Moon, F. C. (1983). Strange attractors and chaos in nonlinear mechanics. *ASME J. Appl. Mech.*, **50**, 1021–1032.

Holmes, P. J. and Rand, D. A. (1976). The bifurcations of Duffing's equation: an application of catastrophe theory. *J. Sound Vib.*, **44**, 237–253.

Holmes, P. J. and Rand, D. A. (1978). Bifurcations of the forced Van der Pol oscillator. *Q. Appl. Math.*, **35**, 495–509.

Holmes, P. J. and Rand, D. A. (1980). Phase portraits and bifurcations of the nonlinear oscillator $\ddot{x} + (a + \gamma x^2)\dot{x} + \beta x + \delta x^3 = 0$. *Int. J. Nonlinear Mech.*, **15**, 449–458.

Holmes, P. J. and Whitley, D. C. (1983). On the attracting set for Duffing's equation. II. A geometrical model for moderate force and damping. *Physica*, **7D**, 111–123.

Holmes, P. J. and Whitley, D. C. (1984). Bifurcations of one-and two-dimensional maps. *Phil. Trans. R. Soc. Lond.*, **A311**, 43–102; Erratum, **A312**, 601–602.

Holmes, P., Lumley, J. L. and Berkooz, G. (1998). *Turbulence, Coherent Structures, Dynamical Systems and Symmetry*. Cambridge University Press: Cambridge.

Hopf, E. (1942). Abzweigung einer periodischen Lösung von einer stationären Lösung eines Differentialsystems. *Ber. Math.-Phys. Klasse Sachs. Akad. Wiss. Leipzig*, **94**, 1–22. English translation in Marsden and McCracken (1976).

Hsu, C. S. (1980a). A theory of index for point mapping dynamical systems. *ASME J. Appl. Mech.*, **47**, 185–190.

Hsu, C. S. (1980b). A theory of cell-to-cell mapping for nonlinear dynamical systems. *ASME J. Appl. Mech.*, **47**, 931–939.

Huberman, B. A., Crutchfield, J. P. and Packard, N. H. (1980). Noise phenomena in Josephson junctions. *Appl. Phys. Lett.*, **37**, 750–752.

Huberman, B. and Crutchfield, J. P. (1979). Chaotic states of anharmonic systems in periodic fields. *Phys. Rev. Lett.*, **43**, 1743–1747.

Hudson, J. L. and Rössler, O. E. (1985). Chaos and complex oscillations in stirred chemical reactors. In *Dynamics of Nonlinear Systems*, V. Hlavacek (ed.) Gordon and Breach: New York.

Hudson, J. L., Hart, M. and Marinko, D. (1979). An experimental study of multiple peak periodic and nonperiodic oscillations in the Belousov–Zhabotinskii reaction. *J. Chem. Phys.*, **71**, 1601–1606.

Hunt, G. W. and Neto, E. L. (1991). Localized buckling in long axially loaded cylindrical shells. *J. Mech. Phys. Solids*, **39**, 881–894.

Hunt, G. W. and Neto, E. L. (1993). Maxwell critical loads for axially loaded cylindrical shells. *J. Appl. Mech.*, **60**, 702–706.

Hunt, G. W. (1981). An algorithm for the nonlinear analysis of compound bifurcation. *Phil. Trans. R. Soc. Lond.*, **A300**, 443–471.

Hunt, G. W., Bolt, H. M. and Thompson, J. M. T. (1989). Structural localization phenomena and the dynamical phase-space analogy. *Proc. R. Soc. Lond.*, **A425**, 245–267.

Huseyin, K. (1978). *Vibrations and Stability of Multiple Parameter Systems*. Noordhoff: Alphen.

Ikeda, K., Daido, H. and Akimoto, O. (1980). Optical turbulence: chaotic behavior of transmitted light from a ring cavity. *Phys. Rev. Lett.*, **45**, 709–712.

Iooss, G. and Joseph, D. D. (1977). Bifurcation and Stability of nT-periodic solutions branching from T-periodic solutions at points of resonance. *Arch. Rat. Mech. Anal.*, **66**, 135–172.

Iooss, G. and Joseph, D. D., (1980). *Elementary Stability and Bifurcation Theory*. Springer-Verlag: New York.

Iooss, G. and Langford, W. F. (1980). Conjectures on the routes to turbulence via bifurcation. In *Nonlinear Dynamics*, R. H. G. Helleman (ed.), pp. 489–505. New York Academy of Sciences: New York.

Iooss, G., Helleman, R. and Stora, R. (eds) (1983). *Chaotic Behavior of Deterministic Systems*. North-Holland: Amsterdam.

Irwin, M. C. (1980). *Smooth Dynamical Systems*. Academic Press: New York.

Jeffries, C. and Perez, J. (1982). Observation of a Pomeau–Manneville intermittent route to chaos in a nonlinear oscillator. *Phys. Rev. A*, **26**, 2117–2122.

Jones, D. S. and Sleeman, B. D. (1983). *Differential Equations and Mathematical Biology*. George Allen and Unwin: London.

Jordan, D. W. and Smith, P. (1977). *Nonlinear Ordinary Differential Equations*. Oxford University Press: Oxford.

Jorna, S. (ed.) (1978). *Topics in Nonlinear Mechanics: A Tribute to Sir Edward Bullard* (*AIP Conference Proceedings*, vol. 46). American Institute of Physics: New York.

Joseph, D. D. (1979). *Stability of Fluid Motions*, vols. 1 and 2. Springer-Verlag: Berlin.

Julia, J. G. (1918). Memoire sur l'iteration des fonctions rationelles. *J. Math., Ser. 7*, **4**, 47–245.

Kadanoff, L. P. (1983). Roads to chaos. *Phys. Today*, **36**, no. 12 (December), 46–53.

Kahn, P. B. and Zarmi, Y. (1997). *Nonlinear Dynamics: Exploration through Normal Forms*. John Wiley & Sons: New York.

Kaneko, K. (1984). Supercritical behavior of disordered orbits of a circle map. *Prog. Theor. Phys.*, **72**, 1089–1103.

Kaneko, K. (1989). Spatiotemporal chaos in one- and two-dimensional coupled map lattices. *Physica*, **37D**, 60–82.

Kaneko, K. and Tsuda, I. (2000). *Complex Systems, Chaos and Beyond: A Constructive Approach with Applications in Life Sciences*. Springer-Verlag: New York.

Kapitaniak, T. and Bishop, S. R. (1999). *A Dictionary of Nonlinear Dynamics*, John Wiley & Sons, Chichester.

Kaplan, D. and Glass, L. (1995). *Understanding Nonlinear Dynamics*, Springer–Verlag: New York.

Kaplan, J. L. and Yorke, J. A. (1979a). Chaotic behavior of multidimensional difference equations. In *Functional Differential Equations and Approximation of Fixed Points*, H. O. Peitgen and H. O. Walther (eds), pp. 228–237 (*Springer Lecture Notes in Mathematics*, vol. 730). Springer-Verlag: New York, Heidelberg and Berlin.

Kaplan, J. L. and Yorke, J. A. (1979b). Preturbulence, a regime observed in a fluid flow model of Lorenz. *Commun. Math. Phys.*, **67**, 93–108.

Katok, A. B. (1980). Lyapunov exponents, entropy and periodic points for diffeomorphisms. *Publ. Math. IHES*, **51**, 137–174.

Keeler, J. D. and Farmer, J. D. (1986). Robust space-time intermittency and $1/f$ noise. *Physica*, **23D**, 413–435.

Kelley, A. (1967). The stable, center stable, center, center unstable and unstable manifolds. *J. Differ. Eq.*, **3**, 546–570.

Kennedy, J. and Yorke, J. A. (1991). Basins of Wada. *Physica*, **51D**, 213–225.

Keolian, R., Turkevich, L. A., Putterman, S. J., Rudnick, I. and Rudnick, J. A. (1981). Subharmonic sequences in the Faraday experiment: departures from period doubling. *Phys. Rev. Lett.*, **47**, 1133–1136.

Kilner, F. A. (1961). Model tests on the motion of moored ships placed on long waves. *Coastal Engineering*, 7th Conference, Hague, 1960, J. W. Johnson (ed.), pp. 723–745.

Kim, J. H. and Stringer, J. (eds) (1992). *Applied Chaos*. John Wiley & Sons: New York.

King, G. and Swinney, H. L. (1983). Limits of stability and irregular flow patterns in wavy vortex flow. *Phys. Rev. A*, **27**, 1240–1243.

Klinker, T., Meyer-Ilse, W. and Lauterborn, W. (1984). Period doubling and chaotic behavior in a driven Toda oscillator. *Phys. Lett.*, **101A**, 371–375.

Koch, B. P., Leven, R. W., Pompe, B. and Wilke, C. (1983). Experimental evidence for chaotic behaviour of a parametrically forced pendulum. *Phys. Lett.*, **96A**, 219–224.

Kolmogorov, A. N. (1957). General theory of dynamical systems and classical mechanics. *Proceedings of the 1954 International Congress of Mathematics*, pp. 315–333. North-Holland: Amsterdam.

Kostelich, E., Kan, I., Grebogi, C., Ott, E. and Yorke, J. A. (1997). Unstable dimension variability: a source of nonhyperbolicity in chaotic systems. *Physica*, **109D**, 81–90.

Krylov, N. M. and Bogoliubov, N. N. (1947). *Introduction to Nonlinear Mechanics*. Princeton University Press: Princeton, NJ.

Kuramoto, Y. (1978). Diffusion-induced chaos in reaction systems. *Prog. Theor. Phys.* (*Suppl.*), **64**, 346–367.

Lai, Y.-C., Grebogi, C. and Yorke, J. A. (1992). *Sudden change in size of* chaotic attractor: how does it occur? Chapter 19 of *Applied Chaos*, J. H. Kim and J. Stringer (eds). John Wiley & Sons: New York.

Lai, Y.-C., Grebogi, C., Yorke, J. A. and Kan, I. (1993). How often are chaotic saddles nonhyperbolic? *Nonlinearity*, **6**, 779–797.

Laing, C. R., McRobie, F. A. and Thompson, J. M. T. (1999). The post-processed Galerkin method applied to non-linear shell vibrations. *Dynamics and Stability of Systems*, **14**, 163–181.

Lanford, O. E. (1982a). The strange attractor theory of turbulence. *Annu. Rev. Fluid. Mech.*, **14**, 347–364.

Lanford, O. E. (1982b). A computer assisted proof of the Feigenbaum conjecture. *Bull. Am. Math. Soc.*, **6**, 427–434.

Lanford, O. E. (1985). A numerical study of the likelihood of phase locking. *Physica*, **14D**, 403–408.

Langford, W. F., Arnéodo, A., Coullet, P., Tresser, C. and Coste, J. (1980). A mechanism for the soft-mode instability. *Phys. Lett. A*, **78**, 11–14.

Lansbury, A. N. and Thompson, J. M. T. (1990). Incursive fractals: a robust mechanism of basin erosion preceding the optimal escape from a potential well. *Phys. Lett.*, **150A**, 355–361.

Lansbury, A. N., Thompson, J. M. T. and Stewart, H. B. (1992). Basin erosion in the twin-well Duffing oscillator: two distinct bifurcation scenarios. *Int. J. Bifurcation and Chaos*, **2**, 505–532.

LaSalle, J. P. and Lefschetz, S. (1961). *Stability by Liapunov's Direct Method with Applications*. Academic Press: New York.

Lauterborn, W. and Cramer, E. (1981). Subharmonic route to chaos observed in acoustics. *Phys. Rev. Lett.*, **47**, 1445–1448.

Laval, G. and Gresillon, D. (eds) (1979). *Intrinsic Stochasticity in Plasmas*. Editions de Physique Courtaboeuf: Orsay.

Lean, G. H. (1971). Subharmonic motions of moored ships subjected to wave action. *R. Inst. Naval Archit. Lond. Suppl. Pap.*, **113**, 387–399.

Lefschetz, S. (1957). *Ordinary Differential Equations: Geometric Theory*. Interscience Publishers: New York. (Reissued in 1977 by Dover: New York.)

Legras, B. and Ghil, M. (1983). Blocking and variations in atmospheric predictability. In *Predictability of Fluid Motions*, G. Holloway and B. J. West (eds), American Institute of Physics: New York.

Leipholz, H. (1970). *Stability Theory*. Academic Press: New York.

Leipnik, R. B. and Newton, T. A. (1981). Double strange attractor in rigid body motion with linear feedback control. *Phys. Lett.*, **86A**, 63–67.

LeTreut, H. and Ghil, M. (1983). Orbital forcing, climatic interactions, and glaciation cycles. *J. Geophys. Res.*, **88**, 5167–5190.

Levi, M. (1981). Qualitative analysis of the periodically forced relaxation oscillations. *Mem. Am. Math. Soc.*, **214**, 1–47.

Levi, M., Hoppensteadt, F. and Miranker, W. (1978). Dynamics of the Josephson junction. *Q. Appl. Math.*, **35**, 167–198.

Levinson, N. (1949). A second-order differential equation with singular solutions. *Ann. Math.*, **50**, 127–153.

Li, T. Y. and Yorke, J. A. (1975). Period three implies chaos. *Am. Math. Monthly*, **82**, 985–992.

Liapunov, A. M. (1949). *Problème Général de la Stabilité du Mouvement* (Annals of Mathematical Studies, vol. 17). Princeton University Press: Princeton, NJ.

Libchaber, A. and Maurer, J. (1982). A Rayleigh–Bénard experiment: helium in a small box. In *Nonlinear Phenomena at Phase Transitions and Instabilities*, T. Riste (ed.), pp. 259–286. Plenum: New York.

Libchaber, A., Fauve, S. and Laroche, C. (1983). Two-parameter study of the routes to chaos. *Physica*, **7D**, 73–84.

Libchaber, A., Laroche, C. and Fauve, S. (1982). Period doubling cascade in mercury: a quantitative measurement. *J. Phys. Lett.*, **43**, L211–L216.

Lichtenberg, A. J. and Lieberman, M. A. (1982). *Regular and Stochastic Motion*. Springer–Verlag: New York, Heidelberg and Berlin.

Liénard, A. (1928). Etude des oscillations entretenues. *Rev. Gen. Electr.*, **23**, 901–912, 946–954.

Linsay, P. (1981). Period doubling and chaotic behavior in a driven anharmonic oscillator. *Phys. Rev. Lett.*, **47**, 1349–1352.

Lorenz, E. N. (1963). Deterministic nonperiodic flow. *J. Atmos. Sci.*, **20**, 130–141.

Lorenz, E. N. (1964). The problem of deducing the climate from the governing equations. *Tellus*, **16**, 1–11.

Lorenz, E. N. (1980). Noisy periodicity and reverse bifurcation. In *Nonlinear Dynamics*, R. H. G. Helleman (ed.), pp. 282–291. New York Academy of Sciences: New York.

Lorenz, E. N. (1984). The local structure of a chaotic attractor in four dimensions. *Physica*, **13D**, 90–104.

Lorenz, E. N. (1993). *The Essence of Chaos*, University of Washington Press: Seattle.

Lozi, R. (1978). Un attracteur étrange? du type attracteur de Hénon. *J. Phys.*, **39**, 9–10.

Lvov, V. S., Prdetechenskii, A. A. and Chernykh, A. I. (1981). Bifurcation and chaos in a system of Taylor vortices: a natural and numerical experiment. *Sov. Phys. JETP*, **53**, 562–573.

Lyubimov, D. V. and Zaks, M. A. (1983). Two mechanisms of the transition to chaos in finite-dimensional models of convection. *Physica*, **9D**, 52–64.

MacDonald, A. H. and Plischke, M. (1983). Study of the driven damped pendulum: application to Josephson junctions and charge-density-wave systems. *Phys. Rev. B*, **27**, 201–211.

MacKay, R. S. and Meiss, J. D. (eds) (1987). *Hamiltonian Dynamical Systems: A Reprint Selection*. Adam Hilger: Bristol.

MacKay, R. S. and Tresser, C. (1984). Transition to chaos for two-frequency systems. *J. Phys. Lett.*, **45**, L741–L746.

MacKay, R. S. and Tresser, C. (1987). Some flesh on the skeleton: the bifurcation structure of bimodal maps. *Physica*, **27D**, 412–422.

MacKay, R. S., Meiss, J. D. and Percival, I. C. (1984). Transport in Hamiltonian systems. *Physica*, **13D**, 55–81.

MacMaster, A. G. and Thompson, J. M. T. (1994). Wave tank testing and the capsizability of hulls. *Proc. R. Soc. Lond.*, **A446**, 217–232.

Macmillen, F. B. J. and Thompson, J. M. T. (1998) Bifurcation analysis in the flight dynamics design process? A view from the aircraft industry. *Phil. Trans. R. Soc. Lond.*, **A356**, 2321–2333.

Malraison, B., Atten, P., Bergé, P. and Dubois, M. (1983). Dimcnsion of strange attractors: an experimental determination for the chaotic regime of two convective systems. *J. Phys. Lett.*, **44**, L897–L902.

Mandelbrot, B. (1983). *The Fractal Geometry of Nature*. W. H. Freeman: San Francisco.

Mankin, J. C. and Hudson, J. L. (1984). Oscillatory and chaotic behavior of a forced exothermic chemical reaction. *Chem. Eng. Sci.*, **39**, 1807–14.

Marcus, P. S. (1981). Effects of truncation in modal representations of thermal convection. *J. Fluid Mech.*, **103**, 241–256.

Markus, L. (1971). *Lectures in Differentiable Dynamics*. American Mathematical Society: Providence, RI.

Marsden, J. E. (1981). *Lectures on Geometric Methods in Mathematical Physics*. SIAM: Philadelphia.

Marsden, J. E. and McCracken, M. (1976). *The Hopf Bifurcation and Its Applications*. Springer-Verlag: New York, Heidelberg and Berlin.

Marzec, C. J. and Spiegel, E. A. (1980). Ordinary differential equations with strange attractors. *SIAM J. Appl. Math.*, **38**, 403–421.

Maurer, J. and Libchaber, A. (1979). Rayleigh–Bénard experiment in liquid helium; frequency locking and the onset of turbulence. *J. Phys. Lett.*, **40**, L419–L423.

Maurer, J. and Libchaber, A. (1980). Effect of the Prandtl number on the onset of turbulence in liquid helium. *J. Phys. Lett.*, **41**, L515–L518.

May, R. M. (1976). Simple mathematical models with very complicated dynamics. *Nature*, **261**, 459–467.

May, R. M. (1979). Bifurcations and dynamic complexity in ecological systems. In *Bifurcation Theory and Applications in Scientific Disciplines*, O. Gurel and O. E. Rössler (eds), pp. 517–529.

Mayer-Kress, G. and Haken, H. (1981). Intermittent behaviour of the logistic system. *Phys. Lett.*, **82A**, 151–155.

Mayer-Kress, G. and Haken, H. (1986). An explicit construction of a class of suspensions and autonomous differential equations for diffeomorphisms in the plane. *Commun. Math. Phys.*, **111**, 63–74.

Maynard-Smith, J. (1971). *Mathematical Ideas in Biology*. Cambridge University Press: Cambridge.

McDonald, S. W., Grebogi, C., Ott, E. and Yorke, J. A. (1985). Fractal basin boundaries. *Physica*, **17D**, 125–153.

McLaughlin, J. B. (1981). Period-doubling bifurcations and chaotic motion for a parametrically forced pendulum. *J. Stat. Phys.*, **24**, 375–388.
McLaughlin, J. B. and Orszag, S. A. (1982). Transition from periodic to chaotic thermal convection. *J. Fluid Mech.*, **122**, 123–142.
McRobie, F. A. (1992a). Birkhoff signature change: a criterion for the instability of chaotic resonance. *Phil. Trans. R. Soc. Lond.*, **A338**, 557–568.
McRobie, F. A. (1992b). Bifurcational precedences in the braids of periodic orbits of spiral 3-shoes in driven oscillators. *Proc. R. Soc. Lond.*, **A438**, 545–569.
McRobie, F. A. and Thompson, J. M. T. (1991). Lobe dynamics and the escape from a potential well. *Proc. R. Soc. Lond.*, **A435**, 659–672.
McRobie, F. A. and Thompson, J. M. T. (1992). Invariant sets of planar diffeomorphisms in nonlinear vibrations. *Proc. R. Soc. Lond.*, **A436**, 427–448.
McRobie, F. A. and Thompson, J. M. T. (1993). Braids and knots in driven oscillators. *Int. J. Bifurcation and Chaos*, **3**, 1343–1361.
McRobie, F. A. and Thompson, J. M. T. (1994). Knot-types and bifurcation sequences of homoclinic and transient orbits of a single-degree-of-freedom driven oscillator. *Dynamics and Stability of Systems*, **9**, 223–251.
McRobie, F. A., Popov, A. A. and Thompson, J. M. T. (1999) Auto-parametric resonance in cylindrical shells using geometric averaging. *J. Sound and Vibration*, **227**, 65–84.
Mees, A. I. (1981). *Dynamics of Feedback Systems*. John Wiley & Sons: Chichester.
Mees, A. I. (ed.) (2000). *Nonlinear Dynamics and Statistics*. Birkhauser: Basel.
Melnikov, V. K. (1963). On the stability of the center for time periodic perturbations. *Trans. Moscow Math. Soc.*, **12**, 1–57.
Metropolis, N., Stein, M. L. and Stein, P. R. (1973). On finite limit sets for transformations on the unit interval. *J. Combin. Theor.*, **A15**, 25–44.
Mielke, A. and Holmes, P. (1988). Spatially complex equilibria of buckled rods. *Arch. Rational Mech. Anal.* **101**, 319–348.
Milnor, J. (1985). On the concept of attractor. *Commun. Math. Phys.*, **99**, 177–195.
Minorsky, N. (1962). *Nonlinear Oscillations*. Van Nostrand: Princeton, NJ.
Mira, C. (1987). *Chaotic Dynamics: from the one-dimensional endomorphism to the two-dimensional diffeomorphism*. World Scientific: Singapore.
Mira, C. and Carcassès, J. P. (1991). On the crossroad area-saddle area and crossroad area-spring area transitions. *Int. J. Bifurcation and Chaos*, **1**, 641–653.
Mira, C., Touzani-Qriouet, M. and Kawakami, H. (1999). Bifurcation structures generated by the nonautonomous Duffing equation. *Int. J. Bifurcation and Chaos*, **9**, 1363–1379.
Miracky, R. F., Clarke, J. and Koch, R. H. (1983). Chaotic noise observed in a resistively shunted self-resonant Josephson tunnel junction. *Phys. Rev. Lett.*, **50**, 856–859.
Misiurewicz, M. (1980). The Lozi mapping has a strange attractor. In *Nonlinear Dynamics*, R. H. G. Helleman (ed.), pp. 348–358. New York Academy of Sciences: New York.
Misiurewicz, M. (1981). Structure of mappings of an interval with zero entropy. *Publ. Math. IHES*, **53**, 5–16.
Mitsui, T., Ueda, Y. and Thompson, J. M. T. (1994). Straddle-orbit location of a chaotic saddle in a high-dimensional realisation of R^{∞}, *Proc. R. Soc. Lond.*, **A445**, 669–677.
Moon, F. C. (1980). Experiments on chaotic motions of a forced nonlinear oscillator: strange attractors. *ASME J. Appl. Mech.*, **47**, 638–644.
Moon, F. C. (1984). Fractal boundary for chaos in a two-state mechanical oscillator. *Phys. Rev. Lett.*, **53**, 962–964.
Moon, F. C. (1992). *Chaotic and Fractal Dynamics*. John Wiley & Sons: New York.
Moon, F. C. (ed) (1997). *Dynamics and Chaos in Manufacturing Processes*, John Wiley & Sons: New York.

Moon, F. C. and Holmes, P. J. (1979). A magnetoelastic strange attractor. *J. Sound Vib.*, **65**, 285–296; Errata, **69**, 339.

Moon, F. C. and Shaw, S. W. (1983). Chaotic vibrations of a beam with nonlinear boundary conditions. *Int. J. Nonlinear Mech.*, **18**, 465–477.

Mori, H. and Kuramoto, Y. (1998). *Dissipative Structures and Chaos*. Springer-Verlag: Berlin.

Moser, J. (1973). *Stable and Random Motions in Dynamical Systems*. Princeton University Press: Princeton, NJ.

Mullin, T. (ed) (1993). *The Nature of Chaos*. Oxford University Press: Oxford.

Murrey, C. D. (1984). Chaotic spinning of hyperion. *Nature*, **311**, 705.

Nagumo, J., Arimoto, S. and Yoshizawa, S. (1962). An active pulse transmission line simulating nerve axon. *Proc. IRE*, **50**, 2061–2070.

Nakatsuka, H., Asaka, S., Itoh, H., Ikeda, K. and Matsuoka, M. (1983). Observation of bifurcation to chaos in an all-optical bistable system. *Phys. Rev. Lett.*, **50**, 109–112.

Nauenberg, M. and Rudnick, J. (1981). Universality and power spectrum at the onset of chaos. *Phys. Rev. B*, **27**, 493–498.

Nayfeh, A. H. and Balachandran, B. (1995). *Applied Nonlinear Dynamics: Analytical, Computational and Experimental Methods*. John Wiley & Sons: New York.

Nayfeh, A. H. and Mook, D. T. (1979). *Nonlinear Oscillations*. John Wiley & Sons: New York.

Newhouse, S. E. (1974). Diffeomorphisms with infinitely many sinks. *Topology*, **13**, 9–18.

Newhouse, S. E. (1979). The abundance of wild hyperbolic sets and non-smooth stable sets for diffeomorphisms. *Publ. Math. IHES*, **50**, 101–152.

Newhouse, S. E. (1980a). Lectures on dynamical systems. In *Dynamical Systems*, CIME Lectures, Bressanone, Italy, June 1978, pp. 1–114 (*Progress in Mathematics*, vol. 8). Birkhauser: Boston.

Newhouse, S. E. (1980b). Asymptotic behavior and homoclinic points in nonlinear systems. In *Nonlinear Dynamics*, R. H. G. Helleman (ed.). New York Academy of Sciences: New York.

Newhouse, S. E., Ruelle, D. and Takens, F. (1978). Occurrence of strange axiom A attractors near quasiperiodic flows on T^m, $m \geq 3$. *Commun. Math. Phys.*, **64**, 35–40.

Nicolis, G. and Prigogine, I. (1977). *Self-Organization in Non-Equilibrium Systems. From Dissipative Structures to Order Through Fluctuations*. John Wiley & Sons: New York.

Nitecki, Z. (1971). *Differentiable Dynamics*. MIT Press: Cambridge, MA.

Nitecki, Z. (1981). Reviews of Simo (1979) and Tresser, Coullet and Arnéodo (1980). *Math. Rev.*, 81j: 58058.

Nitecki, Z. and Robinson, C. (eds) (1980). *Global Theory of Dynamical Systems* (*Springer Lecture Notes in Mathematics*, vol. 819). Springer-Verlag: New York, Heidelberg and Berlin.

Nusse, H. E. and Yorke, J. A. (1998). *Dynamics: Numerical Explorations*, 2nd edn. Springer-Verlag: New York.

Orszag, S. A. and McLaughlin, J. B. (1980). Evidence that random behavior is generic for nonlinear differential equations. *Physica*, **1D**, 68–79.

Oseledec, V. I. (1968). A multiplicative erogodic theorem: Liapunov characteristic numbers for dynamical systems. *Trans. Moscow Math. Soc.*, **19**, 197–231.

Ostlund, S., Rand, D., Sethna, J. and Siggia, E. (1983). Universal properties of the transition from quasiperiodicity to chaos in dissipative systems. *Physica*, **8D**, 303–342.

Ott, E. (1981). Strange attractors and chaotic motions of dynamical systems. *Rev. Mod. Phys.*, **53**, 655–671.

Ott, E. (1993). *Chaos in Dynamical Systems*. Cambridge University Press: New York.

Packard, N. H., Crutchfield, J. P., Farmer, J. D. and Shaw, R. S. (1980). Geometry from a time series. *Phys. Rev. Lett.*, **45**, 712–716.

Palis, J. and deCarmo, M. (eds) (1977). *Geometry and Topology*, Rio de Janeiro, 1978 (*Springer Lecture Notes in Mathematics*, vol. 597). Springer-Verlag: New York, Heidelberg and Berlin.

Palis, J. and deMelo, W. (1982). *Geometric Theory of Dynamical Systems: An Introduction*. Springer-Verlag: New York, Heidelberg and Berlin.

Parker, T. S. and Chua, L. O. (1989). *Practical Numerical Algorithms for Chaotic Systems*. Springer-Verlag: New York.

Parkinson, G. V. and Smith, J. D. (1964). The square prism as an aeroelastic non-linear oscillator. *Q. J. Mech. Appl. Math.*, **17**, 225–239.

Parlitz, U. and Lauterborn, W. (1985). Superstructure in the bifurcation set of the Duffing equation. *Phys. Lett.*, **107A**, 351–355.

Peinke, J., Parisi, J., Rössler, O. E. and Stoop, R. (1992). *Encounter With Chaos: Self-Organized Hierarchical Complexity in Semiconductor Experiments*. Springer-Verlag: New York.

Peixoto, M. M. (1962). Structural stability on two-dimensional manifolds. *Topology*, **1**, 101–120.

Peixoto, M. M. (ed.) (1973). *Dynamical Systems*. Academic Press: New York.

Perez, J. and Jeffries, C. (1982). Direct observation of a tangent bifurcation in a non-linear oscillator. *Phys. Lett.*, **92A**, 82–84.

Pesin, J. B. (1977). Characteristic Lyapunov exponents and smooth ergodic theory. *Russ. Math. Surv.*, **32**, 55–114.

Piangiani, G. and Yorke, J. A. (1979). Expanding maps on sets which are almost invariant: decay and chaos. *Trans. Am. Math. Soc.*, **252**, 351–366.

Pikovsky, A. S. and Rabinovich, M. I. (1981). Stochastic oscillations in dissipative systems. *Physica*, **2D**, 8–24.

Pippard, B. (1982). Instability and chaos: physical models of everyday life. *Interdisc. Sci. Rev.*, **7**, 92–101.

Pismen, L. M. (1982). Bifurcation sequences in a third-order system with a folded slow manifold. *Phys. Lett.*, **89A**, 59–62.

Plykin, R. (1974). Sources and sinks for A-diffeomorphisms. *USSR Math. Sb.*, **23**, 233–253.

Poincaré, H. (1880–90). Mémoire sur les courbes définies par les équations différentielles I–VI, Oeuvre I. Gauthier-Villars: Paris.

Poincaré, H. (1890). Sur les équations de la dynamique et le problème de trois corps. *Acta Math.*, **13**, 1–270.

Poincaré, H. (1899). *Les Methodes Nouvelles de la Mécanique Celeste*, vols 1–3. Gauthier-Villars: Paris.

Pomeau, Y. (1983). The intermittent transition to turbulence. In *Nonlinear Dynamics and Turbulence*, G. I. Barenblatt, G. Iooss and D. D. Joseph (eds) Pitman: London.

Pomeau, Y. and Manneville, P. (1980). Intermittent transition to turbulence in dissipative dynamical systems. *Commun. Math. Phys.*, **74**, 189–197.

Pomeau, Y., Roux, J. C., Rossi, A., Bachelart, S. and Vidal, C. (1981). Intermittent behavior in the Belousov–Zhabotinsky reaction. *J. Phys. Lett.*, **42**, L271–L273.

Popov, A. A., Thompson, J. M. T. and Croll, J. G. A. (1998). Bifurcation analyses in the parametrically excited vibrations of cylindrical panels. *Nonlinear Dynamics*, **17**, 205–225.

Popov, A. A., Thompson, J. M. T. and McRobie, F. A. (1998). Low dimensional models of shell vibrations: parametrically excited vibrations of cylindrical shells. *J. Sound and Vibration*, **209**, 163–186.

Poston, T. and Stewart, I. (1978). *Catastrophe Theory and Its Applications*. Pitman: London.

Preston, C. (1983). *Iterates of Maps on an Interval* (*Springer Lecture Notes in Mathematics*, vol. 999). Springer-Verlag: Berlin.

Prigogine, I. (1980). *From Being to Becoming: Time and Complexity in the Physical Sciences*. W. H. Freeman: San Francisco.

Pustylnikov, L. D. (1978). Stable and oscillating motions in non-autonomous dynamical systems. *Trans. Moscow Math. Soc.*, **14**, 1–101.

Puu, T. (2000). *Attractors, Bifurcations, and Chaos: Nonlinear Phenomena in Economics*. Springer-Verlag: New York.

Rabinovich, M. I. (1978). Stochastic self-oscillations and turbulence. *Sov. Phys. Usp.*, **21**, 443–469.

Rainey, R. C. T. and Thompson, J. M. T. (1991). The transient capsize diagram: a new method of quantifying stability in waves. *J. Ship Research*, **35**, 58–62.

Rand, D. A. (1978). The topological classification of Lorenz attractors. *Math. Proc. Camb. Phil. Soc.*, **83**, 451–460.

Rand, D. A. and Young, L. S. (eds) (1981). *Dynamical Systems and Turbulence* (*Springer Lecture Notes in Mathematics*, vol. 898). Springer-Verlag: New York, Heidelberg and Berlin.

Rand, D. A., Ostlund, S., Sethna, J. and Siggia, E. (1982). A universal transition from quasi-periodicity to chaos in dissipative systems. *Phys. Rev. Lett.*, **49**, 132–135.

Rapp, P. E. (ed.) (1999). *Nonlinear Dynamics and Brain Functioning*. Nova Science Publishers: Commack.

Räty, R., van Boehm, J. and Isomäki, H. M. (1984). Absence of inversion-symmetric limit cycles of even periods and the chaotic motion of Duffing's oscillator. *Phys. Lett.*, **103A**, 289–292.

Rayleigh, J. W. S. (1896). *The Theory of Sound*. Reprinted by Dover: New York.

Rehberg, I. and Ahlers, G. (1985). Experimental observation of a codimension-two bifurcation in a binary fluid mixture. *Phys. Rev. Lett.*, **55**, 500–503.

Reichl, L. E. (1992). *The Transition to Chaos in Conservative and Classical Systems: Quantum Manifestations*. Springer-Verlag: New York.

Rössler, O. (1998). *Endophysics: the world as an interface*. World Scientific: Singapore.

Rössler, O. E. (1976a). An equation for continuous chaos. *Phys. Lett.*, **57A**, 397–398.

Rössler, O. E. (1976b). Different types of chaos in two simple differential equations. *Z. Naturf.*, **31a**, 1664–1670.

Rössler, O. E. (1977). Horseshoe-map chaos in the Lorenz equation. *Phys. Lett.*, **60A**, 392–394.

Rössler, O. E. (1979). Continuous chaos—four prototype equations. In *Bifurcation Theory and Applications in Scientific Disciplines*, O. Gurel and O. E. Rössler (eds), pp. 376–392. New York Academy of Sciences: New York.

Rössler, O. E. (1981). The gluing together principle and chaos. In *Nonlinear problems of Analysis in Geometry and Mechanics*, M. Atteia, D. Bancel and I. Gumowski (eds), pp. 50–56. Pitman: Boston.

Rössler, O. E. (1983). The chaotic hierarchy. *Z. Naturf.*, **38a**, 788–801.

Rössler, O. E., Rössler, R. and Landahl, H. (1978). Arrythmia in a periodically forced excitable system. *Sixth Int. Biophysics Congress*, Kyoto, Abstracts p. 296.

Robbins, K. A. (1979). Periodic solutions and bifurcation structure at high R in the Lorenz model. *SIAM J. Appl. Math.*, **36**, 457–472.

Robinson, A. L. (1982). Physicists try to find order in chaos. *Science*, **218**, 554–556.

Robinson, A. L. (1983). How does fluid flow become turbulent? *Science*, **221**, 140–143.

Robinson, C. (1999). *Dynamical Systems: Stability, Symbolic Dynamics, and Chaos, Second Edition*, CRC Press: Boca Raton.

Robinson, C. (1983). Bifurcation to infinitely many sinks. *Commun. Math. Phys.*, **90**, 433–459.

Rollins, R. W. and Hunt, E. R. (1984). Intermittent transient chaos at interior crises in the diode resonator. *Phys. Rev. A*, **29**, 3327–3334.

Rosen, R. (1970). *Dynamical System Theory in Biology*. Wiley-Interscience: New York.

Roux, J.-C. (1983). Experimental studies of bifurcations leading to chaos in the Belousov–Zhabotinsky reactions. *Physica*, **7D**, 57–68.

Roux, J.-C., Simoyi, R. H. and Swinney, H. L. (1983). Observation of a strange attractor. *Physica*, **8D**, 257–266.

Ruelle, D. (1979). Sensitive dependence on initial conditions and turbulent behavior of dynamical systems. In *Bifurcation Theory and Applications in Scientific Disciplines*, O. Gurel and O. E. Rössler (eds), pp. 408–446. New York Academy of Sciences: New York.

Ruelle, D. (1981). Small random perturbations of dynamical systems and the definition of attractors. *Commun. Math. Phys.*, **82**, 137–151.

Ruelle, D. (1989). *Elements of Differentiable Dynamics and Bifurcation Theory*. Academic Press: San Diego.

Ruelle, D. (1991). *Chance and Chaos*. Princeton University Press: Princeton.

Ruelle, D. and Takens, F. (1971). On the nature of turbulence. *Commun. Math. Phys.*, **20**, 167–192; **23**, 343–344.

Russell, D. A. and Ott, E. (1981). Chaotic (strange) and periodic behavior in instability saturation by the oscillating two-stream instability. *Phys. Fluids*, **24**, 1976–1988.

Russell, R. C. H. (1959). A study of the movement of moored ships subjected to wave action. *Proc. Inst. Civ. Eng.*, **12**, 379–398.

Salam, F. M. A., Marsden, J. E. and Varaiya, P. P. (1983). Chaos and Arnold diffusion in dynamical systems. *IEEE Trans. Circuits Syst.*, **CAS-30**, 697–708.

Saltzman, B. (1962). Finite amplitude free convection as an initial value problem. *J. Atmos. Sci.*, **19**, 329–341.

Sano, M. and Sawada, Y. (1983). Transition from quasi-periodicity to chaos in a system of coupled nonlinear oscillators. *Phys. Lett.*, **97A**, 73–76.

Saperstein, A. and Mayer-Kress, G. (1988). A nonlinear dynamical model of the impact of SDI on the arms race. *J. Conflict Resolution* **32**, 636–670.

Sauer, T., Grebogi, C. and Yorke, J. A. (1997). How long do numerical chaotic solutions remain valid?, *Phys. Rev. Lett.*, **79**, 59–62.

Schaffer, W. M. (1985). Order and chaos in ecological systems. *Ecology*, **66**, 93–106.

Schechter, S. (1987). The saddle-node separatrix loop bifurcation. *SIAM J. Math. Anal.*, **18**, 1142–1156.

Schell, M., Fraser, S. and Kapral, R. (1983). Subharmonic bifurcations in the sine map: an infinite hierarchy of cusp bifurcations. *Phys. Rev. A*, **28**, 373–378.

Schiehlen, W. (ed) (1990). *Nonlinear Dynamics in Engineering Systems*, Proceedings of the IUTAM Symposium, Stuttgart. Springer-Verlag: Berlin.

Schuster, H. G. (1984). *Deterministic Chaos*. Physik-Verlag: Weinheim.

Scott, A. (1999). *Nonlinear Science: Emergence and Dynamics of Coherent Structures*. Oxford University Press: Oxford.

Scott, A. C. (1975). The electrophysics of a nerve fiber. *Rev. Mod. Phys.*, **47**, 487–533.

Seelig, F. F. (1980–83). Unrestricted harmonic balance. *Z. Naturf.*, **35a**, 1054–1061; **38a**, 636–640; **38a**, 729–735.

Segel, L. A. (1984). *Modeling Dynamic Phenomena in Molecular and Cellular Biology*. Cambridge University Press: Cambridge.

Serrin, J. (1959). Mathematical principles of classical fluid mechanics. In *Encyclopedia, of Physics*, S. Flügge (ed.), vol. VIII/1, pp. 125–263. Springer-Verlag: Berlin, Göttingen and Heidelberg.

Shaw, R. (1981). Strange attractors, chaotic behavior, and information flow. *Z. Naturf.*, **36a**, 80–112.

Shaw, R. (1984). *The Dripping Faucet as a Model Chaotic System*. Aerial Press: Santa Cruz, CA.

Shaw, S. W. and Holmes, P. (1983b). A periodically forced piecewise linear oscillator. *J. Sound Vib.*, **90**, 129–144.

Shaw, S. W. and Holmes, P. (1983c). Periodically forced linear oscillator with impacts: chaos and long-period motions. *Phys. Rev. Lett.*, **51**, 623–626.

Shaw, S. W. and Holmes, P. J. (1983a). A periodically forced impact oscillator with large dissipation. *ASME J. Appl. Mech.*, **50**, 849–857.

Shenker, S. J. (1982). Scaling behavior in a map of a circle onto itself: empirical results. *Physica*, **5D**, 405–411.

Shilnikov, L. P. (1965). A case of the existence of a denumerable set of periodic motions. *Sov. Math. Dokl.*, **6**, 163–166.

Shilnikov, L. P. (1967). The existence of a denumerable set of periodic motions in four-dimensional space in an extended neighborhood of a saddle-focus. *Sov. Math. Dokl.*, **8**, 54–58.

Shilnikov, L. P. (1970). A contribution to the problem of the structure of an extended neighborhood of a rough equilibrium state of saddle-focus type. *Math. USSR Sb.*, **10**, 91–102.

Shilnikov, L. P. (1976). Theory of the bifurcation of dynamical systems and dangerous boundaries. *Sov. Phys. Dokl.*, **20**, 674–676.

Shilnikov, L. P., Shilnikov, A. L., Turaev, D. and Chua, L. (2000). *Methods of Qualitative Theory in Nonlinear Dynamics*, Part II, *Nonlinear Science*. World Scientific: Singapore.

Shobu, K., Ose, T. and Mori, H. (1984). Shapes of the power spectrum of intermittent turbulence near its onset point. *Prog. Theor. Phys.*, **71**, 458–473.

Shtern, V. N. (1983). Attractor dimension for the generalized baker's transformation. *Phys. Lett.*, **99A**, 268–270.

Shtern, V. N. and Shumova, L. V. (1984). Metamorphoses of preturbulence. *Phys. Lett.*, **103A**, 167–170.

Simó, C. (1979). On the Hénon-Pomeau attractor. *J. Stat. Phys.*, **21**, 465–494.

Simoyi, R. H., Wolf, A. and Swinney, H. L. (1982). One-dimensional dynamics in a multicomponent chemical reaction. *Phys. Rev. Lett.*, **49**, 245–248.

Sinai, J. G. (1970). Dynamical systems with elastic reflections. *Russ. Math. Surv.*, **25**, 137–189.

Sinai, J. G. and Vul, E. (1981). Hyperbolicity conditions for the Lorenz model. *Physica*, **2D**, 3–7.

Singer, D. (1978). Stable orbits and bifurcations of maps of the interval. *SIAM J. Appl. Math.*, **35**, 260–267.

Skjolding, H., Branner-Jorgensen, B., Christiansen, P. L. and Jenson, H. E. (1983). Bifurcations in discrete dynamical systems with cubic maps. *SIAM J. Appl. Math.*, **43**, 520–534.

Smale, S. (1963). Diffeomorphisms with many periodic points. In *Differential and Combinatorial Topology*, S. S. Cairns (ed.), pp. 63–80. Princeton University Press: Princeton, NJ.

Smale, S. (1967). Differentiable dynamical systems. *Bull. Am. Math. Soc.*, **73**, 747–817.

Smale, S. (1980). *The Mathematics of Time: Essays on Dynamical Systems, Economic Processes and Related Topics*. Springer-Verlag: New York, Heidelberg and Berlin.

Smith, C. W. and Tejwani, M. J. (1983). Bifurcation and the universal sequence for first-sound subharmonic generation in superfluid helium-4. *Physica*, **7D**, 85–88.

Smith, P. (1998). *Explaining Chaos*. Cambridge University Press: Cambridge.

Solari, H. G., Natiello, M. A. and Mindlin, G. B. (1996). *Nonlinear Dynamics: A Two-Way Trip from Physics to Math*. IOP: Bristol.

Soliman, M. S. and Thompson, J. M. T. (1989). Integrity measures quantifying the erosion of smooth and fractal basins of attraction, *J. Sound and Vibration*, **135**, 453–475.

Soliman, M. S. and Thompson, J. M. T. (1991). Basin organisation prior to a tangled saddle-node bifurcation. *Int. J. Bifurcation and Chaos*, **1**, 107–118.

Soliman, M. S. and Thompson, J. M. T. (1992a). Global dynamics underlying sharp basin erosion in nonlinear driven oscillators. *Phys. Rev. A*, **45**, 3425–3431.

Soliman, M. S. and Thompson, J. M. T. (1992b). Indeterminate sub-critical bifurcations in parametric resonance. *Proc. R. Soc. Lond.*, **A438**, 511–518.

Soliman, M. S. and Thompson, J. M. T. (1992c). Indeterminate trans-critical bifurcations in parametrically excited systems. *Proc. R. Soc. Lond.*, **A439**, 601–610.

Soliman, M. S. and Thompson, J. M. T. (1996). Indeterminate bifurcational phenomena in hardening systems. *Proc. R. Soc. Lond.*, **A452**, 487–494.

Soliman, M. S. (1994). Predicting regimes of indeterminate jumps to resonance by assessing fractal boundaries in control space. *Int. J. Bifurcation and Chaos*, **4**, 1645–1653.

Sparrow, C. (1982). *The Lorenz Equations*. Springer-Verlag: New York, Heidelberg and Berlin.

Spyrou, K. J. and Thompson, J. M. T. (eds) (2000). *The Nonlinear Dynamics of Ships*, theme issue. *Phil. Trans. R. Soc. Lond.*, **A358**, 1733–1981.

Stavans, J. Heslot, F. and Libchaber, A. (1985). Fixed winding number and the quasi-periodic route to chaos in a convective fluid. *Phys. Rev. Lett.*, **55**, 596–599.

Stefan, P. (1977). A theorem of Sarkovskii on the existence of periodic orbits of continuous endomorphisms of the real line. *Commun. Math. Phys.*, **54**, 237–248.

Stewart, H. B. (1984). *The Lorenz System* (16 mm film). Aerial Press: Santa Cruz, CA.

Stewart, H. B. (1986). Frequency scaling at the onset of finite amplitude oscillation. *Z. Naturforschung*, **41a**, 1412–1414.

Stewart, H. B. (1987). A chaotic saddle catastrophe in forced oscillators. In *Dynamical Systems Approaches to Nonlinear Problems in Systems and Circuits*, F. Salam and M. Levi (eds), pp. 138–149. SIAM: Philadelphia.

Stewart, H. B. (1991). Application of fixed point theory to chaotic attractors of forced oscillators. *Jpn J. Indust. Appl. Math.*, **8**, 487–504.

Stewart, H. B. (1996). Chaos, dynamical structure, and climate variability, In *Chaos and the Changing Nature of Science and Medicine*, D. Herbert (ed.), AIP Conference Proceedings 346, pp. 80–113. American Instute of Physics: Woodbury.

Stewart, H. B. and Ueda, Y. (1991). Catastrophes with indeterminate outcome. *Proc. R. Soc. Lond.*, **A432**, 113–123.

Stewart, H. B. and Lansbury, A. N. (1992). Forecasting catastrophe by exploiting chaotic dynamics. In *Applied Chaos*, J. H. Kim and J. Stringer (eds), pp 393–410. John Wiley & Sons, New York.

Stewart, H. B., Thompson, J. M. T., Ueda Y. and Lansbury, A. N. (1995). Optimal escape from potential wells: patterns of regular and chaotic bifurcation. *Physica*, **85D**, 259–295.

Stewart, H. B., Ueda, Y., Grebogi, C. and Yorke, J. A. (1995). Double crises in two-parameter dynamical systems. *Phys. Rev. Lett.*, **75**, 2478–2481.

Stewart, H. B., Wiesenfeld, K. and Rossler, O. E. (1990). Unfolding a chaotic bifurcation. *Proc. R. Soc. Lond.*, **A431**, 371–383.

Stewart, I. (1983). Nonelementary catastrophe theory. *IEEE Trans. Circuits Syst.*, **CAS-30**, 663–670.

Stewart, I. (1984). Applications of nonelementary catastrophe theory. *IEEE Trans. Circuits Syst.*, **CAS-31**, 165–174.

Stewart, I. (2000). The Lorenz attractor exists. *Nature*, **406**, 948–949.

Stoker, J. J. (1950). *Nonlinear Vibrations*. John Wiley & Sons: New York.

Stoker, J. J. (1980). Periodic forced vibrations of systems of relaxation oscillators. *Commun. Pure Appl. Math.*, **33**, 215–240.

Strogatz, S. H. (1994). *Nonlinear Dynamics and Chaos*. Addison-Wesley: Reading, MA.

Stutzer, M. J. (1980). Chaotic dynamics and bifurcation in a macro model. *J. Econ. Dyn. Control*, **2**, 353–376.

Swinney, H. L. (1983). Observations of order and chaos in nonlinear systems. *Physica*, **7D**, 3–15. Reprinted as Chapter 17.

Swinney, H. L. and Gollub, J. P. (eds) (1981). *Hydrodynamic Instabilities and the Transition to Turbulence*. Springer-Verlag: Berlin, Heidelberg and New York.

Swinney, H. L. and Roux, J. C. (1984). Chemical chaos. In *Nonequilibrium Dynamics in Chemical Systems*, C. Vidal (ed.). Springer-Verlag: New York.

Takens, F. (1973b). Introduction to global analysis. *Commun. Math. Inst. Rijksuniv. Utrecht*, **2**, 1–111.

Takens, F. (1974). Forced oscillations and bifurcations. *Comm. Math. Inst. Rijkuniversiteit Utrecht*, no. 3, 1–59.

Takens, F. (1980). Detecting strange attractors in turbulence. In *Dynamical Systems and Turbulence*, D. A. Rand and L.-S. Young (eds), pp. 366–381 (*Springer Lecture Notes in Mathematics*, vol. 898). Springer-Verlag: New York, Heidelberg and Berlin.

Takens, F., (1973a). Normal forms for certain singularities of vector fields. *Ann. Inst. Fourier*, **23**, 163–195.

Tavakol, R. K. and Tworkowski, A. S. (1984). On the occurrence of quasiperiodic motion on three tori. *Phys. Lett.*, **100A**, 65–67.

Testa, J., Perez, J. and Jeffries, C. (1982). Evidence of universal chaotic behavior of a driven nonlinear oscillator. *Phys. Rev. Lett.*, **48**, 714–717.

Thom, R. (1975). *Structural Stability and Morphogenesis*. W. A. Benjamin: Reading, MA.

Thompson, J. M. T. (1979). Stability predictions through a succession of folds. *Phil. Trans. R. Soc. Lond.*, **A292**, 1–23.

Thompson, J. M. T. (1982). *Instabilities and Catastrophes in Science and Engineering*. John Wiley & Sons: Chichester.

Thompson, J. M. T. (1983). Complex dynamics of compliant off-shore structures. *Proc. R. Soc. Lond. A*, **387**, 407–427.

Thompson, J. M. T. (1984a). An introduction to nonlinear dynamics. *Appl. Math. Model.*, **8**, 157–168.

Thompson, J. M. T. (1984b). Sir Isaac Newton's pendulum experiments on fluid damping. *Bull. Inst. Math. Appl.*, **20**, 8–11.

Thompson, J. M. T. (1989). Chaotic phenomena triggering the escape from a potential well. *Proc. R. Soc. Lond.*, **A421**, 195–225.

Thompson, J. M. T. (1992). Global unpredictability in nonlinear dynamics: capture, dispersal and the indeterminate bifurcations. *Physica*, **58D**, 260–272.

Thompson, J. M. T. (1994). Basic concepts of nonlinear dynamics. In *Nonlinearity and Chaos in Engineering Dynamics*, J. M. T. Thompson and S. R. Bishop (eds), pp. 1–21. John Wiley & Sons: Chichester.

Thompson, J. M. T. (1996a). Danger of unpredictable failure due to indeterminate bifurcation. *ZAMM*, **S4**, 199–202.

Thompson, J. M. T. (1996b). Global dynamics of driven oscillators: fractal basins and indeterminate bifurcations. In *Nonlinear Mathematics and its Applications*, P. J. Aston (ed.), pp 1–47. Cambridge University Press: Cambridge.

Thompson, J. M. T. (1997). Designing against capsize in beam seas: recent advances and new insights. *Appl. Mech. Rev.*, **50**, 307–325.

Thompson, J. M. T. and Bishop, S. R. (eds) (1994). *Nonlinearity and Chaos in Engineering Dynamics*, Proceedings of the IUTAM Symposium, University College London, July, 1993. John Wiley & Sons: Chichester.

Thompson, J. M. T. and Champneys, A. R. (1996). From helix to localized writhing in the torsional post-buckling of elastic rods. *Proc. R. Soc. Lond.*, **A452**, 117–138.

Thompson, J. M. T. and Chua, L. O. (eds) (1995). *Chaotic Behaviour in Electronic Circuits*, theme issue. *Phil. Trans. R. Soc. Lond.*, **A353**, 1–136.

Thompson, J. M. T. and de Souza, J. R. (1996). Suppression of escape by resonant modal interactions: in shell vibration and heave-roll capsize. *Proc. R. Soc. Lond.*, **A452**, 2527–2550.

Thompson, J. M. T. and Elvey, J. S. N. (1984). Elimination of subharmonic resonances of compliant marine structures. *Int. J. Mech. Sci.*, **26**, 419–425.

Thompson, J. M. T. and Ghaffari, R. (1982). Chaos after period-doubling bifurcations in the resonance of an impact oscillator. *Phys. Lett.*, **91A**, 5–8.

Thompson, J. M. T. and Ghaffari, R. (1983). Chaotic dynamics of an impact oscillator. *Phys. Rev. A*, **27**, 1741–1743.

Thompson, J. M. T. and Gray, P. (eds) (1990). *Chaos and Dynamical Complexity in the Physical Sciences*, theme issue. *Phil. Trans. R. Soc. Lond.*, **A332**, 49–186.

Thompson, J. M. T. and Hunt, G. W. (1973). *A General Theory of Elastic Stability*. John Wiley & Sons: London.

Thompson, J. M. T. and Hunt, G. W. (1984). *Elastic Instability Phenomena*. John Wiley & Sons: Chichester.

Thompson, J. M. T. and Macmillen, F. B. J. (eds) (1998). *Nonlinear Flight Dynamics of High Performance Aircraft*, theme issue. *Phil. Trans. R. Soc. Lond.*, **A356**, no. 1745.

Thompson, J. M. T. and McRobie, F. A. (1993). Indeterminate bifurcations and the global dynamics of driven oscillators. *Proceedings of the 1st European Nonlinear Oscillations Conference*, Hamburg, E. Kreuzer and G. Schmidt (eds), pp. 107–128. Akademie Verlag: Berlin.

Thompson, J. M. T. and Schiehlen, W. (eds) (1992). *Nonlinear Dynamics of Engineering Systems*, theme issue. *Phil. Trans. R. Soc. Lond.*, **A338**, 451–568.

Thompson, J. M. T. and Soliman, M. S. (1990). Fractal control boundaries of driven oscillators and their relevance to safe engineering design. *Proc. R. Soc. Lond.*, **A428**, 1–13.

Thompson, J. M. T. and Soliman, M. S. (1991). Indeterminate jumps to resonance from a tangled saddle-node bifurcation. *Proc. R. Soc. Lond.* A, **432**, 101–111.

Thompson, J. M. T. and Stewart, H. B. (1984). Folding and mixing in the Birkhoff-Shaw chaotic attractor. *Phys. Lett.*, **103A**, 229–231.

Thompson, J. M. T. and Stewart, H. B. (1993). A tutorial glossary of geometrical dynamics. *Int. J. Bifurcation and Chaos*, **3**, 223–239.

Thompson, J. M. T. and Thompson, R. J. (1980). Numerical experiments with a strange attractor. *Bull. Inst. Math. Appl.*, **16**, 150–154.

Thompson, J. M. T. and Ueda, Y. (1989). Basin boundary metamorphoses in the canonical escape equation. *Dynamics and Stability of Systems*, **4**, 285–294.

Thompson, J. M. T. and Virgin, L. N. (1986). Predicting a jump to resonance using transient maps and beats. *Int. J. Nonlinear Mech.*, **21**, 205–216.

Thompson, J. M. T., Bokaian, A. R. and Ghaffari, R. (1983). Subharmonic resonances and chaotic motions of a bilinear oscillator. *IMA J. Appl. Math.*, **31**, 207–234.

Thompson, J. M. T., Bokaian, A. R. and Ghaffari, R. (1984). Subharmonic and chaotic motions of compliant offshore structures and articulated mooring towers. *ASME J. Energy Resources Tech.*, **106**, 191–198.

Thompson, J. M. T., Rainey, R. C. T. and Soliman, M. S. (1990). Ship stability criteria based on chaotic transients from incursive fractals. *Phil. Trans. R. Soc. Lond.*, **A332**, 149–167.

Thompson, J. M. T., Rainey, R. C. T. and Soliman, M. S. (1992). Mechanics of ship capsize under direct and parametric wave excitation. *Phil. Trans. R. Soc. Lond.*, **A338**, 471–490.

Thompson, J. M. T., Stewart, H. B. and Ueda, Y. (1994). Safe, explosive and dangerous bifurcations in dissipative dynamical systems. *Phys. Rev. E*, **49**, 1019–1027.

Tomita, K. (1982). Chaotic response of nonlinear oscillators. *Phys. Rep.*, **86**, 113–167.

Tresser, C. (1983). Nouveaux types de transitions vers une entropie topologique positive. *C. R. Acad. Sci. Paris*, **296**, 729–732.

Tresser, C., Coullet, P. and Arnéodo, A. (1980a). On the existence of hysteresis in a transition to chaos after a single bifurcation. *J. Phys. Lett.*, **41**, L243–L246.

Tresser, C., Coullet, P. and Arnéodo, A. (1980b). Topological horseshoe and numerically observed chaotic behavior in the Hénon mapping. *J. Phys. A.*, **13**, L123–L127

Tsuda, I. (1981). Self-similarity in the Belousov – Zhabotinsky reaction. *Phys. Lett.*, **85A**, 4–8.

Tuchinsky, P. M. (1981). *Man in Competition with the Spruce Budworm: An Application of Differential Equations*. Birkhäuser: Boston.

Tufillaro, N. B., Abbott, T. and Reilly, J. (1992). *An Experimental Approach to Nonlinear Dynamical and Chaos*. Addison-Wesley: Redwood City.

Turcotte, D. L. (1992). *Fractals and Chaos in Geology and Geophysics*. Cambridge University Press: Cambridge.

Turner, J. S., Roux, J.-C., McCormick, W. D. and Swinney, H. L. (1981). Alternating periodic and chaotic regimes in a chemical reaction—experiment and theory. *Phys. Lett.*, **85A**, 9–12.

Ueda, Y. (1979). Randomly transitional phenomena in the system governed by Duffing's equation. *J. Stat. Phys.*, **20**, 181–196.

Ueda, Y. (1980a). Steady motions exhibited by Duffing's equation: a picture book of regular and chaotic motions. In *New Approaches to Nonlinear Problems in Dynamics*, P. J. Holmes (ed.), pp. 311–322. SIAM: Philadelphia. Updated in Ueda (1991).

Ueda, Y. (1980b). Explosion of strange attractors exhibited by Duffing's equation. In *Nonlinear Dynamics*, R. H. G. Helleman (ed.), pp. 422–434. New York Academy of Sciences: New York.

Ueda, Y. (1991). Survey of regular and chaotic phenomena in the forced Duffing oscillator. *Chaos, Solitons, & Fractals*, **1**, 199–231. Reprinted in Ueda (1992).

Ueda, Y. (1992). *The Road to Chaos*. Aerial Press: Santa Cruz.

Ueda, Y. and Akamatsu, N. (1980). Chaotically transitional phenomena in the forced negative-resistance oscillator. *IEEE Trans. Circuits Syst.*, **CAS-28**, 217–224.

Ueda, Y., Akamatsu, N. and Hayashi, C. (1973). Computer simulation of nonlinear ordinary differential equations and non-periodic oscillations. *Electronics and Communications in Japan*, **56A**, 27–34.

Ueda, Y., Yoshida, S., Stewart, H. B. and Thompson, J. M. T. (1990). Basin explosions and escape phenomena in the twin-well Duffing oscillator: compound global bifurcations organising behaviour. *Phil. Trans. R. Soc. Lond.*, **A332**, 169–186.

Ueda, Y., Ueda, Y., Stewart, H. B. and Abraham, R. H. (1998). Nonlinear resonance in basin portraits of two coupled swings under periodic forcing. *Int. J. Bifurcation and Chaos*, **8**, 1183–1197.

Ulam, S. M. and von Neumann, J. (1947). On combinations of stochastic and deterministic processes. *Bull. Am. Math. Soc.*, **53**, 1120.

Ushiki, S. (1982). Central difference scheme and chaos. *Physica*, **4D**, 407–424.

Ushio, T. and Hirai, K. (1983). Chaos in non-linear sampled-data control systems. *Int. J. Control*, **38**, 1023–1033.

Vallée, R., Delisle, C. and Chrostowski, J. (1984). Noise versus chaos is acousto-optic bistability. *Phys. Rev.*, **30**, 336–342.

Vallos, G. K. (1986). El Niño: a chaotic dynamical system. *Science* **232**, 243.

van der Heijden, G. H. M. (2001). The static deformation of a twisted elastic rod constrained to lie on a cylinder. *Proc. R. Soc. Lond.*, **A457**, 695–715.

van der Heijden G. H. M. and Thompson, J. M. T. (1998). Lock-on to tape-like behaviour in the torsional buckling of anisotropic rods, *Physica*, **112D**, 201–224.

van der Heijden, G. H. M. and Thompson, J. M. T. (2000). Helical and localised buckling in twisted rods: a unified analysis of the symmetric case. *Nonlinear Dynamics* **21**, 71–99.

van der Heijden, G. H. M., Champneys, A. R. and Thompson, J. M. T. (1998). The spatial complexity of localised buckling in rods with non-circular cross-section. *SIAM J. Appl. Math.*, **59**, 198–221.

Van der Pol, B. (1926). On relaxation-oscillations. *Phil. Mag. (7)*, **2**, 978–992.

Van der Pol, B. (1927). Forced oscillations in a circuit with nonlinear resistance (reception with reactive triode). *Phil. Mag. (7)*, **3**, 65–80.

Van der Pol, B. and Van der Mark, J. (1927). Frequency demultiplication. *Nature*, **120**, 363–364.

Van der Pol, B. and Van der Mark, J. (1928). The heart beat considered as a relaxation oscillation, and an electrical model of the heart. *Phil. Mag. (7)*, **6**, 763–775.

Verlarde, M. G. and Normand, C. (1980). Convection. *Sci. Am.*, **243**, no. 1, 92–109.

Viana, M. (2000). What's new on Lorenz' strange attractors? *Mathematical Intelligencer*, **22**, 6–19.

Vidal, C. and Pacault, A. (1981). *Nonlinear Phenomena in Chemical Dynamics*. Springer-Verlag: Berlin.

Vidal, C., Roux, J.-C., Bachelart, S. and Rossi, A. (1980). Experimental study of the transition to turbulence in the Belousov–Zhabotinskii reaction. In *Nonlinear Dynamics*, R. H. G. Helleman (ed.), pp. 377–396. New York Academy of Sciences: New York.
Virgin, L. N. (1986a). Parametric studies of the dynamic evolution through a fold. *J. Sound Vibration*. **110**, 99–109.
Virgin, L. N. (1986b). The nonlinear rolling response of a vessel including chaotic motions leading to capsize in regular seas. Submitted to *Applied Ocean Research*.
Virgin, L. N. (2000). *Introduction to Experimental Nonlinear Dynamics: A Case Study in Mechanical Vibration*. Cambridge University Press: Cambridge.
Vyshkind, S. Ya. and Rabinovich, M. I. (1976). The phase stochastization mechanism and the structure of wave turbulence in dissipative media. *Sov. Phys. JETP*, **44**, 292–299.
Walden, R. W., Kolodner, P., Passner, A. and Surko, C. M. (1984). Nonchaotic Rayleigh – Bénard Convection with four and five incommensurate frequencies. *Phys. Rev. Lett.*, **53**, 242–245.
Wiesenfeld, K. (1985). Virtual Hopf phenomenon: a new precursor of period-doubling bifurcations. *Phys. Rev. A*, **32**, 1744–1751.
Wiggins, S. (1988). *Global Bifurcations and Chaos*. Springer-Verlag: New York.
Wiggins, S. (1990). *Introduction to Applied Dynamical Systems and Chaos*, Springer-Verlag: New York.
Williams, R. F. (1967). One-dimensional nonwandering sets. *Topology*, **6**, 473–487.
Williams, R. F. (1974). Expanding attractors. *Publ. Math. IHES*, **43**, 169–203.
Williams, R. F. (1977). The structure of Lorenz attractors. In *Turbulence Seminar Berkeley 1976/77*, P. Bernard and T. Ratiu (eds), pp. 94–112. Springer-Verlag: New York, Heidelberg and Berlin.
Williams, R. F. (1979). The structure of Lorenz attractors. *Publ. Math. IHES*, **50**, 73–99.
Wilson, H. R. and Cowan, J. D. (1972). Excitatory and inhibitory interactions in localized populations of model neurons. *Biophysical J.*, **12**, 1–24.
Wimmer, M. (1981). Experiments on the stability of viscous flow between two concentric rotating cylinders. *J. Fluid. Mech.*, **103**, 117–131.
Winfree, A. T. (1980). *The Geometry of Biological Time*. Springer-Verlag: New York, Heidelberg and Berlin.
Wolf, A., Swift, J. B., Swinney, H. L. and Vastano, J. (1985). Determining Lyapunov exponents from a time series. *Physica*, **16D**, 285–317.
Wolfram, S. (1983). Statistical mechanics of cellular automata. *Rev. Mod. Phys.*, **55**, 601–644.
Yamaguchi, Y. (1983). Chaotic behavior of magnetization in super fluid ^{3}He driven by external periodic field. *Prog. Theor. Phys.*, **69**, 1377–1395.
Yeh, W. J. and Kao, Y. H. (1983). Intermittency in Josephson junctions. *Appl. Phys. Lett.*, **42**, 299–301.
Yeh, W. J., He, D. R. and Kao, Y. H. (1984). Fractal dimension and self-similarity of the devil's staircase in a Josephson-junction simulator. *Phys. Rev. Lett.*, **52**, 480.
Young, L.-S. (1982). Dimension, entropy, and Lyapunov exponents. *Ergodic Theor. Dyn. Syst.*, **2**, 109–124.
Young, L.-S. (1984). Dimension, entropy, and Liapunov exponents in differentiable dynamical systems. *Physica*, **124A**, 639–646.
Zak, M. (1984). Deterministic representation of chaos in classical dynamics. *Phys. Lett.*, **107A**, 125–128.
Zaslavsky, G. M. (1978). The simplest case of a strange attractor. *Phys. Lett.*, **69A**, 145–147.
Zaslavsky, G. M. (1981). Stochasticity in quantum systems. *Phys. Rep.*, **30**, 157–250.
Zaslavsky, G. M. and Chirikov, B. V. (1972). Stochastic instability of non-linear oscillators. *Sov. Phys. Usp.*, **14**, 549–568.

Zeeman, E. C. (1977). *Catastrophe Theory: Selected Papers 1972–1977*. Addison-Wesley: Reading, MA.

Zeeman, E. C. (1981). *1981 Bibliography on Catastrophe Theory*. Mathematics Institute, University of Warwick: Coventry.

Zeeman, E. C. (1982). Bifurcation and catastrophe theory. In *Papers in Algebra, Analysis, and Statistics*, R. Lidl (ed.), p. 207–272. American Mathematical Society: Providence, RI.

Online Resources

- Interactive dynamic computer simulations, and movies of lectures and experiments can be found at

  http://brain.cc.kogakuin.ac.jp/~kanamaru/Chaos/e/Thompson/
  www.culive.org/MichaelThompson

- Frequently asked questions (FAQs) for the USENET newsgroup sci.nonlinear can be found at

  http://amath.colorado.edu/faculty/jdm/faq.html
  http://www.enm.bris.ac.uk/anm/faq/faq-Contents.html

- UK nonlinear news can be found at

  http://www.amsta.leeds.ac.uk/Applied/news.dir/

- Educational resources of Devaney's Dynamical Systems Project can be found at

  http://math.bu.edu/DYSYS/

Index

(See also the Appendix and Glossary)
(Italicized decimals are Figure numbers)

9 780471 87684